AF389411

# Mineralogical Crystallography
# Volume II

# Mineralogical Crystallography Volume II

Editor

**Vladislav V. Gurzhiy**

MDPI • Basel • Beijing • Wuhan • Barcelona • Belgrade • Manchester • Tokyo • Cluj • Tianjin

*Editor*
Vladislav V. Gurzhiy
St. Petersburg State University
Russia

*Editorial Office*
MDPI
St. Alban-Anlage 66
4052 Basel, Switzerland



For citation purposes, cite each article independently as indicated on the article page online and as indicated below:

LastName, A.A.; LastName, B.B.; LastName, C.C. Article Title. *Journal Name* **Year**, *Volume Number*, Page Range.

**ISBN 978-3-0365-5951-3 (Hbk)**
**ISBN 978-3-0365-5952-0 (PDF)**

# Contents

# About the Editor

**Vladislav V. Gurzhiy**

Vladislav V. Gurzhiy is an Associate Professor of the Crystallography Department of the Institute of Earth Sciences at the Saint Petersburg State University, and also holds the position of Chairman of the Institute of Earth Sciences' Scientific Committee. Dr. Gurzhiy was graduated from the St. Petersburg State University in 2007, received a Ph.D. degree in 2009, and a Dr. Habil. degree (Russian Doctor of Science degree in Geology and Mineralogy) in 2022 with a dissertation entitled "The Crystal chemistry of natural and synthetic uranyl sulfates, selenites and selenates". His main research interests are related to the crystal chemistry of minerals and their synthetic analogs, in particular those compounds bearing uranium and transuranium elements, and biominerals. Dr. Gurzhiy is a full Member of the Russian Mineralogical Society and the Saint-Petersburg Society of Naturalists, a member of the Academic Council of the St. Petersburg State University (since 2020) and a member of the Academic Council of the Russian Mineralogical Society (since 2021). He has published over 180 peer-reviewed journal articles and several book chapters. Dr. Gurzhiy was honored with several awards, including the Yu.T. Struchkov Prize and Academia Europaea Prize for research works in the fields of X-ray crystallography and earth sciences.

Editorial

# Mineralogical Crystallography Volume II

**Vladislav V. Gurzhiy**

Department of Crystallography, Institute of Earth Sciences, St. Petersburg State University, University Emb. 7/9, 199034 St. Petersburg, Russia; vladislav.gurzhiy@spbu.ru or vladgeo17@mail.ru

**Citation:** Gurzhiy, V.V. Mineralogical Crystallography Volume II. *Crystals* **2022**, *12*, 1631. https://doi.org/10.3390/cryst12111631

Received: 2 November 2022
Accepted: 10 November 2022
Published: 13 November 2022

**Publisher's Note:** MDPI stays neutral with regard to jurisdictional claims in published maps and institutional affiliations.

The International Mineralogical Association and UNESCO celebrates 2022—the Year of Mineralogy. However, this year was not chosen randomly. Indeed, 2022 is the bicentennial of the death of René Just Haüy (born 1743), who is considered to be one of the founders of crystallography and mineralogy in their modern state. The year 1822 also marks the publication of Haüy's *Traité de minéralogy* and *Traité de cristallographie*. Mineralogy is one of the oldest branches of science, with its origin in at least antic times, but its scientific renaissance started a little more than a century ago, when precise crystallographic studies, such as X-ray structural analysis (mainly, but not only), significantly improved the value of research results. Since the first decade of the XX century, mineralogy and crystallography together have played a key role in our everyday lives.

Various scientific events are being held around the world under the auspices of this landmark event, the Year of Mineralogy, and it is highly satisfying that this Special Issue "Mineralogical Crystallography Volume II" is published in 2022. The first volume of the "Mineralogical Crystallography" Special Issue [1] consisted of such topics as: Discovery of new mineral species; Crystal chemistry of minerals and their synthetic analogs; Behavior of minerals at non-ambient conditions; Biomineralogy; and Crystal growth techniques, and appeared to be very fruitful. The Special Issue "Mineralogical Crystallography Volume II" covers the following topics: Crystal chemistry and properties of minerals and their synthetic analogs; Gemology; Natural-based cement materials; Biomineralogy; and Crystal growth techniques. Additionally, we hope that this continuation will be just as successful, and that the new set of papers will again arouse genuine interest among readers and, perhaps, inspire them in their own successful research. We also believe that with the current collection of papers, we will be able to pay tribute to the union of Mineralogy and Crystallography.

## 1. Crystal Chemistry and Properties of Minerals and Their Synthetic Analogs

Gurzhiy et al. [2] reviewed the state of the art within the structural chemistry of uranyl carbonate minerals and mineral-related synthetic compounds. It was shown that the majority of synthetic analogs of uranyl carbonate minerals were grown from aqueous systems at room temperature, which indicates that the formation of these minerals in nature does not need any specific thermodynamic (increased P and T) conditions, as was assumed for other uranyl-bearing systems.

Kornyakov and Krivovichev [3] report on the crystallization of two novel synthetic phases with structures based on layers of oxocentered ($OCu_4$) tetrahedra, which were previously described in the structure of mineral shchurovskyite.

Huber et al. [4] have studied optical and spectroscopic characteristics of a number of *REE*-bearing (rare earth elements) minerals, such as loparite, lorenzenite, titanite, apatite and others, from massifs of Kola Peninsula (Khibiny, Lovozero, Afrikanda, and Kovdor), which can be regarded as indicator phases in intrusions of ultramafic and alkaline rocks.

The next two papers report on the synthesis and investigation of mineral-related materials with unique sorption characteristics. Samburov et al. [5] have synthesized an analog of the ivanykite mineral and studied its Pb sorption characteristics from model solutions. The maximal sorption capacity reached 400 mg/g at ambient conditions, which is very promising for metal excretion.

Kong and Jiang [6] suggest a protocol for zeolite material preparation from the coal gangue that is enriched by Fe and high quartz content. Technology includes preliminary calcination and leaching, and further activation by sodium carbonate at heating. This procedure appears to be less laborious and with higher product purity.

A further two papers that can be fitted within this chapter deal with the determination of $H_2O$ molecules in structures of minerals. Wu et al. [7] have studied kaolinite and its polytype dickite by the means of near-IR spectroscopy, which helps to distinguish outer Al-OH bending vibrations at c.a. 4600 $cm^{-1}$ and overtones of Al-OH stretching vibrations in the range of 7000–7250 $cm^{-1}$.

Using FTIR, Zuo et al. [8] have found the presence of $H_2O$ molecules in the crystal structure of alkali feldspar within channels with preferable (001) orientation. It is suggested that this water plays an important role in the mechanisms of feldspar alteration and water preservation in nominally anhydrous minerals in Earth's depths.

## 2. Gemology

Wang et al. [9] report that there are two types of $H_2O$ molecules that are arranged in the crystal structure of beryl. It was shown that $Fe^{2+} - Al^{3+}$ hetero-valent substitution shifts color of blue-green beryl from yellow to blue, which is also accompanied by an increase in $H_2O$ content, whose symmetry axis is perpendicular to the *c*-axis.

Zhang et al. [10] have studied apatite of Paraiba-like color from Madagascar. It was shown that studied samples belong to the fluorapatite mineral species. The F/Cl ratio suggests the magmatic origin of apatites, and their fascinating greenish-blue color is caused by the presence of *REE* (Ce and Nd) elements and crystal structure distortion effects.

Peng et al. [11] report on the comprehensive experimental characterization of the natural forsterite crystals from a new locality in Jian forsterite jade (China). This Mg-bearing end member of the olivine group material is remarkable due to the enrichment in B, which comes from H and B substitution for Si.

## 3. Natural-Based Cement Materials

Costafreda and Martín [12] studied bentonites from the San José–Los Escullos deposit (Spain) that can be used in manufacturing durable cements and concretes. The studied rocks contain various clay and phyllosilicate minerals such as smectites of the montmorillonite variety, illite, vermiculite, biotite, muscovite, kaolinite, chlorite, etc. The conducted experiments demonstrated the pozzolanic character of the bentonites, which is very promising in the manufacture of pozzolanic cements.

In the second paper of this workgroup, Martín et al. [13] suggest a protocol for natural and calcined fluorite usage in the manufacture of cements. The authors showed that involvement of local fluorite deposits can help mining companies to reduce $CO_2$ emission and energy costs in the production of cement with good pozzolanic and mechanical properties.

Belmonte et al. [14] describe the pathway of the new cementitious material synthesis that can undergo self-healing, due to the addition of calcium nitrate to Portland cement. This results in the increase in the amount of ettringite crystals and sealing of fissures by them.

## 4. Biomineralogy

Pramono et al. [15] have successfully synthesized hydroxyapatite composite with spinel fittings using bovine bones, beverage Al cans and Mg (as the only commercial reagent). It was shown that the decrease in Al particle size increases hardness and reduces the porosity of the novel perspective biocomposite material.

## 5. Crystal Growth Techniques

Wang et al. [16] have found that syntheses of barite, at room temperature, in the presence of sodium and chlorine ions results in a low amount of rather simple morphologies, while a hydrothermal experiment significantly increased the variety of obtained crystal shapes. In addition, the dendritic type of crystalline barite can be considered as typomorphic for high temperature hydrothermal conditions, with an excess of Ba relative to sulfate ions.

Shakhgildyan et al. [17] have studied the role of gold nanoparticles on multicomponent glass crystallization, microstructure, and optical characteristics. It was shown that thermal precipitation of Au nanoparticles does not affect the crystallization and structure of glass and gahnite nanocrystals within the system, but strongly affects the optical properties of glass-ceramics.

**Funding:** This research received no external funding.

**Data Availability Statement:** Not applicable.

**Acknowledgments:** As the Guest Editor, I would like to acknowledge all the authors for their valuable contribution to this Special Issue, which is expressed in fascinating and inspiring papers.

**Conflicts of Interest:** The author declares no conflict of interest.

## References

1. Gurzhiy, V.V. Mineralogical Crystallography. *Crystals* **2020**, *10*, 805. [CrossRef]
2. Gurzhiy, V.V.; Kalashnikova, S.A.; Kuporev, I.V.; Plášil, J. Crystal Chemistry and Structural Complexity of the Uranyl Carbonate Minerals and Synthetic Compounds. *Crystals* **2021**, *11*, 704. [CrossRef]
3. Kornyakov, I.V.; Krivovichev, S.V. Crystal Chemical Relations in the Shchurovskyite Family: Synthesis and Crystal Structures of $K_2Cu[Cu_3O]_2(PO_4)_4$ and $K_{2.35}Cu_{0.825}[Cu_3O]_2(PO_4)_4$. *Crystals* **2021**, *11*, 807. [CrossRef]
4. Huber, M.; Kamiński, D.; Czernel, G.; Kozlov, E. Optical and Spectroscopic Properties of Lorenzenite, Loparite, Perovskite, Titanite, Apatite, Carbonates from the Khibiny, Lovozero, Kovdor, and Afrikanda Alkaline Intrusion of Kola Peninsula (NE Fennoscandia). *Crystals* **2022**, *12*, 224. [CrossRef]
5. Samburov, G.O.; Kalashnikova, G.O.; Panikorovskii, T.L.; Bocharov, V.N.; Kasikov, A.; Selivanova, E.; Bazai, A.V.; Bernadskaya, D.; Yakovenchuk, V.N.; Krivovichev, S.V. A Synthetic Analog of the Mineral Ivanyukite: Sorption Behavior to Lead Cations. *Crystals* **2022**, *12*, 311. [CrossRef]
6. Kong, D.; Jiang, R. Preparation of NaA Zeolite from High Iron and Quartz Contents Coal Gangue by Acid Leaching—Alkali Melting Activation and Hydrothermal Synthesis. *Crystals* **2021**, *11*, 1198. [CrossRef]
7. Wu, S.; He, M.; Yang, M.; Peng, B. Near-Infrared Spectroscopic Study of OH Stretching Modes in Kaolinite and Dickite. *Crystals* **2022**, *12*, 907. [CrossRef]
8. Zuo, H.; Liu, R.; Lu, A. The Behavior of Water in Orthoclase Crystal and Its Implications for Feldspar Alteration. *Crystals* **2022**, *12*, 1042. [CrossRef]
9. Wang, H.; Shu, T.; Chen, J.; Guo, Y. Characteristics of Channel-Water in Blue-Green Beryl and Its Influence on Colour. *Crystals* **2022**, *12*, 435. [CrossRef]
10. Zhang, Z.-Y.; Xu, B.; Yuan, P.-Y.; Wang, Z.-X. Gemological and Mineralogical Studies of Greenish Blue Apatite in Madagascar. *Crystals* **2022**, *12*, 1067. [CrossRef]
11. Peng, B.; He, M.; Yang, M.; Wu, S.; Fan, J. Natural Forsterite Strongly Enriched in Boron: Crystal Structure and Spectroscopy. *Crystals* **2022**, *12*, 975. [CrossRef]
12. Costafreda, J.L.; Martín, D.A. Bentonites in Southern Spain. Characterization and Applications. *Crystals* **2021**, *11*, 706. [CrossRef]
13. Martín, D.A.; Costafreda, J.L.; Estévez, E.; Presa, L.; Calvo, A.; Castedo, R.; Sanjuán, M.Á.; Parra, J.L.; Navarro, R. Natural Fluorite from Órgiva Deposit (Spain). A Study of Its Pozzolanic and Mechanical Properties. *Crystals* **2021**, *11*, 1367. [CrossRef]
14. Belmonte, I.M.; Soler, M.C.; Saorin, F.J.B.; Costa, C.J.P.; López, C.L.R.; del Pozo Martin, J.; Pacheco, V.M.; Torrano, P.H. Mineralization Reaction of Calcium Nitrate and Sodium Silicate in Cement-Based Materials. *Crystals* **2022**, *12*, 445. [CrossRef]
15. Pramono, A.; Timuda, G.E.; Rifai, G.P.A.; Khaerudini, D.S. Synthesis of Spinel-Hydroxyapatite Composite Utilizing Bovine Bone and Beverage Can. *Crystals* **2022**, *12*, 96. [CrossRef]
16. Wang, C.; Zhou, L.; Zhang, S.; Wang, L.; Wei, C.; Song, W.; Xu, L.; Zhou, W. Morphology of Barite Synthesized by In-Situ Mixing of $Na_2SO_4$ and $BaCl_2$ Solutions at 200 °C. *Crystals* **2021**, *11*, 962. [CrossRef]
17. Shakhgildyan, G.; Durymanov, V.; Ziyatdinova, M.; Atroshchenko, G.; Golubev, N.; Trifonov, A.; Chereuta, O.; Avakyan, L.; Bugaev, L.; Sigaev, V. Effect of Gold Nanoparticles on the Crystallization and Optical Properties of Glass in $ZnO-MgO-Al_2O_3-SiO_2$ System. *Crystals* **2022**, *12*, 287. [CrossRef]

*Review*

# Crystal Chemistry and Structural Complexity of the Uranyl Carbonate Minerals and Synthetic Compounds

Vladislav V. Gurzhiy [1,*], Sophia A. Kalashnikova [1], Ivan V. Kuporev [1] and Jakub Plášil [2]

[1] Crystallography Department, Institute of Earth Sciences, St. Petersburg State University, University Emb. 7/9, 199034 St. Petersburg, Russia; kalashnikova.soff@gmail.com (S.A.K.); st054910@student.spbu.ru (I.V.K.)
[2] Institute of Physics ASCR, v.v.i., Na Slovance 2, 18221 Praha 8, Czech Republic; plasil@fzu.cz
* Correspondence: vladislav.gurzhiy@spbu.ru or vladgeo17@mail.ru

**Abstract:** Uranyl carbonates are one of the largest groups of secondary uranium(VI)-bearing natural phases being represented by 40 minerals approved by the International Mineralogical Association, overtaken only by uranyl phosphates and uranyl sulfates. Uranyl carbonate phases form during the direct alteration of primary U ores on contact with groundwaters enriched by $CO_2$, thus playing an important role in the release of U to the environment. The presence of uranyl carbonate phases has also been detected on the surface of "lavas" that were formed during the Chernobyl accident. It is of interest that with all the importance and prevalence of these phases, about a quarter of approved minerals still have undetermined crystal structures, and the number of synthetic phases for which the structures were determined is significantly inferior to structurally characterized natural uranyl carbonates. In this work, we review the crystal chemistry of natural and synthetic uranyl carbonate phases. The majority of synthetic analogs of minerals were obtained from aqueous solutions at room temperature, which directly points to the absence of specific environmental conditions (increased P or T) for the formation of natural uranyl carbonates. Uranyl carbonates do not have excellent topological diversity and are mainly composed of finite clusters with rigid structures. Thus the structural architecture of uranyl carbonates is largely governed by the interstitial cations and the hydration state of the compounds. The information content is usually higher for minerals than for synthetic compounds of similar or close chemical composition, which likely points to the higher stability and preferred architectures of natural compounds.

**Keywords:** uranyl; carbonate; mineral; crystal structure; topology; structural complexity

**Citation:** Gurzhiy, V.V.; Kalashnikova, S.A.; Kuporev, I.V.; Plášil, J. Crystal Chemistry and Structural Complexity of the Uranyl Carbonate Minerals and Synthetic Compounds. *Crystals* **2021**, *11*, 704. https://doi.org/10.3390/cryst11060704

Academic Editor: Duncan H. Gregory

Received: 18 May 2021
Accepted: 15 June 2021
Published: 19 June 2021

**Publisher's Note:** MDPI stays neutral with regard to jurisdictional claims in published maps and institutional affiliations.

## 1. Introduction

Uranyl carbonate phases play a very important role in all processes related to the nuclear fuel cycle. This conjunction starts from U deposits, where uranyl carbonate minerals form during the direct alteration of primary U-bearing rocks (containing uraninite, etc.) under the influence of groundwaters enriched with $CO_2$, which can be derived from the dissolution of host carbonate rocks or from the atmosphere [1–5]. In dissolved form, uranyl carbonates can play an important role in U release to the environment. And of course, it should not be forgotten that uranyl-carbonate mineralization has been described among the alteration products of the "lavas" that were formed during the accident at the fourth reactor of the Chernobyl nuclear power plant in 1986 [6,7].

There are 40 uranyl carbonate mineral species approved by the International Mineralogical Association as of 1 November 2020, thus making this group one of the most representative among secondary uranium minerals, coming third only after phosphates and sulfates [8,9]. Despite a fairly large number of known compounds, the structural diversity is not as great as one might expect. It is also of interest that about a quarter of approved minerals still have their crystal structures undetermined. The amount of synthetic structurally characterized uranyl carbonates is inferior to natural phases but can give an idea of the crystallization conditions present in the environment.

Current work is devoted to reviewing the topological diversity and growth conditions of natural uranyl carbonates and their synthetic analogs. Information-based complexity measures have been performed to determine contributions of various substructural building blocks and particular topological types into the complexity of the whole structure, which is related to the stability of a crystalline compound.

## 2. Materials and Methods

### 2.1. Structural Data

For the current review, all structural data deposited in the Inorganic Crystal Structure Database (ICSD; version 4.5.0; release February 2020) were selected and supplemented by the data reported in the most recent publications of the author of the current paper (J.P.). Chemical formulae, mineral names, and the basic crystallographic characteristics for all inorganic uranyl carbonates of both natural and synthetic origin are listed in Table 1. In addition, Table 1 contains information on the proposed symmetry and unit cell parameters for the uranyl carbonate minerals with yet undefined crystal structures listed in the IMA Database of Mineral Properties [10].

**Table 1.** Crystallographic characteristics of natural and synthetic uranyl carbonates.

| No. | Chemical Formula | Mineral Name | Sp.Gr. | $a$, Å / $\alpha$, ° | $b$, Å / $\beta$, ° | $c$, Å / $\gamma$, ° | Ref. |
|---|---|---|---|---|---|---|---|
| | | | Finite Clusters | | | | |
| | | | $cc0$-1:2-9 | | | | |
| 1 | $K_4[(UO_2(CO_3)_2(O_2)](H_2O)$ | | $P2_1/n$ | 6.9670(14)/90 | 9.2158(18)/91.511(3) | 18.052(4)/90 | [illegible] |
| 2 | $K_4[(UO_2(CO_3)_2(O_2)](H_2O)_{2.5}$ | | $P2_1/n$ | 6.9077(14)/90 | 9.2332(18)/91.310(4) | 21.809(4)/90 | [illegible] |
| 3 | $(CN_3H_6)_4[UO_2(CO_3)_2(O_2)]·H_2O$ | | $Pca2_1$ | 15.883(1)/90 | 8.788(2)/90 | 16.155(1)/90 | [illegible] |
| | | | $cc0$-1:3-2 | | | | |
| 4 | $Na_4(UO_2)(CO_3)_3$ | Čejkaite | $Cc$ | 9.2919(8)/90 | 16.0991(11)/91.404(5) | 6.4436(3)/90 | [illegible] |
| 4a | $Na_4(UO_2)(CO_3)_3$ | | $P$-3$c$ | 9.3417/90 | 9.3417/90 | 12.824/120 | [illegible] |
| 4b | $Na_4(UO_2)(CO_3)_3$ | Cejkaite old model | $P$-1 | 9.291(2)/90.73(2) | 9.292(2)/90.82(2) | 12.895(2)/120.00(1) | [illegible] |
| 5 | $K_3Na(UO_2)(CO_3)_3$ | | $P$-62$c$ | 9.29(2)/90 | 9.29(2)/90 | 8.26(2)/120 | [illegible] |
| 6 | $K_3Na(UO_2)(CO_3)_3(H_2O)$ | Grimselite | $P$-62$c$ | 9.2507(1)/90 | 9.2507(1)/90 | 8.1788(1)/120 | [illegible] |
| 6a | $Rb_6Na_2((UO_2)(CO_3)_3)_2(H_2O)$ | Rb analogue of Grimselite | $P$-62$c$ | 9.4316(7)/90 | 9.4316(7)/90 | 8.3595(8)/120 | [illegible] |
| 7 | $K_4(UO_2)(CO_3)_3$ | Agricolaite | $C2/c$ | 10.2380(2)/90 | 9.1930(2)/95.108(2) | 12.2110(3)/90 | [illegible] |
| 7a | $K_4UO_2(CO_3)_3$ | | $C2/c$ | 10.247(3)/90 | 9.202(2)/95.11(2) | 12.226(3)/90 | [illegible] |
| 7b | $K_4(UO_2)(CO_3)_3$ | | $C2/c$ | 10.240(7)/90 | 9.198(4)/95.12(4) | 12.222(12)/90 | [illegible] |
| 58 | $Rb_4(UO_2)(CO_3)_3$ | | $C2/c$ | 10.778(5)/90 | 9.381(2)/94.42(3) | 12.509(3)/90 | [illegible] |
| 8 | $Cs_4UO_2(CO_3)_3(H_2O)_6$ | | $P2_1/n$ | 11.1764(4)/90 | 9.5703(4)/96.451(2) | 18.5756(7)/90 | [illegible] |
| 8a | $Cs_4(UO_2(CO_3)_3(H_2O)_6$ | | $P2_1/n$ | 18.723(3)/90 | 9.647(2)/96.84(1) | 11.297(2)/90 | [illegible] |
| 9 | $Cs_4(UO_2(CO_3)_3)$ | | $C2/c$ | 11.5131(9)/90 | 9.6037(8)/93.767(2) | 12.9177(10)/90 | [illegible] |
| 10 | $(NH_4)_4(UO_2(CO_3)_3)$ | | $C2/c$ | 10.679(4)/90 | 9.373(2)/96.43(2) | 12.850(3)/90 | [illegible] |
| 11 | $Tl_4((UO_2)(CO_3)_3)$ | | $C2/c$ | 10.684(2)/90 | 9.309(2)/94.95(2) | 12.726(3)/90 | [illegible] |
| 12 | $Mg_2(UO_2)(CO_3)_3(H_2O)_{18}$ | Bayleyite | $P2_1/a$ | 26.560(3)/90 | 15.256(2)/92.90(1) | 6.505(1)/90 | [illegible] |
| 13 | $CaMg(UO_2)(CO_3)_3(H_2O)_{12}$ | Swartzite | $P2_1/m$ | 11.080(2)/90 | 14.634(2)/99.43(1) | 6.439(1)/90 | [illegible] |
| 14 | $Ca_2(UO_2)(CO_3)_3(H_2O)_{11}$ | Liebigite | $Bba2$ | 16.699(3)/90 | 17.557(3)/90 | 13.697(3)/90 | [illegible] |
| 15 | $Ca_9(UO_2)_4(CO_3)_{13}·28H_2O$ | Markeyite | $Pmmn$ | 17.9688(13) | 18.4705(6) | 10.1136(4) | [illegible] |
| 16 | $Ca_8(UO_2)_4(CO_3)_{12}·21H_2O$ | Pseudomarkeyite | $P2_1/m$ | 17.531(3) | 18.555(3) | 9.130(3)/103.95(3) | [illegible] |
| 17 | $Sr_2UO_2(CO_3)_3(H_2O)_8$ | | $P2_1/c$ | 11.379(2)/90 | 11.446(2)/93.40 (1) | 25.653(4)/90 | [illegible] |
| 18 | $Na_6Mg(UO_2)_2(CO_3)_6·6H_2O$ | Leószilárdite | $C2/m$ | 11.6093(21)/90 | 6.7843(13)/91.378(3) | 15.1058(28)/90 | [illegible] |
| 19 | $Na_2Ca(UO_2)(CO_3)_3(H_2O)_{5.3}$ | Andersonite | $R$-3$m$ | 17.8448(4)/90 | 17.8448(4)/90 | 23.6688(6)/120 | [illegible] |
| 20 | $Na_2Ca_8(UO_2)_4(CO_3)_{13}·27H_2O$ | Natromarkeyite | $Pmmn$ | 17.8820(13) | 18.3030(4) | 10.2249(3) | [illegible] |
| 21 | $Ca_3Na_{1.5}(H_3O)_{0.5}(UO_2(CO_3)_3)_2(H_2O)_8$ | | $Pnnm$ | 18.150(3)/90 | 16.866(6)/90 | 18.436(3)/90 | [illegible] |
| 22 | $K_2Ca(UO_2)(CO_3)_3·6H_2O$ | Braunerite | $P2_1/c$ | 17.6725(12)/90 | 11.6065(5)/101.780(8) | 29.673(3)/90 | [illegible] |
| 23 | $K_2Ca_3[(UO_2)(CO_3)_3]_2·8(H_2O)$ | Linekite | $Pnnm$ | 17.0069(5)/90 | 18.0273(5)/90 | 18.3374(5)/90 | [illegible] |
| 23a | $K_2Ca_3((UO_2)(CO_3)_3)_2(H_2O)_6$ | | $Pnnm$ | 17.015(2)/90 | 18.048(2)/90 | 18.394(2)/90 | [illegible] |
| 24 | $SrMg(UO_2)(CO_3)_3(H_2O)_{12}$ | Swartzite-(Sr) | $P2_1/m$ | 11.216(2)/90 | 14.739(2)/99.48(1) | 6.484(1)/90 | [illegible] |
| 25 | $Na_{0.79}Sr_{1.40}Mg_{0.17}(UO_2(CO_3)_3)(H_2O)_{4.66}$ | | $Pa$-3 | 20.290(3)/90 | 20.290(3)/90 | 20.290(3)/90 | [illegible] |
| 26 | $MgCa_5Cu_2(UO_2)_4(CO_3)_{12}(H_2O)_{33}$ | Paddlewheelite | $Pc$ | 22.052(4)/90 | 17.118(3)/90.474 (2) | 19.354(3)/90 | [illegible] |

**Table 1.** *Cont.*

| No. | Chemical Formula | Mineral Name | Sp.Gr. | $a$, Å/ $\alpha$, ° | $b$, Å/ $\beta$, ° | $c$, Å/ $\gamma$, ° | Ref. |
|---|---|---|---|---|---|---|---|
| 27 | $Na_8[(UO_2)(CO_3)_3](SO_4)_2 \cdot 3H_2O$ | Ježekite | $P\text{-}62m$ | 9.0664(11)/90 | 9.0664(11)/90 | 6.9110(6)/120 | [ ] |
| 28 | $NaCa_3(UO_2)(CO_3)_3(SO_4)F(H_2O)_{10}$ | Schröckingerite | $P\text{-}1$ | 9.634(1)/91.41(1) | 9.635(1)/92.33(1) | 14.391(2)/120.26(1) | [ , ] |
| 29 | $MgCa_4F_2[UO_2(CO_3)_3]_2(H_2O)_{17.29}$ | Albrechtschraufite | $P\text{-}1$ | 13.569(2)/115.82(1) | 13.419(2)/107.61(1) | 11.622(2)/92.84(1) | [ , ] |
| 30 | $Ca_5(UO_2(CO_3)_3)_2(NO_3)_2(H_2O)_{10}$ |  | $P2_1/n$ | 6.5729(9)/90 | 16.517(2)/90.494(3) | 15.195(2)/90 | [ ] |
| 31 | $Ca_6(UO_2(CO_3)_3)_2Cl_4(H_2O)_{19}$ |  | $P4/mbm$ | 16.744(2)/90 | 16.744(2)/90 | 8.136(1)/90 | [ ] |
| 32 | $Ca_{12}(UO_2(CO_3)_3)_4Cl_8(H_2O)_{47}$ |  | $Fd\text{-}3$ | 27.489(3)/90 | 27.489(3)/90 | 27.489(3)/90 | [ ] |
| 33 | $Nd_2Ca[(UO_2)(CO_3)_3](CO_3)_2(H_2O)_{10.5}$ | Shabaite-(Nd) | $P\text{-}1$ | 8.3835(5)/90.058(3) | 9.2766(12)/89.945(4) | 31.7519(3)/90.331(4) | [ , ] |
| 34 | $[C(NH_2)_3]_4[UO_2(CO_3)_3]$ |  | $R3$ | 12.3278(1)/90 | 12.3278(1)/90 | 11.4457(2)/120 | [ ] |
| 35 | $(C_4H_{12}N)_4[UO_2(CO_3)_3] \cdot 8H_2O$ |  | $P2_1/n$ | 10.5377(18)/90 | 12.358(2)/99.343(4) | 28.533(5)/90 | [ ] |
| | $cc$0-1:2-10 | | | | | | |
| 36 | $[C(NH_2)_3]_6[(UO_2)_3(CO_3)_6](H_2O)_{6.5}$ |  | $P\text{-}1$ | 6.941(2)/95.63(2) | 14.488(2)/98.47(2) | 22.374 (2)/101.88(2) | [ ] |
| | **Nanoclusters** | | | | | | |
| 37 | $Mg_8Ca_8(UO_2)_{24}(CO_3)_{30}O_4(OH)_{12}(H_2O)_{138}$ | Ewingite | $I4_1/acd$ | 35.142(2)/90 | 35.142(2)/90 | 47.974(3)/90 | [ ] |
| | **Layers** | | | | | | |
| | $5^44^23^4$ ($\beta\text{-}U_3O_8$) | | | | | | |
| 38 | $CaU(UO_2)_2(CO_3)O_4(OH)(H_2O)_7$ | Wyartite | $P2_12_12_1$ | 11.2706(8)/90 | 7.1055(5)/90 | 20.807(1)/90 | [ - ] |
| 38a | $Ca(U(UO_2)_2(CO_3)_{0.7}O_4(OH)_{1.6})(H_2O)_{1.63}$ | Wyartite dehydrated | $Pmcn$ | 11.2610(6)/90 | 7.0870(4)/90 | 16.8359(10)/90 | [ ] |
| | $6^15^24^23^2$ (*phosphuranylite*) | | | | | | |
| 39 | $Ca(UO_2)_3(CO_3)_2O_2(H_2O)_6$ | Fontanite | $P2_1/n$ | 6.968(3)/90 | 17.276(7)/90.064(6) | 15.377(6)/90 | [ , ] |
| | $6^15^24^23^6$ (*roubaultite*) | | | | | | |
| 40 | $Cu_2(UO_2)_3(CO_3)_2O_2(OH)_2(H_2O)_4$ | Roubaultite | $P\text{-}1$ | 7.767(3)/92.16(4) | 6.924(3)/90.89(4) | 7.850(3)/93.48(4) | [ , ] |
| | $6^13^2$-I (*rutherfordine*) | | | | | | |
| 41 | $(UO_2)(CO_3)$ | Rutherfordine | $Imm2$ | 4.840(1)/90 | 9.273(2)/90 | 4.298(1)/90 | [ , - ] |
| 42 | $Ca(H_2O)_3[(UO_2)_3(CO_3)_{3.6}O_{0.2}]$ | Sharpite | $Cmcm$ | 4.9032(4) | 15.6489(11) | 22.0414(18) | [ ] |
| 42a | $Ca(UO_2)_6(CO_3)_5(OH)_4 \cdot 6H_2O$ | Sharpite | Orth | 21.99(2) | 15.63(2) | 4.487(4) | [ , ] |
| | $6^13^2$-II (*widenmannite*) | | | | | | |
| 43 | $Pb_2[(UO_2)(CO_3)_2]$ | Widenmannite | $Pmmn$ | 4.9350(7)/90 | 9.550(4)/90 | 8.871(1)/90 | [ , - ] |
| | **Layers of Miscellaneous Topology** | | | | | | |
| 44 | $Y_2(UO_2)_4(CO_3)_3O_4 \cdot 14H_2O$ | Kamotoite-(Y) | $P2_1/n$ | 12.3525(5) | 12.9432 (5)/99.857(3) | 19.4409(7) | [ , ] |
| 45 | $[(Y_{4.22}Nd_{3.78})(H_2O)_{25}(UO_2)_{16}O_8(OH)_8(CO_3)_{16}](H_2O)_{14}$ | Bijvoetite-(Y) | $B2_1$ | 21.234(3)/90 | 12.958(2)/90.00(2) | 44.911(7)/90 | [ , ] |
| 46 | $Ca(UO_2)(CO_3)_2 \cdot 5H_2O$ | Meyrowitzite | $P2_1/n$ | 12.376(3) | 16.0867(14)/107.679(13) | 20.1340(17) | [ ] |
| | **Minerals with Undefined Structures** | | | | | | |
| 47 | $Cu_2(Ce,Nd,La)_2(UO_2)(CO_3)_5(OH)_2 \cdot 1.5H2O$ | Astrocyanite-(Ce) | Hex | 14.96(2)/90 | 14.96(2)/90 | 26.86(4)/120 | [ ] |
| 48 | $(UO_2)(CO_3) \cdot H_2O$ | Blatonite | Hex or Trig | 15.79(1)/90 | 15.79(1)/90 | 23.93(3)/120 | [ ] |
| 49 | $(UO_2)(CO_3) \cdot nH_2O$ | Joliotite | Orth | 8.16 | 10.35 | 6.32 | [ ] |
| 50 | $CaGd_2(UO_2)_{24}(CO_3)_8Si_4O_{28} \cdot 60H_2O$ | Lepersonnite-(Gd) | $Pnnm$ or $Pnn2$ | 16.23(3)/90 | 38.74(9)/90 | 11.73(3)/90 | [ ] |
| 51 | $Ca(UO_2)(CO_3)_2 \cdot 3H_2O$ | Metazellerite | $Pbn2_1$ or $Pbnm$ | 9.718(5) | 18.226(9) | 4.965(4) | [ ] |
| 52 | $(UO_2)_2(CO_3)(OH)_2 \cdot 4H_2O$ | Oswaldpeetersite | $P2_1/c$ | 4.1425(6) | 14.098(3)/103.62(1) | 18.374(5) | [ ] |
| 53 | $Ca_3Mg_3(UO_2)_2(CO_3)_6(OH)_4 \cdot 18H_2O$ | Rabbitite | Mon | 32.6(1) | 23.8(1)/~90 | 9.45(5) | [ ] |
| 54 | $Ca(UO_2)_3(CO_3)(OH)_6 \cdot 3H_2O$ | Urancalcarite | $Pbnm$ or $Pbn2_1$, | 15.42(3) | 16.08(4) | 6.970(6) | [ ] |
| 55 | $Ca_2Cu(UO_2)(CO_3)_4 \cdot 6H_2O$ | Voglite | $P2_1$ or $P2_1/m$ | 25.97 | 24.50/104.0 | 10.70 | [ , ] |
| 56 | $Ca(UO_2)(CO_3)_2 \cdot 5H_2O$ | Zellerite | $Pmn2_1$ or $Pmnm$ | 11.220(15) | 19.252(16) | 4.933(16) | [ , ] |
| 57 | $CaZn_{11}(UO_2)(CO_3)_3(OH)_{20} \cdot 4H_2O$ | Znucalite | Orth | 10.72(1) | 25.16(1) | 6.325(4) | [ ] |
| 57a | $CaZn_{12}(UO_2)(CO_3)_3(OH)_{22} \cdot 4H_2O$ | Znucalite | Tricl | 12.692(4)/89.08(2) | 25.096(6)/91.79(2) | 11.685(3)/90.37(3) | [ ] |

## 2.2. Graphical Representation and Anion Topologies

The crystal structures of uranyl carbonate minerals and synthetic compounds discussed in the current review are based on the finite clusters, chains, layers, and even nanoclusters built by the linkage of U-centered coordination polyhedra with each other and carbonate groups. U(VI) atoms make two short $U^{6+} \equiv O^{2-}$ bonds to form approximately linear $UO_2^{2+}$ uranyl cations (*Ur*), which are in turn surrounded in the equatorial plane by other four-to-six O atoms, resulting in the formation of a tetra-, penta-, or hexagonal bipyramid, as a coordination polyhedron of U(VI) atoms. The carbonate group is the simplest oxocarbon anion with C atom arranged in the center of a flat triangle and O atoms occupying all three of its vertices.

Topological analysis of the uranyl carbonate structural complexes is based on the description of *Ur* and $CO_3$ interpolyhedral linkage. The topology of the uranyl carbonate finite clusters and chains can be described using the theory of graphical (nodal) representation, which was first proposed by Hawthorne [110], and subsequently modified by Krivovichev [111,112], in which black and white nodes correspond to *Ur* and $CO_3$ groups, respectively; and the single or double line between the nodes corresponds to a vertex- or edge-sharing way of polyhedra polymerization, respectively (Figure 1a–c). Anion topology approach was proposed by Burns and co-authors [113,114] to describe the U-bearing crystal structures that are based upon sheets with edge-sharing linkage. To obtain anion topology of the layer, all cations, $O_{Ur}$, and O atoms coordinating only one cation should be removed, while the rest of O atoms should be linked via single lines up to c.a. 3.5 Å length (Figure 1d–f). The black-and-white graph is labeled with a special index *cc*D–U:$CO_3$–#, where *cc* means "cation-centered", D specifies dimensionality (0—finite clusters, 1—chains and 2—sheets), U:$CO_3$ ratio, #—registration number of the unit. Sheet-anion topology is indicated by the ring symbol, $p_1^{r_1} p_2^{r_2} \ldots$ , where $p$ is the number of vertices in a topological cycle, and $r$ is the amount of the respective cycle in the reduced section of the sheet.

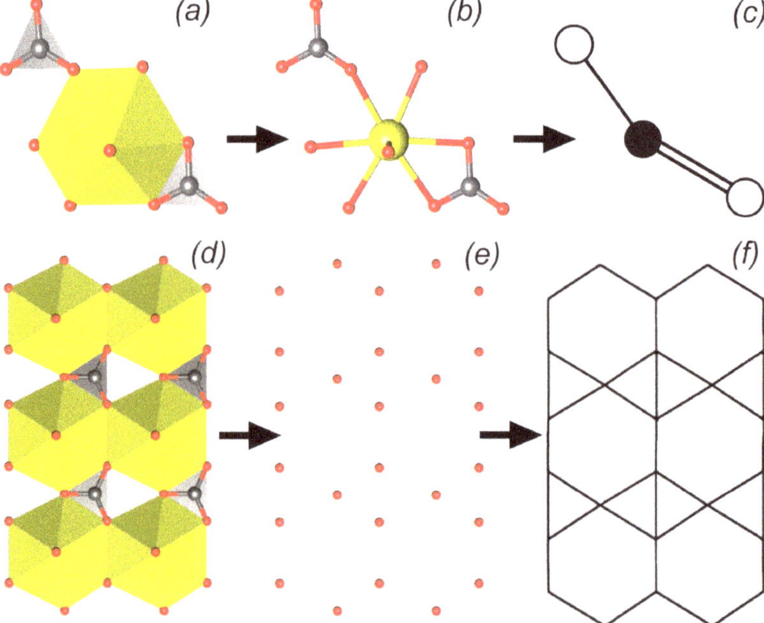



**Figure 1.** Illustration of building topologies. Combination of *Ur* bipyramid with edge- and vertex-shared triangular carbonate groups in polyhedral (**a**), ball-and-stick (**b**) representations, and respective black-and-white graph (**c**). Fragment of the dense uranyl-carbonate layer with edge-sharing interpolyhedral linkage (**d**), O atoms that are involved in linkage with more than one cation (**e**), and the resulted anion topology built on them (**f**). Legend: U-bearing coordination polyhedra = yellow; U atoms = yellow; C atoms = grey; O atoms = red; black nodes = U atoms, white nodes = C atoms; $CO_3$ groups are shown in a ball-and-stick model with grey filling; see Section 2.2 for details.

### 2.3. Complexity Calculations

Structural complexity calculations allow to compare the structures and quantitatively characterize the impact of each substructural unit (i.e., U-bearing complex, interstitial cations, hydration state, etc.) in terms of their information content on the formation of a particular architecture. This approach was recently developed by Krivovichev [115–119], successfully implemented in a number of works (e.g., [120–123]), and is based on the Shannon information content calculations of per atom ($I_G$) and per unit cell ($I_{G,total}$) using the following equations:

$$I_G = -\sum_{i=1}^{k} p_i \, \log_2 \, p_i \quad (\text{bits/atom}) \tag{1}$$

$$I_{G,total} = -v \, I_G = -v \sum_{i=1}^{k} p_i \, \log_2 \, p_i \quad (\text{bits/cell}) \tag{2}$$

where $k$ is the number of different crystallographic orbits (independent sites) in the structure and $p_i$ is the random choice probability for an atom from the $i$-th crystallographic orbit, that is:

$$p_i = m_i/v \tag{3}$$

where $m_i$ is a multiplicity of a crystallographic orbit (i.e., the number of atoms of a specific Wyckoff site in the reduced unit cell), and $v$ is the total number of atoms in the reduced unit cell.

The direct comparison of complexity parameters is possible only for the structures with identical or very close chemical composition (e.g., polymorphs), while changes in hydration state or interstitial complexes could significantly affect the overall complexity behavior. Thus, the structural complexity parameters of various building blocks (uranyl carbonate units, interstitial structure, H-bonding system) were calculated according to the recently suggested algorithm [124–127], so-called "ladders of information", to analyze their contributions into the complexity of the whole structure, and to distinguish which of the blocks plays the most important role, and which has the most influence on symmetry preservation or reduction.

## 3. Results

### 3.1. Uranyl Carbonate Minerals

Uranyl carbonates are one of the most abundant U(VI)-containing minerals at the Earth's surface or near the surface. Their crystallization is possible due to ubiquitous $CO_2$ dissolved in percolating aqueous solutions and can also be amplified by the dissolution of the carbonate minerals in the rock or hydrothermal veins (such as calcite or dolomite). Uranyl-carbonate aqueous species can be important constituents of groundwater under neutral to alkaline conditions as being thermodynamically favored [128,129]. Under such pH conditions, the dominating aqueous species are uranyl monocarbonate, $(UO_2)(CO_3)^0$, uranyl dicarbonate, $(UO_2)(CO_3)_2^{2-}$, and uranyl tricarbonate, $(UO_2)(CO_3)_3^{4-}$, with p$K$ values of 5.5, 7, and 9, respectively [128]. The precipitation of solid phases from the solutions is usually connected with a higher evaporation rate or local oversaturation. Abandoned mining tunnels and adits can serve us as a model for such a situation. Efflorescences of uranyl carbonate minerals precipitating on the mining walls are typical indicators of uranium mineralization in closer or larger distances. Such precipitates have been found hundreds of meters far from the primary U mineralization in the mines (as e.g., in Jáchymov, Czech Republic).

In general, from a genetical point of view and considering the processes that led to the uranyl carbonate formation, we can distinguish among the two groups of uranyl carbonate minerals. The first comprises primary uranyl carbonates, or those that are connected genetically somewhat more with the primary mineralization. The second then comprises a group of uranyl carbonates of the recent or sub-recent origin, usually formed as the precipitates on the walls of the mining adits. Minerals are present, such as sharpite, $Ca[(UO_2)_3O_{0.2}(CO_3)_{3.6}](H_2O)_3$ [88,89], roubaultite, $Cu_2[(UO_2)_3(CO_3)_2(OH)_2](H_2O)_4$ [83] or fontanite, $Ca[(UO_2)_3(CO_3)_2O_2](H_2O)_6$ [81,96]. These minerals are usually tightly spatially connected either with primary minerals (even in relics) or other U(VI) non-carbonate supergene minerals, such as silicates of uranyl-oxides hydroxy-hydrate minerals. As the second, distinctive group, we can consider uranyl carbonate minerals such as grimselite, $K_3Na(UO_2)(CO_3)_3(H_2O)$ [22,23,130], bayleyite, $Mg_2[(UO_2)(CO_3)_3](H_2O)_{18}$ [34,35], or paddlewheelite, $MgCa_5Cu_2[(UO_2)(CO_3)_3]_4(H_2O)_{33}$ [61]. These are usually higher-hydrates oxysalts, tending to appear in powdery aggregates, efflorescence, or curved uneven crystals.

The first uranyl carbonate minerals were discovered in the 19th century. One of the first discovered was mineral liebigite, $Ca_2[UO_2)(CO_3)_3](H_2O)_{~11}$ [40,131], reported in 1848 from Turkey by J.B. Smith and named after German chemist Justus von Liebig. Subsequently, this mineral was more precisely described and characterized from Jáchymov in Czechia (those times a part of the Austro-Hungarian Empire) under the name Uran-calc-carbonat, or Uranothallite, and then all were described to be the same mineral equal to liebigite [132]. Jáchymov became the famous and rich locality for many uranyl-carbonates. Many have been found there for the first time and thus Jáchymov remains one of the richest localities, even for the type uranyl carbonate minerals (from approximately 96 U(VI) supergene minerals, 19 are uranyl carbonates, thus 20% and 10 of them are type-minerals).

Until ca. 2015, there were only U-carbonates from the hydrothermal Variscan-type of U-deposits studied more in detail. Since extensive sampling campaigns at the Red Canyon area (San Juan County, UT, USA) several new particularly interesting U-carbonates with novel structural features and topologies have been discovered [42,98]. As current studies document that at the more detailed scale (micro- to nano-sized), the mineralogy of the localities is of course more diverse than thought, new uranyl carbonates are likely to be discovered. Furthermore, the discovery of U-nano-cages-containing the mineral ewingite [75], has brought up questions about the possible role of nano-scale clusters and cages in the processes of dissolution and formation of uranyl carbonates and uranyl minerals in general. It may be necessary to account for such nano-cages in the geochemical models of particular uranium-bearing systems.

### 3.2. Synthetic Uranyl Carbonates

First of all, it should be noted that the number of synthetic phases for which the structures were determined is significantly inferior to the structurally characterized natural uranyl carbonates in the ratio of 19:32. Whereas for other groups of U(VI)-bearing compounds this proportion is usually opposite [8,127,133]. The first structurally characterized synthetic uranyl carbonate, to our knowledge, was one of the simplest phases **4a**, sodium-bearing $Na_4(UO_2)(CO_3)_3$ [17]. It is of interest that the first crystal structure of the natural uranyl carbonate, rutherfordine (**41**), was reported the year before [85]. The papers of K. Mereiter from the TU Wien (Austria) should be certainly noted in the first row among the works devoted to the synthesis and structural studies of synthetic uranyl carbonates. Then, the studies carried out by the A.M. Fedoseev and I.A. Charushnikova from the Frumkin Institute of Physical chemistry and Electrochemistry RAS (Russian Federation) and by V.N. Serezhkin from the Samara State University (Russian Federation) should be mentioned. A substantial portion of the synthetic uranyl carbonate compounds was synthesized and studied by P. C. Burns and his colleagues from the University of Notre Dame (USA), who significantly contribute to the studies of uranyl carbonate minerals as well.

All synthetic experiments can be roughly divided into two groups. Moreover, the majority of inorganic uranyl carbonates were synthesized by evaporation at room temperature and only a few of them were obtained from hydrothermal conditions. Uranyl nitrate hexahydrate was usually used as the source of U. But in some experiments, more specific reagents were used: $UO_2$ powder (for **1** and **2**), $UO_2(CO_3)$ (for **7a**), $Ag_4[UO_2(CO_3)_3]$ (for **8a**), $\alpha$-$UO_2MoO_4(H_2O)_2$ (for **10**), and $Na_4UO_2(CO_3)_3$ (for **21**). Potassium, sodium, cesium, or thallium carbonates were used as the source of $CO_3$ ions within the synthetic experiments. Compounds **36** and **10** can be considered as exceptions due to the usage of guanidine carbonate and carbamide in the respective syntheses. Several protocols of synthetic experiments deserve special attention. Thus, compound **1** [11] was formed as the result of the dissolution of $UO_2$ powder in the solution of $K_2CO_3$ and $H_2O_2$ at room temperature. Later it was filtered through a 0.45 μm polyamide syringe filter, and an additional 1.5 mL of 35 wt% $H_2O_2$ was added. Afterward it was transferred to a borosilicate scintillation vial and layered with methanol. The compound **2** [12] was obtained similarly, except for the scintillation vial step. The crystals of **4a** [16] were obtained by hydrothermal synthesis at 135 °C in a sealed silica glass tube at about 20 MPa. The compound **7a** [27] was synthesized by evaporation at room temperature, but before being left to evaporate, the dissolution of the precipitate was achieved via heating the solution over a steam bath. The crystals of **7b** [28] were obtained from the solution that was stirred for five days. The compound **11** [33] was synthesized by slow addition of uranyl nitrate solution to the solution of $Tl_2(CO_3)_2$ at the temperature of c.a. 57 °C. The solvent was removed in a vacuum desiccator for two months from the resulting yellow-green solution, and crystals were then taken from the precipitate. To obtain the crystals of **36** [74], the initial solution was stirred vigorously for several days, afterwards, it was centrifuged and the supernatant was removed via pipet. The crystals were obtained from the precipitate, which was slowly cooled to 5 °C under a $CO_2$ atmosphere.

Nine of uranyl carbonate minerals have synthetic analog, which was also obtained mostly by evaporation at room temperature, except for the crystals of **7** [26], which were formed during the two months of evaporation at vacuum-desiccator. The compound **4a** [15] was synthesized by hydrothermal reaction at 220 °C. The crystals of **41** [20] were obtained by purging the solution of $UO_3$ with 70 kPa $CO_2$ at the glove box for 24 h.

Very special attention can be paid to compound **4a** [15], which may be the same sodium uranyl carbonate that was found among alteration products of the "lavas" resulting from the nuclear accident of the Chernobyl nuclear power plant [6,7].

### 3.3. Topological Analysis

The majority of uranyl carbonate crystal structures are based on finite clusters, which are represented by only two topological types (Table 1). The topological variety of layered uranyl carbonate complexes is significantly larger; however, the amount of compounds, which structures are based on the 2D units, is much lower. There are only ten uranyl carbonate compounds known with a layered structure, and all of them are natural phases.

The crystal structures of two synthetic K-bearing uranyl carbonates **1** [11] and **2** [12] differing only in the hydration state and one uranyl carbonate, templated by guanidinium molecules **3** [13], are based on the finite clusters of the *cc0*-1:2-9 topological type (Figure 2a,b). In terms of polyhedral representation, these clusters can be described as a hexagonal bipyramid that shares two of its equatorial edges with $CO_3$ groups spaced by one non-shared equatorial edge. There is a peroxide molecule arranged in the equatorial plane spaced from a carbonate group by another non-shared edge. Such topological type was also described in the structures of two uranyl nitrate compounds, pure inorganic [134] and organically templated [135], in which the peroxide group was replaced by two $H_2O$ molecules. It should be noted that this topological type has an isomer, if two equatorial $CO_3$ groups are *trans*-arranged, being spaced by two non-shared edges, in contrast to the *cis*-arrangement in the structures of **1–3**. The *trans*-isomer is one of the most common types of uranyl nitrate finite clusters [136], while it has not been observed in the structures of uranyl carbonates at all.

Uranyl tricarbonate cluster (UTC), which is shown in Figure 2c, is the most common structural unit among the natural and synthetic uranyl carbonate phases. There are 39 compounds known (Table 1), whose structures are based on these finite clusters, and is in sum 2.5 times more than the amount of all other structurally characterized uranyl carbonates (**15**). The topology of UTC belongs to the *cc0*-1:3-2 type (Figure 2d). This topology can be obtained from the previous *cis-cc0*-1:2-9 type by the replacement of the peroxide molecule by the third $CO_3$ group, resulting in the formation of a triangular cluster with the uranyl hexagonal bipyramid arranged in its core and ideal -6*m*2 point group symmetry.

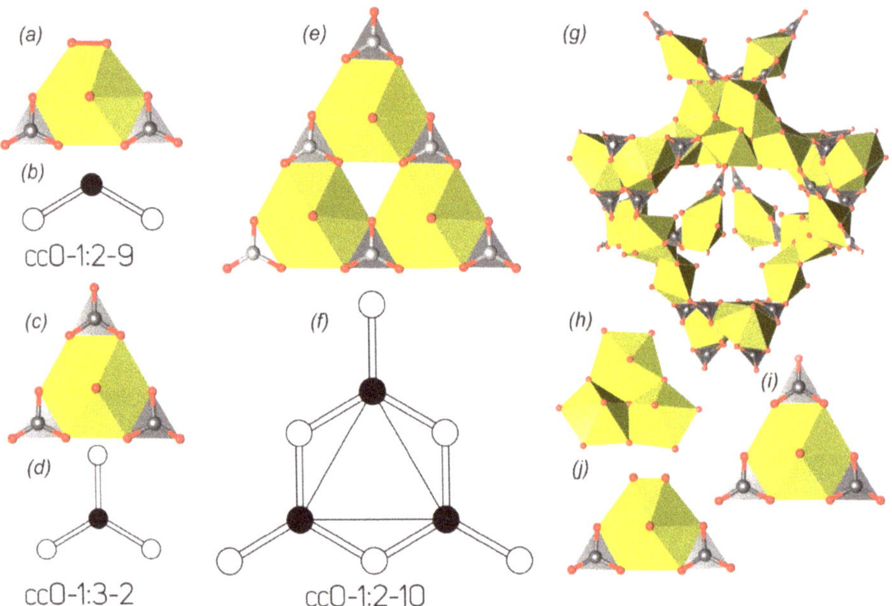

**Figure 2.** Finite clusters in the crystal structures of natural and synthetic uranyl carbonates and their graphical representations (see Table 1 and text for details). Legend: see Figure 1; peroxide molecule is indicated by red bond (**a**); see Section 3.3 for details.

Compound **36** [74], to our knowledge, is the only compound whose structure is based on the triuranyl hexacarbonate finite cluster (Figure 2e). The cluster is built by three vertex-sharing in a cyclic manner uranyl hexagonal bipyramids. Each cavity at the exterior side of such a cycle is occupied by a $CO_3$ group to form a large triangle, each side of which is built by alternating two bipyramids and three carbonate groups. The topology of the uranyl carbonate cluster (Figure 2f) in the structure of **36** belongs to the *cc*0-1:2-10 type. The architecture of this cluster can be also described as trebling of the UTC cluster with keeping triangular motif and ideal -6*m*2 point group symmetry.

Probably the most remarkable structure not only among the uranyl carbonate compounds but among all known minerals, was described in ewingite (**37**) [75]. Ewingite is a calcium-magnesium oxo-hydroxy-hydrate uranyl carbonate natural phase, whose structure is built by 24 uranyl pentagonal and hexagonal bipyramids interlinked with each other and $CO_3$ groups, to form nanoclusters 2.3 nm in diameter (Figure 2g). Three fundamental building units can be distinguished within the uranyl carbonate cluster in ewingite. These are 4 trimers of edge-sharing pentagonal bipyramids, 6 *cis*-isomer of the uranyl bicarbonate unit (*cc*0-1:2-9), and 6 UTC complex (Figure 2h–j). Moreover, linkage of all building units occurs only through the carbonate groups. Mg and Ca ions as well as $H_2O$ molecules are arranged both inside and in between the U-bearing nanoclusters.

The crystal structures of wyartite (**38**) [76–79] and its dehydrated form are based on the similar layered complexes that belong to the so-called β-$U_3O_8$-sheet anion topology (Figure 3a,b). Topology has the $5^4 4^2 3^4$ ring symbol and consists of infinite chains of edge-sharing pentagons that are linked with the neighbor chains through the common vertices and separated by the chains of squares and triangles. The crystal structures of **38** and **38a** are remarkable for being the only U(V)-bearing natural phases. Pentagons are occupied by the $U^{6+}$ ions, squares correspond to the irregular $U^{5+}O_7$ polyhedra, whereas triangles are vacant. Carbonate groups share edges with the U$^{5+}$-centered polyhedra and are arranged towards the interlayer space.

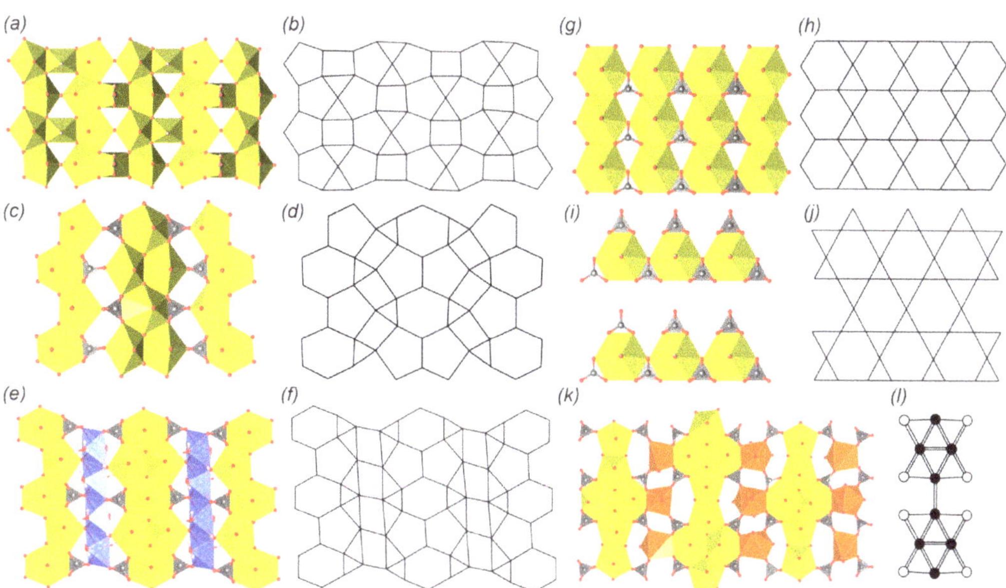

**Figure 3.** 2D complexes in the crystal structures of natural and synthetic uranyl carbonates and their anion topologies or graphical representation (see Table 1 and text for details). Legend: see Figure 1; Cu polyhedra = blue, Nd or Y polyhedra = orange; see Section 3.3 for details.

The crystal structure of fontanite (**39**) [80,81] is based on the layered uranyl carbonate complexes, which correspond to the, so-called, phosphuranylite anion topology [137] (Figure 3c,d) with the $6^1 5^2 4^2 3^2$ ring symbol. The topology consists of two types of alternating infinite chains. The first type of chains is formed by edge-sharing dimers of pentagons that are interlinked by edge-sharing hexagons. The second type of chain is formed by alternating edge-sharing triangles and squares. All hexagons and pentagons are occupied by the uranyl ions, all triangles are occupied by carbonate groups, while all squares are vacant. It should be noted that phosphuranylite anion topology is very common among the U-bearing natural and synthetic phases and is represented by a wide variety of isomers, which differ in the occupancy of polygons. Thus, hexagons may be vacant, and triangles may be occupied by tetrahedral, trigonal pyramidal, or planar trigonal oxyanions (e.g., [8,127]).

Another anion topology that consists of hexagons, pentagons, squares and triangles with the $6^1 5^2 4^2 3^6$ ring symbol (Figure 3e,f) was described in the structure of roubaultite (**40**) [82,83]. Roubaultite anion topology contains the same infinite chains of edge-sharing pentagon dimers linked by edge-sharing hexagons that were observed in the phosphuranylite topology. But in the structure of **40**, these chains are separated by infinite chains of edge-sharing squares decorated with trimers of edge-sharing triangles on the sides. All hexagons and pentagons are also occupied by the uranyl ions, as I was realized in the phosphuranylite topology. Squares are occupied by Cu-centered slightly distorted octahedra. The middle triangle from each trimer is occupied by the carbonate group, leaving the other two triangles vacant.

The simplest uranyl carbonate, at least according to the chemical composition, rutherfordine (UO$_2$)(CO$_3$) [20,84–87], has a layered structure. The anion topology is also rather simple; it consists of parallel chains of edge-sharing hexagons separated by hourglass dimers of edge-sharing triangles (Figure 3g,h). All hexagons are occupied by the uranyl ions, and one triangle from each dimer is occupied by the CO$_3$ group, keeping the second triangle vacant. The crystal structure of sharpite (**42**) is also based on the layered complexes that belong to the same rutherfordine anion topology. However, the polyhedral representation appeared to be much more complex. Thus, the layer can be described as being formed by the 1D modules of rutherfordine topology. Each module represents a triple band of edge-sharing uranyl hexagonal bipyramids, a part of the triangular spaces between which are occupied by carbonate groups. These modules are arranged in the structure at approximate right angles to each other and are linked by the Ca-centered polyhedra, which are arranged on the crests of sawtooth waves (Figure 4a,b). Despite the curvature of such zigzag layers (in contrast to flat layered structure in rutherfordine) and the arrangement of Ca ions in the centers of square antiprisms, projection of such corrugated layers onto the (010) plane corresponds to the rutherfordine topology, with the equatorial arrangement of Ca polyhedra ligands having hexagon shape.

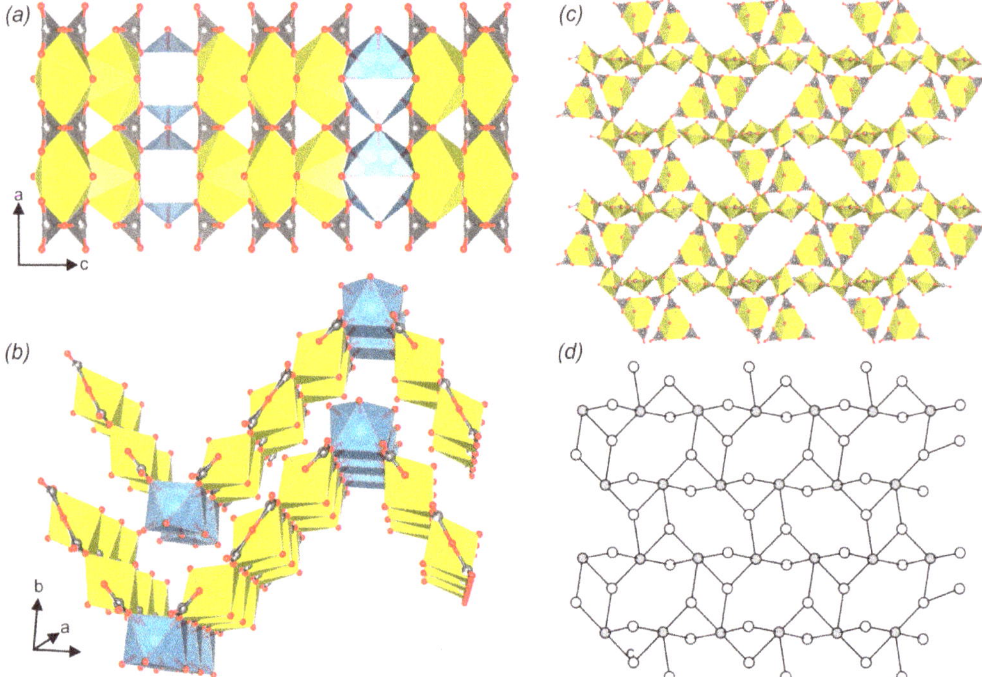

**Figure 4.** The crystal structures of sharpite (**a**,**b**) and meyrowitzite (**c**) with simplified topological representation (**d**). Legend: see Figure 1; Ca polyhedra = light blue.

The crystal structure of widenmannite (**43**) [21,91–93] is based on layered complexes, the topology of which consists of hexagons and triangles with the same $6^1 3^2$ ring symbol as was found in the structure of rutherfordine. However, the arrangement of polygons in both structures is different. Widenmannite anion topology is built by the hexagons linked by vertex-sharing to other six hexagons, while all of its six edges are shared with triangles, thus forming trihexagonal tiling, which was used by Johannes Kepler in his book [138] and is also known under the kagome pattern name. In the ideal structure of the widenmannite (Figure 3i,j) each second row of hexagonal bipyramids should be vacant, but in the real structure the disorder with partial occupancy of the U sites takes place, which results in the occupation of all hexagons by the uranyl ions and half of the triangles oriented in the same direction are occupied by carbonate groups, keeping another half vacant.

The crystal structures of two *REE*-bearing minerals kamotoite-(Y) (**44**) [94,95] and bijvoetite-(Y) (**45**) [96,97] are based on highly remarkable and very rare layered complexes that have not been observed in any other natural or synthetic compound. The topology of the 2D complex is based on infinite chains of alternating dimers of edge-sharing uranyl hexagonal and pentagonal bipyramids (Figure 3k,l). Dimers of pentagonal bipyramids are arranged along the chain's extension, while dimers of hexagonal bipyramids are arranged perpendicularly. Each hexagonal bipyramid shares two of its oblique equatorial edges, not taking part in the linkage between U polyhedra, with $CO_3$ groups. These chains are linked into the 2D structure via irregular $Y^{3+}$- or $Nd^{3+}$-centered coordination polyhedra through the 6th non-shared equatorial edge from one chain and two O atoms of two carbonate groups from the neighbor chain. It should be noted that the resulting U- and *REE*-bearing layers are electroneutral, so the 3D structure formation is provided by the H-bonding system, which involves $H_2O$ molecules from the coordination sphere of *REE* atoms and from the interlayer space.

The final to date topological type, which has been described in the structures of natural and synthetic uranyl carbonates, is observed in the structure of calcium uranyl carbonate mineral meyrowitzite (**46**) [98]. The structure of the layered complex is composed of UTC clusters sharing apical vertices of $CO_3$ groups with uranyl pentagonal bipyramids (Figure 4c). This structural type, unlike the rest of the topologies described herein, is the least dense in terms of the interconnection of U coordination polyhedra, and its topology will become clearer if graphical representation is used for illustration (Figure 4d). Thus, if uranyl pentagonal bipyramid is represented by grey nods, U hexagonal bipyramids by white nods, and the line between nods appearing if pentagonal bipyramid shares an equatorial O atom with the $CO_3$ group from the UTC cluster, the resulting graph of the complex will correspond to one of the most common *cc*1-1:2-4 topological type among the U(VI) bearing structures in general (e.g., [8,127]). Since the concept of graphical representation is violated, this description is more appropriate to use not as a direct interpretation of the topology, but as an approximate model.

### 3.4. Structural and Topological Complexity

Structural complexity measures have been implemented in several stages and the results of calculations are listed in Figures 5 and 6 and Table 2. At the first stage, the topological complexity (**TI**), according to the maximal point (for finite clusters) or layer symmetry group has been determined, as these are the basic structure building units. At the second stage, the structural complexity (**SI**) of the U-bearing complexes has been calculated taking into account its real symmetry. The next informational contribution comes from the stacking (**LS**) of finite clusters and layered complexes (if more than one complex is within the unit cell). The fourth contribution to the total structural complexity is derived by the interstitial structure (**IS**). The last portion of information comes from the interstitial H bonding system (**H**). It should be noted that the H atoms related to the U-bearing clusters and layers were considered as a part of those complexes, but not within the contribution of the H-bonding system. Complexity parameters for the whole structure have been determined using *ToposPro* software [139].

**Table 2.** Structural and topological complexity parameters for the uranyl carbonate compounds.

| No. | Formula | Topology | Sp. Gr. | $\nu$ | $I_G$ | $I_{G,total}$ | Struct. U-C: Point, Rod or Layer Sym. Gr. | $\nu$ | $I_G$ | $I_{G,total}$ | Topol. U-C: Point, Rod or Layer Sym. Gr. | $\nu$ | $I_G$ | $I_{G,total}$ |
|---|---|---|---|---|---|---|---|---|---|---|---|---|---|---|
| | **Finite clusters** | | | | | | | | | | | | | |
| 1 | $K_4(UO_2(O_2)(CO_3)_2)(H_2O)$ | | $P2_1/n$ | 80 | 4.322 | 345.750 | | | | | | | | |
| 2 | $K_4(U(CO_3)_2)O_2(O_2)(H_2O)_{2.5}$ | $cc0$-1:2-9 | $P2_1/n$ | 100 | 4.644 | 464.390 | 1 | 13 | 3.700 | 48.106 | $2mm$ | 13 | 2.777 | 36.101 |
| 3 | $(CN_3H_6)_4[UO_2(CO_3)_2(O_2)]\cdot H_2O$ | | $Pca2_1$ | 204 | 5.672 | 1157.175 | | | | | | | | |
| 4 | $Na_4(UO_2)(CO_3)_3/$**Čejkaite** | | $Cc$ | 38 | 4.248 | 161.420 | | | | | | | | |
| 4b* | $Na_4(UO_2)(CO_3)_3/$**Cejkaite** old model | | $P$-1 | 76 | 5.274 | 400.840 | | | | | | | | |
| 8 | $Cs_4UO_2(CO_3)_3(H_2O)_6$ | | $P2_1/n$ | 148 | 5.209 | 771.000 | | | | | | | | |
| 8a | $Cs_4(UO_2(CO_3)_3)(H_2O)_6$ | | $P2_1/n$ | 148 | 5.209 | 771.000 | | | | | | | | |
| 12 | $Mg_2(UO_2)(CO_3)_3(H_2O)_{18}/$**Bayleyite** | | $P2_1/a$ | 276 | 6.123 | 1689.950 | | | | | | | | |
| 14 | $Ca_2(UO_2)(CO_3)_3(H_2O)_{11}/$**Liebigite** | | $Bba2$ | 200 | 5.664 | 1132.770 | | | | | | | | |
| 16 | $Ca_8(UO_2)_4(CO_3)_{12}\cdot 21H_2O/$**Pseudomarkeyite** | | $P2_1/m$ | 262 | 6.148 | 1610.760 | | | | | | | | |
| 17 | $Sr_2UO_2(CO_3)_3)(H_2O)_8$ | | $P2_1/c$ | 324 | 6.34 | 2054.110 | 1 | 15 | 3.907 | 58.603 | | | | |
| 21 | $Ca_3Na_{1.5}(H_3O)_{0.5}(UO_2(CO_3)_3)_2(H_2O)_8$ | | $Pnnm$ | 484 | 6.059 | 2932.730 | | | | | | | | |
| 22 | $K_2Ca(UO_2)(CO_3)_3\cdot 6H_2O/$**Braunerite** | | $P2_1/c$ | 472 | 6.883 | 3248.610 | | | | | | | | |
| 23 | $K_2Ca_3[(UO_2)(CO_3)_3]_2\cdot 8(H_2O)/$**Linekite** | | $Pnnm$ | 474 | 6.049 | 2867.260 | | | | | | | | |
| 23a | $K_2Ca_3((UO_2(CO_3)_3)_2(H_2O)_6$ | | $Pnnm$ | 462 | 5.99 | 2767.510 | | | | | | | | |
| 25 | $Na_{0.79}Sr_{1.40}Mg_{0.17}(UO_2(CO_3)_3)(H_2O)_{4.66}$ | | $Pa$-3 | 752 | 5.048 | 3795.880 | | | | | | | | |
| 26 | $MgCa_5Cu_2(UO_2)_4(CO_3)_{12}(H_2O)_{33}/$**Paddlewheelite** | | $Pc$ | 616 | 8.267 | 5092.340 | | | | | | | | |
| 28 | $NaCa_3(UO_2)(CO_3)_3(SO_4)F(H_2O)_{10}/$**Schröckingerite** | | $P$-1 | 110 | 5.781 | 635.950 | | | | | $-6m2$ | 15 | 2.106 | 31.584 |
| 29 | $MgCa_4F_2[UO_2(CO_3)_3]_2(H_2O)_{17.29}/$**Albrechtschraufite** | | $P$-1 | 179 | 6.489 | 1161.600 | | | | | | | | |
| 30 | $Ca_5(UO_2(CO_3)_3)_2(NO_3)_2(H_2O)_{10}$ | $cc0$-1:3-2 | $P2_1/n$ | 146 | 5.204 | 759.710 | | | | | | | | |
| 33 | $Nd_2Ca[(UO_2)(CO_3)_3](CO_3)_2(H_2O)_{10.5}/$**Shabaite-(Nd)** | | $P$-1 | 230 | 6.845 | 1574.460 | | | | | | | | |
| 35 | $(C_4H_{12}N)_4[UO_2(CO_3)_3]\cdot 8H_2O$ | | $P2_1/n$ | 428 | 6.741 | 2885.348 | | | | | | | | |
| 7 | $K_4(UO_2)(CO_3)_3/$**Agricolaite** | | $C2/c$ | 38 | 3.406 | 129.420 | | | | | | | | |
| 7a | $K_4UO_2(CO_3)_3$ | | $C2/c$ | 38 | 3.406 | 129.420 | | | | | | | | |
| 7b | $K_4(UO_2)(CO_3)_3$ | | $C2/c$ | 38 | 3.406 | 129.420 | 2 | 15 | 3.107 | 46.603 | | | | |
| 9 | $Cs_4(UO_2(CO_3)_3)$ | | $C2/c$ | 38 | 3.406 | 129.420 | | | | | | | | |
| 10 | $(NH_4)_4(UO_2(CO_3)_3)$ | | $C2/c$ | 70 | 4.215 | 295.050 | | | | | | | | |
| 11 | $Tl_4((UO_2)(CO_3)_3)$ | | $C2/c$ | 38 | 3.406 | 129.420 | | | | | | | | |
| 13 | $CaMg(UO_2)(CO_3)_3(H_2O)_{12}/$**Swartzite** | | $P2_1/m$ | 106 | 4.973 | 527.160 | | | | | | | | |
| 15 | $Ca_9(UO_2)_4(CO_3)_{13}\cdot 28H_2O/$**Markeyite** | | $Pmmn$ | 300 | 5.495 | 1648.650 | | | | | | | | |
| 19 | $Na_2Ca(UO_2)(CO_3)_3(H_2O)_{5.3}/$**Andersonite** | | $R$-3$m$ | 203 | 4.461 | 905.490 | $m$ | 15 | 3.240 | 48.600 | | | | |
| 20 | $Na_2Ca_8(UO_2)_4(CO_3)_{13}\cdot 27H_2O/$**Natromarkeyite** | | $Pmmn$ | 310 | 5.599 | 1735.600 | | | | | | | | |
| 24 | $SrMg(UO_2)(CO_3)_3(H_2O)_{12}/$**Swartzite-(Sr)** | | $P2_1/m$ | 106 | 4.973 | 527.160 | | | | | | | | |
| 18 | $Na_6Mg(UO_2)_2(CO_3)_6\cdot 6H_2O/$**Leószilárdite** | | $C2/m$ | 55 | 4.363 | 239.980 | $m$ | 15 | 3.774 | 56.610 | | | | |
| 31 | $Ca_6(UO_2(CO_3)_3)_2C_{14}(H_2O)_{19}$ | | $P4/mbm$ | 174 | 4.282 | 745.070 | $mm2$ | 15 | 3.107 | 46.603 | | | | |
| 4a | $Na_4(UO_2)(CO_3)_3$ | | $P$-3$c$ | 76 | 3.049 | 231.750 | | | | | | | | |
| 32 | $Ca_{12}(UO_2(CO_3)_3)_4C_{18}(H_2O)_{47}$ | | $Fd$-3 | 438 | 4.397 | 1925.710 | 3 | 15 | 2.639 | 39.585 | | | | |
| 34 | $[C(NH_2)_3]_4[UO_2(CO_3)_3]$ | | $R3$ | 45 | 4.013 | 180.565 | | | | | | | | |
| 5 | $K_3Na(UO_2)(CO_3)_3$ | | $P$-62$c$ | 38 | 2.891 | 109.870 | | | | | | | | |
| 6 | $K_3Na(UO_2)(CO_3)_3(H_2O)/$**Grimselite** | | $P$-62$c$ | 44 | 3.197 | 140.670 | $-6$ | 15 | 2.506 | 37.584 | | | | |

**Table 2.** *Cont.*

| No. | Formula | Topology | Complexity Parameters of the Crystal Structure | | | | Structural Complexity of the U-C Unit | | | | Topological Complexity of the U-C Unit | | | |
|---|---|---|---|---|---|---|---|---|---|---|---|---|---|---|
| | | | Sp. Gr. | $v$ | $I_G$ | $I_{G,total}$ | Point, Rod or Layer Sym. Gr. | $v$ | $I_G$ | $I_{G,total}$ | Point, Rod or Layer Sym. Gr. | $v$ | $I_G$ | $I_{G,total}$ |
| 6a | $Rb_6Na_2((UO_2)(CO_3)_3)_2(H_2O)$ | | $P\text{-}62c$ | 44 | 3.197 | 140.670 | | | | | | | | |
| 27 | $Na_8[(UO_2)(CO_3)_3](SO_4)_2 \cdot 3H_2O$/**Ježekite** | | $P\text{-}62m$ | 42 | 3.498 | 146.930 | $\text{-}6m2$ | 15 | 2.106 | 31.584 | | | | |
| 36 | $[C(NH_2)_3]_6[(UO_2)_3(CO_3)_6](H_2O)_{6.5}$ | $cc0\text{-}1{:}2\text{-}10$ | $P\text{-}1$ | 225 | 6.818 | 1534.100 | 1 | 33 | 5.044 | 166.452 | $\text{-}6m2$ | 33 | 2.914 | 96.162 |
| | **Nanoclusters** | | | | | | | | | | | | | |
| 37 | $Mg_8Ca_8(UO_2)_{24}(CO_3)_{30}O_4(OH)_{12}(H_2O)_{138}$/**Ewingite** | | $I4_1/acd$ | 2508 | 7.311 | 18,335.988 | -4 | 220 | 5.781 | 1271.820 | -4 | 220 | 5.781 | 1271.820 |
| | **Layers** | | | | | | | | | | | | | |
| | $5^4 4^2 3^4$ *($\beta$-$U_3O_8$)* | | | | | | | | | | | | | |
| 38 | $CaU(UO_2)_2(CO_3)O_4(OH)(H_2O)_7$/**Wyartite** | | $P2_12_12_1$ | 156 | 5.285 | 824.52 | $p2_1$ | 40 | 4.322 | 172.880 | $p2_1mn$ | 40 | 3.822 | 152.880 |
| 38a | $Ca(U(UO_2)_2(CO_3)_{0.7}O_4(OH)_{1.6})(H_2O)_{1.63}$ | | $Pmcn$ | 84 | 3.821 | 320.964 | $p2_1mn$ | 36 | 3.614 | 130.103 | | 36 | 3.614 | 130.103 |
| | $6^1 5^2 4^2 3^2$ *(phosphuranylite)* | | | | | | | | | | | | | |
| 39 | $Ca(UO_2)_3(CO_3)_2O_2(H_2O)_6$/**Fontanite** | | $P2_1/n$ | 152 | 5.248 | 797.685 | $p11n$ | 38 | 4.248 | 161.424 | $cmmm$ | 19 | 2.880 | 54.720 |
| | $6^1 5^2 4^2 3^6$ *(roubaultite)* | | | | | | | | | | | | | |
| 40 | $Cu_2(UO_2)_3(CO_3)_2O_2(OH)_2(H_2O)_4$/**Roubaultite** | | $P\text{-}1$ | 37 | 4.236 | 156.750 | $p\text{-}1$ | 23 | 3.567 | 82.042 | $Pa\,mmm$ | 23 | 1.837 | 42.239 |
| | **Finite clusters** | | | | | | | | | | | | | |
| | $6^1 3^2\text{-}1$ *(rutherfordine)* | | | | | | | | | | | | | |
| 41 | $UO_2)(CO_3)$/**Rutherfordine** | | $Imm2$ | 7 | 2.236 | 15.651 | $p2mm$ | 7 | 2.236 | 15.651 | $p2mm$ | 7 | 2.236 | 15.651 |
| 42 | $Ca(H_2O)_3[(UO_2)_3(CO_3)_{3.6}O_{0.2}]$/**Sharpite** | | $Cmcm$ | 72 | 3.837 | 276.235 | $p2/m11$ | 24 | 3.335 | 80.040 | $Pb\,mmm$ | 24 | 3.168 | 76.032 |
| | $6^1 3^2\text{-}II$ *(widenmannite)* | | | | | | | | | | | | | |
| 43 | $Pb_2[(UO_2)(CO_3)_2]$ | | $Pmmn$ | 34 | 3.382 | 114.974 | $p2mm$ | 11 | 2.914 | 32.054 | $Pb\,2mm$ | 11 | 2.914 | 32.054 |
| | *Layers of miscellaneous topology* | | | | | | | | | | | | | |
| 44 | $Y_2(UO_2)_4(CO_3)_3O_4 \cdot 14H_2O$/**Kamotoite-(Y)** | | $P2_1/n$ | 282 | 6.147 | 1733.353 | $p11a$ | 64 | 5.000 | 320.000 | $Pb\,mmm$ | 32 | 3.250 | 104.000 |
| 45 | $[(Y_{4.22}Nd_{3.78})(H_2O)_{25}(UO_2)_{16}O_8(OH)_8(CO_3)_{16}](H_2O)_{14}$/**Bijvoetite-(Y)** | | $B2_1$ | 522 | 8.028 | 4190.567 | $p1$ | 68 | 6.087 | 413.916 | | 34 | 3.382 | 114.988 |
| 46 | $Ca(UO_2)(CO_3)_2 \cdot 5H_2O$/**Meyrowitzite** | | $P2_1/n$ | 320 | 6.322 | 2023.017 | $p2_1/b$ | 132 | 5.044 | 665.808 | $p2_1/b$ | 132 | 5.044 | 665.808 |

*—although the old model of čejkaite is erroneous from the structural point of view, it was decided to keep complexity calculations for it due to presence of structural data in the database.

**Figure 5.** Ladder diagrams showing contributions to structural complexity (bits per unit cell) (**a**) and normalized contributions (in %) (**b**) for the structures based on uranyl carbonate finite clusters. Legend: TI = topological information; SI = structural information; LS = layer stacking; IS = interstitial structure; HB = hydrogen bonding. See Table 2 and text for details.

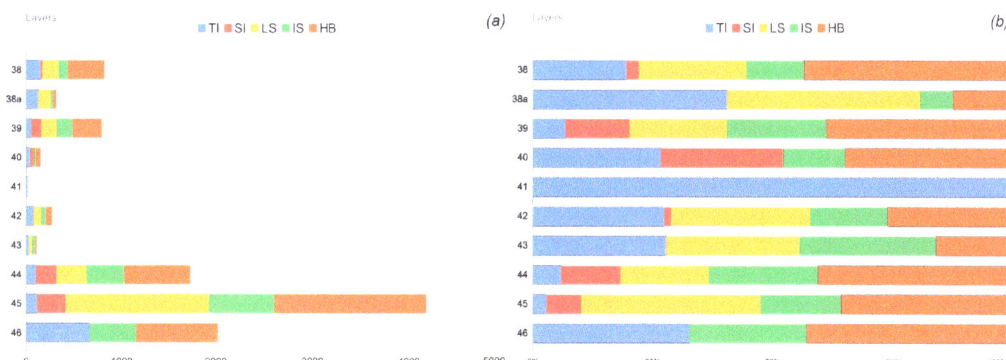

**Figure 6.** Ladder diagrams showing contributions to structural complexity (bits per unit cell) (**a**) and normalized contributions (in %) (**b**) for the structures based on uranyl carbonate layers. Legend: see Figure 5. See Table 2 and text for details.

## 4. Discussion

All three synthetic compounds, whose structures are based on finite clusters of the *cc*0-1:2-9 topological type (Table 2), demonstrate low real (structural) symmetry of the uranyl bicarbonate cluster that corresponds to the point symmetry group 1. However, the topological symmetry of the cluster is much higher and is described by the orthorhombic point symmetry group 2*mm* (Figure 7a). It should be noted that all three compounds have similar $Z = 4$, and thus have equal contribution not only of TI and SI domains, but also from the LS. Complexity of these compounds largely depends on the nature of the interstitial component and hydration state. Substitution of inorganic $K^+$ ions (in **1** and **2**) by the guanidinium cations (in **3**) transfer compound in terms of complexity from intermediate to very complex group.

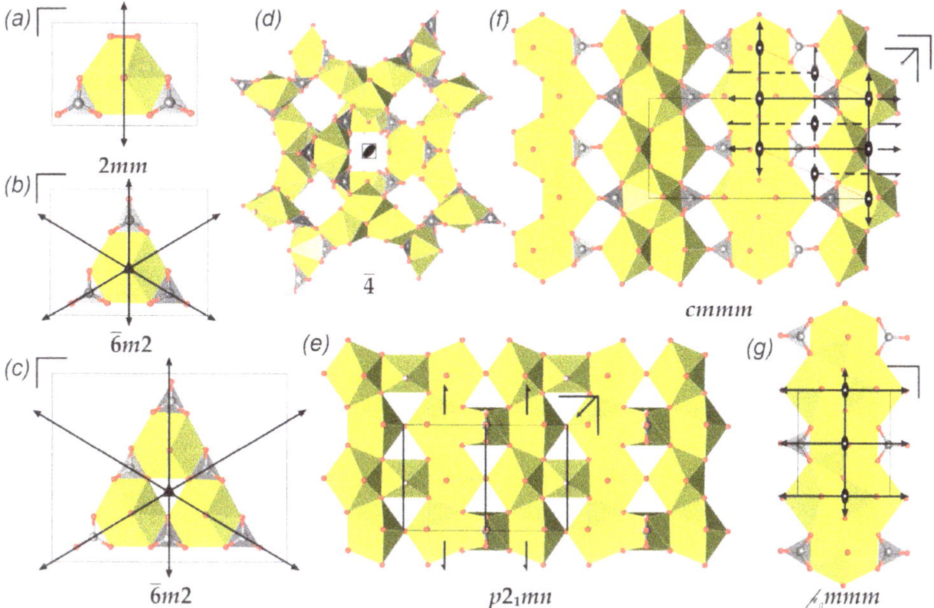

**Figure 7.** Uranyl carbonate complexes and their highest (topological) symmetry groups. Legend: see Figure 1. See text for details.

Topological symmetry of the uranyl tricarbonate cluster (*cc*0-1:3-2 type) is described by the -6*m*2 point group (Figure 7b), and mineral ježekite, **27** [62], and is the only compound whose topological symmetry preserves in a real structure. Another six compounds **4a**, **5**, **6a**, **32**, **34**, including mineral grimselite, **6** [20–23], have rather high structural symmetry of the UTC: trigonal point group 3 (**4a**, **32**, **34**) or hexagonal point group -6 (**5**, **6**, **6a**), which results in a slight increase of SI contribution after TI portion. It should be noted that the cubic symmetry of the whole structures of **4a** and **32** is higher than the topological symmetry of the cluster, which demonstrates that interstitial cations and $H_2O$ molecules can act as symmetry breaking agents, and not only symmetry reducing. It is of interest that all six compounds that have the structural symmetry of the UTC equal to point group *m* are minerals. Moreover, compounds with structural symmetry of the UTC equal to point group 2 are also represented by mineral agricolaite, **7** [25], and its synthetic analogs. Thus, all compounds with monoclinic structural symmetry of the uranyl tricarbonate clusters are related to natural phases. Nearly half of all compounds, whose structures are based on the UTC have the lowest structural symmetry of the cluster, described by the triclinic non-centrosymmetric point group 1, and the distribution between minerals and synthetic compounds here is almost uniform. Second, the most complex compound among the uranyl carbonates is paddlewheelite, **26** [61], whose structure is also based on the UTC units, and information content is 8.267 bits/atom and 5092.340 bits/cell. It should be however mentioned that paddlewheelite has one of the largest information contents per atom among the uranyl carbonate compounds, which is achieved due to a large amount of interstitial cations and $H_2O$ molecules along with the rather high density of their framework arrangement.

The crystal structure of synthetic compound **36** [74] is based on the large triangular tri-uranyl hexacarbonate clusters of *cc*0-1:2-10 topological type, which has the same topological symmetry -6*m*2 as UTC cluster (Figure 7c). But at the same time the structural symmetry of this cluster is the lowest and corresponds to the triclinic point group 1. The unit cell of **36** contains two triuranyl hexacarbonate clusters but their contribution (TI + SI + LS) is comparable with a contribution of four UTC clusters of the same structural symmetry.

Ewingite, **37** [75], is a mineral with the most complex crystal structure known nowadays (Table 2; Figure 7d). According to the structure reported by Olds and coauthors [75], the structural model of ewingite contains 12,684.860 bits/cell, but considering an additional estimated amount of $H_2O$ molecules and respective H atoms in complexity calculations, the total information amount will be presumably equal to c.a. 23,000 bits/cell. Our attempts to get rid of disordered sites and to assign H atoms to $H_2O$ molecules and OH groups resulted in the value of 18,335.988 bits/cell. However, it should be kept in mind that the imperfect quality of measured crystals didn't allow to assign all disordered sites of $H_2O$ molecules in the structural model. Thus, to reach the value of 23,000 bits/cell, an additional 12 fully occupied $H_2O$ sites (48 molecules per formula unit) should appear in the structure, which is somewhat doubtful. Most likely, one could expect the structural complexity of ewingite to be in the range of 19,500–21,500 bits/cell, depending on the variable hydration state. The contribution of TI and SI into the total information content is the smallest (6.9%), stacking is giving another 20.8%, while the contribution of interstitial structure (24.7%) and H-bonding system (47.6%) gives nearly $\frac{3}{4}$ (72.3%) of the total complexity of ewingite crystal structure. Complexity parameters for ewingite are listed in Table 2, but omitted from Figure 5 for clarity.

The crystal structures of wyartite, **38** [76–78], and its dehydrated analog **38a** [79] are based on the layers of β-$U_3O_8$ anion topology, whose topological symmetry is described by the orthorhombic layer symmetry group $p2_1mn$ (Figure 7e). The presence of a significant amount of $H_2O$ molecules in the coordination sphere of interstitial cations reduces the layer's structural symmetry in wyartite, while the release of $H_2O$ and subsequent linkage of U-bearing layers directly through the Ca-centered polyhedra keeps the topological symmetry and breaks the symmetry of the whole structure of **38a**, which results in a decrease of information content by 2.5 times in comparison with the structure of natural compound **38**.

U-bearing layers with phosphuranylite anion topology play an important role in the mineralogy of uranyl selenites as has been recently shown [127]. There is only one uranyl carbonate, mineral fontanite, **39** [80,81], whose structure is based on the layers of this topological type (Figure 7f). However, the structure of uranyl carbonate layers in fontanite differs from that in uranyl selenite minerals and synthetic compounds (Figure 8). Pyramidal selenite groups with lone electron pairs make chains of edge-sharing *Ur* polyhedra arranged as stepped [127] or zig-zag layers (Figure 8c). In the structure of marthozite, as an example of zig-zag layers, equatorial planes of *Ur* bipyramids are arranged parallel to the mean plane of the layer. In the structure of fontanite, uranyl carbonate layer has also a *zig-zag* manner, but equatorial planes of *Ur* bipyramids make an angle of c.a. 135° at the linkage with the neighbor chain of edge-sharing *Ur* polyhedra. Topological symmetry of uranyl carbonate layer (*cmmm*) is higher than that in uranyl selenite marthozite (*pmmm* [127]) because of the flat carbonate groups in comparison to selenite pyramids, but the structural symmetry (*p11n*) and complexity parameters of U-bearing layers in fontanite are the same as were reported for marthozite [127].

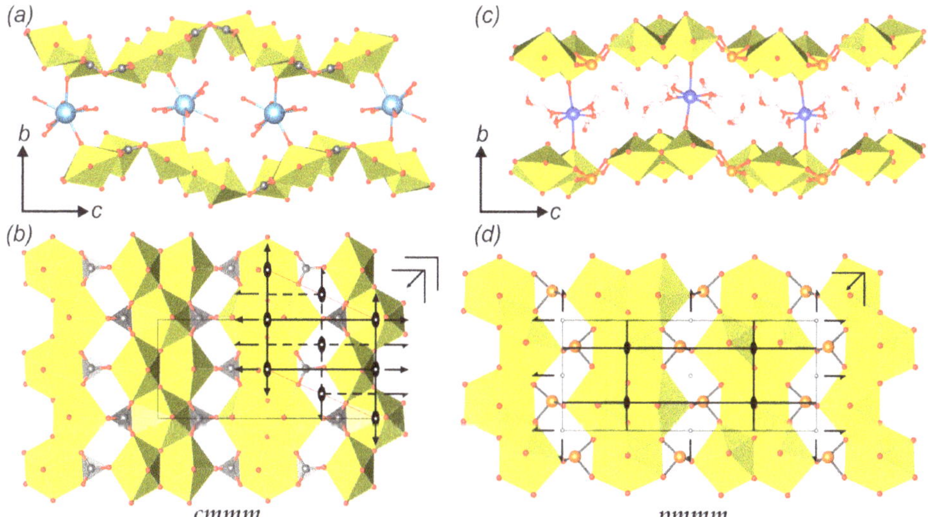



**Figure 8.** The crystal structures of fontanite (**a**) and marthozite (**c**), uranyl carbonate (**b**) and uranyl selenite (**d**) layers of phosphuranylite topology in their structures, and respective topological symmetry groups. Legend: see Figures 1 and 7.

The crystal structure of rutherfordine, **41** [20,84–87], is the simplest among natural and synthetic uranyl carbonates. It is also of special interest because the total amount of information is equal to the topological symmetry contribution (Table 2; Figure 9a). It happens because there is no interstitial structure in rutherfordine, and the body-centered cell is reduced to a single layer.

U-bearing layers of rutherfordine anion topology are also found in the structure of sharpite, **42** [88–90]. However, from a structural point of view, this topology is obtained from uranyl carbonate chains linked by the Ca-centered polyhedra, whose topological symmetry is described by the $p_b mmm$ rod symmetry group (Figure 9b). A similar situation is observed in the structure of roubaultite, **40** [82,83], in which the U-bearing layer can be described in terms of anion topology, but actually is built by the chains of $p_a mmm$ topological symmetry (Figure 7g) similar to those found in phosphuranylite topology, and linked not directly to each other, but through Cu-centered octahedra. Similarly, and as it was described in the Section 3.3, each second row of hexagonal bipyramids in the structure of widenmannite, **43** [93] should ideally be vacant, so the structure can be described as built by chains of the $p_b2mm$ rod symmetry group (Figure 9c).

Bijvoetite-(Y), **45** [96,97], is the most complex compound among the layered uranyl carbonates (Table 2) with a structure based on the uranyl carbonate chains (Figure 9d) linked through the *REE*-centered polyhedra. While meyrowitzite, **46** [98], has the largest TI and SI complexity parameters (Table 2; Figure 9e) among the uranyl carbonate compounds.

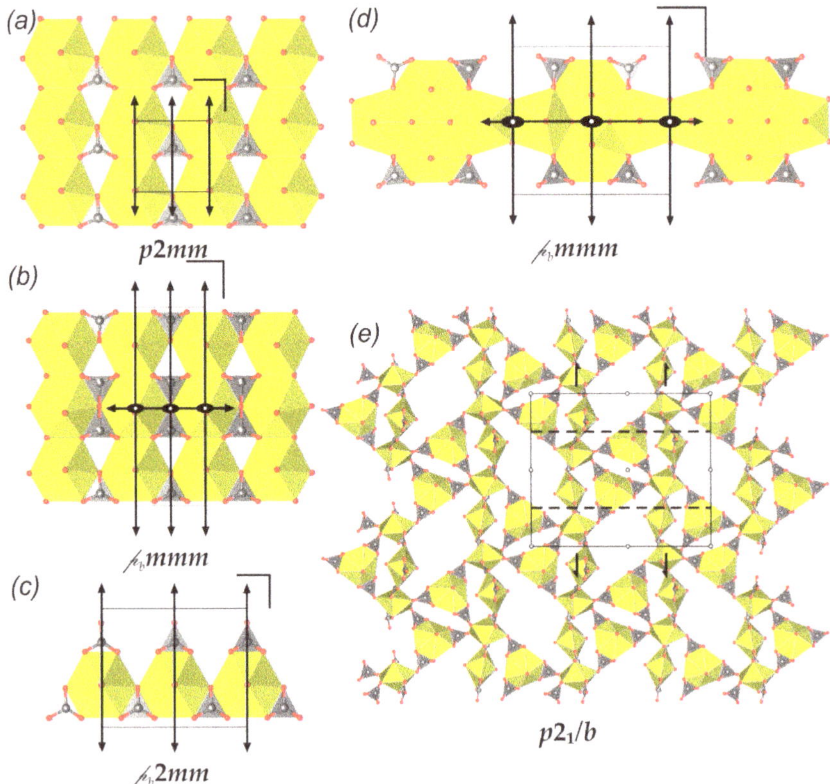


**Figure 9.** Uranyl carbonate complexes and their topological symmetry groups. Legend: see Figure 7. See text for details.

## 5. Conclusions

Comparison of crystal–chemical characteristics of isotypic natural and synthetic compounds can give an answer to the environmental conditions of mineral growth. For instance, the majority of synthetic analogs of minerals were obtained from aqueous solutions at room temperature. This observation allows us to assume that the formation of natural uranyl carbonates does not need any specific environmental conditions (increased P or T), which was recently suggested for uranyl selenites [127] and uranyl sulfates [126].

Structural and topological complexity calculations demonstrate that information content, in general, is usually higher for minerals than for synthetic compounds of similar or close chemical composition, which is likely to point to the higher stability and preferred architectures of natural compounds. However, one interesting feature that was not observed during recent complexity studies of uranyl selenite and sulfate compounds should be noted. A significant portion of compounds have the real symmetry of the structure higher than the structural symmetry of the uranyl carbonate complex. It should be also kept in mind that uranyl carbonate substructural units have average topographical diversity, and mainly include the finite clusters with rigid, edge-sharing manner, of coordination polyhedra polymerization. Thus the structural architecture of uranyl carbonates is largely governed by the interstitial cations and the hydration state of the compounds, while uranyl carbonate complexes play the role of flexible structural skeleton or stone aggregates in concrete.

**Author Contributions:** Conceptualization, V.V.G. and J.P.; Methodology, V.V.G., S.A.K., I.V.K. and J.P.; Investigation, V.V.G., S.A.K., I.V.K. and J.P.; Writing-Original Draft Preparation, V.V.G., S.A.K., I.V.K. and J.P.; Writing-Review & Editing, V.V.G. and J.P.; Visualization, V.V.G. and I.V.K. All authors have read and agreed to the published version of the manuscript.

**Funding:** This research was funded by the Russian Science Foundation (grant 18-17-00018 to V.V.G. and I.V.K.), by the President of Russian Federation grant to leading scientific schools (grant NSh-2526.2020.5 to S.A.K.) and through the Czech Science foundation (project 20-11949S to J.P.).

**Acknowledgments:** We are grateful to reviewers for useful comments.

**Conflicts of Interest:** The authors declare no conflict of interest.

## References

1. Alwan, A.K.; Williams, P.A. The aqueous chemistry of uranium minerals. Part 2. Minerals of the liebigite group. *Mineral. Mag.* **1980**, *43*, 665–667. [CrossRef]
2. Clark, D.L.; Hobart, D.E.; Neu, M.P. Actinide carbonate complexes and their importance in actinide environmental chemistry. *Chem. Rev.* **1995**, *95*, 25–48. [CrossRef]
3. Plášil, J. Oxidation–hydration weathering of uraninite: The current state-of-knowledge. *J. Geosci.* **2014**, *59*, 99–114. [CrossRef]
4. Stefaniak, E.A.; Alsecz, A.; Frost, R.; Mathe, Z.; Sajo, I.E.; Torok, S.; Worobiec, A.; Van Grieken, R. Combined SEM/EDX and micro-Raman spectroscopy analysis of uranium minerals from a former uranium mine. *J. Hazard Mater.* **2009**, *168*, 416–423. [CrossRef] [PubMed]
5. Driscoll, R.J.P.; Wolverson, D.; Mitchels, J.M.; Skelton, J.M.; Parker, S.C. A Raman spectroscopic study of uranyl minerals from Cornwall, UK. *RSC Adv.* **2014**, *4*, 59137–59149. [CrossRef]
6. Teterin, Y.A.; Baev, A.S.; Bogatov, S.A. X-ray photoelectron study of samples containing reactor fuel from "lava" and products growing on it which formed at Chernobyl NPP due to the accident. *J. Electron Spectrosc. Relat. Phenom.* **1994**, *68*, 685–694. [CrossRef]
7. Burakov, B.E.; Strykanova, E.E.; Anderson, E. Secondary uranium minerals on the surface of Chernobyl "Lava". *Mat. Res. Soc. Symp. Proc.* **1996**, *465*, 1309–1311. [CrossRef]
8. Lussier, A.J.; Lopez, R.A.K.; Burns, P.C. A revised and expanded structure hierarchy of natural and synthetic hexavalent uranium compounds. *Can. Mineral.* **2016**, *54*, 177–283. [CrossRef]
9. Gurzhiy, V.V.; Plášil, J. Structural complexity of natural uranyl sulfates. *Acta Crystallogr.* **2019**, *B75*, 39–48. [CrossRef]
10. Lafuente, B.; Downs, R.T.; Yang, H.; Stone, N. The power of databases: The RRUFF project. In *Highlights in Mineralogical Crystallography*; Armbruster, T., Danisi, R.M., Eds.; De Gruyter: Berlin, Germany, 2015; pp. 1–30.
11. Goff, G.S.; Brodnax, L.F.; Cisneros, M.R.; Peper, S.M.; Field, S.E.; Scott, B.L.; Runde, W.H. First identification and thermodynamic characterization of the ternary U(VI) species, $UO_2(O_2)(CO_3)2(4-)$, in $UO_2$-$H_2O_2$-$K_2CO_3$ solutions. *Inorg. Chem.* **2008**, *47*, 1984–1990. [CrossRef]
12. Zehnder, R.; Peper, S.; Brian, L.; Runde, S.; Runde, W. Tetrapotassium dicarbonatodioxoperoxouranium(VI) 2.5-hydrate, $K_4[U(CO_3)_2O_2(O_2)]\cdot2.5H_2O$. *Acta Crystallogr.* **2005**, *C61*, 3–5.
13. Mikhailov, Y.N.; Lobanova, G.M.; Shchelokov, R.N. X-ray structural study of the guanidinium uranyl-peroxo-dicarbonate hydrate $(CN_3H_6)_4UO_2O_2(CO_3)_2\cdot H_2O$. *Zh. Neorg. Khim.* **1981**, *26*, 718–722.
14. Plášil, J.; Fejfarová, K.; Dušek, M.; Škoda, R.; Rohlíček, J. Actinides in Geology, Energy, and the Environment. Revision of the symmetry and the crystal structure of čejkaite, $Na_4(UO_2)(CO_3)_3$. *Am. Mineral.* **2013**, *98*, 549–553. [CrossRef]
15. Li, Y.; Krivovichev, S.V.; Burns, P.C. The crystal structure of $Na_4(UO_2)(CO_3)_3$ and its relationship to schröckingerite. *Mineral. Mag.* **2001**, *65*, 297–304. [CrossRef]
16. Císařová, I.; Skála, R.; Ondruš, P.; Drábek, M. Trigonal $Na_4[UO_2(CO_3)_3]$. *Acta Crystallogr.* **2001**, *E37*, 32–34. [CrossRef]
17. Douglass, M. Tetrasodium uranyl tricarbonate, $Na_4UO_2(CO_3)_3$. *Anal. Chem.* **1956**, *28*, 1635. [CrossRef]
18. Ondruš, P.; Skála, R.; Veselovský, F.; Sejkora, J.; Vitti, C. Čejkaite, the triclinic polymorph of $Na_4(UO_2)(CO_3)_3$—A new mineral from Jachymov, Czech Republic. *Am. Min.* **2003**, *88*, 686–693. [CrossRef]
19. Mazzi, F.; Rinaldi, F. La struttura cristallina del $K_3Na(UO_2)(CO_3)_3$. *Period. Mineral.* **1961**, *30*, 1–21.
20. Kubatko, K.-A.; Helean, K.B.; Navrotsky, A.; Burns, P.C. Thermodynamics of uranyl minerals: Enthalpies of formation of rutherfordine, $UO_2CO_3$, andersonite, $Na_2CaUO_2(CO_3)_3(H_2O)_5$, and grimselite, $K_3NaUO_2(CO_3)_3H_2O$. *Amer. Mineral.* **2005**, *90*, 1284–1290. [CrossRef]
21. Walenta, K. Widenmannit und Joliotit, zwei neue Uranylkarbonatmineralien aus dem Schwarzwald. *Schweiz. Mineral. Petrogr. Mitt.* **1976**, *56*, 167–185.

22. Li, Y.; Burns, P. The crystal structure of synthetic grimselite, $K_3Na[(UO_2)(CO_3)_3](H_2O)$. *Can. Mineral.* **2001**, *39*, 1147–1151. [CrossRef]

23. Plášil, J.; Fejfarová, K.; Skála, R.; Škoda, R.; Meisser, N.; Hloušek, J.; Císařová, I.; Dušek, M.; Veselovský, F.; Čejka, J.; et al. The crystal chemistry of the uranyl carbonate mineral grimselite, $(K,Na)_3Na[(UO_2)(CO_3)_3](H_2O)$, from Jáchymov, Czech Republic. *Mineral. Mag.* **2012**, *76*, 443–453. [CrossRef]

24. Kubatko, K.A.; Burns, P.C. The Rb analogue of grimselite, $Rb_6Na_2((UO_2)(CO_3)_3)_2 (H_2O)$. *Acta Crystallogr. Cryst. Struct. Commun.* **2004**, *C60*, 25–26. [CrossRef] [PubMed]

25. Skála, R.; Ondruš, P.; Veselovský, F.; Císařová, I.; Hloušek, J. Agricolaite, a new mineral of uranium from Jáchymov, Czech Republic. *Mineral. Petrol.* **2011**, *103*, 169–175. [CrossRef]

26. Anderson, A.; Chieh, C.; Irish, D.E.; Tong, J.P.K. An X-Ray crystallographic, Raman, and infrared spectral study of crystalline potassium uranyl carbonate, $K_4UO_2(CO_3)_3$. *Can. J. Chem.* **1980**, *58*, 1651–1658. [CrossRef]

27. Han, J.-C.; Rong, S.-B.; Chen, Q.-M.; Wu, X.-R. The determination of the crystal structure of tetrapotassium uranyl tricarbonate by powder X-ray diffraction method. *Chin. J. Chem.* **1990**, *4*, 313–318.

28. Chernorukov, N.G.; Mikhailov, Y.N.; Knyazev, A.V.; Kanishcheva, A.S.; Zamkovaya, E.V. Synthesis and crystal structure of rubidium uranyltricarbonate. *Russ. J. Coord. Chem.* **2005**, *31*, 364–367. [CrossRef]

29. Charushnikova, I.A.; Fedoseev, A.M.; Perminov, V.P. Synthesis and Crystal Structure of Cesium Actinide(VI) Tricarbonate Complexes $Cs_4AnO_2(CO_3)_3 \cdot 6H_2O$, An(VI) = U., Np, Pu. *Radiochemistry* **2016**, *58*, 578–585. [CrossRef]

30. Mereiter, K. Structure of cesium tricarbonatodioxouranate(VI) hexahydrate. *Acta Crystallogr. Cryst. Struct. Commun.* **1988**, *C44*, 1175–1178. [CrossRef]

31. Krivovichev, S.V.; Burns, P.C. Synthesis and crystal structure of Cs4(UO2(CO3)3). *Radiochemistry* **2004**, *46*, 12–15. [CrossRef]

32. Serezhkin, V.N.; Soldatkina, M.A.; Boiko, N.V. Refinement of the crystal-structure of $(NH_4)_4UO_2(CO_3)_3$. *J. Struct. Chem.* **1983**, *24*, 770–774. [CrossRef]

33. Mereiter, K. Structure of Thallium Tricarbonatodioxouranat (VI). *Acta Crystallogr. Cryst. Struct. Commun.* **1986**, *C42*, 1682–1684. [CrossRef]

34. Axelrod, J.M.; Grimaldi, F.S.; Milton, C.; Murata, K.J. The uranium minerals from the Hillside mine, Yavapai County, Arizona. *Am. Mineral.* **1951**, *36*, 1–22.

35. Mayer, H.; Mereiter, K. Synthetic bayleyite, $Mg_2[UO_2(CO_3)_3] \cdot 18H_2O$: Thermochemistry, crystallography and crystal structure. *Tschermaks Mineral. Petrogr. Mitt.* **1986**, *35*, 133–146. [CrossRef]

36. Mereiter, K. Synthetic swartzite, $CaMg[UO_2(CO_3)_3] \cdot 12H_2O$, and its strontium analogue, $SrMg[UO_2(CO_3)_3] \cdot 12H_2O$: Crystallography and crystal structures. *Neues Jahrb. Mineral. Mon.* **1986**, *1986*, 481–492.

37. Amayri, S.; Arnold, T.; Foerstendorf, H.; Geipel, G.; Bernhard, G. Spectroscopic characterization of synthetic becquerelite, $Ca[UO_2]_6O_4(OH)_6 \cdot 8H_2O$, and swartzite, $CaMg[UO_2(CO_3)_3] \cdot 12H_2O$. *Can. Mineral.* **2004**, *42*, 953–962. [CrossRef]

38. Vochten, R.; Van Haverbeke, L.; Van Springel, K. Synthesis of liebigite and andersonite, and study of their thermal behavior and luminescence. *Can. Mineral.* **1993**, *31*, 167–171.

39. Smith, J.L. Two new minerals—medjidite (sulphate of uranium and lime)—Liebigite (carbonate of uranium and lime). *Am. J. Sci. Arts* **1848**, *5*, 336–338.

40. Mereiter, K. The crystal structure of liebigite, $Ca_2UO_2(CO_3)_3 \cdot {\sim}11H_2O$. *Tschermaks Mineral. Petrogr. Mitt.* **1982**, *30*, 277–288. [CrossRef]

41. Kampf, A.R.; Plášil, J.; Kasatkin, A.V.; Marty, J.; Čejka, J. Markeyite, a new calcium uranyl carbonate mineral from the Markey mine, San Juan County, Utah, USA. *Mineral. Mag.* **2018**, *82*, 1089–1100. [CrossRef]

42. Kampf, A.R.; Olds, T.A.; Plášil, J.; Burns, P.C.; Marty, J. Natromarkeyite and pseudomarkeyite, two new calcium uranyl carbonate minerals from the Markey mine, San Juan County, Utah, USA. *Mineral. Mag.* **2020**, *84*, 753–765. [CrossRef]

43. Mereiter, K. Structure of strontium tricarbonatodioxouranate(VI) octahydrate. *Acta Crystallogr. Cryst. Struct. Commun.* **1986**, *C42*, 1678–1681. [CrossRef]

44. Olds, T.; Sadergaski, L.; Plášil, J.; Kampf, A.; Burns, P.; Steele, I.; Marty, J.; Carlson, S.; Mills, O. Leószilárdite, the first Na,Mg-containing uranyl carbonate from the Markey Mine, San Juan County, Utah, USA. *Mineral. Mag.* **2017**, *81*, 1039–1050. [CrossRef]

45. Gurzhiy, V.V.; Krzhizhanovskaya, M.G.; Izatulina, A.R.; Sigmon, G.E.; Krivovichev, S.V.; Burns, P.C. Structure refinement and thermal stability studies of the uranyl carbonate mineral Andersonite, $Na_2Ca[(UO_2)(CO_3)_3] \cdot (5+x)H_2O$. *Minerals* **2018**, *8*, 586. [CrossRef]

46. Coda, A.; Della Giusta, A.; Tazzoli, V. The structure of synthetic andersonite, $Na_2Ca[UO_2(CO_3)_3] \cdot xH_2O$ (x = 5.6). *Acta Cryst.* **1981**, *B37*, 1496–1500. [CrossRef]

47. Mereiter, K. Neue kristallographische Daten ueber das Uranmineral Andersonit. *Anz. Österr. Akad. Wiss. Mathemat. Naturwiss. Kl.* **1986**, *123*, 39–41.

48. Plášil, J.; Čejka, J. A note on the molecular water content in uranyl carbonate mineral andersonite. *J. Geosci.* **2015**, *60*, 181–187. [CrossRef]

49. Coda, A. Ricerche sulla struttura cristallina dell'Andersonite. *Atti Accad. Naz. Lincei Rend. Cl. Sci. Fis. Mat. Nat. Ser.* **1963**, *34*, 299–304.

50. Čejka, J.; Urbanec, Z.; Čejka, J., Jr. To the crystal chemistry of andersonite. *Neu. Jb. Mineral. Mh.* **1987**, *11*, 488–501.

51. De Neufville, J.P.; Kasdan, A.; Chimenti, R.J.L. Selective detection of uranium by laser-induced fluorescence: A potential remote-sensing technique. 1: Optical characteristics of uranyl geologic targets. *Appl. Opt.* **1981**, *20*, 1279–1296. [CrossRef]

52. Amayri, S.; Arnold, T.; Reich, T.; Foerstendorf, H.; Geipel, G.; Bernhard, G.; Massanek, A. Spectroscopic characterization of the uranium carbonate andersonite Na$_2$Ca[UO$_2$(CO$_3$)$_3$]·6H$_2$O. *Environ. Sci. Technol.* **2004**, *38*, 6032–6036. [CrossRef]

53. Frost, R.L.; Carmody, O.; Ertickson, K.L.; Weier, M.L.; Čejka, J. Molecular structure of the uranyl mineral andersonite—A Raman spectroscopic study. *J. Molec. Struct.* **2004**, *703*, 47–54. [CrossRef]

54. Čejka, J. To the chemistry of andersonite and thermal composition of dioxo-tricarbonatouranates. *Coll. Czech. Chem. Commun.* **1969**, *34*, 1635–1656. [CrossRef]

55. Čejka, J.; Urbanec, Z. Thermal and infrared spectrum analyses of natural and synthetic andersonites. *J. Therm. Anal.* **1988**, *33*, 389–394. [CrossRef]

56. Vochten, R.; van Haverbeke, L.; van Springel, K.; Blaton, N.; Peeters, M. The structure and physicochemical characteristics of a synthetic phase compositionally intermediate between liebigite and andersonite. *Can. Mineral.* **1994**, *32*, 553–561.

57. Plášil, J.; Mereiter, K.; Kampf, A.R.; Hloušek, J.; Škoda, R.; Čejka, J.; Němec, I.; Ederová, J. Braunerite IMA 2015-123. CNMNC Newsletter No. 31. *Mineral. Mag.* **2016**, *80*, 692.

58. Plášil, J.; Čejka, J.; Sejkora, J.; Hloušek, J.; Škoda, R.; Novák, M.; Dušek, M.; Císařová, I.; Němec, I.; Ederová, J. Línekite, K$_2$Ca$_3$[(UO$_2$)(CO$_3$)$_3$]$_2$.8H$_2$O, a new uranyl carbonate mineral from Jáchymov, Czech Republic. *J. Geosci.* **2017**, *62*, 201–213. [CrossRef]

59. Kubatko, K.-A.; Burns, P. The crystal structure of a novel uranyl tricarbonate, K$_2$Ca$_3$[(UO$_2$)(CO$_3$)$_3$]$_2$(H$_2$O)$_6$. *Can. Mineral.* **2004**, *42*, 997–1003. [CrossRef]

60. Effenberger, H.; Mereiter, K. Structure of a cubic sodium strontium magnesium tricarbonatodioxouranate(VI) hydrate. *Acta Crystallogr. Cryst. Struct. Commun.* **1988**, *C44*, 1172–1175. [CrossRef]

61. Olds, T.; Plášil, J.; Kampf, A.; Dal Bo, F.; Burns, P. Paddlewheelite, a New Uranyl Carbonate from the Jáchymov District, Bohemia, Czech Republic. *Minerals* **2018**, *8*, 511. [CrossRef]

62. Plášil, J.; Hloušek, J.; Kasatkin, A.V.; Belakovskiy, D.I.; Čejka, J.; Chernyshov, D. Ježekite, Na$_8$[(UO$_2$)(CO$_3$)$_3$](SO$_4$)$_2$·3H$_2$O, a new uranyl mineral from Jáchymov, Czech Republic. *J. Geosci.* **2015**, *60*, 259–267. [CrossRef]

63. Schrauf, A. Schröckingerit, ein neues Mineral von Joachimsthal. *Tschermaks Mineral. Petrogr. Mitt.* **1873**, *1*, 137–138.

64. Jaffe, H.W.; Sherwood, A.M.; Peterson, M.J. New data on schroeckingerite. *Am. Mineral.* **1948**, *33*, 152–157.

65. Smith, D.K. An X-ray crystallographic study of schroeckingerite and its dehydration product. *Am. Mineral.* **1959**, *44*, 1020–1025.

66. Mereiter, K. Crystal structure and crystallographic properties of a schröckingerite from Joachimsthal. *Tschermaks Mineral. Petrogr. Mitt.* **1986**, *35*, 1–18. [CrossRef]

67. Mereiter, K. The crystal structure of albrechtschraufite, MgCa$_4$F$_2$[(UO$_2$)(CO$_3$)$_3$]$_2$·17H$_2$O. *Acta Crystallogr.* **1984**, *A40*, 247. [CrossRef]

68. Mereiter, K. Description and crystal structure of albrechtschaufite, MgCa$_4$F$_2$[UO$_2$(CO$_3$)$_3$]$_2$·17-18H$_2$O. *Mineral. Petrol.* **2013**, *107*, 179–188. [CrossRef]

69. Li, Y.; Burns, P.C. New structural arrangements in three ca uranyl carbonate compounds with multiple anionic species. *J. Solid State Chem.* **2002**, *166*, 219–228. [CrossRef]

70. Plášil, J.; Škoda, R. Crystal structure of the (REE)–uranyl carbonate mineral shabaite-(Nd). *J. Geosci.* **2017**, *62*, 97–105. [CrossRef]

71. Deliens, M.; Piret, P. La shabaïte-(Nd), Ca(TR)$_2$(U0$_2$)(C0$_3$)$_4$(OH)$_2$.6H$_2$0, nouvelle espéce minérale de Kamoto, Shaba, Zaïre. *Eur. J. Mineral.* **1989**, *1*, 85–88. [CrossRef]

72. Fedosseev, A.M.; Gogolev, A.V.; Charushnikova, I.A.; Shilov, V.P. Tricarbonate complex of hexavalent Am with guanidinium: Synthesis and structural characterization of [C(NH$_2$)$_3$]$_4$[AmO$_2$(CO$_3$)$_3$]·2H$_2$O, comparison with [C(NH$_2$)$_3$]$_4$[*An*O$_2$(CO$_3$)$_3$] (*An*=U, Np, Pu). *Radiochim. Acta* **2011**, *99*, 679–686. [CrossRef]

73. Reed, W.A.; Oliver, A.G.; Rao, L. Tetrakis(tetramethylammonium) tricarbonatodioxidouranate octahydrate. *Acta Crystallogr.* **2011**, *67*, 301–303. [CrossRef]

74. Allen, P.G.; Bucher, J.J.; Clark, D.L.; Edelstein, N.M.; Ekberg, S.A.; Gohdes, J.W.; Hudson, E.A.; Kaltsoyannis, N.; Lukens, W.W.; Neu, M.P.; et al. Multinuclear NMR, Raman, EXAFS, and X ray diffraction studies of uranyl carbonate complexes in near-neutral aqueous solution. X-ray structure of (C(NH2)3)6((UO2)3(CO3)6)·6.5(H2O). *Inorg. Chem.* **1995**, *34*, 4797–4807.

75. Olds, T.; Plášil, J.; Kampf, A.; Simonetti, A.; Sadergaski, L.; Chen, Y.-S.; Burns, P. Ewingite: Earth's most complex mineral. *Geology* **2017**, *45*, 1007–1010. [CrossRef]

76. Guillemin, C.; Protas, J. Ianthinite et wyartite. *Bull. Société Française Minéralogie Cristallogr.* **1959**, *82*, 80–86. [CrossRef]

77. Burns, P.C.; Finch, R.J. Wyartite: Crystallographic evidence for the first pentavalent-uranium mineral. *Am. Mineral.* **1999**, *84*, 1456–1460. [CrossRef]

78. Frost, R.L.; Henry, D.A.; Erickson, K. Raman spectroscopic detection of wyartite in the presence of rabejacite. *J. Raman Spectrosc.* **2004**, *35*, 255–260. [CrossRef]

79. Hawthorne, F.C.; Finch, R.J.; Ewing, R.C. The crystal structure of dehydrated wyartite, Ca(CO$_3$)[U$^{5+}$(U$^{6+}$O$_2$)$_2$O$_4$(OH)](H$_2$O)$_3$. *Can. Mineral.* **2006**, *44*, 1379–1385. [CrossRef]

80. Deliens, M.; Piret, P. La fontanite, carbonate hydraté d'uranyle et de calcium, nouvelle espèce minérale de Rabejac, Hérault, France. *Eur. J. Mineral.* **1992**, *4*, 1271–1274. [CrossRef]

81. Hughes, K.A.; Burns, P.C. A new uranyl carbonate sheet in the crystal structure of fontanite, $Ca[(UO_2)_3(CO_3)_2O_2](H_2O)_6$. *Am. Mineral.* **2003**, *88*, 962–966. [CrossRef]

82. Cesbron, F.; Pierrot, R.; Verbeek, T. La roubaultite $Cu_2(UO_2)_3(OH)_{10} \cdot 5H_2O$, une nouvelle espèce minérale. *Bull. Société Française Minéralogie Cristallogr.* **1970**, *93*, 550–554. [CrossRef]

83. Ginderow, D.; Cesbron, F. Structure de la roubaultite $Cu_2(UO_2)_3(CO_3)_2O_2(OH)_2 \cdot 4H_2O$. *Acta Crystallogr.* **1985**, *41*, 654–657.

84. Marckwald, W. Ueber Uranerze aus Deutsch-Ostafrika. *Zent. Mineral. Geol. Paläontologie* **1906**, *1906*, 761–763.

85. Christ, C.L.; Clark, J.R.; Evans, H.T., Jr. Crystal structure of rutherfordine, $UO_2CO_3$. *Science* **1955**, *121*, 472–473. [CrossRef] [PubMed]

86. Frondel, C.; Meyrowitz, R. Studies of uranium minerals (XIX): Rutherfordine, diderichite, and clarkeite. *Am. Mineral.* **1956**, *41*, 127–133.

87. Finch, R.J.; Cooper, M.A.; Hawthorne, F.C.; Ewing, R.C. Refinement of the crystal structure of rutherfordine. *Can. Mineral.* **1999**, *37*, 929–938.

88. Plášil, J. A unique structure of uranyl-carbonate mineral sharpite: A derivative of the rutherfordine topology. *Z. Krist. Cryst. Mater.* **2018**, *233*, 579–586. [CrossRef]

89. Melón, M.J. La sharpite, nouveau carbonate d'uranyle du Congo belge. *Bull. Séances l'Institut R. Colonial Belg.* **1938**, *9*, 333–336.

90. Cejka, J.; Mrazek, Z.; Urbanec, Z. New data on sharpite, a calcium uranyl carbonate. *Neues Jahrb. Mineral. Monatsh.* **1984**, *1984*, 109–117.

91. Elton, N.J.; Hooper, J.J. Widenmannite from Cornwall, England: The second world occurrence. *Mineral. Mag.* **1995**, *59*, 745–749. [CrossRef]

92. Plášil, J.; Čejka, J.; Sejkora, J.; Škácha, P.; Goliáš, V.; Jarka, P.; Laufek, F.; Jehlička, J.; Němec, I.; Strnad, L. Widenmannite, a rare uranyl lead carbonate: Occurrence, formation and characterization. *Miner. Mag.* **2010**, *74*, 97–110. [CrossRef]

93. Plášil, J.; Palatinus, L.; Rohlíček, J.; Houdková, L.; Klementová, M.; Goliáš, V.; Škácha, P. Crystal structure of lead uranyl carbonate mineral widenmannite: Precession electron-diffraction and synchrotron powder-diffraction study. *Am. Mineral.* **2014**, *99*, 276–282. [CrossRef]

94. Deliens, M.; Piret, P. La kamototïte-(Y), un nouveau carbonate d'uranyle et de terres rares de Kamoto, Shaba, Zaïre. *Bull. Minéralogie* **1986**, *109*, 643–647. [CrossRef]

95. Plášil, J.; Petříček, V. Crystal structure of the (REE)-uranyl carbonate mineral kamotoite-(Y). *Mineral. Mag.* **2017**, *81*, 653–660. [CrossRef]

96. Deliens, M.; Piret, P. Bijvoetite et lepersonnite, carbonates hydrates d'uranyle et de terres rares de Shinkolobwe, Zaire. *Can. Mineral.* **1982**, *22*, 231–238.

97. Li, Y.; Burns, P.C.; Gault, R.A. A new rare-earth element uranyl carbonate sheet in the structure of bijvoetite-(Y). *Can. Mineral.* **2000**, *38*, 153–162. [CrossRef]

98. Kampf, A.R.; Plášil, J.; Olds, T.A.; Nash, B.P.; Marty, J.; Belkin, H.E. Meyrowitzite, $Ca(UO_2)(CO_3)_2 \cdot 5H_2O$, a new mineral with a novel uranyl-carbonate sheet. *Am. Mineral.* **2019**, *104*, 603–610. [CrossRef]

99. Deliens, M.; Piret, P. L'astrocyanite-(Ce), $Cu_2(TR)_2(UO_2)(CO_3)_5(OH)_2 \cdot 1,5\ H_2O$, nouvelle espèce minérale de Kamoto, Shaba, Zaïre. *Eur. J. Mineral.* **1990**, *2*, 407–411. [CrossRef]

100. Vochten, R.; Deliens, M. Blatonite, $UO_2CO_3 \cdot H_2O$, A new uranyl carbonate monohydrate from San Juan County, Utah. *Can. Mineral.* **1998**, *36*, 1077–1081.

101. Coleman, R.G.; Ross, D.R.; Meyrowitz, R. Zellerite and metazellerite, new uranyl carbonates. *Am. Mineral.* **1966**, *51*, 1567–1578.

102. Vochten, R.; Deliens, M.; Medenbach, O. Oswaldpeetersite, $(UO_2)_2CO_3(OH)_2 \cdot 4H_2O$, a new basic uranyl carbonate mineral from the Jomac uranium mine, San Juan County, Utah, U.S.A. *Can. Mineral.* **2001**, *39*, 1685–1689. [CrossRef]

103. Thompson, M.E.; Weeks, A.D.; Sherwood, A.M. Rabbittite, a new uranyl carbonate from Utah. *Am. Mineral.* **1955**, *40*, 201–206.

104. Deliens, M.; Piret, P. L'urancalcarite, $Ca(UO_2)_3CO_3(OH)_6 \cdot 3H_2O$, nouveau minéral de Shinkolobwe, Shaba, Zaïre. *Bull. Minéralogie* **1984**, *107*, 21–24. [CrossRef]

105. Vogl, J.F. Drei neue Mineral-Vorkommen von Joachimsthal. *Jahrb. Kais. Königlichen Geol. Reichsanst.* **1853**, *4*, 220–223.

106. Piret, P.; Deliens, M. New crystal data for Ca, Cu, $UO_2$ hydrated carbonate: Voglite. *J. Appl. Crystallogr.* **1979**, *12*, 616. [CrossRef]

107. Frost, R.L.; Dickfos, M.J.; Cejka, J. Raman spectroscopic study of the uranyl carbonate mineral zellerite. *J. Raman Spectrosc.* **2008**, *39*, 582–586. [CrossRef]

108. Chiappero, P.J.; Sarp, H. Nouvelles données sur la znucalite et seconde occurrence: Le Mas d'Alary, Lodève (Hérault, France). *Arch. Sci.* **1993**, *46*, 291–301.

109. Ondruš, P.; Veselovský, F.; Rybka, R. Znucalite, $Zn_{12}(UO_2)Ca(CO_3)_3(OH)_{22} \cdot 4H_2O$, a new mineral from Príbram, Czechoslovakia. *Neues Jahrb. Mineral. Mon.* **1990**, *1990*, 393–400.

110. Hawthorne, F.C. Graphical enumeration of polyhedral clusters. *Acta Crystallogr.* **1983**, *A39*, 724–736. [CrossRef]

111. Krivovichev, S.V. Combinatorial topology of salts of inorganic oxoacids: Zero-, one- and two-dimensional units with corner-sharing between coordination polyhedra. *Crystallogr. Rev.* **2004**, *10*, 185–232. [CrossRef]

112. Krivovichev, S.V. *Structural Crystallography of Inorganic Oxysalts*; Oxford University Press: Oxford, UK, 2008; p. 303.

113. Burns, P.C.; Miller, M.L.; Ewing, R.C. $U^{6+}$ minerals and inorganic phases: A comparison and hierarchy of structures. *Can. Mineral.* **1996**, *34*, 845–880.

114. Burns, P.C. U$^{6+}$ minerals and inorganic compounds: Insights into an expanded structural hierarchy of crystal structures. *Canad. Mineral.* **2005**, *43*, 1839–1894. [CrossRef]
115. Krivovichev, S.V. Topological complexity of crystal structures: Quantitative approach. *Acta Crystallogr.* **2012**, *A68*, 393–398. [CrossRef]
116. Krivovichev, S.V. Structural complexity of minerals: Information storage and processing in the mineral world. *Mineral. Mag.* **2013**, *77*, 275–326. [CrossRef]
117. Krivovichev, S.V. Which inorganic structures are the most complex? *Angew. Chem. Int. Ed.* **2014**, *53*, 654–661. [CrossRef]
118. Krivovichev, S.V. Structural complexity of minerals and mineral parageneses: Information and its evolution in the mineral world. In *Highlights in Mineralogical Crystallography*; Danisi, R., Armbruster, T., Eds.; Walter de Gruyter: Berlin, Germany, 2015; pp. 31–73.
119. Krivovichev, S.V. Structural complexity and configurational entropy of crystalline solids. *Acta Crystallogr.* **2016**, *B72*, 274–276.
120. Gurzhiy, V.V.; Tyumentseva, O.S.; Izatulina, A.R.; Krivovichev, S.V.; Tananaev, I.G. Chemically Induced Polytypic Phase Transitions in the Mg[(UO$_2$)(TO$_4$)$_2$(H$_2$O)](H$_2$O)$_4$ (*T* = S, Se) System. *Inorg. Chem.* **2019**, *58*, 14760–14768. [CrossRef]
121. Gurzhiy, V.V.; Tyumentseva, O.S.; Belova, E.V.; Krivovichev, S.V. Chemically induced symmetry breaking in the crystal structure of guanidinium uranyl sulfate. *Mendeleev Commun.* **2019**, *29*, 408–410. [CrossRef]
122. Kornyakov, I.V.; Kalashnikova, S.A.; Gurzhiy, V.V.; Britvin, S.N.; Belova, E.V.; Krivovichev, S.V. Synthesis, characterization and morphotropic transitions in a family of *M*[(UO$_2$)(CH$_3$COO)$_3$](H$_2$O)$_n$ (*M* = Na, K, Rb, Cs; n = 0–1.0) compounds. *Z. Kristallogr.* **2020**, *235*, 95–103. [CrossRef]
123. Kornyakov, I.V.; Tyumentseva, O.S.; Krivovichev, S.V.; Gurzhiy, V.V. Dimensional evolution in hydrated K+-bearing uranyl sulfates: From 2D-sheets to 3D frameworks. *Cryst. Eng. Comm.* **2020**, *22*, 4621–4629. [CrossRef]
124. Krivovichev, S.V. Ladders of information: What contributes to the structural complexity in inorganic crystals. *Z. Kristallogr.* **2018**, *233*, 155–161. [CrossRef]
125. Krivovichev, V.G.; Krivovichev, S.V.; Charykova, M.V. Selenium minerals: Structural and chemical diversity and Complexity. *Minerals* **2019**, *9*, 455. [CrossRef]
126. Tyumentseva, O.S.; Kornyakov, I.V.; Britvin, S.N.; Zolotarev, A.A.; Gurzhiy, V.V. Crystallographic insights into uranyl sulfate minerals formation: Synthesis and crystal structures of three novel cesium uranyl sulfates. *Crystals* **2019**, *9*, 660. [CrossRef]
127. Gurzhiy, V.V.; Kuporev, I.V.; Kovrugin, V.M.; Murashko, M.N.; Kasatkin, A.V.; Plášil, J. Crystal chemistry and structural complexity of natural and synthetic uranyl selenites. *Crystals* **2019**, *9*, 639. [CrossRef]
128. Langmuir, D. Uranium solution-mineral equilibria at low temperatures with applications to sedimentary ore deposits. *Geochim. Cosmochim. Acta* **1978**, *A42*, 547–569. [CrossRef]
129. Gorman-Lewis, D.; Burns, P.; Fein, J. Review of uranyl mineral solubility measurements. *J. Chem. Thermodyn.* **2008**, *40*, 335–352. [CrossRef]
130. Walenta, V.K. Grimselit, eln ncuoɛ Kalium-NatriumUranylkarbonat aus dem Grimselgebiet (Oberhasli, Kt. Bern, Schweiz). *Schweiz. Mineral. Petrogr. Mitt.* **1972**, *52*, 93–108.
131. Mereiter, K. Hemimorphy of crystals of liebigite. *Naaes Jahrb. Mineral. Monatsh.* **1986**, *1986*, 325–328.
132. Larsen, E.S. The probable identity of uranothallite and liebigite. *Am. Mineral.* **1917**, *2*, 87.
133. Krivovichev, S.V.; Burns, P.C. Actinide compounds containing hexavalent cations of the VI group elements (S, Se, Mo, Cr, W). In *Structural Chemistry of Inorganic Actinide Compounds*; Krivovichev, S.V., Burns, P.C., Tananaev, I.G., Eds.; Elsevier: Amsterdam, The Netherland, 2007; pp. 95–182.
134. Shuvalov, R.R.; Burns, P.C. A monoclinic polymorph of uranyl dinitrate trihydrate, [UO$_2$(NO$_3$)$_2$(H$_2$O)$_2$]·H$_2$O. *Acta Crystallogr.* **2003**, *C59*, 71–73.
135. Thuéry, P. Uranyl Ion Complexes with Cucurbit[*n*]urils (*n* = 6, 7, and 8): A new family of uranyl-organic frameworks. *Cryst. Growth Des.* **2008**, *8*, 4132–4143. [CrossRef]
136. Gurzhiy, V.V.; Kornyakov, I.V.; Tyumentseva, O.S. Uranyl nitrates: Byproducts of the synthetic experiments or key indicators of the reaction progress? *Crystals* **2020**, *10*, 1122. [CrossRef]
137. Demartin, F.; Diella, V.; Donzelli, S.; Gramaccioli, C.M.; Pilati, T. The importance of accurate crystal structure determination of uranium minerals. I. Phosphuranylite KCa(H$_3$O)$_3$(UO$_2$)$_7$(PO$_4$)$_4$O$_4$·8H$_2$O. *Acta Crystallogr.* **1991**, *B47*, 439–446. [CrossRef]
138. Kepler, J. *Harmonices Mundi Libri V*; Forni: Bologna, Italy, 1619.
139. Blatov, V.A.; Shevchenko, A.P.; Proserpio, D.M. Applied topological analysis of crystal structures with the program package ToposPro. *Cryst. Growth. Des.* **2014**, *14*, 3576–3586. [CrossRef]

*Article*

# Crystal Chemical Relations in the Shchurovskyite Family: Synthesis and Crystal Structures of $K_2Cu[Cu_3O]_2(PO_4)_4$ and $K_{2.35}Cu_{0.825}[Cu_3O]_2(PO_4)_4$

**Ilya V. Kornyakov** [1,2] and **Sergey V. Krivovichev** [1,3,*]

[1]    Department of Crystallography, Institute of Earth Sciences, St. Petersburg State University, University Emb. 7/9, 199034 Saint-Petersburg, Russia; ikornyakov@mail.ru

[2]    Laboratory of Nature-Inspired Technologies and Environmental Safety of the Arctic, Kola Science Centre, Russian Academy of Science, Fesmana 14, 184209 Apatity, Russia

[3]    Nanomaterials Research Center, Federal Research Center–Kola Science Center, Russian Academy of Sciences, Fersmana Str. 14, 184209 Apatity, Russia

*    Correspondence: s.krivovichev@kcs.ru

**Abstract:** Single crystals of two novel shchurovskyite-related compounds, $K_2Cu[Cu_3O]_2(PO_4)_4$ (**1**) and $K_{2.35}Cu_{0.825}[Cu_3O]_2(PO_4)_4$ (**2**), were synthesized by crystallization from gaseous phase and structurally characterized using single-crystal X-ray diffraction analysis. The crystal structures of both compounds are based upon similar Cu-based layers, formed by rods of the $[O_2Cu_6]$ dimers of oxocentered $(OCu_4)$ tetrahedra. The topologies of the layers show both similarities and differences from the shchurovskyite-type layers. The layers are connected in different fashions via additional Cu atoms located in the interlayer, in contrast to shchurovskyite, where the layers are linked by $Ca^{2+}$ cations. The structures of the shchurovskyite family are characterized using information-based structural complexity measures, which demonstrate that the crystal structure of **1** is the simplest one, whereas that of **2** is the most complex in the family.

**Keywords:** shchurovskyite; synthesis; X-ray diffraction; crystal structure; oxocentered tetrahedra

**Citation:** Kornyakov, I.V.; Krivovichev, S.V. Crystal Chemical Relations in the Shchurovskyite Family: Synthesis and Crystal Structures of $K_2Cu[Cu_3O]_2(PO_4)_4$ and $K_{2.35}Cu_{0.825}[Cu_3O]_2(PO_4)_4$. *Crystals* **2021**, *11*, 807. https://doi.org/10.3390/cryst11070807

Academic Editor: Francesco Capitelli

Received: 28 June 2021
Accepted: 9 July 2021
Published: 11 July 2021

**Publisher's Note:** MDPI stays neutral with regard to jurisdictional claims in published maps and institutional affiliations.



## 1. Introduction

The last two decades are marked by the increased interest in mineralogical data from the field of material sciences. Since the discovery of a quantum spin liquid state in herbertsmithite [1,2], the number of mineralogically inspired studies in this field has grown exponentially: from the investigation of magnetic properties of the atacamite-group minerals [3–7], to other Cu minerals such as lindgrenite [8], libethenite [9], dioptase [10,11], volborthite [12], vesigneite [13,14], etc. [15–17]. The most important common feature of all these mineral structures is the presence of $Cu^{2+}$ cations in variable coordination geometries, a consequence of the Jahn–Teller effect [18–20] that results in the existence of at least four most common coordination geometries [21–23] with a diversity of transitional forms. Such a flexibility of $Cu^{2+}$-centered coordination polyhedra leads to the occurrence of a multitude of structure types with interesting physical properties tunable through an interplay between structure and chemical composition.

A number of mineral-related structures are characterized by the presence of 'additional' oxygen atoms that do not participate in the formation of strongly bonded "acid residue" complexes (sulfate, vanadate, phosphate, arsenate, selenite groups, etc.). These structures can be described in terms of anion-centered tetrahedra and attract special attention due to their magnetic properties controlled by the local structure of oxygen-based copper polycations [24–26]. For example, magnetic studies were performed for such anion-centered-based minerals as ilinskite [27], averievite [28,29], yaroshevskite [30], atlasovite [31], etc.

An important trend in the study of magnetic phases is the synthesis of novel mineral-inspired compounds, revealing the dynamics of structural changes and crystal chemical relations in various mineral groups. Recently, we discovered new structure types in the averievite family, $(MX)Cu_5O_2(TO_4)_2$ ($T^{5+}$ = P, V; $M^+$ = K, Rb, Cs, Cu; $X$ = Cl, Br) [32], which demonstrate significant changes in the first coordination spheres of $Cu^{2+}$ cations with the changes induced by the size of alkali metal ions, resulting in significant geometrical changes within the kagomé arrangements of magnetic $Cu^{2+}$ centers.

Aksenov et al. [33] reported on the synthesis and magnetic properties of $Rb_2CaCu_6(PO_4)_4O_2$, belonging to the shchurovskyite family, originally discovered by Pekov et al. [34] in the fumaroles of the Great fissure Tolbachik eruption, Kamchatka peninsula, Russia. Herein, we report on the synthesis and crystal structures of two novel compounds of the shchurovskyite family. Despite the fact that both structures significantly differ from the original shchurovskyite structure, both of them possess shchurovskyite-type Cu-based layers.

## 2. Materials and Methods

### 2.1. Synthesis

Single crystals of **1** ($K_2Cu[Cu_3O]_2(PO_4)_4$) and **2** ($K_{2.35}Cu_{0.825}[Cu_3O]_2(PO_4)_4$) were prepared via gas phase crystallization, successfully employed for the simulation of fumarolic mineral formation in a number of previous experiments [30,35,36]. Stoichiometric amounts of copper oxide (CuO, 99%, Vekton, St. Petersburg, Russia), copper pyrophosphate ($Cu_2P_2O_7$, 99%, Vekton) and potassium chloride (KCl, 99%, Vekton) taken in the 6:2:3 molar ratio were ground in an agate mortar. Due to the hygroscopic nature of potassium chloride [37,38], the resulting mixture was loaded into a porcelain boat and annealed at 250 °C for ~24 h in air. The mixture was further loaded into a fused silica ampule, evacuated to $10^{-2}$ mbar, sealed and placed horizontally in a furnace and heated to 800 °C over a period of 6 h. After two days, the furnace was cooled to 350 °C over a period of 72 h and switched off. The resulting sample contained single crystals of $K_2Cu[Cu_3O]_2(PO_4)_4$ (**1**), $K_{2.35}Cu_{0.825}[Cu_3O]_2(PO_4)_4$ (**2**), $Cu_5O_2(PO_4)_2$ [39] and copper oxide. All the compounds were found in a source zone of the ampule.

### 2.2. Single-Crystal X-ray Diffraction Study

Single crystals of both compounds were selected for data collection under an optical microscope, coated in an oil-based cryoprotectant and mounted on cryoloops. Diffraction data for **1** were collected using a Bruker APEX II DUO X-ray diffractometer (Bruker Co., Billerica, MA, U.S.A.) operated with a monochromated microfocus MoKα tube ($\lambda$ = 0.71073 Å) at 50 kV and 0.6 mA and equipped with a CCD APEX II detector. Diffraction data for **2** were collected using a Rigaku XtaLAB Synergy S X-ray diffractometer (Rigaku Co., Tokyo, Japan) operated with a monochromated microfocus MoKα tube ($\lambda$ = 0.71073 Å) at 50 kV and 1.0 mA and equipped with a CCD HyPix 6000 detector. Exposures were 10 and 74 s per frame for **1** and **2**, respectively. CrysAlisPro software [40] was used for the integration and correction of diffraction data for polarization, background and Lorentz effects as well as for an empirical absorption correction based on spherical harmonics implemented in the SCALE3 ABSPACK algorithm. The unit cell parameters (Table 1) were refined using the least-squares technique. The structures were solved by a dual-space algorithm and refined using the SHELX programs [41,42] incorporated in the OLEX2 program package [43]. The final models include coordinates and anisotropic displacement parameters (Tables A1 and A2).

**Table 1.** Crystallographic data and refinement parameters for **1** and **2**.

| Compound | 1 | 2 |
| --- | --- | --- |
| Formula | $K_2Cu[Cu_3O]_2(PO_4)_4$ | $K_{2.35}Cu_{0.825}[Cu_3O]_2(PO_4)_4$ |
| Space Group | $P$-1 | $P2_1/n$ |
| $a$, Å | 5.7787(3) | 16.7138(4) |
| $b$, Å | 8.2612(4) | 11.2973(3) |
| $c$, Å | 8.3717(4) | 16.8031(4) |
| $\alpha$, ° | 95.813(4) | 90 |
| $\beta$, ° | 103.239(4) | 90.775(2) |
| $\gamma$, ° | 96.821(4) | 90 |
| $V$, Å$^3$ | 382.85(3) | 3172.47(13) |
| $\mu$, mm$^{-1}$ | 10.601 | 10.137 |
| $Z$ | 1 | 8 |
| $D_{calc}$, g/cm$^3$ | 4.055 | 3.925 |
| Color | Light green | Intense green |
| Total reflections | 4126 | 37428 |
| Unique reflections | 1759 | 7219 |
| Reflection with $|F_o|\geq4\sigma_F$ | 1491 | 5761 |
| Angle range $2\theta$, °, MoK$\alpha$ | 5.01 to 54.998 | 6.492 to 55 |
| $R_{int}$, $R_\sigma$ | 0.0286, 0.0359 | 0.0368, 0.0303 |
| $R_1$, $wR_2$ ($|F_o| \geq 4\sigma_F$) | 0.0285, 0.0602 | 0.0394, 0.0847 |
| $R_1$, $wR_2$ (all data) | 0.0368, 0.0634 | 0.0535, 0.0901 |
| GOOF | 1.078 | 1.032 |
| $\rho_{min}$, $\rho_{max}$, $e$/Å$^3$ | 0.69 / −0.64 | 1.57 / −1.94 |
| CSD | 2092265 | 2092266 |

## 3. Results

*Crystal Structure Descriptions*

The crystal structure of **1** contains four symmetrically distinct $Cu^{2+}$ cations, three of which (Cu1, Cu2 and Cu3) form the shchurovskyite-type layer as shown in Figure 1a. Considered in terms of the cation-centered polyhedral, the basic unit of the structure is a rod of edge-sharing $Cu3O_5$ square pyramids (Cu3 · · · Cu3 = 2.856 Å), extended along [100]. The orientation of apical vertices ($O_{ap}$) of square pyramids alternates up (**U**) and down (**D**) relative to the (010) plane, giving the **UDUDUD** sequence within the rod, with the Cu–$O_{ap}$ bond distance equal to 2.239 Å (Figure 1c). Each rod is decorated by $Cu1O_6$ octahedra and $Cu2O_5$ triangular bipyramids from both sides in the (010) plane. The $Cu1O_6$ octahedron shows a typical [4+2] distortion with four short (1.887–2.097 Å) and two long (2.706 and 2.768 Å) Cu–O bonds. It is noteworthy that the Cu1 site has one more elongated (2.936 Å) Cu–O distance to the O atom located near one of the apical ligands (O · · · O = 2.500 Å), which corresponds to the edge of the $(PO_4)^{3-}$ tetrahedron. Such a coordination geometry of $Cu^{2+}$ cations is rather rare and, as far as we know, among all the natural $Cu^{2+}$-containing oxysalts, it was observed in the crystal structures of cesiodymite, cryptochalcite and saranchinaite, which are the products of fumarolic activity of the Tolbachik volcano [44,45]. The $Cu1O_6$ octahedron shares a common edge with the $Cu2O_5$ triangular bipyramid. The equatorial plane of the $Cu2O_5$ bipyramid is formed by one short (1.944 Å) and two elongated (2.237 and 2.299 Å) Cu-O bonds. The apical Cu2–$O_{ap}$ bond lengths are equal to 1.884 and 1.920 Å. The $Cu2O_5$ bipyramid is slightly distorted due to the Berry twist mechanism [46,47]: the $O_{ap}$–Cu2–$O_{ap}$ and $O_{eq}$–Cu2–$O_{eq}$ angles are equal to 166.6 and 124.4°, respectively.

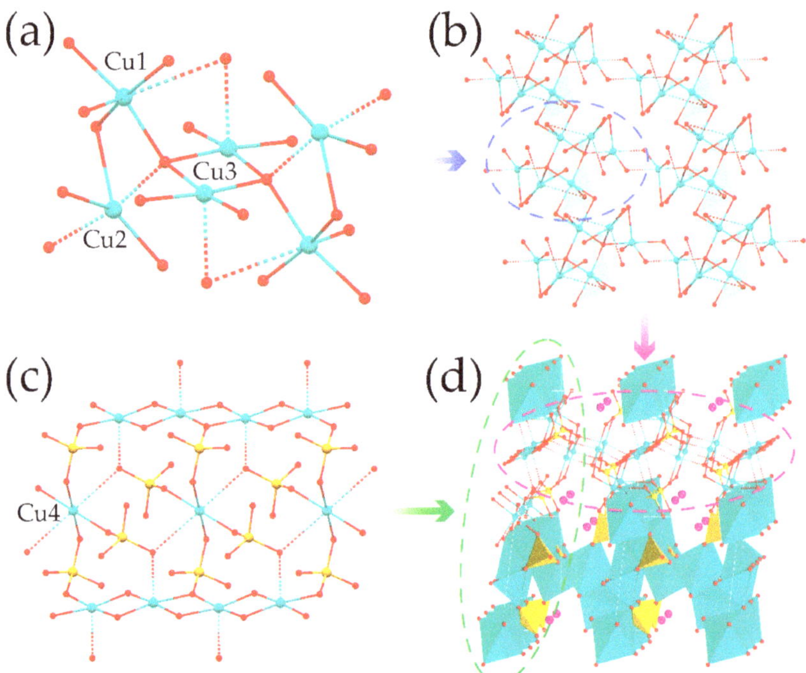

**Figure 1.** The key features of the crystal structure of **1**: (**a**) the linkage of coordination polyhedra of the Cu1, Cu2 and Cu3 atoms; (**b**) the Cu-based layer; (**c**) the rods of $Cu3O_5$ square pyramids and $Cu4O_6$ octahedra, connecting the Cu-based layers; (**d**) lateral view of the crystal structure. Legend: Cu = cyan, O = red, P = yellow, K = purple.

The rods are linked into the Cu-based polyhedral layer parallel to (010) due to the edge-sharing of two adjacent $Cu2O_5$ triangular bipyramids belonging to the neighboring rods (Figure 1b). The connection between the layers in the shchurovskyite-related compounds proceeds via the $PO_4$ tetrahedra. In the crystal structure of **1**, each $P2O_4$ tetrahedron shares common vertices with the $Cu3O_5$ square pyramid and $Cu2O_5$ triangular bipyramid of one layer, and with the $Cu1O_6$ octahedron of the adjacent layer. In opposition, the $P1O_4$ tetrahedron is located within the layer, and shares common vertices with two $Cu2O_5$ triangular bipyramids and one $Cu1O_5$ square pyramid. There is one additional Cu4 site located in between the layers (Figure 1c) and coordinated by six oxygen atoms, belonging to the $PO_4$ tetrahedra, to form [4+2]-distorted octahedron with four short (1.977 ($\times$2) and 1.984 ($\times$2) Å) and two long (2.823 ($\times$2) Å) Cu4-O bonds. Two opposite vertices of the equatorial plane of the $Cu4O_6$ octahedron are common with two $Cu3O_5$ square pyramids of adjacent Cu-based layers. The resulting framework has channels filled by $K^+$ cations (Figure 1d). There is one symmetrically distinct K atom, coordinated by six oxygen atoms (2.633–2.959 Å), with the <K1 $\cdots$ K1> distance equal to 3.865 (3) and 3.732 (3) Å. The calculation of the effective width (e.c.w.) of the channels, by subtracting the ionic diameter of $O^{2-}$ (2.7 Å) from the shortest and longest O $\cdots$ O distances across the channel [48], shows that the channel in the structure of **1** is much smaller than in the structure of shchurovskyite: 1.3 $\times$ 4.6 Å$^2$ in **1** versus 2.8 $\times$ 5.9 Å$^2$ in shchurovskyite.

As with all known shchurovskyite-type minerals and compounds, the crystal structure of **1** contains so-called 'additional' oxygen atoms ($O_{add}$), allowing it to be described in terms of anion-centered tetrahedra [24,25]. The structure of **1** contains one symmetrically distinct 'additional' oxygen atom tetrahedrally coordinated by two Cu3, one Cu1 and one Cu2 atoms (Figure 1a). The $O_{add}$–Cu bond lengths are the shortest among all Cu–O bonds and span the range of 1.881–1.913 Å. Two $OCu_4$ tetrahedra share a common $Cu3 \cdots Cu3$ edge to form a $[O_2Cu_6]^{8+}$ dimer.

The crystal structure description of **2** is far more difficult to describe due to its disordered nature and the enlarged unit cell, which is a consequence of the increased number of symmetrically distinct positions of all atoms. There are two symmetrically distinct rods of $CuO_5$ square pyramids in **2**. The first rod consists of the $Cu3O_5$ and $Cu4O_5$ pyramids, alternating along [010], whereas the alternation of the $Cu8O_5$ and $Cu9O_5$ pyramids builds the second rod (Figure 2a,c,e). The equatorial $Cu–O_{eq}$ bond lengths within square pyramids are in the range of 1.898–1.966 Å. The apical oxygen atom of the $Cu8O_5$ square pyramid is split into two positions, resulting in the $Cu8-O_{ap}$ bond lengths of 2.579 and 2.739 Å; the apical bond lengths in the other square pyramids vary from 2.386 to 2.469 Å. The apical vertices of the square pyramids within the rods are oriented according to the **UUDDUU** sequence, which is different from the sequence **UDUD** observed in **1**.

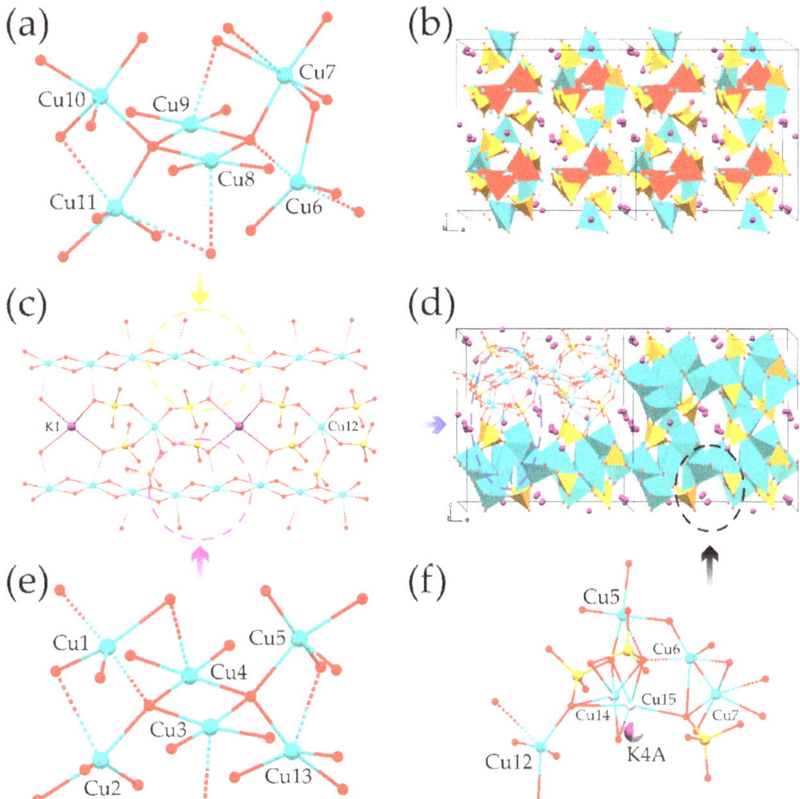


**Figure 2.** The key features of the crystal structure of **2**: (**a**) the fragment of the $Cu9 \cdots Cu8$ rod; (**b**) the arrangement of anion-centered tetrahedra (shown in red); (**c**) two symmetrically distinct rods linked via the $Cu12O_6$ octahedra; (**d**) lateral view of the crystal structure; (**e**) the fragment of the $Cu3 \cdots Cu4$ rod; (**f**) the disordered arrangement of the Cu14 and Cu15 sites, and associated K4A site. Disordered oxygen atoms are shown at most probable positions. Legend: as in Figure 1.

The local coordination of the Cu atoms, involved in the orthogonal connection of the polyhedral rods, is different from those observed in the crystal structure of **1**: instead of one symmetrically distinct $Cu^{2+}$-centered triangular bipyramid, there are four distinct $Cu^{2+}$ polyhedra, forming two types of bridges. The bridge of the first type is formed by the dimer of edge-sharing $Cu2O_5$ and $Cu10O_5$ square pyramids. The average $<Cu–O_{eq}>$ bond distances are equal to 1.953 and 1.966 Å for the $Cu2O_5$ and $Cu10O_5$ pyramids, respectively, whereas the apical bond lengths are 2.597 and 2.305 Å, respectively. Note that the $Cu2O_5$ square pyramid can be considered as a [4+1+1]-distorted octahedron due to the presence of an additional elongated $Cu–O_{ap}$ bond equal to 2.939 Å. The bridge of the second type consists of the $Cu5O_5$ square pyramid and $Cu6O_5$ triangular bipyramid, sharing a common edge. The $Cu5O_5$ square pyramid is similar to the $Cu2O_5$ and $Cu10O_5$ pyramids, with the average $<Cu5–O_{eq}>$ bond length equal to 1.947 Å and the $Cu–O_{ap}$ bond of 2.528 (6) Å. The triangular–bipyramidal coordination geometry of the Cu6 atom is similar to that in **1**, with one short and two elongated (1.88, 2.305 and 2.188 Å) equatorial bonds, and two short (1.965 and 1.880 Å) apical bonds. Taking the crystal structure of **1** as an archetype, the $Cu1O_6$ octahedron in **1** is replaced in **2** by four symmetrically distinct Cu atoms, attached to the rods. The $Cu7O_6$ and $Cu11O_6$ octahedra share common vertices with the $Cu8O_5$ and $Cu9O_5$ square pyramids of the rods (Figure 2a). While the $Cu7O_6$ octahedron shows a typical [4+2]-distortion, the coordination geometry of the Cu11 atom is difficult to assess due to the disorder of one equatorial and one apical ligands. We suppose the overlap of two possible coordination geometries—octahedral and triangular–bipyramidal. The presence of one or another geometry depends on the rotation of the $P5O_4$ tetrahedron that has three split oxygen positions with the site-occupancy factors (S.O.F.) equal to 0.663:0.337. Equatorial bond lengths of the $Cu11O_5$ triangular bipyramid are equal to 1.93, 2.212 and 2.474 Å, whereas the apical bonds are 1.905 and 1.912 Å long. The octahedral coordination geometry of the Cu11 atom shows the [3+1+2]-distortion, with three short (1.905–2.051 Å) and one elongated (2.212 Å) equatorial bonds and two long apical bonds (2.474 and 2.809 Å). The second rod is decorated by the the $Cu13O_6$ octahedra and $Cu1O_5$ triangular bipyramids (Figure 2e). The $Cu13O_6$ octahedron is [4+1+1]-distorted with four short equatorial (1.897–2.052 Å) and two long apical bonds (2.476 and 2.897 Å). The bond distribution in the $Cu1O_5$ polyhedron is different from those in other triangular bipyramids in **1** and **2**: there are two apical bonds (1.906 and 1.911 Å) slightly shorter than the equatorial bonds (2.070–2.172 Å).

The $Cu^{2+}$-centered polyhedra in **2** form the Cu-based layer parallel to the (001) plane. The comparison with the structure of **1** reveals the significant difference between two structures: instead of the $CuO_6$ octahedra connecting the layers in **1**, the crystal structure of **2** shows an alternation of the Cu and K atoms along [010] (Figure 3c). There are two symmetrically distinct fully occupied Cu12 and K1 sites. The Cu12 site is coordinated by five oxygen atoms to form a [4+1]-distorted square pyramid. Taking into account that positions of two neighboring oxygens in the equatorial plane of the $Cu12O_5$ polyhedron are disordered, the average $<Cu12–O_{eq}>$ bond length is 1.972 Å. The coordination geometry of the Cu12 site can also be considered as transitional from square-pyramidal to square-planar due to the long $Cu12–O_{ap}$ bond length (2.731 Å). The K1 atoms are coordinated by eight oxygen atoms of the phosphate groups, with five relatively short (2.711–2.857 Å) and three long (3.375–3.318 Å) K1–O bonds.

The enlargement of the unit cell results in the presence of the second symmetrically distinct intra-framework channel. Both types of channels are filled by highly disordered K atoms (Figure 2d). The first channel contains K2 and K3 atoms, each split into two and three symmetrically distinct partially occupied positions, respectively. The displacement of the K atoms, placed in average positions, is similar to that observed in the structure of **1** due to the zigzag arrangement within the channel, with the $K2_{av}\cdots K3_{av}$ and $K3_{av}\cdots K3_{av}$ distances equal to 3.992 and 4.124 Å, respectively. The effective channel width ($1.3 \times 5.8$ Å$^2$) is close to that observed in the crystal structure of shchurovskyite [34]. The content of the second symmetrically distinct channel is significantly different from that observed in any other shchurovskyite-type structures. Herein, along with the disordered K atoms, additional Cu atoms are present. The Cu site within the channel is split into two symmetrically distinct positions, Cu14 and Cu145, with S.O.F. of 0.5 and 0.15, respectively (Figure 3f). While the most occupied Cu14 position exhibits a typical square-pyramidal coordination geometry

(2.012–2.735 Å), the Cu15 site shows a different bond distribution, having one short (2.033 Å) and four elongated (2.212–2.490 Å) bonds, a consequence of the low occupancy of the disordered configuration. The K atoms are concentrated near to the center of the channel with the K–O bond lengths in the range of 2.564–3.40 Å. Due to the presence of the additional Cu atoms, the effective width of channels is reduced to $1.5 \times 3.0$ Å$^2$.

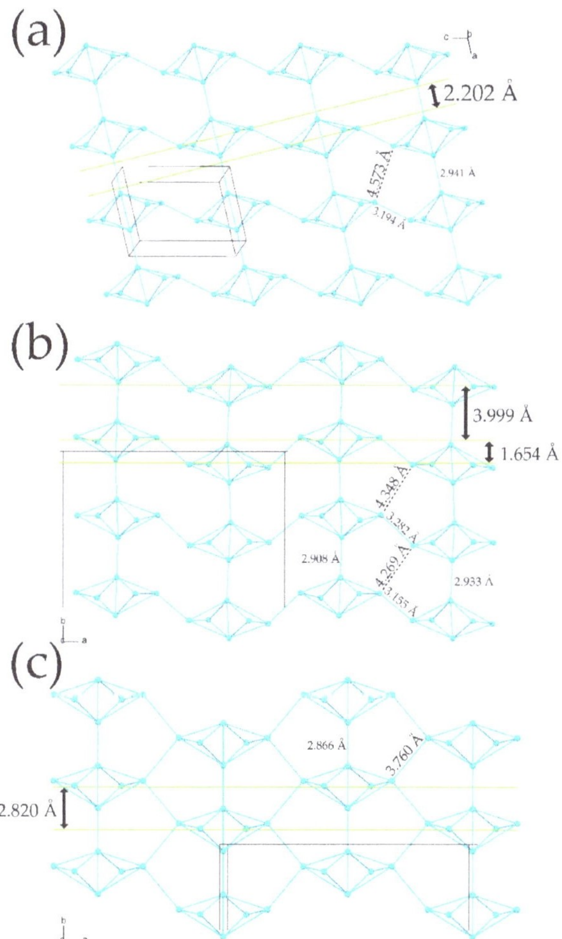


**Figure 3.** Cu-based layers in the crystal structures of **1** (**a**), **2** (**b**) and Rb$_2$Ca[Cu$_3$O]$_2$(PO$_4$)$_4$ (**c**). Thin black lines show boundaries of unit cells. Legend: as in Figure 1. See details in the text.

The crystal structure of **2** contains four symmetrically distinct 'additional' oxygen atoms, tetrahedrally coordinated by copper atoms, with the O$_{add}$–Cu bond lengths varying from 1.881 to 1.928 Å (Figure 2b).

## 4. Discussion

The crystal structures of shchurovskyite, K$_2$Ca[Cu$_3$O]$_2$(AsO$_4$)$_4$, dmisokolovite, K$_3$[Cu$_5$AlO$_2$](AsO$_4$)$_4$, and synthetic Rb$_2$Ca[Cu$_3$O]$_2$(PO$_4$)$_4$ are based upon Cu-based oxo-layers, linked via (PO$_4$)$^{3-}$ tetrahedra [33,34]. The hetero-polyhedral framework contains channels, filled by K$^+$ cations, and one additional position between the copper layers, occupied by Ca$^{2+}$ and K$^+$ atoms in shchurovskyite and dmisokolovite, respectively. Unlike the structure of both minerals, these additional positions in the crystal structures of **1** and **2** are occupied by $Cu^{2+}$ cations, connecting the Cu-based layers, which results in the formation of a Cu-based porous framework, decorated by $(PO_4)^{3-}$ tetrahedra and containing potassium atoms within the channels.

As was mentioned above, all the shchurovskyite-type structures can be described as consisting of Cu-based layers. The simplest way to distinguish the main features of each structure is a topological analysis, which can be performed by means of representation of the connectivity of copper atoms in terms of nets. In order to reduce the influence of chemical composition on crystal chemical comparative analysis, only phosphate members of the shchurovskyite family will be discussed below.

The core of the layers in each structure is a $Cu_6$ cluster, representing a $[O_2Cu_6]$ dimer of two edge-sharing $(OCu_4)$ tetrahedra, with their bases approximately parallel to the planes of the layers. The dimers form linear rods along *a* in **1**, and along *b* in **2** and other shchurovskyite-type structures, with the shortest Cu⋯Cu distances between the adjacent dimers varying from 2.866 (in $Rb_2Ca[Cu_3O]_2(PO_4)_4$), to 2.908 and 2.933 (in **2**), and 2.941 Å (in **1**).

Figure 3a shows the projection of the Cu-based layer onto the (010) plane in **1**, with green lines aligned along terminal Cu atoms of dimers of the neighboring rods. The distance between the lines is 2.202 Å. The rods in the structure of **2** are aligned along two-fold screw axis, which results in the different arrangement (Figure 3b). Herein, each rod is surrounded by two symmetrically related rods shifted by 1.654 Å relative to each other. The shift is maximal in the crystal structure of $Rb_2Ca[Cu_3O]_2(PO_4)_4$, where each rod is shifted by 2.820 Å relative to the adjacent rod (Figure 3c).

The results presented above show the highly flexible nature of the shchurovskyite-type structures. Indeed, at least four different structure types are present: triclinic archetype (**1**), monoclinically distorted $1 \times 1 \times 1$ shchurovskyite and its phosphate synthetic analogue with the *C*2 space groups, monoclinically distorted $1 \times 1 \times 2$ superstructure of dmisokolovite with the *C*2/*c* space group, and monoclinically distorted $2 \times 2 \times 2$ superstructure. Both the archetype (**1**) and $2 \times 2 \times 2$ superstructure (**2**) reported herein were obtained in the same experiment, whereas the phosphate analogue of shchurovskyite does not exhibit any significant difference from its parent structure. At the same time, the $1 \times 1 \times 2$ superstructure of dmisokolovite is due to the partial substitution of Cu by Al cations. Thus, the broad variety of shchurovskyite-type structures most likely depends on the flexibility of the Cu-based framework, and the arrangement of the Cu-based rods in particular. We suppose that the flexibility of the Cu-based framework will lead to the discovery of other novel compounds of the series, including novel mineral species.

Another important feature of the shchurovskyite-type structures is the interlayer space that can accommodate different cations, from $K^+$ (in the structures of dmisokolovite and **2**), to $Ca^{2+}$ (in shchurovskyite and $Rb_2Ca[Cu_3O]_2(PO_4)_4$) and $Cu^{2+}$ (in **1** and **2**). The interlayer sites are responsible for the connection of adjacent Cu-based layers, and it can be assumed that the substitution of non-magnetic cations by $Cu^{2+}$ will lead to significant changes in the magnetic properties of the respective materials.

The differences between the shchurovskyite-related structure types can be easily demonstrated by information-based measures of structural complexity [49–51] (Table 2). In the framework of this approach, the complexity is estimated as the amount of Shannon information contained in a crystal structure according to the following formulas:

$$^{str}I_G = -\sum\nolimits_{i=1}^{k} p_i \log_2 p_i \qquad \text{(bits/atom)} \qquad (1)$$

$$^{str}I_G = -v\sum\nolimits_{i=1}^{k} p_i \log_2 p_i \qquad \text{(bits/cell)} \qquad (2)$$

where *k* is the number of different crystallographic orbits (independent crystallographic Wyckoff sites) in the structure and $p_i$ is the random choice probability for an atom from the $i^{th}$ crystallographic orbit, which is:

$$p_i = m_i/v \qquad (3)$$

where $m_i$ is a multiplicity of a crystallographic orbit (i.e., the number of atoms of a specific Wyckoff site in the reduced unit cell), and $v$ is the total number of atoms in the reduced unit cell.

The crystal structure of **1** (archetype) is the simplest one (3.986 bits/atom and 123.58 bits/cell), whereas the crystal structure of **2** is the most complex in the series (5.974 bits/atom and 1493.446 bits/cell) and belongs to the category of very complex structures [52,53]. The complexity parameters of the crystal structures of shchurovskyite and $Rb_2Ca[Cu_3O]_2(PO_4)_4$ are almost as simple as those for the **1** (4.051 bits/atom and 125.58 bits/cell), whereas the crystal structure of dmisokolovite with the doubled $c$ parameter is twice as complex (4.051 bits/atom and 251.16 bits/cell).

**Table 2.** Selected crystallographic and structural complexity parameters for shchurovskyite-type structures.

| Compound | $K_2Cu[Cu_3O]_2(PO_4)_4$ | $K_{2.35}Cu_{0.825}[Cu_3O]_2(PO_4)_4$ | $Rb_2Ca[Cu_3O]_2(PO_4)_4$ | $K_2Ca[Cu_3O]_2(AsO_4)_4$ | $K_3[Cu_5AlO_2](AsO_4)_4$ |
|---|---|---|---|---|---|
| Space Group | $P$-1 | $P2_1/n$ | $C2$ | $C2$ | $C2/c$ |
| $a$, Å | 5.779 | 16.714 | 16.891 | 17.286 | 17.085 |
| $b$, Å | 8.261 | 11.297 | 5.641 | 5.670 | 5.719 |
| $c$, Å | 8.372 | 16.803 | 8.359 | 8.573 | 16.533 |
| $\alpha, \beta, \gamma$, ° | 95.81, 103.24, 96.82 | 90, 90.77, 90 | 90, 93.92, 90 | 90, 92.95, 90 | 90, 91.72, 90 |
| $V$, Å$^3$ | 382.8 | 3172.5 | 794.6 | 839.2 | 1614.7 |
| $Z$ | 1 | 8 | 2 | 2 | 4 |
| $v$, atoms/cell | 31 | 250 | 31 | 31 | 62 |
| $I_G$, bits/atom | 3.986 | 5.974 * | 4.051 | 4.051 | 4.051 |
| $I_{G,\,total}$, bits/cell | 123.58 | 1493.446 * | 125.58 | 125.58 | 251.16 |
| Reference | This work | This work | 33 | 34 | 34 |

* All the disordered atoms were placed in their average positions.

**Supplementary Materials:** The following are available online at https://www.mdpi.com/article/10.3390/cryst11070807/s1, Crystallographic information files of compounds **1** and **2**.

**Author Contributions:** Conceptualization, I.V.K. and S.V.K.; formal analysis, I.V.K.; investigation, I.V.K.; resources, S.V.K.; writing—original draft preparation, I.V.K.; writing—review and editing, S.V.K.; visualization, I.V.K.; supervision, S.V.K.; project administration, S.V.K.; funding acquisition, S.V.K. All authors have read and agreed to the published version of the manuscript.

**Funding:** This research was funded by the Russian Science Foundation (grant No. 19-17-00038).

**Data Availability Statement:** Supplementary crystallographic data for this paper have been deposited at Cambridge Crystallographic Data Centre (CCDC 2092265 and 2092265 for **1** and **2**, respectively) and can be obtained free of charge via www.ccdc.cam.ac.uk/data_request/cif.

**Acknowledgments:** The XRD measurements were performed at the X-ray Diffraction Centre of St. Petersburg State University.

**Conflicts of Interest:** The authors declare no conflict of interest.

## Appendix A

**Table A1.** Fractional atomic coordinates and isotropic displacement parameters (Å$^2$) for **1**.

| Atom | $x$ | $y$ | $z$ | $U_{eq}$ |
|---|---|---|---|---|
| Cu1 | 0.54313(10) | 0.21854(7) | 0.18319(7) | 0.00826(15) |
| Cu2 | 0.62938(11) | 0.54189(7) | 0.35677(7) | 0.01248(16) |
| Cu3 | 0.25788(9) | 0.51344(7) | 0.00581(7) | 0.00691(14) |
| Cu4 | 1 | 1 | 1 | 0.01075(19) |
| K1 | 1.1877(2) | −0.12671(17) | 0.39996(14) | 0.0251(3) |

**Table A1.** *Cont.*

| Atom | $x$ | $y$ | $z$ | $U_{eq}$ |
|---|---|---|---|---|
| P1 | 0.5789(2) | −0.14031(14) | 0.18292(14) | 0.0066(2) |
| P2 | 0.9306(2) | 0.67348(14) | 0.72047(13) | 0.0053(2) |
| O1 | 0.7258(6) | −0.2413(4) | 0.3016(4) | 0.0146(7) |
| O2 | 0.5839(6) | 0.0261(4) | 0.2838(4) | 0.0102(7) |
| O3 | 0.3192(6) | −0.2209(4) | 0.1123(4) | 0.0134(7) |
| O4 | 0.6982(6) | −0.1156(4) | 0.0350(4) | 0.0106(7) |
| O5 | 0.9594(5) | 0.5576(4) | 0.8563(4) | 0.0093(6) |
| O6 | 1.1450(5) | 0.6696(4) | 0.6400(4) | 0.0074(6) |
| O7 | 0.6911(5) | 0.6036(4) | 0.5917(4) | 0.0093(6) |
| O8 | 0.9150(6) | 0.8492(4) | 0.7897(4) | 0.0120(7) |
| O9 | 0.5292(5) | 0.4366(4) | 0.1373(4) | 0.0066(6) |

**Table A2.** Fractional atomic coordinates, site occupancy factors (S.O.F.) and isotropic displacement parameters ($\text{Å}^2$) for **1**.

| Atom | S.O.F. | $x$ | $y$ | $z$ | $U_{eq}$ |
|---|---|---|---|---|---|
| Cu1 | 1 | 0.84435(4) | 0.42009(6) | 0.88948(4) | 0.01789(15) |
| Cu2 | 1 | 0.92922(4) | 0.41660(6) | 0.73252(4) | 0.01725(15) |
| Cu3 | 1 | 0.75839(4) | 0.54529(5) | 0.73128(4) | 0.01226(14) |
| Cu4 | 1 | 0.74882(4) | 0.30406(5) | 0.75677(4) | 0.01233(14) |
| Cu5 | 1 | 0.57404(4) | 0.39718(6) | 0.75159(4) | 0.01062(13) |
| Cu6 | 1 | 0.43001(4) | 0.59447(7) | 0.74471(5) | 0.02626(19) |
| Cu7 | 1 | 0.33900(4) | 0.56147(6) | 0.89307(4) | 0.01664(15) |
| Cu8 | 1 | 0.25552(4) | 0.44950(5) | 0.73520(4) | 0.01493(15) |
| Cu9 | 1 | 0.24989(4) | 0.69237(5) | 0.76032(4) | 0.01117(14) |
| Cu10 | 1 | 0.07531(4) | 0.59072(5) | 0.75522(4) | 0.01117(13) |
| Cu11 | 1 | 0.16550(4) | 0.58938(6) | 0.60214(4) | 0.01817(15) |
| Cu12 | 1 | 0.25872(5) | 0.82651(7) | 0.49436(5) | 0.0335(2) |
| Cu13 | 1 | 0.34155(4) | 0.58442(6) | 0.39884(4) | 0.01298(14) |
| Cu14 | 0.5 | 0.10290(9) | 0.86289(12) | 0.34216(9) | 0.0188(3) |
| Cu15 | 0.15 | 0.0595(3) | 0.8514(4) | 0.3650(3) | 0.0188(3) |
| K1 | 1 | 0.23938(9) | 0.32812(12) | 0.49506(9) | 0.0313(3) |
| K2A | 0.41 | 0.4229(4) | 0.8501(4) | 0.6448(3) | 0.0718(19) |
| K2B | 0.59 | 0.4599(2) | 0.8680(3) | 0.5471(3) | 0.0613(11) |
| K3A | 0.42 | 0.4539(10) | 0.335(3) | 0.5443(9) | 0.078(6) |
| K3B | 0.36 | 0.4585(5) | 0.3883(10) | 0.5623(9) | 0.0382(19) |
| K3C | 0.22 | 0.445(2) | 0.346(4) | 0.569(2) | 0.069(9) |
| K4A | 0.45 | 0.5346(3) | 0.6385(4) | 0.9684(4) | 0.0725(17) |
| K4B | 0.55 | 0.4942(3) | 0.7322(5) | 1.0182(2) | 0.0804(19) |
| K4C | 0.15 | 0.4597(14) | 0.8276(19) | 1.042(2) | 0.125(12) |
| K4D | 0.55 | 0.4772(3) | 0.9217(6) | 1.0477(3) | 0.0828(17) |
| P1 | 1 | 0.90143(8) | 0.16223(11) | 0.83315(8) | 0.0099(3) |
| P2 | 1 | 0.90101(7) | 0.65707(11) | 0.82032(8) | 0.0080(2) |
| P3 | 1 | 0.40314(7) | 0.33475(11) | 0.81544(8) | 0.0090(2) |
| P4 | 1 | 0.40152(8) | 0.83626(11) | 0.83808(8) | 0.0111(3) |
| P5 | 1 | 0.34316(8) | 0.61861(13) | 0.57888(8) | 0.0144(3) |
| P6A | 0.337(6) | 0.3350(19) | 1.0893(18) | 0.5736(18) | 0.011(3) |
| P6B | 0.663(6) | 0.3349(9) | 1.0612(9) | 0.5739(10) | 0.0129(17) |
| P7 | 1 | 0.16036(8) | 1.05119(12) | 0.42170(8) | 0.0130(3) |
| P8 | 1 | 0.16418(8) | 0.62317(12) | 0.41864(8) | 0.0125(3) |
| O1 | 1 | 0.9538(2) | 0.1157(4) | 0.7643(3) | 0.0240(10) |
| O2 | 1 | 0.8989(3) | 0.0726(4) | 0.9004(3) | 0.0289(11) |
| O3 | 1 | 0.9302(2) | 0.2838(3) | 0.8587(2) | 0.0173(8) |

**Table A2.** *Cont.*

| Atom | S.O.F. | $x$ | $y$ | $z$ | $U_{eq}$ |
|---|---|---|---|---|---|
| O4 | 1 | 0.8155(2) | 0.1732(3) | 0.7957(2) | 0.0161(8) |
| O5 | 1 | 0.9596(2) | 0.5784(3) | 0.7737(2) | 0.0141(8) |
| O6 | 1 | 0.8775(3) | 0.5963(3) | 0.8971(2) | 0.0192(9) |
| O7 | 1 | 0.8277(2) | 0.6751(3) | 0.7635(2) | 0.0152(8) |
| O8 | 1 | 0.9385(2) | 0.7797(3) | 0.8364(2) | 0.0105(7) |
| O9 | 1 | 0.4379(2) | 0.2116(3) | 0.8333(2) | 0.0162(8) |
| O10 | 1 | 0.3249(2) | 0.3197(3) | 0.7645(2) | 0.0167(8) |
| O11 | 1 | 0.4600(2) | 0.4076(3) | 0.7625(2) | 0.0165(8) |
| O12 | 1 | 0.3874(3) | 0.3991(3) | 0.8928(2) | 0.0194(8) |
| O13 | 1 | 0.4321(2) | 0.7117(3) | 0.8570(3) | 0.0210(9) |
| O14 | 1 | 0.4578(2) | 0.9000(4) | 0.7807(3) | 0.0239(10) |
| O15 | 1 | 0.3901(3) | 0.9127(4) | 0.9113(3) | 0.0320(11) |
| O16 | 1 | 0.3197(2) | 0.8238(3) | 0.7932(2) | 0.0159(8) |
| O17 | 1 | 0.3869(2) | 0.6187(4) | 0.5003(2) | 0.0272(10) |
| O18A | 0.337(6) | 0.3132(7) | 0.7541(10) | 0.5773(7) | 0.0199(12) |
| O18B | 0.663(6) | 0.2644(4) | 0.6911(5) | 0.5798(4) | 0.0199(12) |
| O19A | 0.337(6) | 0.4069(9) | 0.6053(15) | 0.6410(9) | 0.021(3) |
| O19B | 0.663(6) | 0.4004(4) | 0.6670(8) | 0.6433(5) | 0.0221(17) |
| O20A | 0.337(6) | 0.2739(8) | 0.5379(11) | 0.5834(8) | 0.0258(13) |
| O20B | 0.663(6) | 0.3189(4) | 0.4881(6) | 0.5983(4) | 0.0258(13) |
| O21 | 1 | 0.4024(3) | 1.0565(5) | 0.6340(3) | 0.0390(13) |
| O22A | 0.337(6) | 0.2806(18) | 0.9835(18) | 0.5575(14) | 0.044(6) |
| O22B | 0.663(6) | 0.3011(7) | 0.9367(8) | 0.5635(7) | 0.030(2) |
| O23 | 1 | 0.2682(3) | 1.1526(5) | 0.6017(3) | 0.0374(12) |
| O24 | 1 | 0.3724(3) | 1.1141(5) | 0.4974(3) | 0.0325(11) |
| O25 | 1 | 0.1884(3) | 0.9209(4) | 0.4246(3) | 0.0269(10) |
| O26 | 1 | 0.2261(2) | 1.1350(4) | 0.3955(2) | 0.0189(8) |
| O27 | 1 | 0.0901(2) | 1.0434(4) | 0.3599(2) | 0.0271(10) |
| O28 | 1 | 0.1255(3) | 1.0924(4) | 0.5000(2) | 0.0209(9) |
| O29 | 1 | 0.1786(3) | 0.4971(3) | 0.3940(2) | 0.0228(9) |
| O30 | 1 | 0.1071(2) | 0.6862(4) | 0.3573(2) | 0.0199(9) |
| O31 | 1 | 0.2454(2) | 0.6921(3) | 0.4201(2) | 0.0147(8) |
| O32 | 1 | 0.1241(2) | 0.6348(4) | 0.5000(2) | 0.0191(8) |
| O33 | 1 | 0.8264(2) | 0.4268(3) | 0.7770(2) | 0.0094(7) |
| O34 | 1 | 0.6801(2) | 0.4241(3) | 0.7134(2) | 0.0092(7) |
| O35 | 1 | 0.3253(2) | 0.5669(3) | 0.7799(2) | 0.0110(7) |
| O36 | 1 | 0.1812(2) | 0.5730(3) | 0.7141(2) | 0.0116(7) |

## References


1. Han, T.-H.; Singleton, J.; Schlueter, J.A. Barlowite: A spin-1212 antiferromagnet with a geometrically perfect kagome motif. *Phys. Rev. Lett.* **2014**, *113*, 227203. [CrossRef]
2. Norman, M.R. Herbertsmithite and the search for the quantum spin liquid. *Rev. Mod. Phys.* **2016**, *88*, 041002. [CrossRef]
3. Zheng, X.G.; Kawae, T.; Kashitani, Y.; Li, C.S.; Tateiwa, N.; Takeda, K.; Yamada, H.; Xu, C.N.; Ren, Y. Unconventional magnetic transitions in the mineral clinoatacamite $Cu_2Cl(OH)_3$. *Phys. Rev. B Condens. Matter Mater. Phys.* **2005**, *71*, 052409. [CrossRef]
4. Zheng, X.G.; Mori, T.; Nishiyama, K.; Higemoto, W.; Yamada, H.; Nishikubo, K.; Xu, C.N. Antiferromagnetic transitions in polymorphous minerals of the natural cuprates atacamite and botallackite $Cu_2Cl(OH)_3$. *Phys. Rev. B Condens. Matter Mater. Phys.* **2005**, *71*, 174404. [CrossRef]
5. Li, Y.-S.; Zhang, Q.-M. Structure and magnetism of S = 1/2 kagome antiferromagnets $NiCu_3(OH)_6Cl_2$ and $CoCu_3(OH)_6Cl_2$. *J. Phys. Condens. Matter.* **2013**, *25*, 026003. [CrossRef]
6. Malcherek, T.; Mihailova, B.; Welch, M.D. Structural phase transitions of clinoatacamite and the dynamic Jahn-Teller effect. *Phys. Chem. Miner.* **2017**, *44*, 307–321. [CrossRef]
7. Malcherek, T.; Welch, M.D.; Williams, P.A. The atacamite family of minerals—A testbed for quantum spin liquids. *Acta Crystallogr.* **2018**, *B74*, 519–526. [CrossRef]
8. Vilminot, S.; André, G.; Richard-Plouet, M.; Bourée-Vigneron, F.; Kurmoo, M. Magnetic Structure and Magnetic Properties of Synthetic Lindgrenite, $Cu_3(OH)_2(MoO_4)_2$. *Inorg. Chem.* **2006**, *45*, 10938–10946. [CrossRef] [PubMed]
9. Belik, A.A.; Koo, H.-J.; Whangbo, M.-H.; Tsujii, N.; Naumov, P.; Takayama-Muromachi, E. Magnetic Properties of Synthetic Libethenite $Cu_2PO_4OH$: A New Spin-Gap System. *Inorg. Chem.* **2007**, *46*, 8684–8689. [CrossRef] [PubMed]

10. Janson, O.; Tsirlin, A.A.; Schmitt, M.; Rosner, H. Large quantum fluctuations in the strongly coupled spin-1/2 chains of green dioptase $Cu_6Si_6O_{18} \cdot 6H_2O$. *Phys. Rev. B Condens. Matter Mater. Phys.* **2010**, *82*, 014424. [CrossRef]

11. Podlesnyak, A.; Prokhorenko, O.; Nikitin, S.E.; Kolesnikov, A.I.; Matsuda, M.; Dissanayake, S.E.; Prisk, T.R.; Nojiri, H.; Díaz-Ortega, I.F.; Kidder, M.K.; et al. magnetic ground state and magnetic excitations in black dioptase $Cu_6Si_6O_{18}$. *Phys. Rev. B Condens. Matter Mater. Phys.* **2019**, *100*, 184401. [CrossRef]

12. Hiroi, Z.; Ishikawa, H.; Yoshida, H.; Yamaura, J.-I.; Okamoto, Y. Orbital Transitions and Frustrated Magnetism in the Kagome-Type Copper Mineral Volborthite. *Inorg. Chem.* **2019**, *58*, 11949–11960. [CrossRef] [PubMed]

13. Yoshida, H.; Michiue, Y.; Takayama-Muromachi, E.; Isobe, M. β-Vesignieite $BaCu_3V_2O_8(OH)_2$: A structurally perfect $S = 1/2$ kagomé antiferromaget. *J. Matter. Chem.* **2012**, *22*, 18793–18796. [CrossRef]

14. Boldrin, D.; Knight, K.; Wills, A.S. Orbital frustration in the $S = 1/2$ kagome magnet vesignieite, $BaCu_3V_2O_8(OH)_2$. *J. Matter. Chem. C* **2016**, *4*, 10315–10322. [CrossRef]

15. Sun, W.; Huang, Y.-X.; Pan, Y.; Mi, J.-X. Synthesis and magnetic properties of centennialite: A new $S = 1/2$ Kagomé antiferromagnet and comparison with herbertsmithite and kapellasite. *Phys. Chem. Miner.* **2016**, *43*, 127–136. [CrossRef]

16. Vilminot, S.; Richard-Plouet, M.; André, G.; Swierczynski, D.; Guillot, M.; Bourée-Vigneron, F.; Drillon, M.; André, G.; Swierczynski, D.; Guillot, M.; et al. Magnetic structure and properties of $Cu_3(OH)_4SO_4$ made of triple chains of spins $s = 1/2$. *J. Solid State Chem.* **2003**, *170*, 255–264. [CrossRef]

17. Brandão, P.; Rocha, J.; Reis, M.S.; dos Santos, A.M.; Jin, R. Magnetic properties of $KNaMSi_4O_{10}$ compounds ($M$ = Mn, Fe, Cu). *J. Solid State Chem.* **2009**, *182*, 253–258. [CrossRef]

18. Jahn, H.A.; Teller, E. Stability of polyatomic molecules in degenerate electronic states. *Proc. R. Soc. Ser. A* **1937**, *161*, 220–235.

19. Hathaway, B.J. A new look at the stereochemistry and electronic properties of complexes of the copper(II) ion. In *Complex Chemistry. Structure and Bonding*; Emsley, J., Ernst, R.D., Hathaway, B.J., Warren, K.D., Eds.; Springer: Berlin/Heidelberg, Germany, 1984; Volume 57, pp. 55–118.

20. Burns, P.C.; Hawthorne, F.C. Static and dynamic Jahn-Teller effects in $Cu^{2+}$ oxysalt minerals. *Can. Mineral.* **1996**, *34*, 1089–1105.

21. Effenberger, H. Contribution to the Stereochemistry of Copper. The Transition from a Tetragonal Pyramidal to a Trigonal Bipyramidal $Cu(II)O_5$ Coordination Figure with a Structure Determination of $PbCu_2(SeO_3)_3$. *J. Solid State Chem.* **1988**, *73*, 118–126. [CrossRef]

22. Eby, R.K.; Hawthorne, F.C. Structural Relations in Copper Oxysalt Minerals. I. Structural Hierarchy. *Acta Crystallogr. Sect. B Struct. Sci. Cryst. Eng. Mater.* **1993**, *49*, 28–56. [CrossRef]

23. Burns, P.C.; Hawthorne, F.C. Coordination-geometry structural pathways in $Cu^{2+}$ oxysalt minerals. *Can. Mineral.* **1995**, *33*, 889–905.

24. Krivovichev, S.V.; Filatov, S.K. Structural principles for minerals and inorganic compounds containing anion-centered tetrahedra. *Am. Mineral.* **1999**, *84*, 1099–1106. [CrossRef]

25. Krivivochev, S.V.; Mentré, O.; Siidra, O.I.; Colmont, M.; Filatov, S.K. Anion-Centered Tetrahedra in Inorganic Compounds. *Chem. Rev.* **2013**, *113*, 6459–6535. [CrossRef]

26. Volkova, L.M.; Marinin, D.V. Frustrated Antiferromagnetic Spin Chains of Edge-Sharing Tetrahedra in Volcanic Minerals $K_3Cu_3(Fe_{0.82}Al_{0.18})O_2(SO_4)_4$ and $K_4Cu_4O_2(SO_4)_4MeCl$. *J. Supercond. Nov. Magn.* **2017**, *30*, 959–971. [CrossRef]

27. Badrtdinov, D.I.; Kuznetsova, E.S.; Verchenko, V.Y.; Berdonosov, P.S.; Dolgikh, V.A.; Mazurenko, V.V.; Tsirlin, A.A. magnetism of coupled spin tetrahedra in ilinskite-type $KCu_5O_2(SeO_3)_2Cl_3$. *Sci. Rep.* **2018**, *8*, 2379. [CrossRef]

28. Botana, A.S.; Zheng, H.; Lapidus, S.H.; Mitchell, J.F.; Norman, M.R. Averievite: A copper oxide kagome antiferromagnet. *Phys. Rev. B* **2018**, *98*, 054421. [CrossRef]

29. Winiarski, M.J.; Tran, T.T.; Chamorro, J.R.; McQueen, T.M. $(CsX)Cu_5O_2(PO_4)_2$ ($X$ = Cl, Br, I): A Family of $Cu^{2+}$ $S = \frac{1}{2}$ Compounds with Capped-Kagomé Networks Composed of $OCu_4$ Units. *Inorg. Chem.* **2019**, *58*, 4328–4336. [CrossRef]

30. Siidra, O.I.; Vladimirova, V.A.; Tsirlin, A.A.; Chukanov, N.V.; Ugolkov, V.L. $Cu_9O_2(VO_4)_4Cl_2$, the first copper oxychloride vanadate: Mineralogically inspired synthesis and magnetic behavior. *Inorg. Chem.* **2020**, *59*, 2136–2143. [CrossRef] [PubMed]

31. Fujihala, M.; Morita, K.; Mole, R.; Mitsuda, S.; Tohyama, T.; Yano, S.-I.; Yu, D.; Sota, S.; Kuwai, T.; Koda, A.; et al. Gapless spin liquid in a square-kagome lattice antiferromagnet. *Nat. Commun.* **2020**, *11*, 3429. [CrossRef]

32. Kornyakov, I.V.; Vladimirova, V.A.; Siidra, O.I.; Krivovichev, S.V. Expanding the Averievite Family, $(MX)Cu_5O_2(T^{5+}O_4)_2$ ($T^{5+}$ = P, V; M = K, Rb, Cs, Cu; $X$ = Cl, Br): Synthesis and Single-Crystal X-ray Diffraction Study. *Molecules* **2021**, *26*, 1833. [CrossRef]

33. Aksenov, S.M.; Borovikova, E.Y.; Mironov, V.S.; Yamnova, N.A.; Volkov, A.S.; Ksenofontov, D.A.; Gurbanova, O.A.; Dimitrova, O.V.; Deyneko, D.V.; Zvereva, E.A.; et al. $Rb_2CaCu_6(PO_4)_4O_2$, a novel oxophosphate, with a shchurovskyite-type topology: Synthesis, structure, magnetic properties and crystal chemistry of rubidium copper phosphate. *Acta Crystallogr.* **2019**, *B75*, 903–913. [CrossRef]

34. Pekov, I.V.; Zubkova, N.V.; Belakovskiy, D.I.; Yapaskurt, V.O.; Vigasina, M.F.; Sidorov, E.G.; Pushcharovsky, D.Y. New arsenate minerals from the Arsenatnaya fumarole, Tolbachik volcano, Kamchatka, Russia. IV. Shchurovskyite, $K_2CaCu_6O_2(AsO_4)_4$ and dmisokolovite, $K_3Cu_5AlO_2(AsO_4)_4$. *Mineral. Mag.* **2015**, *79*, 1737–1753. [CrossRef]

35. Kornyakov, I.V.; Krivovichev, S.V.; Gurzhiy, V.V. Oxocentered Units in Three Novel Rb-Containing Copper Compounds Prepared by CVT Reactions Method. *Z. Anorg. Allg. Chem.* **2018**, *644*, 77–81. [CrossRef]

36. Kovrugin, V.M.; Siidra, O.I.; Colmont, M.; Mentré, O.; Krivovichev, S.V. Emulating exhalative chemistry: Synthesis and structural characterization of ilinskite, $Na[Cu_5O_2](SeO_3)_2Cl_3$, and its K-analogue. *Miner. Petrol.* **2015**, *109*, 421–430. [CrossRef]

37. Jing, B.; Peng, C.; Wang, Y.; Liu, Q.; Tong, S.; Zhang, Y.; Ge, M. Hygroscopic properties of potassium chloride and its internal mixtures with organic compounds relevant to biomass burning aerosol particles. *Sci. Rep.* **2015**, *7*, 43572. [CrossRef]

38. Giamarelou, M.; Smith, M.; Papapanagiotou, E.; Martin, S.; Biskos, G. Hygroscopic properties of potassium-halide nanoparticles. *Aerosol Sci. Technol.* **2018**, *52*, 536–545. [CrossRef]

39. Brunel-Laügt, M.; Guitel, J.-C. Structure crystalline de $Cu_5O_2(PO_4)_2$. *Acta Crystallogr.* **1977**, *B33*, 3465–3468. [CrossRef]

40. *CrysAlisPro Software System*, version 1.171.38.46; Rigaku Oxford Diffraction: Oxford, UK, 2015.

41. Sheldrick, G.M. SHELXT—Integrated space-group and crystal structure determination. *Acta Crystallogr.* **2015**, *A71*, 3–8. [CrossRef]

42. Sheldrick, G.M. Crystal structure refinement with SHELXL. *Acta Crystallogr.* **2015**, *C71*, 3–8.

43. Dolomanov, O.V.; Bourhis, L.J.; Gildea, R.J.; Howard, J.A.K.; Puschmann, H. OLEX2: A complete structure solution, refinement and analysis program. *J. Appl. Cryst.* **2009**, *42*, 339–341. [CrossRef]

44. Pekov, I.V.; Zubkova, N.V.; Agakhanov, A.A.; Pushcharovsky, D.Y.; Yapaskurt, V.O.; Belakovskiy, D.I.; Vigasina, M.F.; Sidrov, E.G.; Britvin, S.N. Cryptochalcite, $K_2Cu_5O(SO_4)_5$, and cesiodymite, $CsKCu_5O(SO_4)_5$, two new isotypic minerals and the K-Cs isomorphism in this solid-solution series. *Eur. J. Mineral.* **2018**, *30*, 593–607. [CrossRef]

45. Siidra, O.I.; Lukina, E.A.; Nazarchuk, E.V.; Depmeier, W.; Bubnova, R.S.; Agakhanov, A.A.; Avdontseva, E.Y.; Filatov, S.K.; Kovrugin, V.M. Saranchinaite, $Na_2Cu(SO_4)_2$, a new exhalative mineral from Tolbachik volcano, Kamchatka, Russia, and a product of the reversible dehydration of kröhnkite, $Na_2Cu(SO_4)_2(H_2O)_2$. *Mineral. Mag.* **2018**, *82*, 257–274. [CrossRef]

46. Berry, R.S. Correlation of Rates of Intramolecular Tunneling Processes, with Application to Some Group V Compounds. *J. Chem. Phys.* **1960**, *32*, 933–938. [CrossRef]

47. Pasquarello, A.; Petri, I.; Salmon, P.S.; Parisel, O.; Car, R.; Tóth, E.; Powell, D.H.; Fischer, H.E.; Helm, L.; Merbach, A.E. First Solvation Shell of the Cu(II) Aqua Ion: Evidence for Fivefold Coordination. *Science* **2001**, *291*, 856–859. [CrossRef]

48. McCusker, L.B.; Liebau, F.; Engelhardt, G. Nomenclature of structural and compositional characteristics of ordered microporous and mesoporous materials with inorganic hosts: (IUPAC recommendations 2001). *Microporous Mesoporous Mater.* **2003**, *58*, 3–13. [CrossRef]

49. Krivovichev, S.V. Topological complexity of crystal structures: Quantitative approach. *Acta Crystallogr.* **2012**, *A68*, 393–398. [CrossRef] [PubMed]

50. Krivovichev, S.V. Structural complexity of minerals: Information storage and processing in the mineral world. *Miner. Mag.* **2013**, *77*, 275–326. [CrossRef]

51. Krivovichev, S.V. Structural complexity and configurational entropy of crystalline solids. *Acta Crystallogr.* **2016**, *B72*, 274–276.

52. Krivovichev, S.V. Ladders of information: What contributes to the structural complexity in inorganic crystal. *Z. Kristallogr.* **2018**, *233*, 155–161. [CrossRef]

53. Krivovichev, S.V. Which inorganic structures are the most complex? *Angew. Chem. Int. Ed.* **2014**, *53*, 654–661. [CrossRef] [PubMed]

*Article*

# Optical and Spectroscopic Properties of Lorenzenite, Loparite, Perovskite, Titanite, Apatite, Carbonates from the Khibiny, Lovozero, Kovdor, and Afrikanda Alkaline Intrusion of Kola Peninsula (NE Fennoscandia)

Miłosz Huber [1,*], Daniel Kamiński [2], Grzegorz Czernel [3] and Evgeni Kozlov [4]

[1] Department of Geology, Soil Science and Geoinformacy, Faculty of Earth Science and Spatial Management, Maria Curie-Skłodowska University, 2d/107 Kraśnickie Rd, 20-718 Lublin, Poland
[2] Department of General Chemistry, Coordination Chemistry and Crystallography, Faculty of Chemistry, Maria Curie-Skłodowska University, 2/508 Maria Curie-Skłodowska Sq, 20-031 Lublin, Poland; daniel.kaminski@mail.umcs.pl
[3] Department of Biophysics, University of Life Sciences in Lublin, Akademicka 13, 20-950 Lublin, Poland; grzegorz.czernel@up.lublin.pl
[4] Geological Institute, Kola Science Centre, Russian Academy of Sciences, 14, Fersman Street, 184209 Apatity, Russia; kozlov_e.n@mail.ru
* Correspondence: milosz.huber@mail.umcs.pl; Tel.: +48-692283572

Citation: Huber, M.; Kamiński, D.; Czernel, G.; Kozlov, E. Optical and Spectroscopic Properties of Lorenzenite, Loparite, Perovskite, Titanite, Apatite, Carbonates from the Khibiny, Lovozero, Kovdor, and Afrikanda Alkaline Intrusion of Kola Peninsula (NE Fennoscandia). *Crystals* **2022**, *12*, 224. https://doi.org/10.3390/cryst12020224

Academic Editors: Dinadayalane Tandabany and Alessandro Chiasera

Received: 12 December 2021
Accepted: 30 January 2022
Published: 4 February 2022

**Publisher's Note:** MDPI stays neutral with regard to jurisdictional claims in published maps and institutional affiliations.



**Abstract:** This manuscript deals with the analysis of significant rare earth elements (REE) minerals such as eudialyte, lorenzenite, loparite, perovskite, titanite, apatite, and carbonates. These minerals are found in the rocks of the Khibiny, Lovozero, Afrikanda, and Kovdor massifs (the Paleozoic hotspot activity in the Kola-Karelian Alkaline Province is estimated at about 100,000 km$^2$). Performed microscopic analyses that demonstrated their structure and optical features (dimming, interference colors, relief). Single-crystal analysis using XRD methods, SEM-EDS, and spectroscopic (FTIR) studies allowed the characteristics of described minerals: Lorenzenite in Lovozero probably crystalized after loparite have small additions of Nb, La, Ce, Pr, and Nd. Loparite and perovskite have the addition of Ce, Nb, and Ta. The same dopants have titanite probably crystalized after perovskite. Calcite in these massifs had the addition of Ce and Sr, the same as in fluorapatite, which was found in these rocks too. All of the analyzed minerals are REE-bearing and can be considered as deposits

**Keywords:** lorenzenite; perovskite; loparite; titanite; apatite; carbonates; Khibiny; Lovozero; Kovdor; Afrikanda; alkaline intrusions; NF Fennoscandia

## 1. Introduction

The Kola-Karelian Alkaline Province is a unique world complex containing about forty different intrusions of ultramafic alkaline rocks with rare rock and mineral types [1–4]. In the Khibiny, about 10% [5] of all known minerals have been discovered. The minerals lorenzenite, loparite, perovskite, titanite, apatite, and carbonate, among others, are found in these rocks and are admixtures in the discussed rocks. Rare earth inclusions have been found in these minerals, which both enrich their crystal structure and form inclusions and distinct phases. Lorenzenite is a mineral that occurs in the alkaline rocks of the Khibiny and the Lovozero intrusions. Loparite, on the other hand, is found in the urtite formations of the Lovozero and Kovdor rocks. Besides these minerals, apatite is also found in whole rocks of described massifs and is accompanied by minerals such as titanite containing REE admixtures. In addition to these minerals, perovskite and carbonate (contain admixtures of Sr and Ce) are found in Afrikanda in ultramafic-alkaline rocks [6–12]. Lorenzenite, loparite, and Perovskite are relatively rare, and their occurrences in the world are limited, and the optical and spectroscopic characteristics of these minerals are relatively unknown. The authors in this paper decided to introduce the properties of these minerals in this paper by presenting the original results of SEM-EDS, FTIR, and single-crystal X-ray studies of these minerals. The purpose is to present the properties of the minerals mentioned above found in the Murmansk District. Below are the characteristics of the study region, the individual intrusions from which samples were taken, and the properties of the minerals in question.

## 2. Study Area

The NE Fennoscandia region is where more than a dozen diverse intrusions of alkaline-basic rocks are located. This is one of many intrusions connected with the hotspots in the early Paleozoic [1–19], of the alkaline continental magmatism such as the Khibiny, Lovozero, Turiy, Niva, and Salmagora massifs, and many others, and it contains the most original cumulate magma [1–3]. These intrusions form several concentric belts exposed in the region of Murmansk District, Karelia, Finland (and several in the Arkhangelsk region). The outermost of them are composed of ultramafic-alkaline rocks (Afrikanda, Kolvitza region, Lesnaya Varaka, and Ozernaya Varaka) [4,6] of carbonatites (Kovdor, Sokli, and Patyan Vaara) [9,10] and alkaline rocks (Turiy, Khibiny, Lovozero, and Selbyavr) [12,13]. The complex of these intrusions also includes numerous vein formations [1] and kimberlites of the Terski Bereg and Arkhangelsk regions [20–23]. These rocks are intruded in Archean and Paleoproterozoic formations of magmatic and metamorphic character, which are the basement of the Baltic Shield of the Eastern European Platform [24,25]. They were formed due to the impact of a hotspot that occurred between 465–364 million years ago [18], forming the Kola-Karelian Alkaline Province (Figure 1). Some of these intrusions are under the Pleistocene sediments [26,27].

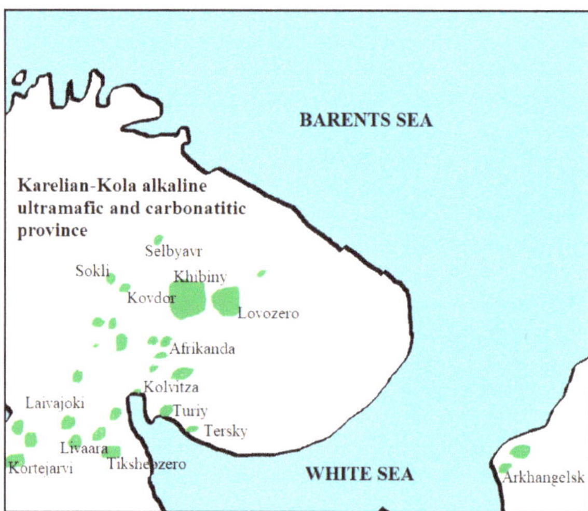


**Figure 1.** Kola-Karelian Alkaline Province with intrusions localizations (after [22], changed by author).

Most of them are central intrusions usually composed of concentric rings of rocks forming the metasomatized contact zone and internal zones of magmatic derivatives. These sequences are more numerous in large massifs such as Khibiny and Lovozero. An ore zone (Khibiny) [28–30] and zones of pegmatites and vein formations (Khibiny, Lovozero) usually appear there [31–37]. There are elevated amounts of rare earth elements (Ce) [31–38]. The share of these elements is variable, but their occurrences sometimes cause enrichment in these elements of the rock-forming minerals of the discussed massifs or the form of solid inclusions and separate phases.

### 3. Geology of the Khibiny, Lovozero, Afrikanda, and Kovdor Massifs

The Khibiny massif is a concentric Early Paleozoic intrusion of alkaline rocks, along with the Lovozero intrusion. Khibiny is the largest alkaline rock massif in the NE Fennoscandia region, having an area of 1300 km$^2$ (Lovozero 587 km$^2$). The age of these massifs is 385–375 Ma [31–37]. The Khibiny Massif is built of concentrically arranged rock formations in the form of belts, which can be divided into four sequences [25,38–45]. The outermost is composed of so-called massive khibinites or alkali-feldspathic syenites. The inner one (Figure 2) is composed of similarly formed porphyritic foyaites with the presence of biotite, phlogopite, and pegmatites, seen in the nested form [31–37]. Pegmatites are often characterized by the presence of many rare minerals [31,33].

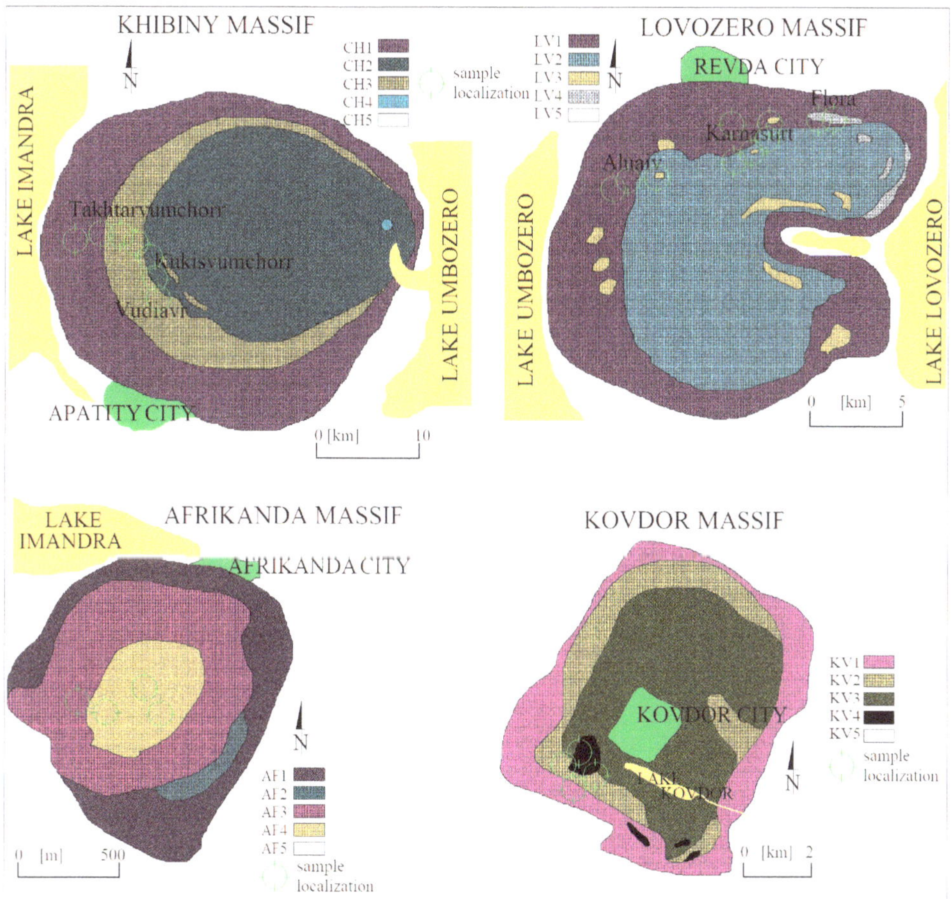


**Figure 2.** Generalized geological sketch of the Khibiny, Lovozero, Afrikanda, and Kovdor massifs (after [11,39] changed by authors). Rocks abbreviations: Khibiny: CH1-massive syenites, CH2-rischorites, CH3-ore zone, CH4-carbonatites, CH5-surrounded rocks; Lovozero: LV1-syenites, LV2-lujavrites, jovites, LV3-ijolites, LV4-porphyrites, LV5-surrounded rocks; Afrikanda: AF1-contact zone phenites, AF2-melteigites, AF3-pyroxenites, AF4-olivinites with perovskite-magnetite ore, AF5-surrounded rocks; Kovdor: KV1-contact zone phenites, KV2-foidoites, KV3-pyroxenites and olivinites, KV4-magnetite ore, KV5-Precambrian bSasement.

Between these formations is an ore zone (third sequence) formed by titanitic nephelinites. The fourth sequence is formed by veins and dikes cutting the above concentric system, including porphyrites, syenite pegmatites, and melteigites [29]. Carbonatites are also present in the eastern region of the massif [38]. The formations of the zones mentioned above in the area of Lake Vudiavr, Vudiavrchorr hills, Tachtarvumchorr, Yudychvumchorr, Kukisvumchorr, and Malaya Belaya Valley were investigated by the authors (Figure 2, [1,5]).

The Lovozero massif is an intrusion of rocks similar to Khibiny, localized several kilometers to the east. However, they differ significantly in their composition and prevalence of accessory minerals as well as in the directional texture of the rocks dominant in the Lovozero massif. It is an intrusion with an age of 370–359 Ma constituting the second-largest massif in the NE Fennoscandia region after Khibiny [40–44]. This massif is composed of several zones of alkaline rocks, forming concentrically arranged circles. The first outer one is mainly made of trachytes and metasomatized porphyrites. The second, inner one is made of lujavrites accompanied by other foyaites. These rocks are interspersed with various vein rocks and pegmatites forming the third zone of the discussed massif. The rocks were sampled in the western (in the Aluaiv Mt. area) and northern part of the massif (in the Karnasurt and Flora Mts. area, Figure 3) [42].

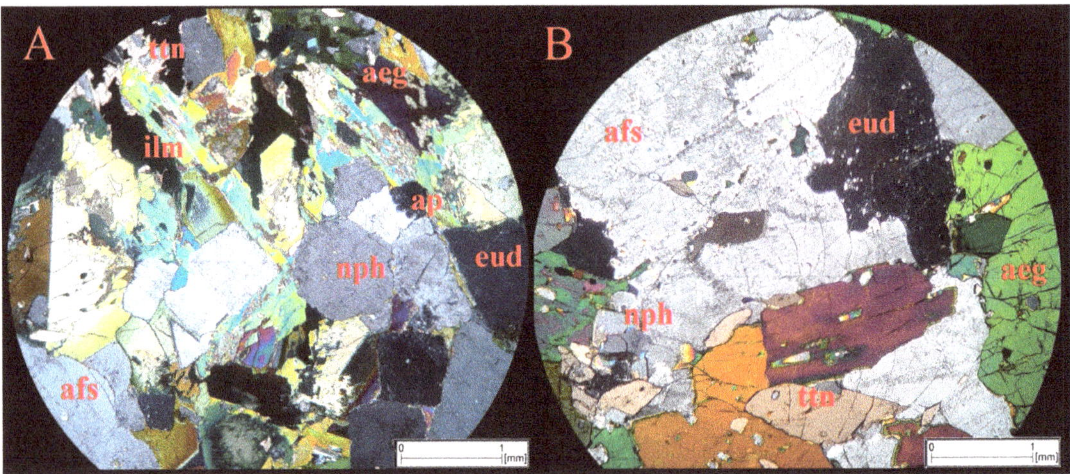


**Figure 3.** Macro photographs of the typical rocks from Khibiny: Massive syenite with eudialyte (eud) crystals with nepheline (nph), orthoclase (afs), apatite (ap), aegirine (aeg), ilmenite (ilm), and titanite (ttn) (**A**); Melasyenite (malinite) with titanite, aegirine, eudialyte, and nepheline (**B**).

Afrikanda intrusion is located in the southern part of the Kola Peninsula in the NE part of the Scandinavian Peninsula [45–48]. It is located on the southern shore of Lake Imandra, 35 km south of Khibiny and about 10 km from Polarnye Zori town. It constitutes a single hill with several peaks. The massif has a small area of 6.4 km$^2$ [11]. It is a Paleozoic intrusion dated by the U-Pb method at 379 ± 5 Ma [1]. It shows similar age to other intrusions in the Kola Peninsula region [8]. It shows an isometric shape with numerous apophyses of pyroxenites. It is characterized by a distinct zone-ring structure. Analyzing the geological structure of the Afrikanda massif (Figure 4), two zones can be distinguished in it, the outer and the inner. The outer ring is made of melteigites. The inner zone comprises various alkaline and ultramafic rocks: olivinites, pyroxenites, and carbonatites (Figure 4). The discussed intrusion represents a complex of ultramafic and alkaline formations, which show the occurrence of clinopyroxene-perovskite association with olivines, orthopyroxenes, as well as magnetite, titanomagnetite, rutile, and titanite. Olivines usually have a significant proportion of magnesium and iron and minor admixtures of nickel. As a result of residual crystallization, pegmatites were formed, in which garnet-schorlomite occurs. Along with the subsequent stages of formation of these rocks, there are processes of their metasomatic transformation (phenitization, amphibolization, and saussuritization). Phenitization causes the appearance of apatite, nepheline. These rocks also show the effects of hydrothermal interactions highlighted by the presence of Cu, Zn, Pb sulfides, and sulfates such as barite [45].

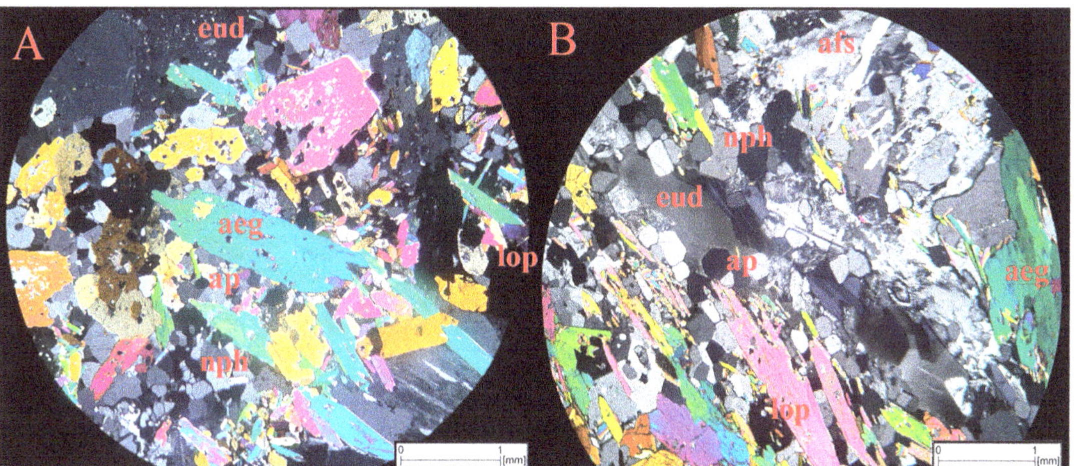


**Figure 4.** Microphotograph of typical rocks from the Lovozero massif. Massive syenite with visible aegirine crystals and eudialyte, apatite, nepheline, and loparite (**A**); Foyaite with visible apatite, zonal eudialyte, albite and orthoclase, and associated aegirine and loparite (**B**).

The Kovdor massif occurs in the southwestern part of the Murmansk District, in the vicinity of the Finnish border. It is an intrusion composed of concentrically arranged rocks, with an area of 25 km$^2$ [49–54]. The age of this massif is determined to be 382 $\pm$ 3 Ma [55]. The massif includes Lake Kovdor and a mining town of the same name. This intrusion is composed of ultramafic and alkaline rocks, which are arranged in several circles. The outermost one is formed by syenites and melilites found in the marginal zone. In the next one, these rocks transition into pyroxenites, among which olivine is also found [49–54]. These rocks are located in the central part of the massif. They are interspersed with carbonatites (calcite phoscorites, dolomite phoscorites, calcite-dolomite phoscorites, and manetitites) in which magnetite is the main component. In these rocks, apart from carbonates and magnetite, there are also diopside, olivine, nepheline, apatite, and many rare minerals such as loparite, perovskite, baddeleyite, and others. Accompanying these rocks are phlogopite crystals, the aggregates of which sometimes reach considerable sizes. In weathering zones, phlogopite transforms into vermiculite. In the zones of fractures and tectonic faults, there are breccias cemented with apatite [54]. Like Afrikanda, this intrusion is a complex of ultramafic and alkaline rocks. In this intrusion, there are also zones of derivatives crystallizing in the form of large-crystalline pegmatites with phlogopite, diopside, accompanied by magnetite, and carbonates. In the contact zone of the intrusion, one can observe strong phenitization of rocks forming transition zones where apatite and melilites can be found. In the discussed intrusion, there are also numerous dispersed sulfides forming polymetallic phases of multistage hydrothermal crumbling. However, the proportion of these phases is relatively small.

## 4. Materials and Methods

Rock samples were collected by the author during numerous field investigations between 1999 and 2021. During this time, the discussed massifs were visited, and geological documentation of the samples was also done. In the Khibiny massif, the Malaya Belaya valley regions, Takhtarvumchorr region, Kukisvumchorr, and the massifs surrounding Lake Vudiavr were studied. In the Lovozero region, these were the Aluaiv and Karnasurt Mts. slopes in the western part of the massif and the Flora region in the northern part. In the Afrikanda area, rocks were sampled in the main quarry and several smaller ones in this intrusion. In the Kovdor area, samples were taken corresponding to the most important rock types discussed in the main quarry area. The selected rocks were targeted for thin section preparations to determine further the characteristics of the minerals discussed. Subsequently, these minerals were subjected to analyses using a Leica DM2500P polarizing optical microscope and examination with a Scanning Electron Microscope Hitachi SU6600, with an EDS attachment. These samples were analyzed under low vacuum (10 Pa), 15 kV beam diameter of 02 μm. A total of 3711 mineral analyses were performed in the microprobe (at Khibiny 676, at Lovozero 1338, at Kovdor 888, and Afrikanda 869, respectively). Next, the selected minerals were separated and analyzed with single-crystal X-ray diffraction. Data were collected on a Rigaku diffractometer (Rigaku, Tokyo, Japan) with CuK$\alpha$ radiation ($\lambda$ = 1.54184 Å) at 293 K. Crystallographic refinement and data collection as well as data reduction, and analysis was performed with the CrysAlisPro 1.171.39.27b (Rigaku Oxford Diffraction, Tokyo, Japan) [56]. Selected single crystals were mounted on the nylon loop with oil. Structures were solved by applying direct methods using the SHELXS-86 program and refined with SHELXL$-$2018/3 [56–60] in Olex2 software [57]. These samples were also examined with the FTIR technique. Samples were measurement using Shimadzu FTIR spectrophotometer equipped with an ATR accessory, QATR-S (wide-band diamond crystal). Apodization Function: Happ-Genzel, Detector: DLATGS. Every single spectrum was obtained from 25 records at 2 cm$^{-1}$ resolution ranging from 250 to 2000 cm$^{-1}$. The elaboration of a spectrum was carried out using Grams 9.1 program. Optical and microscopic studies were performed in the Department of Geology, Soil Science, and Geoinformation of the Institute of Earth and Environmental Sciences, and crystal-chemistry studies were performed in the Department of Crystallochemistry, Faculty of Chemistry, Maria Curie-Skłodowska University in Lublin. FTIR analysis was performed in the Department of Biophysics, University of Life Sciences in Lublin.

## 5. Results

### *5.1. Petrology of Selected Massifs*

To better describe the mineralogy of the phases discussed in the title, the main features of the most important rocks of the massifs are briefly presented: Khibiny, Lovozero, Afrikanda, and Kovdor.

### 5.1.1. Khibiny

In the Khibiny massif, these are massive syenites, melasyenites, porphyritic foyaites, and apatite-nepheline ores. Massive syenites-khibinites are usually grayish-greenish rocks with a coarse crystalline structure and compact, disorderly texture (Figure 3A). Large crystals of orthoclase and albite are visible along with other minerals of the rock, filling the space, touching each other. They are accompanied by needle-shaped aegirine crystals, which form sheaves, radial clusters, sometimes interspersed with feldspar. Next to these minerals, riebeckite appears. Along with these minerals, nepheline with characteristic cube-shaped crystals and apatite forming an admixture occurs in the rock. Large crystals of eudialyte, usually cherry red, can also be seen in the rock. Arfvedsonite, astrophyllite, and lorenzenite are found as accessory minerals. Ore minerals are represented by ilmenite and titanite (up to 10% by volume) less frequently, loparite, pyrite, magnetite. Melasyenites (malinites) are a variety of rocks that form vein formations and usually have fine crystalline structures and linear textures (Figure 3B). They are also distinguished from the above syenites by the increased presence of mafic minerals. There is much more aegirine in the matrix of these rocks, but sometimes augite aegirine, arfvedsonite, aenigmatite, ilmenite, and titanite are found. Occasionally there is also a large admixture of hematite. Porphyritic foyaites, on the other hand, are gray-pink rocks with a coarse-crystalline, porphyritic structure and a compact, disorderly texture. The matrix of these rocks consists of orthoclase, nepheline, and apatite crystals.

Next to these minerals, eudialyte is sometimes visible in the rocks. It is interspersed with needles of aegirine, which also occurs in the spaces between the discussed crystals. It is accompanied by biotite. Aenigmatite, arfvedsonite, and phlogopite are also present. Ore minerals are represented by titanite, ilmenite, as well as pyrite and titan-rich magnetite. Nepheline-apatite-titanite ores are brown-greenish-pink rocks with a coarse crystalline structure and compact, disorderly texture. The matrix in this rock are crystals of apatite and nepheline in different proportions co-occurring with each other. They are also accompanied by titanite, sometimes forming a matrix mineral in the rock. Titanite in these rocks usually shows a zonal structure. Along with these minerals, aegirine and ilmenite are also leading minerals. The accessories minerals such as aenigmatite and arfvedsonite were also found.

### 5.1.2. Lovozero

The Lovozero Massif contains mainly massive syenites, lujavrites, foyaites, jovites, and porphyrites. Massive syenites are gray-green rocks with a fine crystalline structure and a compact, disorderly texture (Figure 4A). Larger phenocrysts of microcline accompanied by plaques of albite and apatite embedded in a finer matrix (or groundmass matrix) of the rock are visible. Between these crystals, one can see nepheline and aegirine. Next to them, the eudialyte is also visible. Accessory minerals visible are titanite, loparite, lovozerite, and many other minerals. They are accompanied by ore minerals such as titanite, ilmenite, and pyrite. Apatite is usually enriched in cerium, while titanite is in niobium, thorium, and cerium. Lujavrites are gray-green rocks with red speckles with a medium crystalline structure, an extremely linear texture, which is emphasized by femic minerals. The rock matrix is formed by aegirine crystals, giving the rock its linearity. They are accompanied by augite and aenigmatite. Between these crystals, apatite, albite, and eudialyte are visible, forming red "dots" in the rock. Between these minerals, there is also loparite and ilmenite. Accessory minerals are present in small additions, including pyrite, chalcopyrite, and zircon. Foyaites are grayish rocks of medium crystalline structure with a compact and linear texture. In this rock, the matrix is formed by aegirine needles between nepheline and microcline crystals (Figure 4B). Next to these minerals apatite, albite sometimes single crystals of small eudialyte are visible. They are accompanied by ilmenite, loparite, riebeckite. Also, single crystals of zircon can be seen. There are varieties rich also in murmanite, a characteristic pink mineral that occurs in these rocks. Jovites are rocks with a greenish color, porphyritic structure, and linear texture. These rocks are similar to the foyaites described above, with a larger amount of leucocratic minerals occurring in them. The rock matrix consists of microcline, nepheline forming large crystals interlayered with aegirine. Next to them, eudialyte and ilmenite may also occur. In these rocks may occur loparite and titanite. Porphyries are gray-green rocks with porphyritic structure and compact, disorderly texture. The rock matrix is built by aegirine needles together with albite, microcline, nepheline. These are generally fine-crystalline minerals. Next to them, there are large, euhedral crystals of eudialyte, lorenzenite, epistolite, astrophyllite, loparite, and narsarsukite. As a rule, these crystals have a skeletal character; except for a certain border made of the phase that builds the discussed minerals, the interior is filled with numerous fractures forming the rock matrix. Small druses filled with euhedral crystals of eudialyte, lorenzenite, and aegirine can be found in these rocks.

### 5.1.3. Afrikanda

In the Afrikanda massif, there are pyroxenites and olivinite accompanied by magnetitites as well as melteigites, pegmatites, and carbonatites. Pyroxenites and olivine pyroxenites are black or black-greenish rocks, with a coarse and very coarse crystalline structure, compact, disorderly, less frequently miarolitic texture. The rock matrix is formed by large crystals of pyroxene (diopside, augite, and aegirine), sometimes reaching several centimeters in size; next to these crystals, there are also large aggregates of phlogopite reaching several centimeters in size. Between these minerals, the common hornblende is seen accessory. They are accompanied by biotite, apatite, magnetite, and perovskite. The accessory is also epidote, chlorite, sulfides (pyrite, chalcopyrite, chalcopyrite, sphalerite), and sulfates (barite). Sometimes, there are druses filled by euhedral diopside, magnetite, perovskite, and carbonates in the discussed rocks. Olivinites are black-green rocks with a fine crystalline structure, disorderly texture (Figure 5B). The matrix of these rocks is built by olivines, often serpentinized, and next to them, one can see small admixtures of diopside, nepheline, and apatite. Magnetite-perovskite magnetites are metallic-black, coarse-crystalline, granular rocks with a compact, disorderly texture (Figure 5A). The matrix of these rocks is formed by magnetite, which is close together with large crystals reaching up to 1 cm in size. It is accompanied by perovskite with polysynthetic associations, sometimes with an admixture of Ce-rich perovskite (knopite). Apart from these minerals, sometimes single crystals of orthopyroxene, carbonates, titanite are found in the discussed rocks. Accessories are pyrite, chalcopyrite, pentlandite. Melteigites: these are pink-greenish rocks with a coarse crystalline structure, compact, disorderly structure. The matrix of these rocks consists of orthopyroxene crystals (bronzite), sometimes accompanied by olivine (serpentinized). These minerals occur in the matrix of nepheline and sodalite. Apatite and microcline titanite are also visible next to them. Carbonatites are light gray rocks, occurring mainly in the form of veins, druses, and small accumulations accompanying magnetite-perovskite ores and other rocks (e.g., pyroxenites and peridotites). They have a coarse crystalline structure, compact, disorderly texture. They are composed mainly of calcite and dolomite and, in smaller amounts, strontianite. Between these crystals, there sometimes appear very well-formed (self-shaped) titanites, perovskite, magnetite, and also diopside, hornblende, phlogopite. Pegmatites are rocks distinguished primarily by their large crystalline structure of the rock. The matrix of these rocks consists of albite with microcline and also phlogopite, epidote, and chlorite. The feldspars are often sericitized. In these rocks, there is a rare schoolmite garnet, sometimes reaching several cm in size. The accessory is titanite, magnetite, and ilmenite.

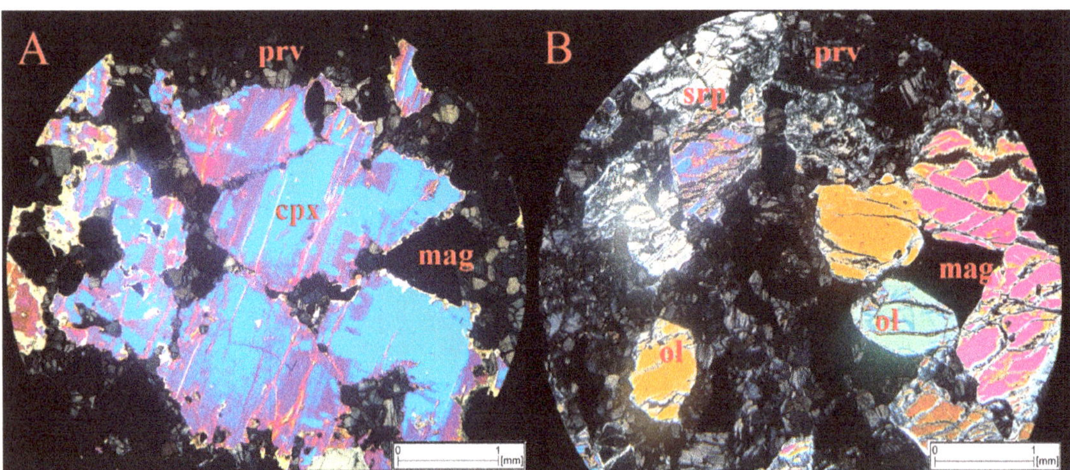


**Figure 5.** Microphotography of typical rocks from the Afrikanda massif. Perovskite magnetite magnetitites with diopside (cpx), magnetite (mag), and perovskite (prv) (**A**); olivinite with olivine (ol) serpentinized crystals (srp), magnetite, and perovskite (**B**).

### 5.1.4. Kovdor

In the Kovdor massif, syenites are the rocks of the contact zone. They have a coarsely crystalline, poikilitic, compact, disorderly texture. The rock matrix is filled by aggregates of aegirine and diopside accompanied by large crystals of nepheline and orthoclase. They are accompanied by apatite and small amounts of magnetite. Melilites are also found among these rocks. These rocks have a coarse crystalline structure, compact texture, sometimes porphyritic. The matrix of these rocks consists of melilite crystals in contact with each other. Next to these minerals, there is nepheline and apatite. Single magnetite crystals, phlogopite aggregates, and carbonates can also be seen in the rock. In the central part, these rocks transform into pyroxenites. Pyroxenites have a coarse-crystalline, porphyritic structure with a compact, disorderly texture. They are dominated by diopside accompanied by hypersthene. These minerals are usually large, compactly contacting each other. In the spaces between them, magnetite has crystallized, subordinately also hercynite. Single crystals of olivine and small aggregates of phlogopite can also be seen in these rocks. Between these phases, apatite and calcite are also sporadically present in the rock. Olivinites are rocks that accompany pyroxenites. They have a fine crystalline structure and a compact and disorderly texture (Figure 6A). They are composed mostly of olivine, sometimes bearing traces of serpentinization visible in the marginal zones of the crystals. In their vicinity, one can see magnetite, usually of small size. Next to olivine, there are also crystals of diopside and aggregates of phlogopite. Sometimes single specimens of nepheline, apatite, and carbonates can be found in the spaces between these minerals. Phoscorites and associated magnetitites can be found in the central part of the intrusion. Phoscorites are coarsely crystalline rocks with a compact, disorderly texture (Figure 6B). They are characterized by the presence of various carbonates. These are crystals of calcite, dolomite, or carbonates doped with iron and other elements (e.g., strontium). In addition to these carbonates, the rock contains euhedral crystals of diopside, which occur together with aggregates of phlogopite. Magnetite crystals are also found in these rocks, sometimes in significant amounts forming transitions to magnetitites. They are also accompanied by baddeleyite, monazite, and loparite in small amounts. Next to mafic minerals, one can also see small crystals of apatite and nepheline. Hematite and sulfides (pyrite and chalcopyrite) also occur accessorily. In tectonic zones, there is also a breccia cemented with apatite. Next to apatite and carbonates, there is vermiculite, sometimes forming quite large aggregates. It is accompanied by magnetite, nepheline, and hematite.

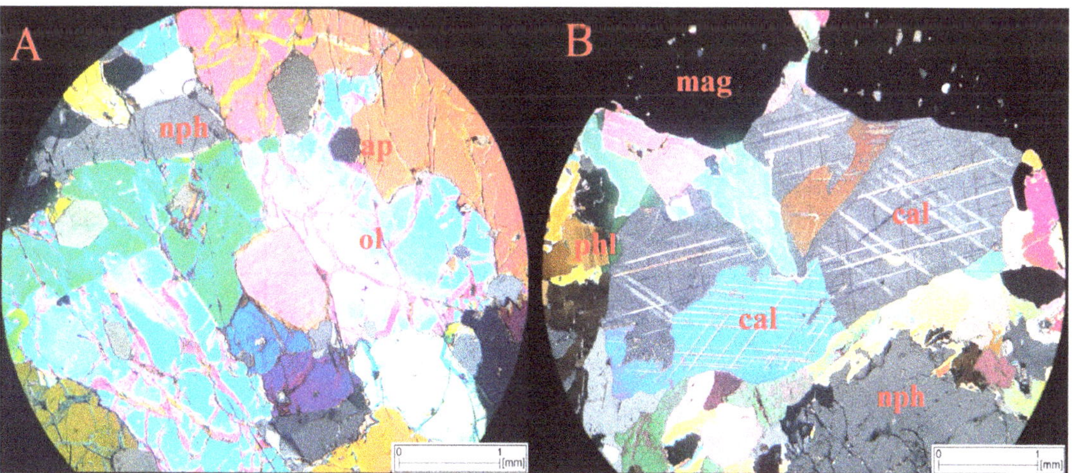


**Figure 6.** Microphotographs of typical rocks from the Kovdor massif. Olivinite with visible crystals of olivine and apatite and nepheline (**A**); phoscorites with visible calcite (cal), phlogopite (phl) nepheline, and crystals of magnetite (**B**).

### 5.2. Structures of the Discussed Minerals

*Lorenzenite* is chocolate brown, forming nice crystals up to several cm in size. Both single euhedral specimens of 2–3 cm in size and smaller, usually darker, can be observed. These are usually euhedral crystals found in porphyrites.

These minerals in the Lovozero massif usually have a skeletal character (Figure 7A). Their outer boundaries are formed properly, while in the middle, they have numerous inclusions of ambient minerals cemented by the substance of the phase in question. In the Khibiny massif, lorenzenite is found less frequently among syenite rocks. It forms single small crystals in the vicinity of titanite and other mafic minerals.

**Figure 7.** Macro photographs of the minerals discussed: lorenzenite (**A**), loparite (**B**), perovskite (**C**), titanite (**D**), apatite (**E**), and eudialyte (**F**).

Loparite in the discussed rocks forms small crystals with a metallic silver-steel luster. It usually forms polysynthetic twins (Figure 7C). This is evident both in the shape of the discussed crystals showing twinned adhesions. In the rocks of the Lovozero massif, these crystals are euhedral similarly in the Khibiny massif. The Kovdor rocks show far-reaching corrosion and have a strongly developed boundary. In Afrikanda, they rarely occur, accompanying perovskites by filling the spaces between these minerals, crystallizing on their walls as a form of epitaxy.

Perovskite macroscopically forms crystals with a metallic luster and steel-gray color, up to 1 cm in size. It forms euhedral crystals with numerous twinned adhesions, occurring as admixtures in pyroxenites and other Afrikanda rocks (Figure 7B). In megnetitites, they form nested ore-like accumulations, filling 80% of the volume of these rocks. At Kovdor, they are usually heavily corroded, as is loparite, accompanying magnetites in phoscorites. At Khibiny, they infrequently occur in the vicinity of femic minerals such as titanite and pyroxene.

Titanite usually has strongly elongated crystals reaching several mm in size, colored in chocolate brown (Figure 7D). In the Khibiny massif, it forms large crystals, sometimes with a visible sector structure. In syenitic rocks, it forms numerous admixtures together with ilmenite. In the ore zone, it forms rock-forming accumulations, together with nepheline, apatite, and femic minerals, forming admixtures. In these rocks, it is often twinned. In malinites, it occurs in anhedral form, crystallizing between crystals of aegirine, arfvedsonite, nepheline, and apatite. In the Lovozero massif, titanite occurs less frequently, accompanying femic minerals and often having an anhedral shape. In the Afrikanda massif, titanite occurs adjacent to perovskite, where it forms xenomorphic crystals together with carbonates. At Kovdor, titanite occurs accessorily in the form of small admixtures in the rocks in question.

Apatite usually has sugary light green or pink crystals. It forms minor accessory admixtures in all the discussed rocks of the Khibiny, Lovozero, Afrikanda, and Kovdor massifs. These minerals usually have a euhedral shape, are small in size, and co-occur with other phases in the discussed rocks (Figure 7E).

Carbonates form anhedral crystals, forming carbonatite veins cutting the Afrikanda pyroxenites. They are also present in minor admixtures in Afrikanda pyroxenite rocks, occurring between clinopyroxene and magnetite crystals. Small admixtures of carbonates were also found in the Khibiny massif stuck in the space between aegirine and titanite, nepheline. At Lovozero, calcite is much less common. In addition, at Khibiny, there are zones of carbonatite breccias in the northeastern part of the massif. At Kovdor, the carbonates form complexes of phoscorites rocks, where they occur in significant amounts co-occurring with magnetite, apatite, pyroxenes, olivine, phlogopite, and other minerals. These carbonates are represented by calcite, dolomite, less frequently ankerite, siderite, and strontianite. The latter often occurs as admixtures and inclusions in other carbonates, also accompanying phosphates.

Eudialyte crystals usually have a cherry-red to light blood-red color (Figure 7F). In rocks of the Khibiny massif, they usually form anhedral crystals crystallizing between nepheline, apatite, and femic minerals. In the vein varieties of malinites (melasyenites) they usually form crystals that have a sector structure. In the Lovozero Massif, these crystals are more intensely red and sometimes have euhedral forms (in porphyrites) where they crystallize similarly to Lorenzenite forming euhedral, skeletal crystals with numerous fractures in the center.

*5.3. Optical Properties of the Discussed Minerals*

Lorenzenite in the thin section is a mineral with visible positive relief, this relief is the highest of all the minerals discussed. It has straw-brown colors showing dichroism. In polarized light, it has interference colors on the borderline of the 2nd order (Figure 8A,B).

Loparite in a thin section has the same high positive relief as lorenzenite (Figure 8C,D). It shows a brown-orange coloration, and under polarized light, the sector structure and polysynthetic twins are visible. Interference colors are of the 3rd order, but because this mineral is dark, these colors are not well visible and on the matrix of other minerals forming more intense interference colors.

Perovskite also has high positive relief, in the thin section forming brown colored crystals, in polarized light, it shows a gray-blue color on the border between the colors of 3rd order. Polysynthetic twinning can be seen in it (Figure 8E,F).

**Figure 8.** Microphotographs of minerals by polarization microscopy in transmitted light (1N: (**A,C,E,G,J,M**)) and polarized light (Nx: (**B,D,F,H,I,K,L,N,O**)): lorenzenite (**A,B**); loparite (**C,D**); perovskite (**E,F**); titanite (**G–I**); eudialyte (**J–L**); apatite (**M,N**); and calcite (**O**).

Titanite is a well-known mineral. In the rocks of the discussed massif, it has a positive relief although much smaller than the minerals discussed above. In a thin section, it shows a trace color in straw-brown tones. In polarized light, it dims obliquely and has interference colors of the 3rd order of a brown character, showing sector structure and twins (especially in Khibiny rocks, Figure 8G–I).

Carbonates are mainly represented by calcite, less frequently ankerite dolomite, siderite, or strontianite. They form small crystals with positive relief. In a thin section, they are colorless, while in polarized light, they show interference colors of the third-order (Figure 8O). These minerals form their rocks in the Kovdor massif, where they are a rock-forming component of calcite and dolomitic phocorites. In the thin section, a domain structure is often visible, emphasized by a change in the direction of optical alignment in the carbonate. In Afrikanda rocks, they form small nests in pyroxenites and adjacent to perovskite in magnetitites. They also form mineral veins, where calcite is mainly present and is sometimes admixed with strontianite. In the Khibiny and Lovozero massifs, they usually represent a minor admixture of strontianite, although, at Khibiny, they also form carbonatite breccias in the northeastern region of the intrusion.

Eudialyte in the thin section is usually colorless, although it shows a slight pinkish-cherry coloration in some rocks of Khibiny or Lovozero (Figure 8J–L). This crystal under polarized light is colored dark gray, sometimes showing a slight pleochroism in the colors of the borderline 1st order (dark purple-blue). Numerous twining, zonal, or sector structures of these crystals are then visible.

Apatite forms crystals with a small positive relief, in polarized light, it is usually dark gray. In most of the discussed rocks, it forms euhedral crystals occurring both in the vicinity of leucocratic minerals (with nepheline and feldspar) and in the form of inclusions accompanying femic minerals. In the Kovdor and Afrikanda massifs, it accompanies pyroxenes and olivines in the Khibiny and Lovozero massifs it occurs in titanite (Figure 8M,N).

### 5.4. Spectroscopic Properties of the Discussed Minerals

All the minerals in question were examined spectroscopically. Lorenzenite was examined from the 'Flora' exposure area in the northern part of Lovozero, loparite from the Aluaiv area in the northwestern parts of Lovozero, perovskite was examined from rocks at Afrikanda, titanite from the ore zone at Khibiny. Calcite was examined from the Afrikanda carbonatite veins, eudialyte was examined from the Kukisvumchorr area in Khibiny and the Aluaiv area in Lovozero. Fluoroapatite, on the other hand, as well as titanite, are from the Khibiny ore zone. FTIR spectroscopic spectra are shown in Figure 9. In the spectrum of lorenzenite, bands at 893 and 829 $cm^{-1}$ were observed, which correspond to stretching vibrations of the Si-O group, bands at 728, 688, and 308 $cm^{-1}$, which can be related to stretching vibrations of the Ti-O group, vibrations at 509 $cm^{-1}$ corresponding to bending vibrations of the Na-O group, and vibrations at 417 $cm^{-1}$ correspond to bending vibrations of the Si-O group (Figure 9A). In the case of loparite (Figure 9B) and perovskite (Figure 9B), some similarity spectra were found in the region of 433–441 $cm^{-1}$ indicating bending vibrations, in the case of loparite, the bending vibrations of the Ti-O group were noted at 326 $cm^{-1}$, while for loparite, it was 367 $cm^{-1}$. Ti-O stretching vibrations are seen in the region at 664 $cm^{-1}$ for the examined perovskite, while in loparite, these vibrations were not noted. In the group of minerals, perovskite-loparite-tausonite group of minerals, such types of vibrations may not be evident everywhere as described in [61]. In loparite vibrations in the region, 503 $cm^{-1}$ are visible, which can be connected with torsional vibrations for Nb-O.

Vibrations in the region of 710 $cm^{-1}$ in perovskite can be connected with Ca-O stretching vibrations and in loparite 811 and 727 $cm^{-1}$ for Ce-O group and Ca-O, Na-O groups. Unexplained remains a band in Perovskite in the region of 1245 $cm^{-1}$, which can come from carbonate suffix in this mineral. The examined titanite crystal (Figure 9C) has a band in the region of 845 $cm^{-1}$, which may be related to Si-O stretching vibrations, and in the region of 688 $cm^{-1}$, characteristic of stretching vibrations of $TiO_6$ octahedra. In the region of 554 cm$^{-1}$ and 372 cm$^{-1}$, bands corresponding to Si-O bending vibrations were found, and for 334 cm$^{-1}$, bending vibrations of the Ti-O group. Some deviations from the standard results [62] may be related to the substitution of small amounts of Th, Nb, Ce in place of Ca, Ti. In the case of the calcite spectra (Figure 9D), three large peaks were found at 1215 cm$^{-1}$, characteristic of the stretching vibrations of the carboxyl group, 660 cm$^{-1}$, and 490 cm$^{-1}$, and a small peak at 875 cm$^{-1}$ probably related to the vibrations of the Ca-O bridge [63] and possible substitutions of Sr-O, and Ce-O, which occur in small amounts in the Afrikanda crystals discussed. Spectra of Eudialyte (Figure 9E) have been overlaid on spectra of samples from Khibiny and Lovozero. They show great similarity to each other. There are bands in the range of 764 cm$^{-1}$ and 708 cm$^{-1}$, which correspond to tetrahedral vibrations. The bands in the range 519 cm$^{-1}$, 474 cm$^{-1}$, and 432 cm$^{-1}$ are from the stretching vibrations of Zr-O and Fe-O bridges [64]. From 301 to 311 cm$^{-1}$, there are stretching vibrations of tetrahedral groups. The slight differences in the vibrations of the eudialyte from Khibiny and Lovozero may be related to slight differences in the substitutions with Mn, Fe, Na, and Ca ions. For hydroxyl and chlorine groups, vibrations are recorded from 3200 cm$^{-1}$ to 3700 cm$^{-1}$ [64], which was not captured during the measurement. The examined fluorapatite from Khibiny (Figure 9F) shows a band at 890 cm$^{-1}$, characteristic of bending vibrations between oxygen and fluorine, and bands in the range 821 cm$^{-1}$, 755 cm$^{-1}$ associated with Ca-O stretching vibrations. The bands at 474 cm$^{-1}$, 371 cm$^{-1}$, and 340 cm$^{-1}$ may be responsible for the bending vibrations of the Ca-O group and the phosphate group [65].

### 5.5. Single Crystals Analysis Results

The same crystal samples as in the spectroscopic analyses were examined, along with an eudialyte from Khibiny. Lorenzenite crystallizes in an orthorhombic—dipyramidal H-M Symbol (2/m 2/m 2/m) Space Group: P bcn [66]. The investigated crystal cell parameters from the 'flora' exposure in the Lovozero massif are a = 8.7035(9) Å, b = 5.2266(5) Å, c = 14.476(1) Å, Z = 4; V = 658.51 Å3 Den(Calc) = 3.44. Some minor differences in the cell parameters may be related to the La, Ce, Nd substitutions found in the micro-area studies. The crystal structure of lorenzenite is shown in Figure 10A. Some minor differences in the cell parameters may be related to the bad quality of the Lorenzenite crystals. Therefore, it cannot be excluded that the structure contains trace amounts of La, Ce, and Nd substitutions found in the micro-area studies.

The analysis of loparite showed that it crystallizes in isometric-hexoctahedral H-M system Symbol (4/m -3 2/m) Space Group: P m-3m [67]. The studied parameters of the Aluaiv slope crystal from the Lovozero massif are: a = 3.891(1) Å, Z = 1; V = 58.94 Å3 Den(Calc) = 4.79. The deviations from the standard data are related to the loparte-perovskite-tausonite crystallization system. The loparite-tausonite member crystallizes in a similar arrangement, while the loparite-perovskite cell arrangement differs (see text below). The substitutions of Nb, Ti, and Ce, Ca in the crystal structure modify the arrangement in which this mineral crystallizes. The crystal structure of the examined loparite is shown in Figure 10B. Both micro-area studies and spectroscopic analyses demonstrated this.

The situation is similar for perovskite. These crystals crystallize in an orthorhombic—dipyramidal H-M Symbol (2/m 2/m 2/m) Space Group: P nma [68]. The elemental cell parameters of the studied crystal from Afrikand are: a = 5.4530(5), b = 7.660(3) Å, c = 5.397(1) Å, Z = 4; V = 225.45 Å3, Den(Calc) = 4.03. The differences in the studied values and the benchmark values are explained similarly for loparite, since Afrikanda is Ce-rich perovskite and, in addition, some Nb doping and probably also Fe were also found in the perovskite structure. An image of the crystal structure of perovskite is shown in Figure 10C.

Titanite crystallizes in the monoclinic—prismatic H-M system Symbol (2/m) Space Group: P 21/a [69]. The elemental cell parameters of the studied titanite from the Khibiny massif (ore zone) are as follows: a = 6.5605(5) Å, b = 8.7123(6) Å, c = 7.0698(5) Å, Z = 4; beta = 113.885° V = 369.48 Å3 Den(Calc) = 3.55.

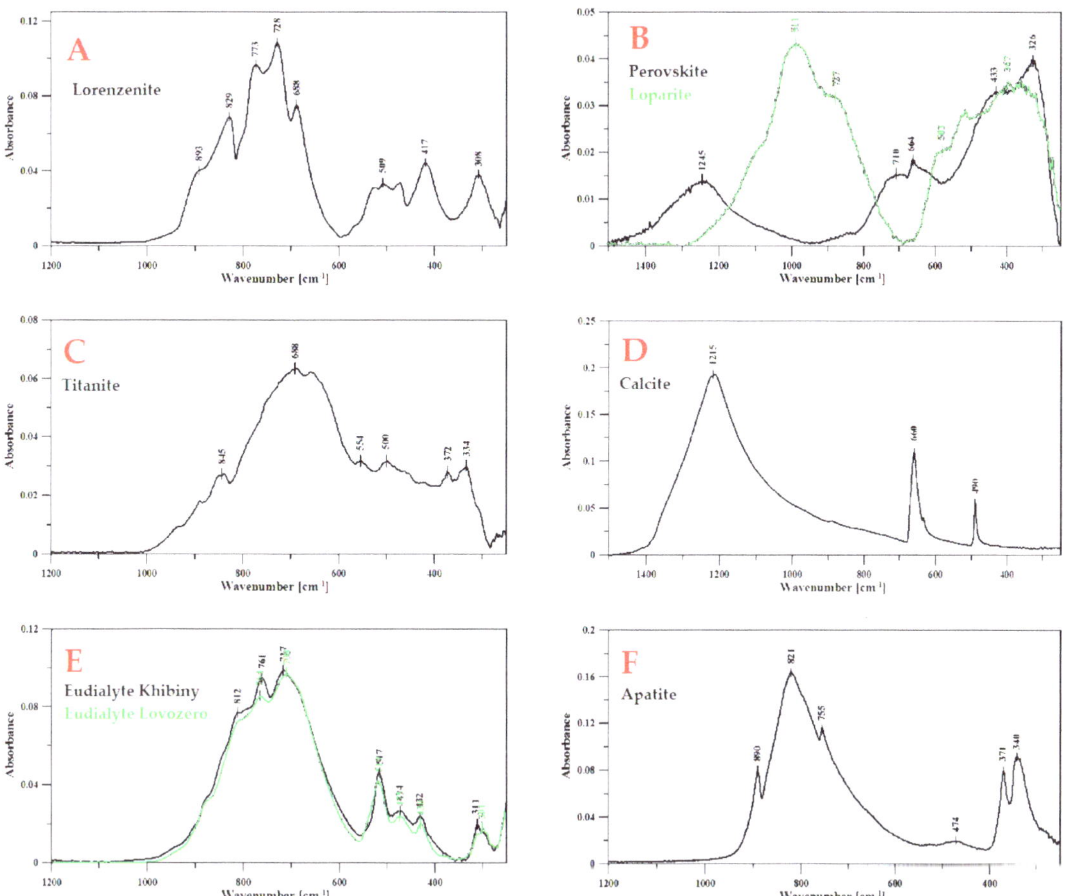

**Figure 9.** FTIR spectra of the described minerals: Lorenzenite (**A**), Perovskite (black, (**B**)), Loparite (green, (**B**)), Titanite (**C**), Calcite (**D**), Eudialyte from Khibiny (black, (**E**)), Eudialyte from Lovozero (green, (**E**)), and apatite (**F**).

Some slight differences in the cell constant c and angle β in the studied crystal compared to the standard are probably due to substitutions of Nb and W, in place of Ti, found in these crystals. The crystal structure of titanite from Khibiny is illustrated in Figure 10D.

Calcite from carbonatite veins in the Afrikanda massif was studied. Calcite crystallizes in a trigonal-hexagonal Scalenohedral H-M Symbol (3 2/m) Space Group: R 3c [70]. Some admixtures of strontianite (crystallizing in the Orthorhombic-Dipyramidal H-M Symbol (2/m 2/m 2/m) Space Group: Pmcn [70]) and Ce admixtures in the calcite crystal structure found in the micro-area. The crystal structure of calcite is visualized in Figure 10E.

The mineral eudialyte is a part of the eudialyte group, in which numerous substitutions of Na, Ce, La, Sr, Mn, Ti, and Fe [64] can occur, this also influences the nature of the elementary cell of this crystal. The studied eudialyte crystallizes in trigonal-hexagonal scalenohedral H-M Symbol (3 2/m) Space Group: R -3m [71]. Its elemental cell parameters are: a = 14.2508(2) Å, c = 30.0846(5) Å, Z = 12; V = 5291.19 Å3, Den(Calc) = 3.67. And are consistent within error with the reference structures from [71]. The eudialyte crystals studied in the microarray, along with their given formulas, are in the text below. An illustrated eudialyte elemental cell from the Khibiny massif is illustrated in Figure 10G.

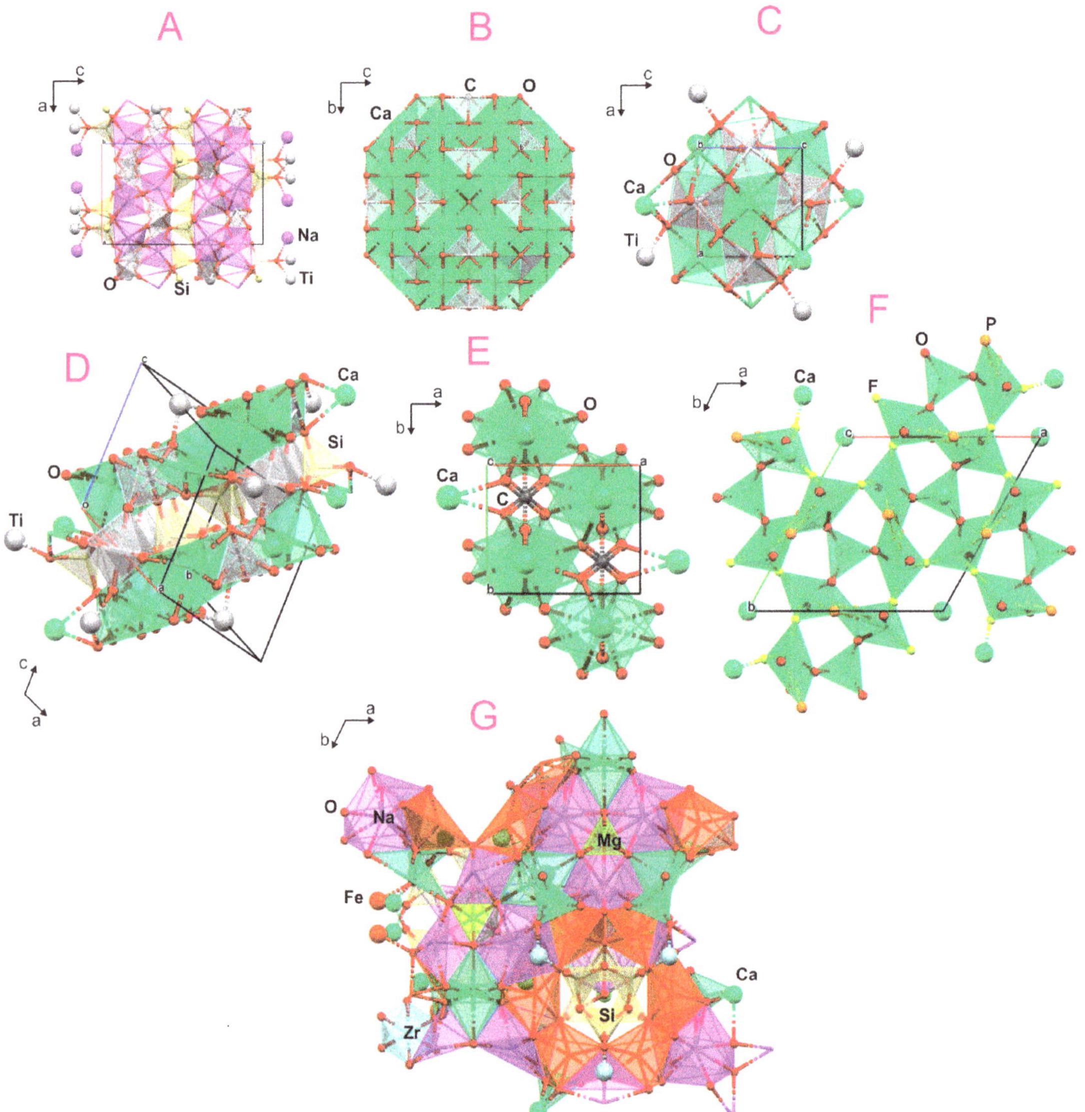

**Figure 10.** Results of the crystal structure of described minerals: lorenzenite (**A**), loparite (**B**), perovskite (**C**), titanite (**D**), calcite (**E**), apatite (**F**), and eudialyte (**G**).

Apatites are minerals that have numerous substitutions. It may be hydroxyapatite, fluorapatite, chlorapatite, carbonate-rich apatite, while Ca be replaced by La and Ce. In the Khibiny massif, fluorapatite is the most common (visible in the examined structure), although carbonate and chlorine admixtures are also found. The studied apatite crystal from the ore zone in the Khibiny massif crystallizes in the Hexagonal—Dipyramidal H-M Symbol (6/m) Space Group: P 63/m [72], and has the following elemental cell parameters: a = 10.0103(3) Å, c = 8.3924(3) Å, Z = 2; V = 728.30 Å3 Den(Calc) = 3.20. The sizes in the c direction deviate significantly from the reference structure. The crystal structure of Khibiny fluorapatite is illustrated in Figure 10F.

### 5.6. Results of SEM-EDS Analyses

#### 5.6.1. Lorenzenite

This mineral has been found in rocks of the Lovozero and Khibiny massifs [73–77]. At Lovozero, it occurs mostly among porphyrites, where it forms euhedral crystals together with eudialyte, astrophyllite, epistolite, and murmanite. Analyses in the SEM-EDS (Figure 11A) showed that this mineral contains small amounts of niobium. In the boundary zones of the crystal, numerous small inclusions of phosphates containing La, Ce, Pr, Nd, and loparite, also containing some of U, were found. In the Khibiny rocks, it was found as an admixture, accompanying rocks of the ore zone; these minerals co-occur with titanite and pyroxenes. Lorenzenite in these rocks has an anhedral shape and is generally small and strongly corroded. Details are shown in Table A1 in the Appendix A.

#### 5.6.2. Loparite

This mineral has been found in all the discussed massifs, although it occurs in them in different proportions [78–85]. In the Lovozero massif, it is relatively common, occurs as an accessory mineral, and is a rock-forming mineral in some pegmatites. In Lovozero rocks, it usually has a euhedral shape and is visible against other minerals. The results of the analysis in the Lovozero loparite SEM-EDS showed that it contains admixtures of uranium and strontium. The mineral also has niobium in it. In the Khibiny Massif, loparite occurs much less frequently. It has been found, inter alia, in the ore zone, where it accompanies apatite-nepheline rocks in the vicinity of titanite. It is typically small in size and corroded. Analysis in the SEM-EDS (Figure 11B) showed that loparite from Khibiny is sometimes slightly doped with strontium. In the Kovdor rocks, loparite is relatively common as a minor admixture. It co-occurs with magnetite, which is often intergrown with perovskite, and with femic minerals such as pyroxene and phlogopite. SEM-EDS analysis has shown that some of these minerals are admixed with Ta and U, and also Th and Cd. Small admixtures of barite are also found in their vicinity. In Afrikanda rocks, loparite is a rare mineral, co-occurring with perovskite and Ce -rich perovskite as minor associated phases.

The results of the SEM-EDS analyses for loparite are shown in Table A2 in the Appendix A. The results of the analyses are shown in Table A3 in the Appendix A.

#### 5.6.3. Perovskite

This mineral has been found in all massifs discussed [46–48,81]. In Afrikanda, it occurs mostly in rocks (derivatives in magnetitites and pyroxenites). The crystals found there are numerous, large, euhedral. Studies in the SEM-EDS (Figure 11C) have shown that it is doped with Ce, Th, Nb, and sometimes W. In the Kovdor massif, perovskite is also relatively abundant accompanying magnetites, growing on these crystals and in the form of admixtures in femic minerals. In the Khibiny massif, it rarely occurs, as a small admixture in rocks of the ore zone co-occurring with titanite. The same is true in the Lovozero massif, where small amounts of this crystal have been studied in jovite rocks (04LV12). These minerals in this massif have numerous admixtures of Nd, Th, and sometimes also Ce and Th [86].

#### 5.6.4. Titanite

This mineral occurs in all discussed massifs [73–75,86,87]. The Khibiny massif sometimes forms significant accumulations, especially in the ore zone, where it can occur as a rock-forming mineral together with arfvedsonite and aegirine in intercalations with apatite-nepheline ore. In other rocks, it is often present along with ilmenite and less often with magnetite. The titanite specimens examined in the SEM-EDS (Figure 11D) often show admixtures of Nb, Ta, and, less frequently, W. Moreover, inclusions of lanthanide-containing phases are common in these minerals. The same is true for titanite from the Lovozero massif, where it is less common, although it occurs along with ilmenite in syenites, lujavrites, and jovites. When examined in the SEM-EDS, it has abundant admixtures of Nb, Th, Ce, and occasionally Nd, as evidenced by SEM-EDS analyses. Among the Afrikanda rocks, titanite sometimes co-occurs with perovskite, forming a phase of a secondary character, while in the rocks of the Kovdor massif, it is a rare phase. The results of analyses in the SEM-EDS of this mineral can be found in Table A4 in the Appendix A.

### 5.6.5. Calcite

Calcite occurs in all of the discussed massifs in varying proportions. Most of these minerals are found in Kovdor, where they root their rocks (phoscorites) [4,10–13,49–52,88–93]. In this massif, it is usually calcite, sometimes admixed with iron and manganese. Along with calcite, there is also dolomite forming dolomite phoscorites. In the Afrikanda massif, carbonates are represented mainly by calcite. It forms carbonatite veins and nests among pyroxenites and other rocks. Studies of the SEM-EDS showed that there are numerous admixtures of lanthanum and cerium, and less frequently strontium (Figure 11G). In the Khibiny massif, calcite is an accessory mineral and is less common in alkaline rocks except in zones in the northeastern part of the massif where carbonatites and carbonatite breccias occur. The studied calcite from Khibiny has a large admixture of strontium (strontianite). In the Lovozero massif, as in Khibiny, carbonates generally occur less frequently as accessory minerals in some alkalic rock types. The studied rock samples are dominated by a large admixture of strontium similar to Khibiny. The results of the analyses in the carbonate micro-area can be found in Table A7 in the Appendix A.

### 5.6.6. Eudialyte

Eudialyte occurs in rocks of the Khibiny and Lovozero massifs [94–97]. At Khibiny, eudialyte is an accessory mineral to many different rocks, as it is at Lovozero. Detailed analyses in the SEM-EDS (Figure 11F) showed little variation in the occurrence of zircon, manganese, and occasionally doped titanium contents in these minerals (can be found in Table A6 in the Appendix A).

### 5.6.7. Apatite

Apatite crystals co-occur in rocks in all discussed massifs [52,77,89]. At Khibiny, they form rock-forming accumulations in the ore zone, where nepheline-apatite rocks occur. In addition to these rocks, apatite is also present in all varieties of agpatite rocks found there, usually as an accessory mineral. Analyses in the SEM-EDS (Figure 11E) showed that hydroxyapatite sometimes dominates with a small admixture of fluorine less frequently chlorite. The same is true for the Lovozero massif, where apatite also occurs frequently as an accessory mineral. Studies in the SEM-EDS also indicate the occurrence of hydroxyapatite admixed with fluorine, rarely chlorine. In addition, admixtures of lanthanum, cerium, strontium less frequently silver are found in this mineral. In the Afrikanda massif, apatite forms an accessory mineral representing as above hydroxyapatite and fluorapatite. In the case of the Kovdor rocks, apatite sometimes forms their breccias, filling their spaces. These are usually phases of apatite, growing in a concentric form on crumbs of phoscorite breccia. Apart from these zones, apatite also occurs as accessory crystals in rocks, usually as hydroxyapatite and less frequently as fluorapatite. Micro-area studies have shown that barite inclusions are also sometimes present in this mineral. The results of micro-area analyses of this mineral can be found in Table A5 in the Appendix A.

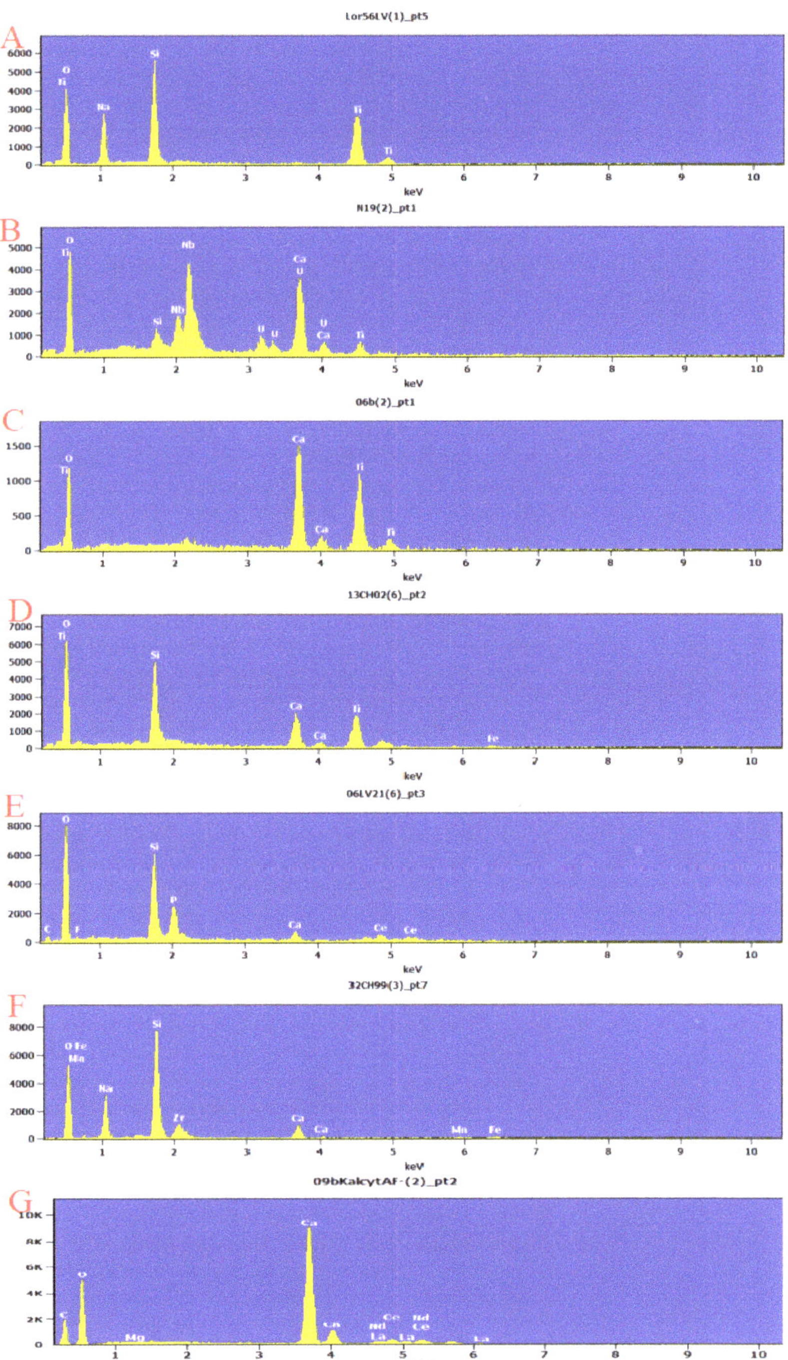

**Figure 11.** Example EDS spectra of the studied minerals: Lorenzenite (**A**), loparite (**B**), perovskite (**C**), titanite (**D**), apatite (**E**), eudialyte (**F**), and carbonates (**G**). The elemental contributions to each phase are discussed in the text.

## 6. Discussion

The discussed minerals are important components of rocks in the Khibiny, Lovozero, Kovdor, and Afrikanda massifs. Their presence indicates multi-stadial crystallization processes occurring in the discussed intrusions [98–106]. Perovskite is the mineral of the earliest crystallization that occurs mainly in the Afrikanda massif, less frequently in Kovdor, and is a rare mineral in the Khibiny and Lovozero massifs. In contrast, REE-rich phases such as knopite and loparite are phases that crystallized somewhat later in the crystallization phase of the contaminated melt. It is evidenced by the petrographic features, where in the microscopic image, knopite occurs mostly in the outer parts of perovskite or the crack zones of this crystal. SEM-EDS studies showed numerous Nb and Sr substitutions in perovskite, which was also confirmed by single crystals analysis and FTIR studies. In turn, loparite occurs in rocks in an accessory form, crystallizing together with alkaline phases such as eudialyte and lorenzenite. In carbonatite rocks, loparite has a corroded character which may mean that under changing conditions, it was disintegrated. This hypothesis may be supported by the fact that loparite and perovskite occur at Khibiny, mainly as an admixture among femic minerals, most often together with titanite. In addition, analysis of titanite from Khibiny and Lovozero indicates that it often has Nb admixtures as in loparite. The crystallizing phases in the Loparite-Perovskite-Tausonite series have many dopants indicating some contribution of these members to the loparite structure. Their microanalysis and single-crystals, as well as FTIR studies, indicate a small admixture of Nb and Ti, and Ce and Ca in the crystal structure modify the arrangement in which this mineral crystallizes. Titanite occurs much more frequently at Khibiny and Lovozero than at Afrikanda and Kovdor. It is a secondary mineral to perovskite and loparite, which passed to titanite during the impact of secondary processes. In zones of titanite occurrence, one can see relics of perovskite and loparite, especially in Khibiny. Due to the nature of titanite (usually a secondary mineral after perovskite or loparite), REE dopants are also found in this mineral. Apatite crystallized probably at the same time as loparite. This is evidenced especially by apatites in the ore zone rocks, forming euhedral crystal inclusions along with loparite and perovskite surrounded by titanite. Apatite from the discussed massifs is also rich in REE elements which were identified using SEM-EDS, single-crystals, and FTIR analysis. Minerals such as lorenzenite and eudialyte crystallized in the residual or hydrothermal phase [40,88,93]. This is particularly evident in the porphyrites of the Lovozero massif, where these minerals form skeletal crystals with numerous inclusions of surrounding phases. In the other rocks of the Lovozero and Khibiny massifs, eudialyte usually forms minerals of anhedral character filling the voids in the rocks between mafic phases as well as feldspars. Lorenzenite is present only in a few hydrothermal impact areas like the porphyrites mentioned above. Crystal structure analyses indicated a model crystal structure. However, FTIR studies showed some slight differences between the eudialyte from Khibiny and that from Lovozero. Carbonates in the discussed massifs are usually present in the form of admixtures, although, in Afrikanda, they form their nests and Kovdor rocks. Their position has been the subject of dispute for many researchers, who have emphasized their features indicating both crystallization during the mixing phase of alkali melts with supracrustal material and later [86]. Isotopic studies of these minerals indicate that they were generally formed by fluid crystallization from the Earth's mantle, although some crystallize later, as also evidenced by $\delta^{13}C$ values in carbonates [11,89,107] and $\delta^{34}S$ sulfides [108–111]. This is also confirmed by data obtained by other researchers. According to Arzamastsev, the age of perovskite from Khibiny is 383 ± 7 Ma, while the age of apatite-nepheline phases is 370 ± 3 Ma, respectively [22]. This indicates the order of crystallization of these phases. Sr-Nd [89] isotopic data indicate a mantle origin of the components of the intrusions in question and subsequent contamination with supracrustal material at the migration stage of these magmas towards the Earth surface and their further evolution during the formation of the intrusions in question. The Afrikanda and Kovdor intrusions have the most primordial character, and the Khibiny and Lovozero intrusions are the most contaminated [112]. This hypothesis is also supported by the geochemical data and results of helium isotope studies [113,114].

## 7. Conclusions

The discussed minerals are indicator phases in intrusions of ultramafic and alkaline rocks. Their presence is associated with different phases of the evolution of the melts from which the intrusions were formed in the Kola-Karelian Alkaline Province [7,20,21,115,116]. Their presence and petrographic character indicate the processes of their crystallization, corrosion, and phase formation at the expense of these minerals (e.g., titanite after loparite and perovskite). These minerals in the discussed massifs are carriers of REE elements. They are interesting from a mineralogical as well as deposit point of view. Their optical, spectroscopic features are closely related to the internal structure, optical properties of these minerals, and the form of crystallization. Their presence in alkaline rocks arouses interest and is an important genetic indicator of the studied rocks. The occurrence of these phases in the Khibiny, Lovozero, Afrikanda, and Kovdor intrusions is unique in the world. These massifs are relatively easy to access; moreover, the Khibiny and Lovozero massifs are large, showing the multi-zone nature of the different rocks. This provides the opportunity to observe these minerals macroscopically and study them with modern methods of analysis.

**Author Contributions:** Conceptualization, M.H.; methodology, M.H., D.K. and G.C.; software, M.H.; validation, M.H., D.K., G.C. and E.K.; formal analysis, M.H.; investigation, M.H.; resources, M.H. and E.K.; data curation, M.H.; writing—original draft preparation, M.H., D.K. and G.C.; writing—review and editing, M.H.; visualization, supervision, and project administration, M.H.; funding acquisition, M.H. and E.K. All authors have read and agreed to the published version of the manuscript.

**Funding:** This research received no external funding.

**Institutional Review Board Statement:** Not applicable.

**Informed Consent Statement:** Not applicable.

**Data Availability Statement:** This statement if the study did not report any data.

**Acknowledgments:** Authors would like to acknowledge of V. V. Balaganskiy and F. P. Mintrofanov from Kola Science Center of Russian Academy of Sciences, for help us collect minerals and needed informations.

**Conflicts of Interest:** The authors declare no conflict of interest.

## Appendix A

**Table A1.** Results of the microanalysis of the lorenzenite crystals.

| Sample | O | Na | Si | Ca | Ti | Nb | U |
|---|---|---|---|---|---|---|---|
| CH20(3)_pt1 | 46.96 | 4.58 | 7.92 | | 9.76 | | |
| CH20(6)_pt12 | 45.19 | 8.11 | 12.01 | | 15.97 | | |
| CH20(6)_pt13 | 44.33 | 8.57 | 12.67 | | 15.31 | | |
| CH20(6)_pt16 | 43.83 | 6.99 | 11.36 | | 16.78 | | |
| CH20(6)_pt6 | 43.37 | 8.7 | 13.62 | | 17.85 | | |
| CH20(6)_pt7 | 45.57 | 8.84 | 12.75 | | 14.32 | | |
| CH20(6)_pt8 | 45.44 | 7.6 | 11.18 | | 13.89 | | |
| CH20(6)_pt9 | 44 | 7.42 | 12.06 | | 16.46 | | |
| 06LV12-(2)_pt1 | 49.25 | 11.83 | 16.07 | | 12.37 | | |
| 47LV00(2)_pt3 | 49.67 | 17.16 | 15.53 | | 15.86 | 0.81 | |
| 47LV00(5)_pt1 | 27.81 | 11.61 | 13.59 | | 18.25 | 0.79 | |
| 51LV00(2)_pt1 | 22.33 | 9.09 | 15.72 | 0.88 | 42.28 | | |
| 51LV00(2)_pt2 | 23.77 | 8.21 | 16.06 | 0.92 | 43.73 | | |
| 51LV00(2)_pt3 | 23.52 | 8.25 | 16.29 | | 40.73 | | |
| 51LV00(2)_pt4 | 24.26 | 8.46 | 15.76 | 0.83 | 40.12 | | |

**Table A1.** *Cont.*

| Sample | O | Na | Si | Ca | Ti | Nb | U |
|---|---|---|---|---|---|---|---|
| 51LV00(2)_pt5 | 22.72 | 8.67 | 15.84 | | 42.43 | | |
| 51LV00(3)_pt1 | 35.42 | 1.15 | 20.31 | 3.07 | 6.03 | 13.98 | |
| 51LV00(3)_pt2 | 22.94 | 0.59 | 14.66 | 2 | 6.15 | 19.22 | 14.5 |
| 52LV00(3)_pt3 | 52.27 | 17.13 | 16.93 | | 12.29 | | |
| 53LV00(1)_pt1 | 30.47 | 9.83 | 16.06 | 0.6 | 40.66 | | |
| 53LV00(1)_pt2 | 26.33 | 10.15 | 16.14 | | 39.04 | | |
| 53LV00(1)_pt3 | 30.68 | 10.28 | 16.44 | 0.65 | 38.9 | | |
| 53LV00(1)_pt4 | 30.44 | 10.06 | 16.42 | | 37.48 | | |
| 53LV00(1)_pt5 | 28.73 | 9.11 | 16.46 | 0.71 | 31.4 | | |
| 53LV00(1)_pt6 | 30.67 | 10.1 | 16.99 | 0.94 | 38.87 | | |
| 53LV00(2)_pt2 | 35.12 | 4.94 | 20.98 | 2.44 | 7.61 | 8.79 | 13.46 |
| 53LV00(3)_pt4 | 34.92 | 8.35 | 29.79 | 0.83 | 3.40 | | |
| 54LV00(1)_pt1 | 29.24 | 3.19 | 11.99 | | 9.2 | 25.71 | 15.67 |
| 54LV00(1)_pt2 | 31.32 | 4.16 | 16.95 | | 6.01 | 16.34 | 16.16 |
| 54LV00(1)_pt3 | 36.36 | 8.63 | 32.68 | 1.18 | 2.87 | | |
| 54LV00(1)_pt5 | 30.8 | 9.29 | 19.44 | | 36.45 | | |
| 54LV00(1)_pt6 | 30.07 | 9.32 | 18.94 | | 37.45 | | |
| 54LV00(1)_pt7 | 29.96 | 9.18 | 19.12 | | 37.5 | | |
| 54LV00(1)_pt8 | 31.76 | 9.12 | 19.27 | | 35.88 | | |
| 54LV00(2)_pt1 | 38.19 | 4.55 | 19 | | 9.1 | 22.32 | |
| 54LV00(2)_pt3 | 38.6 | 4.54 | 20.33 | | 9.48 | 19.49 | |
| 54LV00(2)_pt6 | 32.57 | 9.45 | 21.16 | | 33.52 | | |
| 54LV00(2)_pt7 | 31.82 | 9.77 | 19.98 | | 35.07 | | |
| 54LV00(2)_pt8 | 32.3 | 9.43 | 20.38 | | 34.39 | | |
| 54LV00(3)_pt1 | 30.76 | 8.92 | 17.87 | 0.77 | 38.38 | | |
| 54LV00(3)_pt2 | 32.53 | 8.79 | 19.91 | | 34.6 | | |
| 54LV00(3)_pt3 | 33.52 | 9.17 | 19.17 | | 33.47 | | |
| 54LV00(3)_pt4 | 31.21 | 9.06 | 17.94 | 0.65 | 37.71 | | |
| 54LV00(3)_pt5 | 36.63 | 4.39 | 19 | 3.53 | 9.04 | 17.37 | |
| 54LV00(3)_pt6 | 30.06 | 3.29 | 12.79 | | 8.99 | 22.61 | 16.23 |
| 54LV00(3)_pt7 | 30.34 | 3.95 | 14.03 | | 6.82 | 23.96 | 15.32 |
| 54LV00(3)_pt8 | 30.29 | 3.94 | 13.18 | | 7.34 | 23.38 | 15.8 |
| 55LV00(1)_pt1 | 25.93 | 7.85 | 16.5 | | 43.65 | | |
| 55LV00(1)_pt1 | 25.93 | 7.85 | 16.5 | | 43.65 | | |
| 55LV00(1)_pt2 | 26.34 | 7.82 | 17.54 | | 41.46 | | |
| 55LV00(1)_pt2 | 26.34 | 7.82 | 17.54 | | 41.46 | | |
| 55LV00(1)_pt3 | 27.29 | 8.22 | 16.96 | | 42.55 | | |
| 55LV00(1)_pt3 | 27.29 | 8.22 | 16.96 | | 42.55 | | |
| 55LV00(1)_pt4 | 24.82 | 7.78 | 16.93 | 1.16 | 36.58 | | |
| 55LV00(1)_pt4 | 24.82 | 7.78 | 16.93 | 1.16 | 36.58 | | |
| 55LV00(1)_pt5 | 28.82 | 7.69 | 18.67 | | 38.07 | | |
| 55LV00(1)_pt5 | 28.82 | 7.69 | 18.67 | | 38.07 | | |
| 56LV00(1)_pt10 | 21.57 | 7.3 | 14.06 | 0.69 | 46.98 | | |
| 56LV00(1)_pt11 | 20.03 | 6.9 | 13.77 | 0.73 | 46.19 | | |
| 56LV00(1)_pt2 | 24.91 | 7.8 | 15.18 | 0.83 | 44.61 | | |
| 56LV00(1)_pt3 | 24.74 | 7.86 | 14.58 | 0.63 | 45.68 | | |
| 56LV00(1)_pt4 | 23.51 | 7.17 | 14.47 | | 48.08 | | |
| 56LV00(1)_pt5 | 22.99 | 6.7 | 14.06 | | 49.06 | | |
| 56LV00(1)_pt6 | 20.96 | 7.27 | 14.75 | | 48.88 | | |
| 56LV00(1)_pt7 | 20.58 | 7.41 | 14.62 | | 47.65 | | |
| 56LV00(1)_pt8 | 20.77 | 7.06 | 14.39 | 0.68 | 45.01 | | |
| 56LV00(1)_pt9 | 21.88 | 7.19 | 14.34 | 0.73 | 46.99 | | |
| 56LV00(2)_pt1 | 30.35 | 8.79 | 14.13 | 0.56 | 38.7 | | |
| 56LV00(2)_pt10 | 31.24 | 8.81 | 14.57 | | 37.83 | | |
| 56LV00(2)_pt2 | 30.72 | 8.97 | 14.32 | 0.63 | 37.65 | | |

**Table A1.** *Cont.*

| Sample | O | Na | Si | Ca | Ti | Nb | U |
|---|---|---|---|---|---|---|---|
| 56LV00(2)_pt3 | 33.41 | 9.5 | 13.02 | 0.89 | 32.02 | | |
| 56LV00(2)_pt4 | 33.75 | 9.23 | 12.22 | 0.94 | 29.6 | | |
| 56LV00(2)_pt5 | 33.58 | 8.8 | 12.17 | 1.39 | 29.16 | | |
| 56LV00(2)_pt6 | 33.78 | 9.04 | 12.18 | 1.06 | 29.58 | | |
| 56LV00(2)_pt7 | 33.58 | 9.23 | 12.78 | 0.71 | 30.33 | | |
| 56LV00(2)_pt8 | 34.45 | 9.66 | 13.06 | 0.41 | 32.79 | | |
| 56LV00(2)_pt9 | 30.68 | 9.34 | 14.16 | | 39.07 | | |

**Table A2.** Results of the microanalysis of the loparite crystals.

| Sample | O | Na | Ca | Ti | Nb | Pb | U | Ag | Ce | Ta |
|---|---|---|---|---|---|---|---|---|---|---|
| 13AF15(10)_pt1 | 21.28 | 2.83 | 13.43 | 32.92 | 3.71 | | | | 19.09 | |
| 06CH(22)_pt1 | 37.69 | | 1.94 | 6.65 | 16.73 | | | | | |
| 06CH(22)_pt2 | 35.29 | | 2.11 | 7.18 | 17.94 | | | | | |
| 06CH(22)_pt3 | 43.3 | | 1.94 | 6.81 | 16.43 | | | | | |
| 10CH02(17)_pt1 | 37.75 | 6.12 | 4.43 | 31.36 | 9.15 | | | | | |
| 10CH02(17)_pt2 | 46.28 | 6.58 | 3.44 | 23.78 | 7.9 | | | | | |
| 10CH02(19)_pt1 | 33.5 | 3.68 | 5.1 | 37.94 | 5.11 | | | | | |
| 01KV(10)_pt1 | 23.62 | 2.45 | 22.56 | 4.19 | 33.38 | | | | | |
| 01KV(11)_pt1 | 23.73 | 1.18 | 27.87 | | 39.13 | | | | | |
| 01KV(11)_pt2 | 31.51 | | 37.88 | 4.98 | 18.30 | | | | | |
| 01KV(11)_pt3 | 26.77 | | 32.57 | 4.42 | 23.82 | | | | | 8.25 |
| 01KV(12)_pt1 | 27.13 | 0.56 | 26.42 | 3.55 | 29.58 | | | | | |
| 01KV(12)_pt2 | 25.62 | 0.66 | 24.91 | 3.99 | 33.94 | | | | | |
| 01KV(12)_pt3 | 25.11 | 0.80 | 22.65 | 3.54 | 37.39 | | | | | |
| 01KV(12)_pt4 | 25.50 | | 24.67 | 3.61 | 37.55 | | | | | |
| 01KV(12)_pt5 | 30.28 | | 29.99 | | 33.96 | | | | | |
| 01KV(13)_pt1 | 23.89 | 1.65 | 27.75 | 3.59 | 33.26 | | | | | |
| 01KV(13)_pt2 | 24.54 | 1.01 | 26.01 | 2.07 | 32.01 | | | | | 5.15 |
| 01KV(13)_pt3 | 24.11 | 0.64 | 27.20 | 3.07 | 29.7 | | | | | 7.74 |
| 01KV(13)_pt4 | 25.89 | 1.01 | 31.96 | 3.18 | 27.43 | | | | | 5.24 |
| 01KV(13)_pt5 | 25.20 | 1.02 | 32.56 | 1.50 | 26.81 | | | | | 4.53 |
| 01KV(14)_pt1 | 23.82 | | 13.29 | 0.78 | 18.48 | | | | | |
| 01KV(14)_pt2 | 28.47 | | 8.43 | 3.47 | 17.01 | | | | | |
| 01KV(14)_pt3 | 31.22 | 0.54 | 17.79 | 6.53 | 15.79 | | | | | |
| 01KV(14)_pt4 | 26.09 | | 18.01 | 2.57 | 16.65 | | | | | |
| 01KV(14)_pt5 | 28.75 | | 17.76 | 8.1 | 21.35 | | | | | |
| 01KV(14)_pt6 | 28.53 | | 18.46 | 5.08 | 21.86 | | | | | |
| 01KV(15)_pt1 | 26.89 | | 12.26 | | 24.16 | | | | | |
| 01KV(15)_pt2 | 25.57 | | 10.64 | 1.31 | 21.61 | | | | | |
| 01KV(15)_pt3 | 26.6 | | 11.4 | | 18.7 | | | | | |
| 01KV(15)_pt4 | 31.92 | | 13.06 | 10.26 | 17.58 | | | | | |
| 01KV(15)_pt5 | 28.45 | | 10.72 | 4.22 | 17.61 | | | | | |
| 01KV(16)_pt1 | 25.17 | | 15.97 | 2.95 | 20.11 | | | | | |
| 01KV(16)_pt2 | 26.15 | | 18.93 | 2.96 | 19.76 | | | | | |
| 01KV(16)_pt3 | 23.05 | | 15.42 | | 18.44 | | | | | |
| 01KV(16)_pt4 | 25.96 | | 18.37 | | 21.01 | | | | | |
| 01KV(16)_pt5 | 30.96 | | 20.74 | 4.19 | 22.62 | | | | | |
| 01KV(18)_pt1 | 29.11 | | 14.66 | 7.71 | 20.16 | | | | | |
| 01KV(18)_pt2 | 27.52 | | 15.12 | 3.06 | 17.59 | | | | | |
| 01KV(2)_pt1 | 24.72 | 0.98 | 27.74 | 3.95 | 32.68 | | | | | |
| 01KV(2)_pt2 | 26.42 | | 27.16 | | 28.17 | | | | | 9.66 |
| 01KV(23)_pt1 | 25.49 | | 23.58 | | 18.44 | | | | | |
| 01KV(23)_pt2 | 26.27 | | 24.4 | 3.05 | 18.34 | | | | | |

**Table A2.** *Cont.*

| Sample | O | Na | Ca | Ti | Nb | Pb | U | Ag | Ce | Ta |
|---|---|---|---|---|---|---|---|---|---|---|
| 01KV(23)_pt3 | 25.06 | | 24.65 | 3.66 | 16.3 | | | | | |
| 01KV(3)_pt1 | 25.29 | 0.84 | 23.7 | 3.79 | 31.36 | | | | | 8.87 |
| 01KV(3)_pt2 | 25.17 | 0.53 | 27.18 | 2.93 | 28.2 | | | | | 8.12 |
| 01KV(3)_pt3 | 24.39 | 1.18 | 24.65 | 3.85 | 31.22 | | | | | 7.5 |
| 01KV(3)_pt4 | 25.04 | 0.78 | 25.96 | 3.32 | 32.59 | | | | | 4.94 |
| 01KV(3)_pt5 | 23.99 | 1.97 | 28.4 | 3.51 | 30.55 | | | | | 5.27 |
| 01KV(3)_pt6 | 24.11 | 1.26 | 27.5 | 2.97 | 30.61 | | | | | 6.77 |
| 01KV(4)_pt1 | 25.01 | | 24.64 | 2.55 | 24.6 | | | | | 0.95 |
| 01KV(4)_pt2 | 23.66 | 1.54 | 25.7 | 3.41 | 31.31 | | | | | 6.34 |
| 01KV(4)_pt3 | 25.97 | 0.47 | 25.73 | 5.33 | 29.8 | | | | | 6.34 |
| 01KV(4)_pt4 | 26.06 | 0.33 | 27.11 | 3.59 | 28.06 | | | | | 8.95 |
| 01KV(4)_pt5 | 24.23 | 0.82 | 25.29 | 4.87 | 34.97 | | | | | |
| 01KV(6)_pt1 | 24.25 | 2.25 | 27.33 | 2.65 | 35.66 | | | | | |
| 01KV(6)_pt2 | 24.96 | 1.1 | 26.61 | 3.9 | 30.44 | | | | | 5.56 |
| 01KV(6)_pt3 | 26.68 | 0.58 | 27.09 | 3.93 | 32.98 | | | | | |
| 01KV(6)_pt4 | 26.59 | 0.71 | 27.31 | 4.2 | 31.58 | | | | | 2.41 |
| 01KV(6)_pt5 | 24.3 | 1.12 | 27.13 | 3.01 | 27.89 | | | 2.65 | | 2.12 |
| 01KV(6)_pt6 | 26.04 | 1.67 | 30.96 | 3.52 | 28.69 | | | | | |
| 01KV(8)_pt1 | 27.13 | 0.53 | 27.95 | 4.94 | 29.86 | | | | | |
| 01KV(8)_pt2 | 27.5 | | 27.4 | 4.28 | 28.66 | | | | | |
| 01KV(8)_pt3 | 28.51 | | 27.01 | 5.07 | 29.23 | | | | | |
| 01KV(8)_pt4 | 23.91 | | 23.26 | | 25.71 | | | | | |
| N19(1)_pt2 | 24.62 | 2.25 | 29.44 | 5.15 | 17.06 | | 13.89 | | | |
| N19(2)_pt1 | 21.00 | | 23.27 | 6.79 | 23.89 | | 14.09 | | | |
| N19(2)_pt2 | 21.91 | | 21.93 | 5.72 | 26.24 | | 12.64 | | | |
| 06LV12-(21)_pt1 | 32.43 | | | 3.77 | 34.27 | | 18.98 | | | |
| 06LV12-(21)_pt2 | 32.43 | | | 4.45 | 32.09 | | 19.2 | | | |
| 06LV12-(21)_pt3 | 33.79 | | | 7.23 | 38.38 | | 16.16 | | | |
| 06LV12-(21)_pt4 | 33.8 | | | 7.41 | 34.33 | | 20.82 | | | |

**Table A3.** Results of the microanalysis of the perovskite crystals.

| Sample | O | Na | Ca | Ti | Nb | Ce | Nd | W | Th |
|---|---|---|---|---|---|---|---|---|---|
| 13AF15(1)_pt1 | 32.07 | | 58.87 | 4.86 | | | | | |
| 13AF15(1)_pt10 | 25.11 | | 28.21 | 32.92 | | | | | |
| 13AF15(1)_pt11 | 31.67 | | 53.28 | 10.37 | | | | | |
| 13AF15(1)_pt2 | 28.43 | | 54.18 | 13.41 | | | | | |
| 13AF15(1)_pt3 | 25.17 | | 28.69 | 32.61 | | | | | |
| 13AF15(1)_pt4 | 17.78 | 2.19 | 14.77 | 42.11 | | 21.75 | | | |
| 13AF15(1)_pt5 | 24.67 | | 21.32 | 17.31 | | 25.54 | | | |
| 13AF15(1)_pt6 | 27.18 | | 20.31 | 20.98 | | 24.03 | | 2.79 | |
| 13AF15(1)_pt7 | 19.39 | 0.39 | 32.62 | 45.74 | | | | | |
| 13AF15(1)_pt8 | 19.07 | | 32.66 | 47.17 | | | | | |
| 13AF15(1)_pt9 | 24.81 | | 27.43 | 35.05 | | | | | |
| 13AF15(10)_pt2 | 22.79 | 0.56 | 29.16 | 42.2 | | | | | |
| 13AF15(10)_pt3 | 23.46 | 0.56 | 28.32 | 41.93 | | | | | |
| 13AF15(10)_pt4 | 29.79 | 0.89 | 24.47 | 21.98 | | | | | |
| 13AF15(3)_pt1 | 30.15 | | 18.78 | 19.38 | | 22.36 | | | |
| 13AF15(3)_pt10 | 19.61 | 0.31 | 31.21 | 46.16 | | | | | |
| 13AF15(3)_pt11 | 19.83 | | 30.74 | 47.43 | | | | | |
| 13AF15(3)_pt12 | 18.71 | 0.39 | 31.31 | 47.7 | | | | | |

**Table A3.** *Cont.*

| Sample | O | Na | Ca | Ti | Nb | Ce | Nd | W | Th |
|---|---|---|---|---|---|---|---|---|---|
| 13AF15(3)_pt13 | 19.58 | 0.3 | 31.24 | 46.96 | | | | | |
| 13AF15(3)_pt14 | 19.83 | 0.38 | 30.75 | 46.85 | | | | | |
| 13AF15(3)_pt15 | 20.67 | | 31.08 | 44.92 | | | | | |
| 13AF15(3)_pt2 | 24.71 | | 16.02 | 19.36 | | 31.11 | | | |
| 13AF15(3)_pt3 | 27.29 | | 20.31 | 25.9 | | 18.74 | | | |
| 13AF15(3)_pt4 | 24.97 | | 21.48 | 21.23 | | 21.99 | | | |
| 13AF15(3)_pt5 | 23.44 | | 19.29 | 18.16 | | 29.99 | | | |
| 13AF15(3)_pt6 | 41.4 | | 23.43 | 27.48 | | | | | |
| 13AF15(3)_pt7 | 27.39 | | 22.47 | 19.88 | | 22.44 | | | |
| 13AF15(3)_pt8 | 23.15 | | 21.29 | 18.37 | | 27.57 | | | |
| 13AF15(3)_pt9 | 23.48 | | 20.58 | 19.46 | | 26.48 | | | |
| 13AF15(4)_pt1 | 18.37 | | 14.51 | | | 29.14 | | | |
| 13AF15(4)_pt10 | 28.44 | | 16.14 | 11.16 | | 33.06 | | 7.63 | |
| 13AF15(4)_pt11 | 26.78 | | 13.17 | 12.35 | | 31.83 | | 13.48 | |
| 13AF15(4)_pt12 | 28.37 | | 16.36 | 13.41 | | 31.67 | | 6.13 | |
| 13AF15(4)_pt13 | 25.22 | | 19.8 | 19.04 | | 25.08 | | | |
| 13AF15(4)_pt14 | 20.29 | 0.38 | 31.26 | 45.69 | | | | | |
| 13AF15(4)_pt15 | 19.84 | | 31.31 | 46.75 | | | | | |
| 13AF15(4)_pt16 | 19.79 | 0.24 | 32.86 | 45.29 | | | | | |
| 13AF15(4)_pt17 | 19.64 | | 31.62 | 46.49 | | | | | |
| 13AF15(4)_pt18 | 19.4 | 0.41 | 32.08 | 46.05 | | | | | |
| 13AF15(4)_pt2 | 25.23 | | 18.28 | 15.82 | | 29.49 | | | |
| 13AF15(4)_pt3 | 28.9 | | 17.29 | 17.67 | | 25.49 | | | |
| 13AF15(4)_pt4 | 24.98 | | 15.11 | 15.88 | | 31.83 | | | 2.08 |
| 13AF15(4)_pt5 | 25.25 | | 20.62 | 17.03 | | 26.25 | | | |
| 13AF15(4)_pt6 | 26.82 | | 18.34 | 18.39 | | 30 | | | |
| 13AF15(4)_pt7 | 28.42 | | 16.46 | 13.71 | | 34.58 | | | |
| 13AF15(4)_pt8 | 29.5 | | 15.23 | 12.81 | | 29.65 | | 9.09 | |
| 13AF15(4)_pt9 | 26.6 | | 15.27 | 9.66 | | 37.61 | | 7.47 | |
| 13AF15(5)_pt1 | 24.54 | | 20.81 | 10.26 | | 33.49 | | | |
| 13AF15(5)_pt2 | 25.05 | | 23.75 | 10.71 | | 27.92 | | | |
| 13AF15(5)_pt3 | 24.17 | | 18.67 | 16.51 | | 32.38 | | | |
| 13AF15(5)_pt4 | 26.45 | | 18.97 | 17.1 | | 27.51 | | | |
| 13AF15(7)_pt4 | 29.01 | 0.73 | 22.83 | 21.92 | | | | | |
| 13AF15(8)_pt10 | 21.63 | 0.73 | 31.98 | 43.49 | | | | | |
| 13AF15(8)_pt5 | 24.39 | 2.23 | 20.48 | 46.76 | 3.63 | | | | |
| 13AF15(8)_pt6 | 29.33 | | 26.63 | 7.44 | | 23 | | 10.65 | |
| 13AF15(8)_pt7 | 31.44 | | 53.69 | 9.73 | | | | | |
| 13AF15(8)_pt8 | 32.79 | 2.27 | 29.35 | 6.63 | | | | | |
| 13AF15(8)_pt9 | 26.05 | | 31.75 | 28.7 | | | | | |
| 06bCH(2)_pt1 | 20.43 | 1.1 | 29.97 | 42.43 | | | | | |
| 06bCH(2)_pt2 | 19.46 | 0.83 | 29.77 | 42.61 | | | | | |
| 06bCH(5)_pt1 | 19.15 | 0.51 | 30.73 | 45.46 | | | | | |
| 06bCH(5)_pt2 | 21.21 | 0.42 | 32.14 | 41.23 | | | | | |
| 06bCH(5)_pt3 | 19.65 | 0.43 | 31.6 | 43.23 | | | | | |
| 06bCH(5)_pt4 | 19.04 | 0.66 | 31.76 | 43.72 | | | | | |
| 06bCH(6)_pt1 | 22.34 | 1.5 | 28.64 | 37.86 | | | | | |
| 06bCH(6)_pt1 | 22.34 | 1.5 | 28.64 | 37.86 | | | | | |
| 06bCH(6)_pt2 | 20.66 | 0.53 | 32.93 | 37.13 | | | | | |
| 11KV(1)_pt1 | 23.55 | 0.82 | 29.33 | 31.09 | | | | | |
| 11KV(1)_pt2 | 24.81 | 0.68 | 28.62 | 30.15 | | | | | |
| 11KV(2)_pt3 | 26.05 | 0.63 | 30.76 | 29.89 | | | | | |
| 11KV(2)_pt4 | 23.49 | 0.75 | 29.97 | 34.86 | | | | | |
| 11KV(2)_pt5 | 22.07 | 0.79 | 34.15 | 35.53 | | | | | |
| 11KV(2)_pt6 | 24.02 | 0.86 | 31.88 | 33.38 | | | | | |
| 7bKV(3)_pt1 | 22.11 | 0.63 | 29.12 | 37.56 | | | | | |

**Table A3.** *Cont.*

| Sample | O | Na | Ca | Ti | Nb | Ce | Nd | W | Th |
|---|---|---|---|---|---|---|---|---|---|
| 7bKV(3)_pt2 | 23.11 | 0.49 | 28 | 37.96 | | | | | |
| 7bKV(3)_pt3 | 20.17 | 0.62 | 30.34 | 35.1 | | | | | |
| 7bKV(3)_pt5 | 27.34 | 0.45 | 23.57 | 29.01 | | | | | |
| 7bKV(3)_pt7 | 24.8 | 0.43 | 24.96 | 33.97 | | | | | |
| 7bKV(4)_pt1 | 24.08 | 0.51 | 25.2 | 30.27 | | | | | |
| 7bKV(4)_pt2 | 25.54 | 0.63 | 24.88 | 30.51 | | | | | |
| 7bKV(4)_pt3 | 23.87 | | 27.44 | 32.44 | | | | | |
| 7bKV(5)_pt4 | 24.43 | 0.55 | 26.91 | 31.46 | | | | | |
| 7bKV(5)_pt5 | 26.39 | 0.59 | 25.62 | 31.7 | | | | | |
| 7bKV(5)_pt6 | 24.19 | 0.4 | 26.38 | 34.05 | | | | | |
| N09kv03(2)_pt2 | 15.05 | | 6.28 | 9.05 | | | | | |
| N09kv03(2)_pt3 | 15.64 | | 8.61 | 7.67 | | | | | |
| N09kv03(2)_pt4 | 12.99 | | 6.24 | 10.08 | | | | | |
| N09kv03(2)_pt6 | 14.13 | | 7.86 | 8.3 | | | | | |
| 04LV12-(29)_pt1 | 32.69 | | | 24.16 | | 13.01 | 13.54 | | 4.88 |
| 04LV12-(29)_pt2 | 33.6 | | | 22.18 | | 10.96 | 16.26 | | 5.51 |
| 04LV12-(29)_pt3 | 46.39 | | 1.47 | 19.98 | | | 16.73 | | |
| 04LV12-(29)_pt6 | 37.87 | | | 24.42 | | | 16.52 | | 8.12 |

**Table A4.** Results of the microanalysis of the titanite crystals.

| Sample | O | Si | Ca | Ti | Fe | Nb | Ce | Nd | Th |
|---|---|---|---|---|---|---|---|---|---|
| 13AF15(3)_pt16 | 24.83 | 10.69 | 28.92 | 33.19 | | | | | |
| 13AF15(3)_pt17 | 24.84 | 10.18 | 27.17 | 35 | 1.38 | | | | |
| 13AF15(4)_pt19 | 25.79 | 11.79 | 27.16 | 33.37 | | | | | |
| 13AF15(4)_pt20 | 25.08 | 11.08 | 28.4 | 33.65 | | | | | |
| 13AF15(5)_pt10 | 25.81 | 10.06 | 33.2 | 26.32 | | | | | |
| 13AF15(5)_pt5 | 25.51 | 10.55 | 31.27 | 29.43 | | | | | |
| 13AF15(5)_pt6 | 25.91 | 11.17 | 30.18 | 29.56 | | | | | |
| 13AF15(5)_pt7 | 25.85 | 11.09 | 28.31 | 31.32 | | | | | |
| 13AF15(5)_pt8 | 25.95 | 10.37 | 29.66 | 31.2 | | | | | |
| 13AF15(5)_pt9 | 24.94 | 10.37 | 30.52 | 31.72 | | | | | |
| 13AF15(6)_pt1 | 24.11 | 10.1 | 29.32 | 35.01 | | | | | |
| 13AF15(6)_pt2 | 24.86 | 11.94 | 28.05 | 34.02 | | | | | |
| 06bCH(5)_pt5 | 23.4 | 10.29 | 31.8 | 30.53 | | | | | |
| 06bCH(6)_pt4 | 37.43 | 24.43 | 15.4 | 6.72 | | | | | |
| 06bCH(6)_pt5 | 33.62 | 28.59 | 11.08 | 3.06 | 1.03 | | | | |
| 10CH02(6)_pt4 | 26.2 | 11.89 | 17.06 | 23.89 | | | | | |
| 10CH02(7)_pt1 | 24.11 | 13.4 | 24.63 | 35.42 | | | | | |
| 10CH02(7)_pt2 | 24.95 | 12.92 | 24.61 | 34.91 | | | | | |
| 10CH02(7)_pt3 | 25.68 | 12.97 | 23.8 | 34.2 | | | | | |
| 10CH02(7)_pt4 | 25.18 | 13.3 | 24.46 | 33.83 | | | | | |
| 10CH02(7)_pt5 | 25.49 | 13.87 | 24.03 | 33.13 | | | | | |
| 10CH02(8)_pt4 | 26.9 | 14.63 | 20.28 | 31.11 | | | | | |
| 10CH02(8)_pt5 | 26.31 | 14.54 | 23.5 | 31.03 | | | | | |
| 10CH02(8)_pt6 | 26.77 | 14.32 | 22.46 | 31.86 | | | | | |
| 20aCH02(7)_pt4 | 29.41 | 15.66 | 18.89 | 25.72 | | | | | |
| c(2)_pt1 | 63.09 | 13.06 | 9.33 | 6.83 | | | | | |
| CH20(2)_pt2 | 64.6 | 12.15 | 8.2 | 5.73 | | | | | |
| CH20(2)_pt3 | 33.93 | 15.93 | 21.74 | 20.54 | | | | | |
| CH20(2)_pt4 | 50.18 | 15.98 | 12.41 | 11.68 | | | | | |

**Table A4.** *Cont.*

| Sample | O | Si | Ca | Ti | Fe | Nb | Ce | Nd | Th |
|---|---|---|---|---|---|---|---|---|---|
| 7bKV(5)_pt1 | 34.91 | 18.78 | 17.38 | 2.63 | 9.07 | | | | |
| 7bKV(5)_pt2 | 34.06 | 17.54 | 16.42 | 4.09 | 10.62 | | | | |
| 7bKV(5)_pt3 | 35.23 | 18.32 | 17.59 | 3.21 | 7.46 | | | | |
| 03LV12-(16)_pt1 | 39.35 | 13.36 | 14.69 | 4.12 | | 2.63 | | | 1.98 |
| 03LV12-(16)_pt2 | 36.34 | 14.62 | 16.02 | 4.52 | | 1.71 | 5.19 | | 3.84 |
| 03LV12-(17)_pt3 | 38.43 | 14.17 | 14.93 | 5.23 | | | | 3.86 | |
| 03LV12-(17)_pt4 | 40.96 | 13.97 | 14.32 | 4.51 | | 2.25 | 3.95 | | 2.87 |
| 03LV12-(17)_pt6 | 33.64 | 15.2 | 12.6 | 4.97 | | 2.33 | 6.73 | | 2.45 |
| Lov1(4)_pt1 | 54.06 | 14.01 | 10.73 | 14 | | 0.85 | | | |
| Lov1(4)_pt2 | 48.82 | 14.87 | 13.19 | 17.82 | | | | | |
| LV12(18)_pt3 | 50.59 | 13.52 | 12.03 | 16.92 | | | | | |
| LV12(18)_pt4 | 41.13 | 12.42 | 15.4 | 22.04 | | | | | |
| LV12(20)_pt1 | 41.66 | 13.84 | 15.89 | 21.66 | 0.36 | 0.86 | | | |
| LV12(20)_pt2 | 50.81 | 12.34 | 2.33 | 15.51 | | | | | |
| LV12(5)_pt2 | 41.98 | 14.22 | 15.64 | 21.92 | | | | | |
| LV12(5)_pt4 | 46.45 | 14.09 | 14.31 | 19.26 | | | | | |
| LV12(7)_pt8 | 44.45 | 14.22 | 13.87 | 18.83 | | | | | |

**Table A5.** Results of the microanalysis of the apatite crystals.

| Sample | C | O | F | P | Cl | Ca | Sr | Nb | U | La | Ce |
|---|---|---|---|---|---|---|---|---|---|---|---|
| 13AF15(7)_pt1 | 2.11 | 27.12 | 0.83 | 11.98 | | 48.21 | | | | | |
| 13AF15(7)_pt2 | 1.8 | 27.33 | 0.16 | 12.3 | | 49.51 | | | | | |
| 13AF15(7)_pt3 | 1.98 | 26.85 | 1.22 | 12.23 | | 48.35 | | | | | |
| 13AF15(8)_pt1 | 2 | 28.15 | 0.77 | 11.63 | | 48.89 | | | | | |
| 13AF15(8)_pt2 | 1.79 | 27.17 | 0.89 | 12.32 | | 49.16 | | | | | |
| 13AF15(8)_pt3 | 1.94 | 28.19 | 0.27 | 12.08 | | 49.43 | | | | | |
| 13AF15(8)_pt4 | 1.9 | 25.92 | 0.55 | 11.92 | | 48.48 | | | | | |
| 10CH02(11)_pt3 | 2.15 | 30.66 | 1.09 | 12.03 | | 37.24 | | | | | |
| 10CH02(12)_pt1 | 4.84 | 30.23 | | 12.55 | | 41.7 | | | | | |
| 10CH02(12)_pt2 | 4.23 | 29.33 | 1.01 | 12.91 | | 38.25 | | | | | |
| 10CH02(12)_pt3 | 6 | 28.05 | 1.66 | 13.67 | | 41.76 | | | | | |
| 10CH02(13)_pt1 | 3.17 | 28.39 | 1.48 | 14.16 | | 44.79 | | | | | |
| 10CH02(13)_pt2 | 2.33 | 28.44 | 1.38 | 14.34 | | 45.15 | | | | | |
| 10CH02(13)_pt3 | 3.33 | 28.55 | 1.41 | 14.46 | | 44.97 | | | | | |
| 10CH02(13)_pt4 | 1.94 | 29.36 | 1.52 | 13.76 | | 43.23 | | | | | |
| 10CH02(13)_pt5 | 2.08 | 29.73 | 1.15 | 13.13 | | 43.93 | | | | | |
| 10CH02(13)_pt6 | 1.61 | 27.92 | 1.59 | 14.59 | | 46.91 | | | | | |
| 10CH02(13)_pt7 | 11.76 | 26.47 | 1.19 | 12.42 | | 40.15 | | | | | |
| 10CH02(14)_pt1 | 1.05 | 29.26 | 1.25 | 13.95 | | 42.43 | | | | | |
| 10CH02(14)_pt2 | 1 | 28.96 | 1.36 | 13.73 | | 42.82 | | | | | |
| 10CH02(14)_pt3 | 1.13 | 28.97 | 1.44 | 14.02 | | 43.55 | | | | | |
| 10CH02(14)_pt4 | 1.42 | 28.94 | 1.81 | 13.88 | | 44.21 | | | | | |
| 10CH02(14)_pt5 | 1.54 | 29.88 | 1.6 | 14.13 | | 42.27 | | | | | |
| 10CH02(15)_pt1 | 1.52 | 29 | 1.35 | 13.56 | | 42.57 | | | | | |
| 10CH02(15)_pt2 | 1.52 | 28.63 | 1.15 | 14.75 | | 43.17 | | | | | |
| 10CH02(15)_pt3 | 2.17 | 30.19 | 2.16 | 12.72 | | 41.92 | | | | | |
| 10CH02(15)_pt4 | 2.13 | 29.12 | 1.17 | 13.74 | | 43.75 | | | | | |
| 10CH02(2)_pt4 | 2.41 | 30.99 | 0.98 | 11.95 | | 36.07 | | | | | |
| 10CH02(2)_pt5 | 3.03 | 30.02 | 0.99 | 11.72 | | 35.96 | | | | | |

**Table A5.** *Cont.*

| Sample | C | O | F | P | Cl | Ca | Sr | Nb | U | La | Ce |
|---|---|---|---|---|---|---|---|---|---|---|---|
| 10CH02(3)_pt6 | 3.97 | 29.22 | 0.58 | 12.04 | | 35.26 | | | | | |
| 10CH02(3)_pt7 | 2.33 | 30.72 | 1.14 | 12.42 | | 36.78 | | | | | |
| 10CH02(3)_pt8 | 1.73 | 30.2 | 0.93 | 11.84 | | 37.19 | | | | | |
| 10CH02(5)_pt4 | 1.92 | 27.79 | 1.54 | 12.96 | | 43.28 | | | | | |
| 10CH02(5)_pt5 | 3.46 | 28.04 | 0.94 | 12.27 | | 38.4 | | | | | |
| 10CH02(6)_pt3 | 4.87 | 27.75 | 0.89 | 13.06 | | 41.22 | | | | | |
| 10CH02(8)_pt1 | 2.07 | 30.67 | 1.34 | 11.89 | | 38.22 | | | | | |
| 10CH02(8)_pt2 | 2.05 | 30.49 | 1.34 | 12.21 | | 37.33 | | | | | |
| 20aCH02(6)_pt1 | 0.85 | 32.96 | 1.86 | 12.42 | | 37.73 | | | | | |
| 20aCH02(6)_pt2 | 0.91 | 33.35 | 2.7 | 12.4 | | 38.06 | | | | | |
| 20aCH02(6)_pt3 | 1.89 | 32.66 | | 6.71 | 0.9 | 25.53 | | | | | |
| CH20(1)_pt1 | 2.54 | 46.35 | 5.26 | 14.1 | | 31.32 | | | | | |
| CH20(1)_pt2 | 2.89 | 47.36 | 5.07 | 14.03 | | 30.66 | | | | | |
| CH20(1)_pt3 | 4.23 | 29.84 | 2.62 | 17.18 | | 45.95 | | | | | |
| CH20(1)_pt4 | 1.95 | 43.21 | 5.2 | 14.31 | | 35.34 | | | | | |
| CH20(1)_pt5 | 6.57 | 21.97 | 1.36 | 13.47 | | 55.41 | | | | | |
| CH20(1)_pt6 | 3.6 | 32.73 | 3.41 | 15.72 | | 44.26 | | | | | |
| CH20(2)_pt1 | 45.54 | | 5.25 | 13.17 | | 32.51 | | | | | |
| CH20(2)_pt2 | 48.1 | | 2.33 | 9.99 | | 25.77 | | | | | |
| CH20(2)_pt3 | 49.89 | | 5.71 | 13.75 | | 26.79 | | | | | |
| CH20(2)_pt4 | 54.92 | | 7.18 | 13.64 | | 17.75 | | | | | |
| CH20(2)_pt5 | 59.65 | | 10.47 | 11.6 | | 5.99 | | | | | |
| CH20(3)_pt1 | 2.25 | 14.5 | 0.87 | 17.46 | | 64.92 | | | | | |
| CH20(3)_pt2 | 1.32 | 19.43 | 1.69 | 17.09 | | 60.48 | | | | | |
| CH20(3)_pt3 | 12.6 | 18.22 | 0.79 | 12.28 | | 53.39 | | | | | |
| CH20(3)_pt5 | 2.06 | 42.86 | 5.16 | 13.96 | | 35.96 | | | | | |
| CH20(3)_pt6 | 1.12 | 37.44 | 3.62 | 13.33 | | 44.5 | | | | | |
| CH20(3)_pt7 | 3.28 | 40.4 | 2.13 | 12.46 | | 32.27 | | | | | |
| 01KV(10)_pt3 | 4.72 | 29.1 | 1 | 15.42 | | 49.09 | | | | | |
| 01KV(11)_pt4 | 2.85 | 27.85 | | 14.22 | | 55.08 | | | | | |
| 01KV(12)_pt6 | | 25.13 | | 14.17 | | 57.33 | | | | | |
| 01KV(14)_pt10 | 5.22 | 27.97 | 0.88 | 13.07 | | 40.12 | | | | | |
| 01KV(14)_pt7 | 4.09 | 28.8 | 0.87 | 14.41 | | 43.33 | | | | | |
| 01KV(14)_pt8 | 5.01 | 29.66 | | 15 | | 42.45 | | | | | |
| 01KV(14)_pt9 | 4.35 | 30.54 | | 14.21 | | 42.7 | | | | | |
| 01KV(16)_pt6 | 7.31 | 30.08 | 0.63 | 12.6 | | 39.82 | | | | | |
| 01KV(18)_pt4 | 4.49 | 33.41 | 3.07 | 13.34 | | 42.69 | | | | | |
| 01KV(19)_pt3 | 4.48 | 25.64 | 1.07 | 13.15 | | 38.46 | | | | | |
| 01KV(19)_pt4 | 3.4 | 26.71 | 0.31 | 13.31 | | 42.43 | | | | | |
| 01KV(19)_pt5 | 3.75 | 24.41 | 0.57 | 12.47 | | 39.22 | | | | | |
| 01KV(19)_pt6 | 7.21 | 22.86 | 1.23 | 8.36 | | 27.78 | | | | | |
| 01KV(19)_pt7 | 4.08 | 27.54 | | 12.8 | | 37.23 | | | | | |
| 01KV(19)_pt8 | 4.26 | 28.73 | | 12.43 | | 38.36 | | | | | |
| 01KV(19)_pt9 | 6.14 | 25.85 | 0.54 | 10.92 | | 33.05 | | | | | |
| 01KV(2)_pt4 | 2.77 | 26.78 | | 13.76 | | 56.69 | | | | | |
| 01KV(21)_pt2 | 3.02 | 26.38 | 0.83 | 13.45 | | 43.82 | | | | | |
| 01KV(21)_pt3 | 3.96 | 26.36 | | 13.63 | | 41.58 | | | | | |
| 01KV(21)_pt5 | 3.84 | 26.02 | | 12.71 | | 44.65 | | | | | |
| 01KV(21)_pt6 | 4.31 | 28.27 | | 10.72 | | 43.25 | | | | | |
| 01KV(4)_pt8 | | 28.51 | | 14.82 | | 52.79 | | | | | |
| 01KV(7)_pt1 | 4.27 | 26.26 | 1.17 | 14.4 | | 50.67 | | | | | |
| 01KV(8)_pt5 | | 28.34 | | 14.57 | | 53.62 | | | | | |
| N09kv03(2)_pt38 | | 17.44 | 1.63 | 12.88 | | 28.44 | | | | | |
| N09kv03(2)_pt39 | | 19.21 | 1.63 | 13.84 | | 29.23 | | | | | |
| N09kv03(2)_pt40 | | 15.79 | 0.92 | 13.22 | | 28.61 | | | | | |
| N09kv03(2)_pt41 | | 16.94 | | 14.12 | | 29.56 | | | | | |

**Table A5.** *Cont.*

| Sample | C | O | F | P | Cl | Ca | Sr | Nb | U | La | Ce |
|---|---|---|---|---|---|---|---|---|---|---|---|
| N09kv03(2)_pt42 | | 18.11 | 0.49 | 11.91 | | 27.99 | | | | | |
| N09kv03(2)_pt43 | | 17.83 | | 13.16 | | 27.12 | | | | | |
| N09kv03(2)_pt45 | | 17.52 | 1.48 | 13.86 | | 29.42 | | | | | |
| N09kv03(2)_pt46 | | 17.9 | | 13.03 | | 27.66 | | | | | |
| N09kv03(2)_pt47 | | 17.64 | | 12.56 | | 27.81 | | | | | |
| N09kv03(2)_pt48 | | 18.19 | 1.43 | 13.08 | | 27.91 | | | | | |
| N09kv03(2)_pt49 | | 18.65 | 0.97 | 13.06 | | 28.16 | | | | | |
| N09kv03(2)_pt50 | | 18.04 | 1.42 | 13.35 | | 29.67 | | | | | |
| N09kv03(2)_pt51 | | 17.17 | | 12.93 | | 30.27 | | | | | |
| N09kv03(2)_pt52 | | 18.99 | 1.64 | 12.51 | | 29.79 | | | | | |
| N09kv03(2)_pt53 | | 17.42 | | 14.4 | | 32.05 | | | | | |
| N09kv03(2)_pt54 | | 23.99 | | 6.21 | | 22.24 | | | | | |
| N09kv03(2)_pt55 | | 18.23 | | 16.02 | | 30.71 | | | | | |
| N09kv03(2)_pt56 | | 21.2 | | 15.29 | | 30.84 | | | | | |
| N09kv03(2)_pt57 | | 19.38 | 0.96 | 14.25 | | 28.81 | | | | | |
| N09kv03(2)_pt58 | | 19.6 | | 12.41 | | 30.18 | | | | | |
| N09kv03(2)_pt59 | | 17.9 | | 14.12 | | 29.35 | | | | | |
| N09kv03(2)_pt60 | | 20.71 | 1.56 | 14.9 | | 29.92 | | | | | |
| N09kv03(2)_pt61 | | 16.36 | 2.25 | 12.78 | | 28.51 | | | | | |
| N09kv03(2)_pt62 | | 19.05 | | 15.28 | | 31.69 | | | | | |
| N09kv03(2)_pt63 | | 17.97 | | 16.37 | | 27.86 | | | | | |
| N09kv03(2)_pt64 | | 17.36 | 1.05 | 13.29 | | 28.33 | | | | | |
| N09kv03(2)_pt65 | | 17.86 | | 13.71 | | 30.06 | | | | | |
| N09kv03(2)_pt66 | | 16.66 | | 13.41 | | 29.45 | | | | | |
| N09kv03(2)_pt67 | | 19.6 | | 16.75 | | 32.04 | | | | | |
| N09kv03(2)_pt68 | | 17.3 | | 13.59 | | 27.64 | | | | | |
| N09kv03(2)_pt69 | | 16.65 | | 13.63 | | 27.39 | | | | | |
| N09kv03(2)_pt70 | | 18.4 | 1.06 | 12.26 | | 29.31 | | | | | |
| N8KV(1)_pt1 | 2.26 | 27.97 | 0.14 | 13.87 | | 53.19 | | | | | |
| N8KV(1)_pt2 | 2.34 | 28.86 | 0.47 | 12.96 | | 52.84 | | | | | |
| N8KV(1)_pt3 | 2.29 | 28.49 | 0.35 | 13.14 | | 55.73 | | | | | |
| N8KV(1)_pt4 | 2.78 | 28.57 | 0.18 | 12.87 | | 55.34 | | | | | |
| 06LV21(10)_pt1 | 1.67 | 32.03 | | 11.46 | | 5.94 | | | | | 24.24 |
| 06LV21(10)_pt2 | 3.61 | 36.51 | 0.41 | 6.7 | | 8.32 | | | | | |
| 06LV21(10)_pt3 | 2.07 | 32.87 | 0 | 7.29 | | 4.84 | | | | | 26.96 |
| 06LV21(10)_pt4 | 2.92 | 34.56 | | 7.36 | 0.66 | 5.1 | | | | | 21.08 |
| 06LV21(10)_pt5 | 2.95 | 34.36 | | 11.74 | | 5.28 | | | | | 22.13 |
| 06LV21(10)_pt6 | 2.28 | 31.2 | 0.52 | 8.63 | | 4.66 | | | | | 29.7 |
| 06LV21(10)_pt7 | 1.82 | 33.28 | 0 | 8.79 | | 5.58 | | | | | 25.18 |
| 06LV21(6)_pt1 | 1.8 | 31.4 | | 9.28 | | 5.77 | | | | | 25.6 |
| 06LV21(6)_pt2 | 1.59 | 30.36 | 0.03 | 8.55 | 0.73 | 6.61 | | | | | 27.43 |
| 06LV21(6)_pt3 | 1.54 | 29.78 | 0 | 8.84 | | 6.05 | | | | | 27.89 |
| 06LV21(6)_pt4 | 2.07 | 30.57 | 0 | 10.04 | | 6.47 | | | | | 24.96 |
| 06LV21(6)_pt5 | 2.63 | 30.68 | 0 | 8.61 | | 6.42 | | | | | 27.27 |
| 06LV21(9)_pt1 | 1.48 | 32.93 | | 11.01 | | 5.19 | | | | | 23.55 |
| 06LV21(9)_pt2 | 0.95 | 31.95 | 0 | 8.11 | | 4.56 | | | | | 26.42 |
| 06LV21(9)_pt3 | 1.13 | 32.46 | | 10.86 | | 5.41 | | | | | 25.37 |
| 06LV21(9)_pt4 | 1.28 | 31.65 | | 10.42 | | 5.38 | | | | | 25.78 |
| 06LV21(9)_pt5 | 1.61 | 33.28 | | 7.5 | | 4.84 | | | | | 22.21 |
| 06LV21(9)_pt6 | 1.24 | 33.47 | 0 | 9.24 | | 5.18 | | | | | 24.03 |
| 06LV21(9)_pt7 | 1.36 | 32.78 | 0 | 8.86 | | 5.07 | | | | | 25.82 |
| 06LV21(9)_pt8 | 1.28 | 31.17 | 0 | 8.3 | | 3.6 | | | | | 28.88 |
| LV12(11)_pt12 | 4.81 | 50.48 | | 12.79 | | 20.54 | 3.2 | | | | |
| LV12(11)_pt5 | 4.5 | 49.28 | | 12.97 | | 23.2 | 2.13 | | | | |
| LV12(15)_pt2 | 5.2 | 42.21 | 5.6 | 14.49 | | 25.96 | 5.37 | | | | |
| LV12(14)_pt3 | 7.28 | 42.38 | 4.48 | 11.95 | | 20.49 | 2.55 | | | | |
| 55LV00(1)_pt6 | 3.65 | 30.53 | 1.41 | 10.18 | | 28.04 | | | | | |

**Table A5.** *Cont.*

| Sample | C | O | F | P | Cl | Ca | Sr | Nb | U | La | Ce |
|---|---|---|---|---|---|---|---|---|---|---|---|
| 55LV00(1)_pt7 | 2.95 | 28.56 | 1.71 | 9.15 | | 31.17 | | | | | |
| 55LV00(1)_pt6 | 3.65 | 30.53 | 1.41 | 10.18 | | 28.04 | | | | | |
| 55LV00(1)_pt7 | 2.95 | 28.56 | 1.71 | 9.15 | | 31.17 | | | | | |
| 55LV00(2)_pt1 | 5.52 | 41.9 | | 10.02 | | 7.33 | | | | | |
| 55LV00(2)_pt2 | 4 | 29.93 | 0.61 | 8.63 | | 15.12 | | | | | |
| 55LV00(3)_pt1 | 1.99 | 18.81 | | 5.55 | | 4.45 | | | | 13.29 | 28.74 |
| 55LV00(3)_pt3 | 2.51 | 22.66 | | 8.72 | | 7.22 | | | | 13.27 | 23.05 |
| 55LV00(3)_pt4 | 5.42 | 30.7 | | 7.19 | 0.63 | 6.94 | | | | | 16.07 |
| 55LV00(3)_pt7 | 2.84 | 34.04 | | 8.81 | | 26.24 | | | | | |
| 55LV00(3)_pt8 | 4.28 | 32.9 | | 3.74 | | 10.47 | | | | | |

**Table A6.** Results of the microanalysis of the eudialyte crystals.

| Sample | O | Na | Si | Ca | Mn | Fe | Zr |
|---|---|---|---|---|---|---|---|
| 32CH99(3)_pt1 | 29.31 | 8.65 | 26.33 | 10.38 | 9.72 | 2.13 | 9.19 |
| 32CH99(3)_pt2 | 29.69 | 8.09 | 24.92 | 10.75 | 6.91 | 6.99 | 8.94 |
| 32CH99(3)_pt3 | 28.38 | 8.57 | 25.6 | 10.08 | 8.23 | 6.3 | 9.43 |
| 32CH99(3)_pt4 | 28.93 | 8.66 | 26.73 | 10.27 | 9.25 | 2.57 | 9.41 |
| 32CH99(3)_pt5 | 28.05 | 7.72 | 25.12 | 10.02 | 10.07 | 6.47 | 8.02 |
| 32CH99(3)_pt6 | 27.3 | 8.9 | 26.45 | 11.04 | 10.66 | 1.86 | 10.73 |
| 32CH99(3)_pt7 | 27.99 | 8.8 | 27.02 | 11.27 | 6.43 | 3.67 | 10.57 |
| 32CH99(5)_pt1 | 26.75 | 7.79 | 22.82 | 9.45 | 5.46 | 8.1 | 8.04 |
| 32CH99(5)_pt2 | 27.67 | 8.56 | 25.94 | 11.74 | 7.48 | 2.81 | 9.16 |
| 32CH99(5)_pt3 | 31.25 | 8.83 | 28.7 | 12.38 | | 2.78 | 9.94 |
| 32CH99(5)_pt4 | 28.12 | 8.71 | 27.27 | 12.38 | 6.11 | 2.92 | 10.22 |
| 32CH99(5)_pt5 | 31.75 | 9.19 | 28.29 | 12.13 | | 2.6 | 10.83 |
| 32CH99(5)_pt6 | 27.96 | 8.69 | 27.18 | 11.45 | 7 | 2.33 | 9.98 |
| 06LV21(10)_pt8 | 42.69 | 9 | 27.46 | 2.05 | | | 3.56 |
| 06LV21(14)_pt1 | 32.48 | 5.14 | 28.47 | 8.57 | 9.02 | 1.61 | 9.67 |
| 06LV21(14)_pt2 | 32.11 | 5.23 | 28.95 | 9.14 | 8.23 | 0.52 | 10.53 |
| 06LV21(14)_pt3 | 30.94 | 4.19 | 28.5 | 9.77 | 8.86 | | 11.83 |
| 06LV21(14)_pt4 | 32.99 | 5.34 | 28.89 | 7.17 | 7.1 | 0 | 12.37 |
| 06LV21(7)_pt1 | 32.37 | 4.64 | | 8.53 | 8 | 0.04 | 9.03 |
| 06LV21(7)_pt10 | 32.67 | 4.92 | | 8.88 | 6.74 | | 8.43 |
| 06LV21(7)_pt2 | 30.91 | 3.96 | | 7.93 | 8.3 | 0.6 | 8.65 |
| 06LV21(7)_pt3 | 32.7 | 4.83 | | 8.31 | 8.13 | 0.77 | 9.18 |
| 06LV21(7)_pt4 | 32.01 | 4.42 | | 8.55 | 6.52 | | 10.09 |
| 06LV21(7)_pt5 | 29.54 | 4.17 | | 8.91 | 6.55 | 1.26 | 11.31 |
| 06LV21(7)_pt6 | 31.21 | 4.27 | | 8.93 | 7.29 | 0.14 | 9.83 |
| 06LV21(7)_pt7 | 31.74 | 4.64 | | 8.33 | 7.01 | 1.3 | 10.32 |
| 06LV21(7)_pt8 | 31.6 | 4.61 | | 7.98 | 8.78 | 0 | 9.55 |
| 06LV21(7)_pt9 | 31.17 | 4.07 | | 7.63 | 8.25 | | 9.43 |
| 06LV21(8)_pt3 | 32.03 | 5.15 | 27.66 | 8.21 | | | 10.37 |
| 06LV21(8)_pt4 | 31.07 | 5.02 | 28.79 | 8.96 | | 1.38 | 11.08 |
| 53LV00(1)_pt10 | 34.19 | 5.67 | 28.24 | 6.61 | 9.18 | | 9.2 |
| 53LV00(1)_pt11 | 32.53 | 8.38 | 27.52 | 7.09 | 11.61 | | 7.23 |
| 53LV00(1)_pt7 | 34.32 | 5.23 | 24.31 | 6.73 | 6.44 | 0 | 7.51 |
| 53LV00(1)_pt8 | 31.88 | 7.79 | 24.3 | 6.07 | 6.7 | 0 | 7.76 |
| 53LV00(1)_pt9 | 32.88 | 7.36 | 27.97 | 6.3 | 9.3 | | 10.27 |
| 54LV00(2)_pt4 | 32.59 | 4.11 | 16.4 | 1.84 | 22.26 | | 9.68 |
| 54LV00(2)_pt5 | 31.53 | 7.9 | 26.5 | 6.92 | 9.29 | 6.85 | 6.2 |
| 55LV00(3)_pt10 | 31.13 | 6.32 | 29.76 | 9.32 | | | 8.94 |
| 55LV00(3)_pt2 | 33.13 | 2.15 | 27.49 | 6.84 | | 1.05 | 15.05 |
| 55LV00(3)_pt5 | 31.46 | 6.01 | 28.05 | 8.18 | 6.04 | | 8.91 |
| 55LV00(3)_pt6 | 32.42 | 3.93 | 26.64 | 6.19 | 7.03 | | 13.83 |

**Table A7.** Results of the microanalysis of the carbonate crystals.

| Sample | C | O | Mg | Ca | Fe | Nb | Ta | Sr | La | Ce |
|---|---|---|---|---|---|---|---|---|---|---|
| 13AF15(10)_pt5 | 4.92 | 35.52 | 1.33 | 50.73 | | | | | | |
| 13AF15(5)_pt11 | 3.15 | 32.66 | | 55.68 | | | | | | |
| 13AF15(5)_pt12 | 3.45 | 31.68 | | 55.76 | | | | | | |
| 13AF15(6)_pt3 | 3.31 | 31.7 | | 57.56 | | | | | | |
| 13AF15(6)_pt4 | 1.01 | 22.07 | 0.24 | 15.95 | | 5.23 | 7.87 | | | |
| 13AF15(9)_pt1 | 4.65 | 23.95 | 4.63 | 6.5 | 0.34 | | | | 12.5 | 34.89 |
| 13AF15(9)_pt2 | 4.26 | 22.55 | 4.11 | 5.23 | 5.06 | | | | 12.92 | 35.31 |
| AF-cc(1)_pt1 | 1.48 | 16.74 | | 25.86 | | | | | 15.56 | 35.84 |
| AF-cc(1)_pt2 | 2.21 | 27.56 | | 27.95 | | | | | 11.05 | 24.87 |
| AF-cc(1)_pt3 | 1.84 | 22.93 | | 29.86 | | | | | 12.76 | 27.84 |
| AF-cc(1)_pt4 | 2.12 | 21.99 | | 55.63 | | | | | | 19.46 |
| AF-cc(1)_pt5 | 2.39 | 27.19 | | 53.91 | | | | | | 15.74 |
| AF-cc(1)_pt6 | 1.45 | 21.59 | | 76.95 | | | | | | |
| AF-cc(1)_pt7 | 3.42 | 38.56 | | 58.02 | | | | | | |
| AF-cc(1)_pt8 | 3.45 | 36.33 | | 60.22 | | | | | | |
| AF-cc(1)_pt9 | 2.73 | 33.94 | | 63.33 | | | | | | |
| AF-cc(2)_pt1 | 1.89 | 20.13 | | 25.84 | | | | | 12.26 | 33.98 |
| AF-cc(2)_pt2 | 1.49 | 14.39 | | 29.52 | | | | | 13.69 | 38.01 |
| AF-cc(2)_pt3 | 2.89 | 32.74 | | 45.52 | | | | | | 16.17 |
| AF-cc(2)_pt4 | 2.18 | 26.65 | | 71.17 | | | | | | |
| AF-cc(2)_pt5 | 3.19 | 37.94 | | 58.87 | | | | | | |
| AF-cc(2)_pt6 | 2.35 | 30.49 | | 67.16 | | | | | | |
| AF-cc(2)_pt7 | 3.22 | 34.38 | | 62.4 | | | | | | |
| AF-cc(3)_pt1 | 1.42 | 15.78 | | 27.65 | | | | | 12.25 | 39.06 |
| AF-cc(3)_pt2 | 1.87 | 17.47 | | 31.98 | | | | | 12.13 | 33.1 |
| AF-cc(3)_pt3 | 2.32 | 27.49 | | 32.4 | | | | | | 32 |
| AF-cc(3)_pt4 | 1.97 | 30.04 | | 65.57 | | | | | | |
| AF-cc(4)_pt1 | 2.35 | 24.09 | | 49.75 | | | | | | 21.4 |
| AF-cc(4)_pt10 | 2.6 | 31.4 | | 65.62 | | | | | | |
| AF-cc(4)_pt11 | 3.24 | 34.63 | | 62.12 | | | | | | |
| AF-cc(4)_pt15 | 2.38 | 30.22 | | 67.41 | | | | | | |
| AF-cc(4)_pt16 | 3.44 | 37.72 | | 58.84 | | | | | | |
| AF-cc(4)_pt2 | 2 | 21.15 | | 50.13 | | | | | | 24.8 |
| AF-cc(4)_pt3 | 1.43 | 19.19 | | 19.88 | | | | | 14.01 | 37.99 |
| AF-cc(4)_pt4 | 1.44 | 13.2 | | 27.53 | | | | | 15.81 | 38.2 |
| AF-cc(4)_pt5 | 1.99 | 27.68 | | 69.73 | | | | | | |
| AF-cc(4)_pt6 | 1.53 | 18.35 | | 21.19 | | | | | 13.07 | 38.68 |
| AF-cc(4)_pt7 | 2.31 | 27.8 | | 69.88 | | | | | | |
| AF-cc(4)_pt8 | 3.48 | 38.43 | | 58.09 | | | | | | |
| AF-cc(4)_pt9 | 2.85 | 33.68 | | 63.15 | | | | | | |
| 10CH(4)_pt2 | 5.05 | 34.55 | | 3.63 | | | | 13.66 | 25.36 | 15.73 |
| 01KV(11)_pt5 | 3.51 | 38.03 | | 58.46 | | | | | | |
| 01KV(11)_pt6 | 3.99 | 35.85 | | 60.16 | | | | | | |
| 01KV(13)_pt7 | 2.62 | 27.58 | | 55.67 | | | | | | |
| 01KV(13)_pt8 | 3.89 | 36.27 | 0.98 | 58.86 | | | | | | |
| 01KV(13)_pt9 | 3.82 | 35.92 | 5.58 | 45.08 | | | | | | |
| 01KV(4)_pt8 | 3.27 | 28.51 | 0.62 | 52.79 | | | | | | |
| 01KV(4)_pt9 | 3.58 | 34.9 | 1.31 | 60.22 | | | | | | |
| 01KV(7)_pt2 | 3.49 | 36.3 | 1.15 | 59.07 | | | | | | |
| 01KV(9)_pt3 | 3.72 | 27.67 | 0.96 | 51.35 | | | | | | |
| 01KV(9)_pt5 | 2.64 | 26.32 | 0.84 | 52.45 | | | | | | |
| LV12(1)_pt12 | 10.87 | 24.58 | | 41.25 | | | | 6.37 | | |
| LV12(1)_pt7 | 3.82 | 24.99 | | 34.49 | | | | 28.86 | | |

## References

1. Arzamastsev, A.A. *Unique Paleozoic Intrusions of the Kola Peninsula*; Russian Academy of Science Publ.: Apatity, Russia, 1994; p. 79.
2. Huber, M. *Mineralogical-Petrographic Characteristic of the Selected Alkaline Massifs of the Kola Peninsula*; TMKarpinski Publisher: Suchy Las, Poland, 2015; p. 187.

3. Kogarko, L.N.; Lahaye, Y.; Brey, G.P. Plume-related mantle source of super-large rare metal deposits from the Lovozero and Khibiny massifs on the Kola Peninsula, Eastern part of Baltic Shield: Sr, Nd and Hf isotope systematics. *Miner. Petrol.* **2010**, *98*, 197–208. [CrossRef]

4. Kukharenko, A.A. *Caledonian Complex Ultrabasic Alkaline Rocks and Carbonatites of the Kola Peninsula and Northern Karelia*; Nedra: Moskov, Russia, 1965; p. 215. (In Russian)

5. Pozhilienko, V.I.; Gavrilenko, B.V.; Zhirov, C.V.; Zhabin, S.V. *Geology of Mineral Areas of the Murmansk Region*; Kola Scientific Center, Russian Academy of Sciences: Moscow, Russia, 2002; p. 360. (In Russian)

6. Fanasev, B.V. *Mineral Resources of the Alkaline—Ultrabasic Massifs of the Kola Peninsula*; Roza Vetrov: Apatity, Russia, 2011; p. 151. (In Russian)

7. Akimenko, M.I.; Kogarko, L.N.; Sorokhtina, N.V.; Kononkova, N.N.; Mamontov, V.P. A New Occurrence of Alkaline Magmatism on the Kola Peninsula: An Agpaitic Dyke in the Kandalaksha Region. *Dokl. Earth Sci.* **2014**, *458*, 1125–1128. [CrossRef]

8. Chukanov, N.V.; Pekov, I.V.; Sokolov, S.V.; Nekrasov, A.N.; Ermolaeva, V.N.; Naumova, I.S. On the Problem of the Formation and Geochemical Role of Bituminous Matter in Pegmatites of the Khibiny and Lovozero Alkaline Massifs, Kola Peninsula, Russia. *Geochem. Int.* **2006**, *44*, 715–728. [CrossRef]

9. Kogarko, L.N. *Alkaline Rocks of the Eastern of Baltic Shield (Kola Peninsula)*; Special Publications; Geological Society: London, UK, 1987; Volume 30, pp. 531–544.

10. Al Ani, T.; Sarapää, O. Mineralogical and geochemical study on carbonatites and fenites from the Kaulus drill cores, southern side of the Sokli Complex, NE Finland. *Geol. Surv. Finl.* **2013**, *145*, 1–58.

11. Kozlov, E.; Fomina, E.; Sidorov, M.; Shilovskikh, V.; Bocharov, V.; Chernyavsky, A.; Huber, M. The Petyayan-Vara Carbonatite-Hosted Rare Earth Deposit (Vuoriyarvi, NW Russia). Mineralogy and Geochemistry. *Minerals* **2020**, *10*, 73. [CrossRef]

12. Huber, M. Mineralogy and petrographic characteristic of carbonate samples from Kovdor (Kola Peninsula, N Russia). *JBES* **2015**, *5*, 49–58.

13. Kogarko, L.N.; Kononova, V.A.; Orlova, M.P.; Woolley, A.R. *Alkaline Rocks and Carbonatites of the World: Former USSR*; Chapman & Hall: London, UK, 1995; p. 22.

14. Huber, M. Petrology of the Africanda ultrabasic ingenious rocks (Kola Peninsula, N Russia). *Geo-Sci. Educ. J.* **2017**, *5*, 44–52.

15. Arzamastsev, A.; Yakovenchuk, V.; Pakhomovsky, Y.; Ivanyuk, G. The Khibiny and Lovozero alkaline massifs: Geology and unique mineralization. In Proceedings of the 33rd International Geological Congress Excursion, Oslo, Norway, 22 July–2 August 2008.

16. Kogarko, L.N.; Williams, C.T.; Woolley, A.R. Chemical evolution and petrogenetic implications of loparite in the layered, agpaitic Lovozero complex, Kola Peninsula, Russia. *Mineral. Petrogr.* **2002**, *74*, 1–24.

17. Zartman, R.E.; Kogarko, L.N. Lead isotopic evidence for interaction between plume and lower crust during emplacement of peralkaline Lovozero rocks and related rare-metal deposits, East Fennoscandia, Kola Peninsula, Russia. *Contrib. Mineral. Petrol.* **2017**, *172*, 32. [CrossRef]

18. Bayanova, T.B.; Pozhylienko, V.I.; Smolkin, V.F.; Kudryshov, N.M.; Kaulina, T.V.; Vetrin, V.R. *Katalogue of the Geochronology Data of N-E part of the Baltic Shield*; Kola Science Center of the Russian Academy of Sciences Publisher: Apatity, Russia, 2002; p. 53. (In Russian)

19. Huber, M.A. *Geology of the Kandalaksha region of Lapland Granulite Belt (Kola Peninsula, Russia)*; TMKarpinski Publisher: Suchy Las, Poland, 2014; p. 135.

20. Arzamastsev, A.A.; Glaznev, V.N. Plume-Lithosphere Interaction in the Presence of an Ancient Sublithospheric Mantle Keel: An Example from the Kola Alkaline Province. *Dokl. Earth Sci.* **2008**, *419A*, 384–387. [CrossRef]

21. Arzamastsev, A.A.; Arzamastseva, L.V. Geochemical Indicators of the Evolution of the Ultrabasic—Alkaline Series of Paleozoic Massifs of the Fennoscandian Shield. *Petrology* **2013**, *21*, 249–279. [CrossRef]

22. Kramm, U.; Kogarko, L.N.; Kononova, V.A.; Vartiainen, H. The Kola alkaline province of the CIS and Finland: Precise Rb–Sr ages define 380–360 Ma age range for all magmatism. *Lithos* **1993**, *30*, 33–44. [CrossRef]

23. Britvin, S.N.; Ivanov, G.Y.; Yakuvenchuk, V.N. *Mineralogical Accessory on the Kola Peninsula*; World of Stones: Moskov, Russia, 1995; pp. 5–6. (In Russian)

24. Mitrofanov, A.F. *Geological Characteristics of Kola Peninsula*; Russian Academy of Science Publ.: Apatity, Russia, 2000; p. 166.

25. Glebovitsky, V.A. *Early Precambrian of the Baltic Shield*; Nauka: St Petersburg, Russia, 2005; p. 710.

26. Evzerov, V.Y. Glacier Deposits As Unique Formations in the Loose Cover of the Baltic Shield. *Lithol. Miner. Resour.* **2001**, *36*, 109–115. [CrossRef]

27. Korsakova, O.P. Pleistocene marine deposits in the coastal areas of Kola Peninsula (Russia). *Quat. Int.* **2009**, *206*, 3–15. [CrossRef]

28. Dudkin, O.B.; Sandimirov, S.S. Geochemistry of Water-Rock Interaction in the Area of the Khibiny Alkaline Massif. *Geochem. Int.* **2007**, *45*, 1103–1110. [CrossRef]

29. Huber, M. Melteigite rocks from the Khibiny massif analysis using SEM-EDS and ICP. *JBES* **2015**, *5*, 82–99.

30. Huber, M.A. Microanalysis of alkaline rocks from the Khibiny massif using SEM-EDS. *JBES* **2015**, *5*, 74–81.

31. Konoplevaa, N.G.; Ivanyuka, G.Y.; Pakhomovskya, Y.A.; Yakovenchuka, V.N.; Mikhailova, Y.A. Loparite-(Ce) from the Khibiny Alkaline Pluton, Kola Peninsula. *Russia. Geol. Ore Depos.* **2017**, *59*, 729–737. [CrossRef]

32. Ogorodova, L.P.; Mel'chakova, L.V.; Vigasina, M.F.; Olysich, L.V.; Pekov, I.V. Cancrinite and Cancrisilite in the Khibiny–Lovozero Alkaline Complex: Thermochemical and Thermal Data. *Geochem. Int.* **2009**, *47*, 260–267. [CrossRef]

33. Pakhomovsky, Y.A.; Yakovenchuk, V.N.; Ivanyuk, G.Y. Kalsilite of the Khibiny and Lovozero Alkaline Plutons, Kola Peninsula. *Geol. Ore Depos.* **2009**, *51*, 822–826. [CrossRef]

34. Rastsvetaeva, R.K.; Zaitsev, V.A.; Pekov, I.V. Crystal Structure of Niobium-Rich Lomonosovite with Symmetry P1 from the Khibiny Massif (Kola Peninsula). *Crystallogr. Rep.* **2020**, *65*, 422–427. [CrossRef]
35. Sindern, S.; Zaitsev, A.N.; Deme´ny, A.; Bell, K.; Chakmouradian, A.R.; Kramm, U.; Moutte, J.; Rukhlov, A.S. Mineralogy and geochemistry of silicate dyke rocks associated with carbonatites from the Khibiny complex (Kola, Russia)—Isotope constraints on the genesis and small-scale mantle sources. *Mineral. Petrol.* **2004**, *80*, 215–239. [CrossRef]
36. Yakovleva, O.S.; Pekov, I.V.; Bryzgalov, I.A. Chromium Mineralization in the Khibiny Alkaline Massif (Kola Peninsula, Russia). *Mosc. Univ. Geol. Bull.* **2009**, *64*, 220–229. [CrossRef]
37. Chukanov, N.V.; Pekov, I.V.; Olysych, L.V.; Massa, W.; Yakubovich, O.V.; Zadov, A.E.; Rastsvetaeva, R.K.; Vigasina, M.F. Kyanoxalite, a New Cancrinite Group Mineral Species with Extraframework Oxalate Anion from the Lovozero Alkaline Pluton, Kola Peninsula. *Geol. Ore Depos.* **2010**, *52*, 778–790. [CrossRef]
38. Ermolaeva, V.N.; Mikhailova, A.V.; Kogarko, L.N.; Kolesov, G.M. Leaching Rare-Earth and Radioactive Elements from Alkaline Rocks of the Lovozero Massif, Kola Peninsula. *Geochem. Int.* **2016**, *54*, 633–639. [CrossRef]
39. Boruckiy, B.E. *Rock-Forming Minerals of the High-Alkaline Complexes*; Nauka: Moscov, Russia, 1989; p. 214. (In Russian)
40. Ermolaeva, V.N.; Pekov, I.V.; Chukanov, N.V.; Zadov, A.E. Thorium Mineralization in Hyperalkaline Pegmatites and Hydrothermalites of the Lovozero Pluton, Kola Peninsula. *Geol. Ore Depos.* **2007**, *49*, 758–775. [CrossRef]
41. Evseev, A. Minerals of the Lovoziero Massif. 2014. Available online: http://geo.web.ru/druza/l-Lovozero.htm (accessed on 12 December 2021).
42. Huber, M. *Microanalysis of Alkaline Rocks from the Lovoziero Massif Using SEM-EDS Methods*; TMKarpinski Publisher: Suchy Las, Poland, 2015; p. 125.
43. Kotel'nikov, A.R.; Ogorodova, L.P.; Mel'chakova, L.V.; Vigasina, M.F. Ussingite from the Lovozero Alkaline Massif: Calorimetric, Thermal, and IR Spectroscopic Study. *Geochem. Int.* **2010**, *48*, 183–186. [CrossRef]
44. Veselovskiy, R.V.; Arzamastsev, A.A.; Demina, L.I.; Travin, A.V.; Botsyun, S.B. Paleomagnetism, geochronology, and magnetic mineralogy of Devonian dikes from the Kola alkaline province (NE Fennoscandian Shield). *Izv. Phys. Solid Earth* **2013**, *49*, 526–547. [CrossRef]
45. Huber, M. *Ultrabasic-Alkaline Intrusion in Afrikanda (N Russia), Petrological-Geochemistry Study*; Sciences Publisher: Lublin, Poland, 2017; p. 87.
46. Potter, N.J.; Kamenetsky, V.S.; Chakhmouradian, A.R.; Kamenetsky, M.B.; Goemann, K.; Rodemann, T. Polymineralic inclusions in oxide minerals of the Afrikanda alkaline-ultramafic complex: Implications for the evolution of perovskite mineralization. *Contrib. Mineral. Petrol.* **2020**, *175*, 18. [CrossRef]
47. Potter, N.J.; Ferguson, M.R.M.; Kamenetsky, V.S.; Chakhmouradian, A.R.; Sharygin, V.V.; Thompson, J.M.; Goemann, K. Textural evolution of perovskite in the Afrikanda alkaline–ultramafic complex, Kola Peninsula, Russia. *Contrib. Mineral. Petrol.* **2018**, *173*, 100. [CrossRef]
48. Reguir, E.P.; Camacho, A.; Yang, P.; Chakhmouradian, A.R.; Kamenetsky, V.S.; Halden, N.M. Trace-element study and uranium-lead dating of perovskite from the Afrikanda plutonic complex, Kola Peninsula (Russia) using LA-ICP-MS. *Miner. Petrol.* **2010**, *100*, 95 103. [CrossRef]
49. Huber, M.; Hałas, S.; Mokrushin, A.V.; Neradovsky, Y.N.; Lata, L.; Sikorska-Jaworowska, M.; Skupiński, S. Chemical composition of Ca-Mg-Sr carbonates and the stable isotope $\delta 13C$ study, the Kovdor massif showcase (Kola Region, NW Russia). *Pr. Fersmanowskiej Ses. Nauk. Inst. Geol. Kolsk. Cent. Nauk. Ros. Akad. Nauk.* **2018**, *15*, 383–385. [CrossRef]
50. Kalashnikov, A.O.; Ivanyuk, G.Y.; Mikhailova, J.A.; Sokharev, V.A. Approach of automatic 3D, geological mapping: The case of, the Kovdor phoscorite-carbonatite complex, NW Russia. *Sci. Rep.* **2017**, *7*, 6893. [CrossRef]
51. Nivin, V.A.; Treloar, P.J.; Konopleva, N.G.; Ikorskya, S.V. A review of the occurrence, form and origin of C-bearing species in the Khibiny Alkaline Igneous Complex, Kola Peninsula, NW Russia. *Lithos* **2005**, *85*, 93–112. [CrossRef]
52. Lapin, A.V.; Lyagushkin, A.P. The Kovdor Apatite–Francolite Deposit as a Prospective Source of Phosphate Ore. *Geol. Ore Depos.* **2014**, *56*, 61–80. [CrossRef]
53. Rass, I.T.; Petrenko, D.B.; Koval'chuk, E.V.; Yakushev, A.I. Phoscorites and Carbonatites: Relations, Possible Petrogenetic Processes, and Parental Magma, with Reference to the Kovdor Massif, Kola Peninsula. *Geochem. Int.* **2020**, *58*, 753–778. [CrossRef]
54. Sorokhtin, N.V.; Zaitsev, V.A.; Petrov, S.V.; Kononkovaa, N.N. Estimation of Formation Temperature of the Noble Metal Mineralization of the Kovdor Alkaline-Ultrabasic Massif (Kola Peninsula). *Geochem. Int.* **2021**, *59*, 474–490. [CrossRef]
55. Arzamastsev, A.A.; Wu, F.Y. U-Pb Geochronology and Sr-Nd Isotopic Systematics of Minerals from the Ultrabasic–Alkaline Massifs of the Kola Province. *Petrology* **2014**, *22*, 462–479. [CrossRef]
56. Rigaku. *CRYSALIS Software System*; Rigaku: Oxford, UK, 2016.
57. Sheldrick, G.M. A short history of SHELX. *Acta Crystallogr.* **2008**, *A64*, 112–122. [CrossRef]
58. Sheldrick, G.M. SHELXT—Integrated space-group and crystal-structure determination. *Acta Crystallogr.* **2015**, *A71*, 3–8. [CrossRef]
59. Sheldrick, G.M. Crystal structure refinement with SHELXL. *Acta Crystallogr.* **2015**, *C71*, 3–8.
60. Dolomanov, O.V.; Bourhis, L.J.; Gildea, R.J.; Howard, J.A.K.; Puschmann, H.J. A Complete Structure Solution, Refinement and Analysis Program. *Appl. Cryst.* **2009**, *42*, 339–341. [CrossRef]
61. Perry, C.H.; Khan, D.B.N. Infrared Studies of Perovskite Titanates. *Phys. Rev.* **1964**, *135*, A408–A412. [CrossRef]
62. Zhang, M.; Salje, E.K.H.; Bismayer, U.; Groat, L.A.; Malcherek, T. Metamictization and recrystallization of titanite: An infrared spectroscopic study. *Am. Mineral.* **2002**, *87*, 882–890. [CrossRef]

63. Gunasekaran, S.; Anbalagan, G.; Pandi, S. Raman and infrared spectra of carbonates of calcite structure. *J. Raman Spectrosc.* **2006**, *37*, 892–899. [CrossRef]

64. Rastsvetaeva, R.K.; Chukanov, N.V.; Pekov, I.V.; Schäfer, C.; Van, K.V. New Data on the Isomorphism in Eudialyte-Group Minerals. 1. Crystal Chemistry of Eudialyte-Group Members with Na Incorporated into the Framework as a Marker of Hyperagpaitic Conditions. *Minerals* **2020**, *10*, 587. [CrossRef]

65. Fleet, M.E. Infrared spectra of carbonate apatites: n2-Region bands. *Biomaterials* **2009**, *30*, 1473–1481. [CrossRef]

66. Sundberg, M.R.; Lehtinen, M.; Kivekäs, R. Refinement of the crystal structure of ramsayite (lorenzenite). *Am. Mineral.* **1987**, *72*, 173–177.

67. Mitchell, R.H.; Burns, P.C.; Chakhmouradian, A.R. The crystal structures of loparite-(Ce). *Can. Mineral.* **2000**, *38*, 145–152. [CrossRef]

68. Ali, R.; Yashima, M. Space group and crystal structure of the Perovskite $CaTiO_3$ from 296 to 1720 K. *J. Solid State Chem.* **2005**, *178*, 2867–2872. [CrossRef]

69. Speer, J.A.; Gibbs, G.V. The crystal structure of synthetic titanite, $CaTiOSiO_4$, and the domain textures of natural titanites. *Am. Mineral.* **1976**, *61*, 238–247.

70. Smyth, J.R.; Arhens, T.J. The crystal structure of calcite III. *Geophys. Res. Lett.* **1997**, *24*, 1595–1598. [CrossRef]

71. Golyshev, V.M.; Simonov, V.I.; Belov, N.V. Crystal structure of eudialyte. *Sov. Phys. Crystallography.* **1971**, *16*, 70–74.

72. Ivanova, T.I.; Frank-Kamenetskaya, O.V.; Kol'tsov, A.B.; Ugolkov, V.L. Crystal Structure of Calcium-Deficient Carbonated Hydroxyapatite. Thermal Decomposition. *J. Solid State Chem.* **2001**, *2*, 340–349. [CrossRef]

73. Arzamastsev, A.A.; Arzamastseva, L.V.; Zhirova, A.M.; Glaznev, V.N. Model of Formation of the Khibiny–Lovozero Ore-Bearingvolcanic-Plutonic Complex. *Geol. Ore Depos.* **2013**, *55*, 341–356. [CrossRef]

74. Huber, M. Preliminary characteristic of the rocks—Veins from the Malaya Belaya Valley in Khibiny. *J. Biol. Earth Sci.* **2013**, *3*, 1–11. (In Polish)

75. Kogarko, L.N. Geochemistry of Fractionation of Coherent Elements (Zr and Hf) during the Profound Differentiation of Peralkaline Magmatic Systems: A Case Study of the Lovozero Complex. *Geochem. Int.* **2016**, *54*, 1–6. [CrossRef]

76. Huber, M. LA-ICP MS Characteristic of geochemical composition of veins rocks from the center of the Khibiny Massive in the Kola Peninsula, Northern Russia. *J. Chem. Eng.* **2014**, *8*, 364–370.

77. Kogarko, L.N. Chemical Composition and Petrogenetic Implications of Apatite in the Khibiny Apatite-Nepheline Deposits (Kola Peninsula). *Minerals* **2018**, *8*, 532. [CrossRef]

78. Arzamastsev, A.A.; Arzamastseva, L.V.; Zaraiskii, G.P. Contact Interaction of Agpaitic Magmas with Basement Gneisses: An Example of the Khibiny and Lovozero Massifs. *Petrology* **2011**, *19*, 109–133. [CrossRef]

79. Lebedev, V.N.; Rudenko, A.V. Processing of Rare-Earth Elements from Loparite. *Radiochemistry* **2005**, *47*, 80–85. [CrossRef]

80. Maiorov, V.G.; Nikolaev, A.I.; Kopkov, V.K. Processing of Precipitates of Hydrated Tantalum(V), Niobium(V), and Titanium(IV) Oxides in Loparite Technology. *Russ. J. Appl. Chem.* **2003**, *76*, 1720–1723. [CrossRef]

81. Mitchell, R.H.; Chakhmouradian, A.R.; Woodward, P.M. Crystal chemistry of perovskite-type compounds in the tausonite-loparite series, $(Sr_{1-2x}Na_xLa_x)TiO_3$. *Phys. Chem. Miner.* **2000**, *27*, 583–589. [CrossRef]

82. Zubkova, N.V.; Arakcheeva, A.V.; Pushcharovskii, D.Y.; Semenov, E.I.; Atencio, D. Crystal structure of Loparite. *Crystallogr. Rep.* **2000**, *45*, 210–214. [CrossRef]

83. Pakhomovsky, Y.A.; Ivanyuk, G.Y.; Yakovenchuk, V.N. Loparite[1](Ce) in rocks of the Lovozero Layered Complex at Mt. Karnasurt and Mt. Kedykvyrpakhk. *Geol. Ore Depos.* **2014**, *56*, 685–698. [CrossRef]

84. Semenov, E.I.; Kochemasov, G.G.; Bykova, A.V. Zirkelite and Rozenbushite from Metasomatic Rocks of Contact in the Lovozero Massif. *Tr. IMGRE* **1963**, *15*, 106–109.

85. Dekusar, V.M.; Kolesnikova, M.S.; Nikolaev, A.I.; Maiorov, V.G.; Zilberman, B.Y. Mineral reserves of naturally radioactive thorium-bearing raw materials. *At. Energy* **2012**, *111*, 185–194. [CrossRef]

86. Krivovichev, V.G.; Charykova, M.V. Mineral systems, their types, and distribution in nature. *I. Khibiny Lovozero Mont Saint-Hilaire Geol. Ore Depos.* **2016**, *85*, 551–558. [CrossRef]

87. Ghobadia, M.; Gerdesa, A.; Kogarkoc, L.N.; Breya, G. New Data on the Composition and Hafnium Isotopes of Zircons from Carbonatites of the Khibiny Massif. *Dokl. Earth Sci.* **2012**, *446*, 1083–1085. [CrossRef]

88. Nivin, A.V. Molecular–Mass Distribution of Saturated Hydrocarbons in Gas of the Lovozerskii Nepheline–Syenite Massif. *Dokl. Earth Sci.* **2009**, *429A*, 1580–1582. [CrossRef]

89. Zaitsev, A.; Bell, K. Sr and Nd isotope data of apatite, calcite, and dolomite as indicators of source, and the relationships of phoscorites and carbonatites from the Kovdor massif, Kola peninsula, Russia. *Contrib. Miner. Pet.* **1995**, *121*, 324–335. [CrossRef]

90. Kogarko, L.N. Conditions of Accumulation of Radioactive Metals in the Process of Differentiation of Ultrabasic Alkaline-Carbonatite Rock Associations. *Geol. Ore Depos.* **2014**, *56*, 230–238. [CrossRef]

91. Jones, A.P.; Genge, M.; Carmody, L. Carbonate Melts and Carbonatite. *Rev. Mineral. Geochem.* **2013**, *75*, 289–322. [CrossRef]

92. Kogarko, L.N. Zirconium and Hafnium Fractionation in Differentiation of Alkali Carbonatite Magmatic Systems. *Geol. Ore Depos.* **2016**, *58*, 173–181. [CrossRef]

93. Nivin, V.A. Variations in the Composition and Origin of Hydrocarbon Gases from Inclusions in Minerals of the Khibiny and Lovozero Plutons, Kola Peninsula, Russia. *Geol. Ore Depos.* **2011**, *53*, 699–707. [CrossRef]

94. Lebedev, V.N.; Shchur, T.E.; Maiorov, D.V.; Popova, L.A.; Serkova, R.P. Specific Features of Acid Decomposition of Eudialyte and Certain Rare-Metal Concentrates from the Kola Peninsula. *Russ. J. Appl. Chem.* **2003**, *76*, 1191–1196. [CrossRef]
95. Ivanyuk, G.Y.; Pakhomovsky, Y.A.; Yakovenchuk, V.N. Eudialyte Group Minerals in Rocks of Lovozero Layered Complex at Mt. Karnasurt and Mt. Kedykvyrpakhk. *Geol. Ore Depos.* **2015**, *57*, 600–613. [CrossRef]
96. Zakharov, V.I.; Maiorov, D.V.; Alishkin, A.R.; Matveev, V.A. Causes of Insufficient Recovery of Zirconium during Acidic Processing of Lovozero Eudialyte Concentrate. *Russ. J. Non-Ferr. Met.* **2011**, *52*, 423–428. [CrossRef]
97. Kogarko, L.N. Geochemistry of Rare Earth Metals in the Ore Eudialyte Complex of the Lovozero Rare Earth Deposit. *Dokl. Earth Sci.* **2020**, *491*, 231–234. [CrossRef]
98. Arzamastsev, A.A.; Mitrofanov, F.P. Paleozoic Plume-Lithospheric Processes in Northeastern Fennoscandia: Evaluation of the Composition of the Parental Mantle Melts and Magma Generation Conditions. *Petrology* **2009**, *17*, 300–313. [CrossRef]
99. Terekhov, E.N.; Baluev, A.S.; Przhiyalgovsky, E.S. Structural Setting and Geochemistry of Devonian Dikes in the Kola Peninsula. *Geotectonics* **2012**, *46*, 69–84. [CrossRef]
100. Bogatyreva, E.V.; Ermilov, A.G.; Khokhlova, O.V. X-ray Crystal Analysis to Forecast Efficiency of Mechanical Pre-Activation of Loparite Concentrate. *J. Min. Sci.* **2013**, *49*, 664–669. [CrossRef]
101. Kozlov, E.N.; Arzamastsev, A.A. Petrogenesis of metasomatic rocks in the fenitized zones of the Ozernaya Varaka alkaline ultrabasic complex, Kola Peninsula. *Petrology* **2015**, *23*, 45–67. [CrossRef]
102. Kogarko, L.N. A New Geochemical Criterion for Rare-Metal Mineralization of High-Alkalic Magmas (Lovozero Deposit, Kola Peninsula). *Dokl. Earth Sci.* **2019**, *487*, 922–924. [CrossRef]
103. Arzamastsev, A.A.; Fedotov, Z.A.; Arzamastseva, L.V. *Dyke Magmatism of the N-E of Baltic Shield*; Nauka: Sankt Petersburg, Russia, 2009; p. 383.
104. Ivanyuk, G.Y.; Pakhomovsky, Y.A.; Konopleva, N.G.; Kalashnikov, A.O.; Korchak, Y.A.; Selivanova, E.A.; Yakovenchuk, V.N. Rock-Forming Feldspars of the Khibiny Alkaline Pluton, Kola Peninsula, Russia. *Geol. Ore Depos.* **2010**, *52*, 736–747. [CrossRef]
105. Arzamastsev, A.A.; Arzamastsev, L.V.; Travin, A.V.; Belyatsky, B.V.; Shamatrina, A.M.; Antonov, A.V.; Larionov, A.N.; Rodionov, N.V.; Sergeev, S.A. Duration of Formation of Magmatic System of Polyphase Paleozoic Alkaline Complexes of the Central Kola: U-Pb, Rb-Sr, Ar-Ar Data. *Dokl. Earth Sci.* **2006**, *413A*, 432–436. [CrossRef]
106. Arzamastsev, A.A.; Fedotova, Z.A.; Arzamastseva, L.V.; Travin, A.V. Paleozoic Tholeiite Magmatism in the Kola Igneous Province: Spatial Distribution, Age, Relations with Alkaline Magmatism. *Dokl. Earth Sci.* **2010**, *430*, 205–209. [CrossRef]
107. Huber, M.; Hałas, S.; Sikorska, M. Evolution of prehnite-albite-calcite veins in Metamorphic rocks from the Lapland Granulite Belt (Kandalaksha region of Kola Peninsula). *Geologija* **2007**, *57*, 1–7.
108. Hałas, S.; Pieńkos, T. *Simultaneous Determination of* $\delta^{34}S$ *and* $\delta^{36}S$ *on* $SO_2$ *Gas*; XIV Workshop of the European Society for Isotope Research: Balle Govora, Romania, 2017; pp. 25–29.
109. Farquhar, J.; Bao, H.; Thiemens, M. Atmospheric influence of Earth's earliest sulfur cycle. *Science* **2000**, *289*, 756–758. [CrossRef]
110. Farquhar, J.; Wing, B.A. Multiple sulfur isotopes and the evolution of the atmosphere. *Earth Planet. Sci. Lett.* **2003**, *213*, 1–13. [CrossRef]
111. Huber, M.; Hałas, S.; Serov, P.A.; Ekimova, N.A.; Bayanova, T.B. Stable isotope geochemistry and Sm-Nd, U-Pb dating of sulfides from layered intrusions in the northern part of Baltic Shield. *Cent. Eur. Geol.* **2013**, *56*, 134–135.
112. Downes, H.; Balaganskaya, E.; Beard, A.; Liferovich, R.; Demaiffe, D. Petrogenetic processes in the ultramafic, alkaline and carbonatitic magmatism in the Kola Alkaline Province: A review. *Lithos* **2005**, *85*, 48–75. [CrossRef]
113. Nivin, V.A. Helium and Argon Isotopes in Rocks and Minerals of the Lovozero Alkaline Massif. *Geochem. Int.* **2008**, *46*, 482–502. [CrossRef]
114. Kogarko, L.N. Fractionation of Zirconium and Hafnium during Evolution of a Highly Alkaline Magmatic System, Lovozero Massif, Kola Peninsula. *Dokl. Earth Sci.* **2015**, *463*, 792–794. [CrossRef]
115. Suk, N.I.; Kotel'nikov, A.R.; Koval'skii, A.M. Mineral Thermometry and the Composition of Fluids of the Sodalite Syenites of the Lovozero Alkaline Massif. *Petrology* **2007**, *15*, 441–458. [CrossRef]
116. Arzamastsev, A.A.; Arzamastseva, L.V.; Bea, F.; Montero, P. Trace Elements in Minerals as Indicators of the Evolution of Alkaline Ultrabasic Dike Series: LA-ICP-MS Data for the Magmatic Provinces of Northeastern Fennoscandia and Germany. *Petrology* **2009**, *17*, 46–72. [CrossRef]

*Article*

# A Synthetic Analog of the Mineral Ivanyukite: Sorption Behavior to Lead Cations

Gleb O. Samburov [1,*], Galina O. Kalashnikova [2], Taras L. Panikorovskii [1], Vladimir N. Bocharov [3], Aleksandr Kasikov [4], Ekaterina Selivanova [2,5], Ayya V. Bazai [2,5], Daria Bernadskaya [6], Viktor N. Yakovenchuk [2,4] and Sergey V. Krivovichev [2]

[1] Laboratory of Nature-Inspired Technologies and Environmental Safety of the Arctic, Kola Science Centre, Russian Academy of Sciences, 184209 Apatity, Russia; t.panikorovskii@ksc.ru

[2] Nanomaterials Research Centre, Kola Science Centre, Russian Academy of Sciences, 184209 Apatity, Russia; g.kalashnikova@ksc.ru (G.O.K.); selivanova@geoksc.apatity.ru (E.S.); bazai@geoksc.apatity.ru (A.V.B.); yakovenchuk@geoksc.apatity.ru (V.N.Y.); s.krivovichev@ksc.ru (S.V.K.)

[3] Geo Environmental Centre "Geomodel", Saint-Petersburg State University, 198504 St. Petersburg, Russia; bocharov@molsp.phys.spbu.ru

[4] I.V. Tananaev Institute of Chemistry and Technology of Rare Elements and Mineral Raw Materials, Kola Science Centre, Russian Academy of Sciences, 184209 Apatity, Russia; a.kasikov@ksc.ru

[5] Geological Institute, Kola Science Centre, Russian Academy of Sciences, 184209 Apatity, Russia

[6] Institute of the North Industrial Ecology Problems, Kola Science Centre, 184209 Apatity, Russia; d.bernadskaya@ksc.ru

* Correspondence: g.samburov@ksc.ru; Tel.: +7-921-030-88-75

**Citation:** Samburov, G.O.; Kalashnikova, G.O.; Panikorovskii, T.L.; Bocharov, V.N.; Kasikov, A.; Selivanova, E.; Bazai, A.V.; Bernadskaya, D.; Yakovenchuk, V.N.; Krivovichev, S.V. A Synthetic Analog of the Mineral Ivanyukite: Sorption Behavior to Lead Cations. *Crystals* **2022**, *12*, 311. https://doi.org/10.3390/cryst12030311

Academic Editor: Francesco Capitelli

Received: 31 January 2022
Accepted: 21 February 2022
Published: 23 February 2022

**Publisher's Note:** MDPI stays neutral with regard to jurisdictional claims in published maps and institutional affiliations.

**Abstract:** The production of electrolytic nickel includes the stage of leaching of captured firing nickel matte dust. The solutions formed during this process contain considerable amounts of Pb, which is difficult to extraction due to its low concentration upon the high-salt background. The sorption of lead from model solutions with various compositions by synthetic and natural titanosilicate sorbents (synthetic ivanyukite-Na-$T$ (SIV), ivanyukite-Na-$T$, and AM-4) have been investigated. The maximal sorption capacity of Pb is up to 400 mg/g and was demonstrated by synthetic ivanyukite In solutions with the high content of $Cl^-$ (20 g/L), extraction was observed only with a high amount of Na (150 g/L). Molecular mechanisms and kinetics of lead incorporation into ivanyukite were studied by the combination of single-crystal and powder X-ray diffraction, microprobe analysis, and Raman spectroscopy. Incorporation of lead into natural ivanyukite-Na-$T$ with the $R3m$ symmetry by the substitution $2Na^+ + 2O^{2-} \leftrightarrow Pb^{2+} + \square + 2OH^-$ leds to its transformation into the cubic $P-43m$ Pb-exchanged form with the empirical formulae $Pb_{1.26}[Ti_4O_{2.52}(OH)_{1.48}(SiO_4)_3]\cdot3.32(H_2O)$.

**Keywords:** ivanyukite; lintisite; SIV; AM-4; synthesis; sorption; lead; ion-exchange; titanosilicate; Arctic

## 1. Introduction

Minerals of the ivanyukite group were discovered in 2009 by Yakovenchuk and co-authors in a pegmatite vein of the Koashva apatite mine, Khibiny Massif, Kola Peninsula, Russia [1]. Ivanyukite group minerals have been found in several variaties including ivanyukite-Na-$T$ ($T$—rigonal form), ivanyukite-Na-$C$ ($C$—cubic form), ivanyukite—K, and ivanyukite—Cu. The observed chemical diversity was assigned to the high ion-exchange and sorption capacities of the minerals.

In contrast to othertitanosilicates discovered in the Khibiny Massif or Lovozero Massif (Kola Peninsula, Russia)—such as lintisite, zorite and sitinakite [2–7]—the synthetic of ivanyukite was was known prior to its mineralogical discovery, being obtained by D. Chapman and A. Roe in 1990 [8] under the name name 'grace titanium silicate' (GTS). For their synthesis process, the authors had used $Ti(OC_2H_5)_4$ as a source of Ti. The synthetic titanosilicate AM-4 was first obtained by M.S. Dadachov in 1997 using $TiCl_3$ as a starting reagent [9].

In this work, the synthesis of ivanyukite and AM-4 was done using different Ti sources. For example, semi-product (TiCl$_4$) of hydrochloric acid loparite ore treatment (JSC "Solikamsk Magnesium Works", Solikamsk, Russia) [10] and a semi-product (ammonium titanylsulfate (NH$_4$)$_2$TiO(SO$_4$)$_2$·H$_2$O) of sulfuric acid titanite ore treatment (mining JSC "Apatit", PhosAgro, Russia) [11,12] were used extensively, providing a link between industrially available compounds and a new material with potential innovative applications.

There are human health risks associated with the lead exposure and are important for industrial areas and surrounding cities. For example, in the Monchegorsk city area (Kola peninsula) the nickel and copper pollution is by 6–1500 times higher than the European background levels [13]. Analysis of trace metals by atomic adsorption spectroscopy [14] showed that the highly Pb-polluted waters of the Kola Peninsula reached 425-times the background concentration (8 μg/L) in areas of anthropogenic impact compared to the background level of the Murmansk region [14]. The concentration of Pb is more than 19-times higher than the reference recommended values of maximum tolerable concentrations of this element in soils around copper–nickel metallurgical smelters [15–17].

Since this work was carried out taking into account the principle of integrated nature management, one of the subjects of research was both models and real samples of solutions of the nickel–copper plant JSC "Severonickel KMM" (Monchegorsk, Russia). These solutions are characterized by a low concentrations of Ag and Pb cations (0.3 g/L and up to 1 g/L, respectively) with the high concentrations of NaCl or NaSO$_4$ (up to 200 g/L).

For the purification of silver-containing solutions, a new technique has been developed and a patent application RU 2021124566A has been filed [16]. In this work, we propose new conditions (molar ratio, temperature, and volume) for the synthesis of the AM-4 and SIV titanosilicates and test their sorption properties with respect to Pb for different solutions. The molecular mechanism and kinetics of Pb incorporation into ivanyukite structure has been studied here in detail.

## 2. Materials and Methods

### 2.1. Materials and Their Abbreviations

SIV—is a synthetic titanosilicate material with the composition Na$_4$(TiO)$_4$(SiO$_4$)$_3$·nH$_2$O that possesses a microporous crystal structure similar to that observed in ivanyukite-group minerals [1].

AM-4—is a synthetic analog of the mineral lintisite that is used in this work for comparison of its sorption properties with those of SIV was obtained according to the methodology described in detail in [3,18,19].

SL3—is the protonated form of AM-4 obtained according to the methods described in [3,18].

STA—is ammonium sulfate oxytitanium, (NH$_4$)$_2$TiO(SO$_4$)$_2$·H$_2$O, the pre-product of titanite concentrate reprocessing (PJSC "PhosAgro", Apatity, Russia) [20].

### 2.2. Reagents

Titanium tetrachloride (JSC "Solikamsk Magnesium Works", Solikamsk, Russia) and (NH$_4$)$_2$TiO(SO$_4$)$_2$·H$_2$O (PJSC PhosAgro, Apatity, Russia) were of technical grade quality; sodium hydroxide (Aldrich, Moscow, Russia) was of ACS grade quality; sodium metasilicate, lead nitrate, and sodium chloride produced by the Neva Reactive were of U.S.P. grade quality. The solutions 1–7 were model and real solutions after processing (leaching) of fine dusts of nickel production generated by the pyrometallurgical processes at metallurgical plants and captured by electrostatic precipitators during dry cleaning of dust and gas phases.

### 2.3. Synthesis

Sodium metasilicate and titanium tetrachloride (or ammonium sulfate oxytitanium) were taking according to they molar ratios of Na$_2$O:SiO$_2$:TiO$_2$:H$_2$O = 4.5:4:1:160 [9,16] and 5.6:3.1:1:128, for the synthesis of SIV and of AM-4, respectively. The mixture of initial reagents was dissolved in a distilled water at 298 K and stirred with the mixing speed of

200 revolutions per minute during 4 h to obtain SIV powders. The autoclaves with the mixtures were maintained under hydrothermal conditions (453 K, 1.5 MPa) for 96 h (in the case of $TiCl_4$) and 48 h (in the case of ammonium sulfate oxytitanium). The product were filtered by vacuum flask, washed by distilled water (1/5 of the total mixture volume), and dried at 348 K.

Autoclaves with the volumes of 40 $cm^3$ and 450 $cm^3$ (production of FRC Kola Science Centre, Apatity, Russia) were used for the hydrothermal synthesis of the SIV. The autoclave with volume of 7 L (Parr Instrument Company, Moline, IL, USA) was used for SIV synthesis only. TheSIV powders were firstly obtained with the initial molar ratio of $Na_2O:SiO_2:TiO_2:H_2O$ = 5.6:3.1:1:128 under 373 K.

The high precision universal drying oven SNOL (SNOL-TERM, Utena, Lithuania), with the operating temperature range from 323 to 573 K and programmable thermostat, was used for heating autoclaves and for power drying.

The synthesized powders were separated from the mother liquor by vacuum filtration using a diaphragm pump (Technology for Vacuum system, Vacuubrand, Wertheim, Germany).

In all cases where mixing was necessary, a magnetic stirrer IKA RT 5 (Germany) was used.

The drying process of the powder was carried out for 4 h at 348 K using the SNOL drying oven (SNOL-TERM, Utena, Lithuania).

Sartorius—ED224S-RCE (Sartorius, Goettingen, Germany) was the first class accuracy analytical balance used for weighing synthetic titanosilicate powders and reagents.

*2.4. Composition*

Investigation of morphology of the Pb-exchanged ivanyukite samples was carried out using a scanning electron microscope LEO-1450 (Carl Zeiss Microscopy, Oberkochen, Germany) and chemical composition was studied with an Oxford Instruments Ultim Max 100 analyzer at 20 kV, 500–1000 pA, 1–3 μm beam diameter (Geological Institute of Kola Science Centre, Apatity, Russia).

The content of Pb in the samples was determined by the inductively coupled plasma atomic emission spectrometry (ICP AES, Institute of the North Industrial Ecology Problems of Kola Science Centre, Apatity, Russia). To construct the calibration curve, a multi-element standard sample of $Pb^{2+}$ ions GSO 7877-2000 was used. The samples were diluted with 2% nitric acid in a ratio of 1:100. The 2% nitric acid was obtained by mixing purified water—18.2 M$\Omega$ × cm (Ultra Clear TP UV UF TM, EVOQUA, Pittsburgh, PA, USA) with distilled nitric acid obtained by an acid distillation unit (SPK-2, Saint-Petersburg, Russia).

The obtained solutions were analyzed by an "Optima 2100 DV" (PerkinElmer, Waltham, MA, USA) an inductively coupled plasma atomic emission spectrometer with the following settings: wavelength 220.353 nm; axial torch viewing position; nebulizer gas flow 0.8 L/min; flow rate 1.5 mL/min; power 1300 Watts (Institute of the North Industrial Ecology Problems of Kola Science Centre, Apatity, Russia). The method of standard addition was used to estimate the validity of the analysis.

*2.5. Pb Sorption from the Simulated Model Solution*

The possibility of extracting Pb from its solutions by employing titanosilicates was studied using nitrate salt. For this purpose, the $Pb(NO_3)_2$ solutions containing 0.39, 0.88, and 1.79 g/L of Pb (initial solutions 1–3) were prepared by dissolving in 1000 $cm^3$ of distilled water 0.62, 1.4, and 2.87 g of $Pb(NO_3)_2$ respectively. The sorption experiments were conducted by the immersion of titanosilicate powders into solutions at 298 K for 4 h with a constant stirring. The ratio of liquid and solid phases was V:m = 50:0.5 (mL:g) in each experiment. After sorption, the powders were filtered by vacuum filtration, washed with distilled water in three steps (50 mL for each step), and dried at 348 K. For each experiment, a parallel control experiment was conducted.

*2.6. Pb Sorption from the Dust Leaching Solution (Model Solution)*

Before starting work with a real solution of leaching of nickel Feinstein firing dusts, an experiment was conducted on a model solution, simulating the composition of industrial solutions. In order to prepare the latter, 60 mL of NaCl solution with the concentration 333 g/L (20 g NaCl dissolved in 60 mL of distilled water) and 20 mL of $Pb(NO_3)_2$ solution with a concentration 6.5 g/L (0.13 g $Pb(NO_3)_2$ dissolved in 20 mL of distilled water) were added with a small portion of 20 mL $AgNO_3$ solution at a concentration of 0.8 g/L (0.016 g $AgNO_3$ dissolved in 20 mL of distilled water); the resulting solution was diluted in 400 mL of $Na_2SO_4$ solution with a concentration of 95 g/L. The experimental conditions agreed well with the conditions described for the experiments with a model solution $Pb(NO_3)_2$ with a difference in a liquid–solid ratio, which in this case was equal to V:m = 25:0.25 ($cm^3$:g). The SL3 sorbent was not used in the experiments with the first model solutions because of its low efficiency in recent studies. For each experiment, a parallel control experiment was conducted.

*2.7. Pb Sorption from the Dust Leaching Solution (Real Solution)*

Two different solutions of leaching of fine dusts of nickel production firing dusts from copper-nickel plant JSC "Severonickel KMM" were used in this study as provided by the industrial plant laboratory. The first one, solution-6, was obtained by the treatment the fine-grained dust of nickel production with a solution of sodium chloride with the concentration of 200 g/L and precipitation of Pb with a solution of sodium sulphate with the concentration 200 g/L. The second one, solution-7, was a product of the dusts of nickel production treatment similar to solution-6, except for that leaching was carried out with a solution of sulfuric acid with the concentration of 200 g/L. The conditions of the sorption experiments were same as those implemented for model solutions.

For each experiment, a second parallel experiment was conducted for control purposes.

The cation-exchange capacity of the sorbent (mg/g) in each experiment was calculated by Equation (1)

$$q = (C_0 - C_e)\cdot(V/m) \tag{1}$$

where $C_0$—is the initial concentration of the element in the solution (mg/L); $C_e$—is the equilibrium concentration of the element in the solution after sorption (mg/L); V—is the volume of solution (L); m—is the mass of the sorbent (g).

The extent of extraction of the element (%) was calculated by Equation (2)

$$A = \frac{C_0 - C_e}{C_0}\cdot 100 \tag{2}$$

where $C_0$—the initial concentration of the element in the solution (mg/L); $C_e$—the equilibrium concentration of the element in the solution after sorption (mg/L).

*2.8. Raman Spectroscopy*

The Raman spectra (RS) of SIV, SIV-Pb, and Pb-exchanged form of ivanyukite collected from uncoated individual grains were recorded with a Horiba Jobin-Yvon LabRAM HR800 spectrometer (Horiba, Kyoto, Japan) equipped with an Olympus BX-41 microscope (Olympus Corporation, Tokyo, Japan) in backscattering geometry (Saint-Petersburg State University, Saint-Petersburg, Russia). Raman spectra were excited by a solid-state laser (532 nm) with an actual power of 2 mW under the 50× objective (NA 0.75). The spectra were obtained in a range of 70–4000 $cm^{-1}$ at a resolution of 2 $cm^{-1}$ at room temperature. To improve the signal-to-noise ratio, the number of acquisitions was set to 15. The spectra were processed using the algorithms implemented in Labspec (Horiba, Kyoto, Japan) and OriginPro 8.1 (OriginLab Corporation, Northampton, MA, USA) software packages.

## 2.9. Powder and Single-Crystal X-ray Diffraction

The synthetic products were investigated by means of powder X-ray diffraction using a Bruker D2 Phaser diffractometer (Bruker Corporation, Billerica, MA, USA) (Cu$K\alpha$ radiation, 30 kV/10 mA) (XRD Research Center, St. Petersburg State University, Saint-Petersburg, Russia). The experiments were carried out in the $2\theta$ range 5–65 ($°$) with a step of $0.02°$, and the exposure at each point was 1 s.

The crystal-structure study of Pb-exchanged ivanyukite was carried out at the X-ray Diffraction Resource Centre of St. Petersburg State University (Saint-Petersburg, Russia) by means of the Synergy S single-crystal diffractometer equipped with a Hypix (Rigaku Corporation, Tokyo, Japan) detector using monochromatic Mo$K\alpha$ radiation ($\lambda = 0.71069$ Å) at room temperature. More than a half of the diffraction sphere was collected with scanning step of $1°$, and an exposure time of 30 s. The data were integrated and corrected by means of the CrysAlis (Rigaku Corporation, Tokyo, Japan) program package, which was also used to apply empirical absorption correction using spherical harmonics, implemented in the SCALE3 ABSPACK scaling algorithm [21]. The structure was refined using the SHELXL software package [21]. The crystal structure was drawn using the VESTA 3 program [22]. Occupancies of the cation sites were calculated from the experimental site-scattering factors (except for the low-occupied sites) in accordance with the empirical chemical composition. Hydrogen sites could not be located.

The SCXRD data are deposited in CCDC under entry no. 2130913. Crystal data, data collection information, and refinement details are given in Table 1. Atom coordinates and isotropic parameters of atomic displacements are given in Table S1, interatomic distances in Table S2, and the anisotropic parameters of atomic displacements are given in Table S3.

**Table 1.** Crystal data, data collection information, and refinement details Pb-exchanged form ivanyukite.

| Temperature/K | 293 (2) |
| --- | --- |
| Crystal system | cubic |
| Space group | $P$-43m |
| $a = b = c$/Å | 7.8037 (7) |
| $\alpha = \beta = \gamma$/$°$ | 90 |
| Volume/Å$^3$ | 475.23 (13) |
| Z | 1 |
| $\rho_{calc}$g/cm$^3$ | 2.956 |
| $\mu$/mm$^{-1}$ | 13.003 |
| F (000) | 388.0 |
| Crystal size/mm$^3$ | $0.12 \times 0.12 \times 0.12$ |
| Radiation | Mo K$\alpha$ ($\lambda = 0.71073$) |
| $2\theta$ range for data collection/$°$ | 7.384 to 52.998 |
| Index ranges | $-6 \leq h \leq 9, -9 \leq k \leq 4, -9 \leq l \leq 9$ |
| Reflections collected | 612 |
| Independent reflections | 200 [$R_{int} = 0.0486$, $R_{sigma} = 0.0310$] |
| Data/restraints/parameters | 200/0/23 |
| Goodness-of-fit on F$^2$ | 1.116 |
| Final R indexes [$I \geq 2\sigma$ (I)] | $R_1 = 0.0497$, w$R_2 = 0.1186$ |
| Final R indexes [all data] | $R_1 = 0.0644$, w$R_2 = 0.1257$ |
| Largest diff. peak/hole/e Å$^{-3}$ | $0.71/-0.64$ |
| Flack parameter | 0.008(19) |

## 3. Results

### 3.1. Composition

Natural ivanyukite contains numerous intergrowths of aegirine (Figure 1a). Pb-exchanged natural ivanyukite contains up to 0.04 apfu of $Fe^{2+}$ and 0.12 apfu of $Nb^{5+}$, whereas its synthetic counterpart may contain up to 0.09 apfu of $Al^{3+}$. The SIV aggregates are represented by a fine-grained matrix with nanometer-sized crystallites (Figure 1b). AM-

4 forms rosette-like aggregates with the size of individual crystals less than $100 \times 100 \times 10$ µm (Figure 1f). Both samples of Pb-exchanged SIV and AM-4 (Figure 1c,e) contained an admixtures of prismatic crystals of $Pb_3O_4$ (Figure 1d).

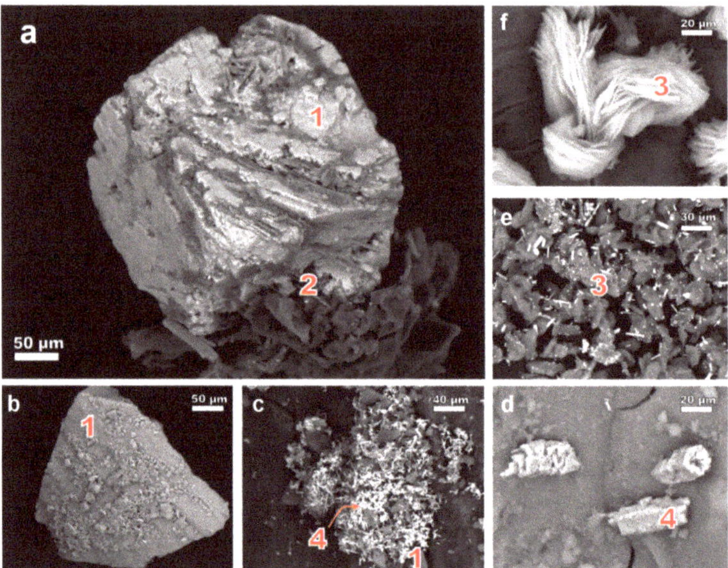

**Figure 1.** Backscattered images of (**a**) Pb-exchanged natural ivanyukite (1) with aegirine (2) inclusion; (**b**) Pb-exchanged SIV; (**c**) formation of $Pb_3O_4$ (4) crystals on the surface Pb-exchanged SIV; (**d**) individual prismatic crystals of $Pb_3O_4$; (**e**) Pb-exchanged AM-4 (3); (**f**) AM-4 rosette-like aggregates.

The Pb-exchanged SIV contains residual 0.06–0.36 apfu of $K^+$. Both synthetic and natural ivanyukite demonstrate exceptional ion-exchange properties close to the ideal substitution scheme $2Na^+ \leftrightarrow 1Pb^{2+} + 1\square$. Table 2 provides analytical results for natural and synthetic ivanyukite and their Pb-exchanged forms.

**Table 2.** Chemical composition of Pb-exchanged SIV and Pb-exchanged natural ivanyukite.

| Constituent | Pb-Exchanged SIV | | | Pb-Exchanged Natural Ivanyukite | |
|---|---|---|---|---|---|
| $SiO_2$ | 18.11 | 19.08 | 19.40 | 18.23 | 19.23 |
| $TiO_2$ | 30.74 | 30.01 | 31.64 | 31.65 | 32.50 |
| $Al_2O_3$ | 0.45 | | 0.23 | | |
| FeO | | | | 0.22 | 0.31 |
| $Nb_2O_5$ | | | | 1.64 | 1.68 |
| $K_2O$ | 0.44 | 0.32 | 1.85 | | |
| PbO | 38.60 | 38.40 | 36.03 | 37.59 | 34.87 |
| $H_2O$ * | 9.70 | 10.00 | 11.00 | 10.10 | 10.10 |
| Total | 98.04 | 97.81 | 100.15 | 99.43 | 98.69 |
| $Si^{4+}$ | 3.00 | 3.00 | 3.00 | 3.00 | 3.00 |
| $Ti^{4+}$ | 3.83 | 3.55 | 3.68 | 3.92 | 3.81 |
| $Al^{3+}$ | 0.09 | | 0.04 | | |
| $Fe^{2+}$ | | | | 0.03 | 0.04 |
| $Nb^{5+}$ | | | | 0.12 | 0.12 |
| Sum O | 3.92 | 3.55 | 3.72 | 4.07 | 3.97 |

**Table 2.** *Cont.*

| Constituent | Pb-Exchanged SIV | | | Pb-Exchanged Natural Ivanyukite | |
|---|---|---|---|---|---|
| $K^+$ | 0.09 | 0.06 | 0.36 | | |
| $Pb^{2+}$ | 1.72 | 1.63 | 1.50 | 1.67 | 1.46 |
| Sum A | 1.81 | 1.69 | 1.86 | 1.67 | 1.46 |
| $OH^-$ | 10.72 | 10.49 | 11.35 | 11.09 | 10.51 |

* The content of $H_2O$ was calculated according to the ivanyukite formula (4 apfu) and the content of OH according to the charge-balance requirements.

### 3.2. Pb Sorption from the Simulated Model Solution

The concentrations of Pb in the model solutions were determined by atomic emission spectrometry and are given in Table 3.

**Table 3.** Results of sorption experiments of Pb from $Pb(NO_3)_2$ solutions.

| | C(Pb), g/L | q, mg/g | A, % |
|---|---|---|---|
| Initial solution 1 | 0.39 | | |
| SIV | 0.0089 | 38.11 | 97.72 |
| SL3 | 0.31 | 8 | 20.51 |
| AM-4 | 0.0008 | 38.92 | 99.79 |
| Initial solution 2 | 0.88 | | |
| SIV | 0.004 | 87.6 | 99.55 |
| SL3 | 0.80 | 8 | 9.09 |
| AM-4 | 0.0009 | 87.91 | 99.90 |
| Initial solution 3 | 1.79 | | |
| SIV | 0.0026 | 178.74 | 99.85 |
| AM-4 | 0.0009 | 178.91 | 99.95 |
| Initial solution 4 | 1.78 | | |
| SIV | 0.97 | 403 | 45.28 |
| AM-4 | 1.73 | 26 | 2.92 |
| Initial solution 5 * | 0.16 | | |
| SIV | 0.016 | 14.4 | 90 |
| AM-4 | 0.017 | 14.3 | 89 |
| Initial solution 6 * | 23.70 | | |
| SIV | 0.11 | 2.4 | 99.5 |
| AM-4 | <0.03 | 2.4 | 99.9 |
| Initial solution 7 * | 56.23 | | |
| SIV | 55.04 | 0.1 | 2.1 |
| AM-4 | 55.68 | 0.1 | 1.0 |

* Data on the sorption of silver from solutions are not given due to the registration of the patent RU 2021124566 [17].

The cation-exchange capacity of the sorbents with respect to Pb was up to 178 mg/g for SIV and AM-4 with the extent of extraction of up to 99.85% and 8 mg/g for SL3 in each solution (initial solutions 1–3). In all cases the extent of extraction of Pb by the SIV and AM-4 was more than 99%.

It was found that, under given conditions for Pb sorption from the industrial solutions, both sorbents demonstrated excellent sorption properties with respect to Pb (Table 3, initial solution 5). The cation-exchange capacity for Pb was 14.4 and 14.3 mg/g for SIV and AM-4, respectively, with the extents of extraction equal to 90%.

The extent of Pb extraction from the real dust leaching solution-6 was 99.55% with the cation-exchange capacity for SIV equal to of 13 mg/g. The results of the experiment are presented in Table 3.

### 3.3. Raman Spectroscopy

The Raman spectra of SIV (ivanyukite-Na-*T*), SIV-Pb, and Pb-exchanged natural ivanyukite are shown in Figure 2. The assignments of the absorption bands were made by analogy with structurally related pharmacosiderite type compounds [23–25] and titanosilicates [26–29].

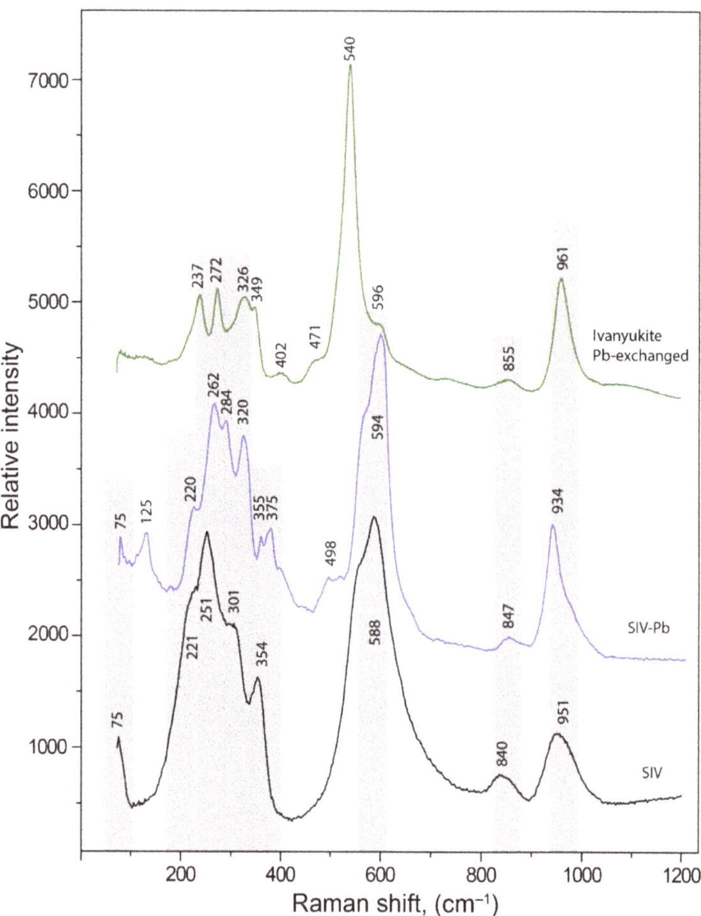


**Figure 2.** Raman spectra of initial SIV, SIV-Pb, and Pb-exchanged natural ivanyukite. The main bands in the initial SIV spectra are indicated by gray lines.

The intense vibrational bands at 951, 934, and 961 cm$^{-1}$ can be attributed to asymmetric stretching vibrations of $SiO_4$ tetrahedra, while the bands at 840, 847, and 855, cm$^{-1}$ are assigned to symmetric vibration modes involving the same bonds [24,28]. The most intense bands at 588 and 594 cm$^{-1}$ for rhombohedral SIV are significantly shifted (by ~60 cm$^{-1}$) in comparison with the band at 540 cm$^{-1}$ observed in the spectrum of Pb-exchanged natural ivanyukite and are related to the asymmetric bending vibrations of Si–O bonds or overlapping stretching vibrations of Ti–O bonds [25,28]. The same shift was observed previously for natural ivanyukite and corresponded to the transition from rhombohedral to cubic form [27]. The bands in the range of 350–500 cm$^{-1}$ correspond to symmetric bending vibrations of O–Si–O bonds and overlapping stretching vibrations of Ti–O bonds [25–27]. The bands at 498 and 471 cm$^{-1}$ were assigned to different modes of Ti–O stretching vibrations. Bands of different intensities in the region of 200–350 cm$^{-1}$ belong to different bending vibration modes of the Ti–O bonds in TiO$_6$ octahedra [23,24]. The bands with the wave numbers below 200 cm$^{-1}$ belong to translational vibrations. The bands at 75 and 126–128 cm$^{-1}$ are observed in the Raman spectra of synthetic samples and are absent in the Raman spectrum of the Pb-exchanged natural ivanyukite and probably respond to translation modes of K.

In general, the spectrum of Pb-exchanged natural ivanyukite corresponds to the cubic form, whereas the spectrum of SIV-Pb has characteristic features corresponding to both trigonal and cubic forms. It contains bands near 498 cm$^{-1}$ that are characteristic of cubic Pb-exchanged natural ivanyukite and at the same time has bands near 220 and 350 cm$^{-1}$ that are characteristic of trigonal SIV. This can be explained by the presence of domains with both trigonal and cubic symmetries within the single grains of SIV-Pb.

### 3.4. Powder Diffraction

The initial XRD pattern of SIV is in good agreement with that of synthetic ivanyukite-Na-*T*, (PDF Card No. 00-052-1204). Intensity of the (101) reflection significantly decreases after Pb sorption (Figure 3). The appearance of (440) reflection at 68.18°2θ (1.3743 Å) in SIV-Pb instead two peaks at 67.15 and 68.80°2θ observed in SIV pattern indicates the R3m to P−43m symmetry change. The observed pattern for the Pb-exchanged SIV agrees with that of synthetic ivanyukite-Na-C (PDF Card No.00-047-0042). The increase in the Pb concentration results in the decrease of intensities of main reflections and is probably connected with the increase of absorption coefficient induced by the increases amounts of Pb in the structure. When the maximum sorption capacity is reached, the Pb$_3$O$_4$ lead oxide (PDF Card No. 01-071-0562) starts to crystallize. The presence of Pb$_3$O$_4$ was confirmed by our chemical data (Figure 1d) and the presence of additional reflections at 24.82°2θ (3.584 Å), 25.49°2θ (3.492 Å), and 43.48°2θ (2.079 Å).

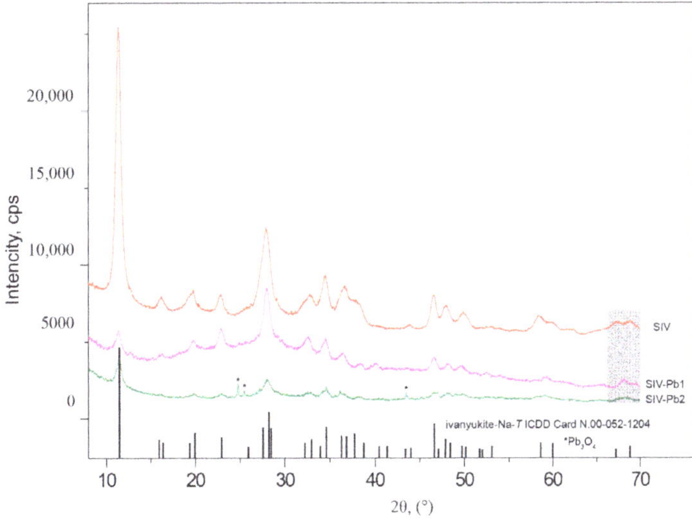


**Figure 3.** Diffraction patterns of SIV (red curve) and its Pb-exchanged forms with the S:L ratios of 1:500 (pink) and 1:133 (green). Asterisks indicate peaks of Pb$_3$O$_4$.

### 3.5. Single-Crystal XRD

The crystal structure of Pb-exchanged ivanyukite was refined in the P−43m space group to R$_1$ = 0.049 for 200 (R$_{int}$ = 0.049, R$_{sigma}$ = 0.030) independent reflections with F$_o$ > 4σ(F$_o$) using the model of ivanyukite-K [27]. The crystal structure of Pb-exchanged ivanyukite (Figure 4a) possesses a pharmacosiderite structural topology and is based upon a topologically identical three-dimensional framework [28]. The main structural feature of ivanyukite-group minerals is the presence of cubane-like $[Ti_4O_4]^{8+}$ clusters formed by four edge-sharing $TiO_6$ octahedra [29]. The $[Ti_4O_4]^{8+}$ clusters are connected by sharing corners with $SiO_4$ tetrahedra to form a negatively charged $[(TiO)_4(SiO_4)_3]_{4-}$ framework. The framework has a three-dimensional system of channels defined by 8-membered rings (8-MRs) with a free (suitable for migration) crystallographic diameter of ~3.5 Å [30]. The channels are occupied by extra-framework cations (e.g., $Na^+$, $K^+$) and $H_2O$ molecules.

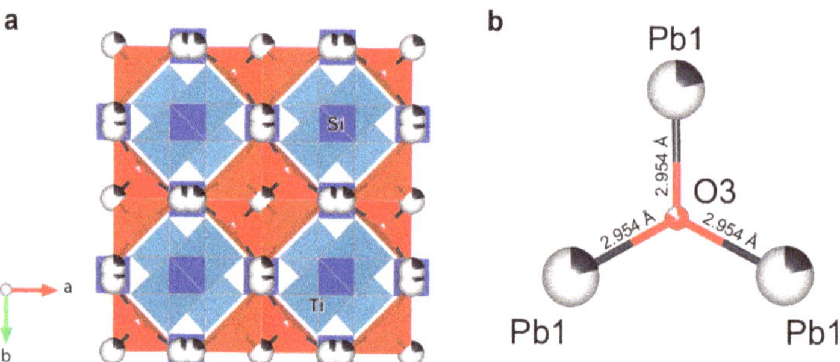


**Figure 4.** Crystal structure of Pb-exchanged ivanyukite in general projection (**a**); local coordination of Pb in the crystal structure of Pb-exchanged ivanyukite (**b**).

In the crystal structure of Pb-exchanged ivanyukite, as in other cubicpharmacosiderite-supergroup minerals [31], there is one symmetrically independent Si1 and one Ti1 site coordinated by the O1 and O2 atoms. The mean bond lengths for $SiO_4$ tetrahedra $TiO_6$ octahedra are, <1.642> Å and <1.961> Å, respectively, in good agreement with their full occupancies. The splitted Pb1 site is situated at the center of the 8-MR with the site occupation factor (s.o.f.) = of 0.21 and the Pb−Pb distance of 0.62 Å. The Pb1 site is bonded to two $O3(H_2O)$ sites.

The structural formula of Pb-exchanged ivanyukite determined from the structure refinement can be written as $Pb_{1.26}[Ti_4O_{2.52}(OH)_{1.48}(SiO_4)_3]\cdot3.32(H_2O)$.

## 4. Discussion

Incorporation of Pb into natural ivanyukite-Na-*T* by the substitution scheme $2Na^+ + 2O^{2-} \leftrightarrow Pb^{2+} + \square + 2OH^-$ results in its transformation into the cubic Pb-exchanged form. The same transition with increasing symmetry from $R3m$ to $P-43m$ was observed for the synthetic material and confirmed by powder X-ray diffraction (Figure 3) and Raman spectroscopy (Figure 2). During the Pb sorption, the significant decrease of the (101) reflection intensity and the appearance of additional peak at 68.18°2θ, instead of two peaks at 67.15 and 68.80°2θ, was observed. In the Raman spectra the shift for the most intense band at 580–600 cm$^{-1}$ to 500–540 cm$^{-1}$ related to the asymmetric bending vibrations of Si–O bonds or overlapping stretching vibrations of Ti–O bonds was observed. Such a shift was previously described for similar transitions in the ivanyukite-group minerals [27].

According to our chemical and sorption data, the maximal sorption capacity in relation to Pb was demonstrated by synthetic ivanyukite and reaches 400 mg/g at ambient conditions with synthetic ivanyukite. The maximal Pb content reaches 1.71 apfu. In the crystals structure of Pb-exchanged ivanyukite, the Pb1 site is situated at the center of the 8-MR and participates in the $OPb_3$ triangles with the O−Pb distance of 2.954 Å (Figure 4b). Oxocentred triangles $OA_3$ (A = metal) are typical structural units in bismuth oxysalts with additional oxygen atoms [31,32], and was observed in Pb oxysalts as well [33].

There is a certain balance between the amounts of Pb in precipitates and the remaining solution. At the same time, there is an observed dependence of these amount from the NaCl content that is usually introduced at the beginning of the dust leaching process. When $Na^+$ concentration is up to 200 g/L, the precipitation of Pb is seemingly incomplete. When the

Na$^+$ concentration is more than 200 g/L, no additional precipitation of trace amounts of Pb was observed.

In the case of the studied solutions, strongly acidic medium (pH < 1) the fast protonation reactions in SIV and AM-4 titanosilicates is favored. In this case, Na$^+$ cations should be quickly leaching into solution and two concurrent processes are observed described by the equations $2Na^+ + O^{2-} \leftrightarrow 2\square + OH$ (protonation) and $2Na^+ \leftrightarrow Pb^{2+}$ (ion-exchange).

It is known that Pb$^{2+}$ cations in acidic chloride-sulfate solutions of metallurgical industries can form stable anionic complexes—such as $[PbCl_3]^-$, $[PbCl_4]^{2-}$, $[PbCl_6]^{4-}$, among others—which can be extracted from a solution on anion exchangers (e.g., AMP or AM-2b) with $-CH_2N^+$, $-CH_2N(CH_3)_2$, and $-CH_2N(CH_3)_3$ serving the role of exchange groups [34,35]. The sorption of Pb from acid mine drainage also includes its association to organic hydroxyl and carboxyl, $SiO_4^{2-}$, and $-R-O-H$ functional groups [36,37]. The presence of carboxyl groups in some clay minerals also improves the adsorption of Pb [38]. In the case of titanosilicates used in this work, the key-role is presumably played by extra-framework sodium cations, which displacement may significantly affect the change in the surface charge of the samples. The SL3 material in this case is a good example of the formation of an intermediate between the product and the initial AM-4 phase leading to the change in the nature of the active centers of the material (decrease in the acidic strength of the active centers as a result of protonation) [19]. Since SL3 did not prove to be a good sorbent for Pb in our studies, we can assume that titanosilicate samples containing Na$^+$ are required to decrease the stability of lead chloride complexes. In solutions with the content of Cl$^-$ (20 g/L), extraction is observed with a high background of sodium in the solution (150 g/L) only. For this reason, it is important to determine and fix the mechanisms of the reactions taking place in natural ivanyukite crystals.

## 5. Conclusions

Occurrence of extra framework Na$^+$ cations in the structure of titanosilicate sorbents plays a key role in Pb sorption. Both AM-4, $Na_2Ti_2[Si_4O_{13}(OH)]\cdot2H_2O$ and SIV, $Na_3Ti_4(SiO_4)_3O_4\cdot4H_2O$ titanosilicates with extra framework Na$^+$ cations demonstrate 90% to 99% of Pb extraction from different solutions. At the same time SL3, $Ti_2[Si_4O_{10}(OH)_4]$—protonated modification of AM-4 does not demonstrate good sorption properties. The maximal sorption capacity with respect to Pb is demonstrated by SIVand reaches 400 mg/g at ambient conditions. Both AM-4 and SIV may be recommended for the Pb sorption from the Pb-bearing dust leaching solution.

In solutions with the content of Cl$^-$ (20 g/L), extraction was observed only with the high background of Na (150 g/L). It seems that Na$^+$ ions neutralize Cl$^-$, which catalyzes the exchange reaction.

The molecular mechanism of incorporation of Pb$^{2+}$ into ivanyukite structure by the substitution scheme $2Na^+ + 2O^{2-} \leftrightarrow Pb^{2+} + \square + 2OH^-$ result in the structure transition with increasing symmetry from $R3m$ to $P-43m$. The symmetry transition has been confirmed by the SC XRD, PXRD, and Raman data. In the crystal structure of Pb-substituted form, Pb$^{2+}$ cations occupy the centers of the 8-MRs and forms oxo-centered OPb$_3$ units.

## 6. Patents

Nikolaev, A.I.; Samburov, G.O.; Kalashnikova, G.O.; Kasikov, A.G.; Panikorovskii, T.L.; Bazai, A.V. Method of silver extraction from pyrometallurgical wastes. Patent application RU 2021124566A, 16 February 2022.

**Supplementary Materials:** The following are available online at https://www.mdpi.com/article/10.3390/cryst12030311/s1, Table S1: Fractional Atomic Coordinates ($\times10^4$) and Equivalent Isotropic Displacement Parameters ($\mathring{A}^2 \times 10^3$) for Pb-exchanged ivanyukite. U$_{eq}$ is defined as 1/3 of the trace of the orthogonalised U$_{IJ}$ tensor; Table S2: Anisotropic Displacement Parameters ($\mathring{A}^2 \times 10^3$) for Pb-exchanged ivanyukite. The Anisotropic displacement factor exponent takes the form: $-2\pi^2[h^2a \times {}^2U_{11} + 2hka \times b \times U_{12} + \ldots]$; Table S3: Bond Lengths for Pb-exchanged ivanyukite.

**Author Contributions:** Conceptualization, V.N.Y. and A.K.; Methodology, G.O.S., G.O.K., A.V.B., E.S. and D.B.; Software, T.L.P. and V.N.B.; Validation, G.O.S., G.O.K., T.L.P., E.S., A.V.B., D.B. and A.K.; Formal analysis, E.S., D.B., T.L.P. and V.N.B.; Investigation, G.O.S., G.O.K., T.L.P. and E.S.; Resources, A.K. and V.N.Y.; Data curation, A.K. and S.V.K.; Writing—original draft preparation, G.O.S., T.L.P., G.O.K. and S.V.K.; Writing—review and editing, A.K., E.S., A.V.B., D.B. and S.V.K.; Visualization, T.L.P. and A.V.B.; Supervision, V.N.Y. and S.V.K.; Project administration, A.K.; Funding acquisition, G.O.S. and G.O.K. All authors have read and agreed to the published version of the manuscript.

**Funding:** This research was funded by the Ministry of Education and Science of the Murmansk Region, grant no. 244 at 2 September 2021 (synthesis of titanosilicates), by the Russian Foundation for Basic Research grant no. 20-33-90326 (sorption experiments) and Russian Science Foundation, project No. 21-77-10103 (SCXRD study).

**Institutional Review Board Statement:** Not applicable.

**Informed Consent Statement:** Not applicable.

**Data Availability Statement:** The CIF filecontains data about the ivanyukite-Pb structure, Fractional Atomic Coordinates ($\times 10^4$) and Equivalent Isotropic Displacement Parameters ($\text{Å}^2 \times 10^3$) for Pb-exchanged ivanyukite. $U_{eq}$ is defined as 1/3 of the trace of the orthogonalised $U_{IJ}$ tensor, Anisotropic Displacement Parameters ($\text{Å}2 \times 10^3$) for Pb-exchanged ivanyukite. The Anisotropic displacement factor exponent takes the form: $-2\pi^2[h^2a \times {}^2U_{11} + 2hka \times b \times U_{12} + \dots]$ and Bond Lengths for Pb-exchanged ivanyukite can be found at the link: https://doi.org/10.5281/zenodo.6232628 (accessed on 30 January 2022).

**Acknowledgments:** The team of authors would like to acknowledge our professors from I.V. Tananaev Institute of Chemistry and Technology of Rare Elements and Mineral Raw Materials of the KSC RAS and Saint-Petersburg University—Britvin S.N., Gerasimova L.G., and Korovin V.N.—for their useful consultations, comments, and valuable advice regarding questions of hydrothermal synthesis of titanosilicates and features of chloride ligands in the complexes of platinum metals. The authors are also grateful to the X-ray Diffraction Centre, Geo Environmental Centre "Geomodel" of Saint-Petersburg State University for their help in our experimental studies.

**Conflicts of Interest:** The authors declare no conflict of interest.

# References


1. Yakovenchuk, V.N.; Nikolaev, A.P.; Selivanova, E.A.; Pakhomovsky, Y.A.; Korchak, J.A.; Spiridonova, D.V.; Zalkind, O.A.; Krivovichev, S.V. Ivanyukite-Na-T, ivanyukite-Na-C, ivanyukite-K, and ivanyukite-Cu: New microporous titanosilicates from the Khibiny massif (Kola Peninsula, Russia) and crystal structure of ivanyukite-Na-T. *Am. Mineral.* **2009**, *94*, 1450–1458. [CrossRef]
2. Popa, K.; Pavel, C.C.; Bilba, N.; Cecal, A. Purification of waste waters containing 60Co$^{2+}$, 115mCd$^{2+}$ and 203Hg$^{2+}$ radioactive ions by ETS-4 titanosilicate. *J. Radioanal. Nucl. Chem.* **2006**, *269*, 155–160. [CrossRef]
3. Kalashnikova, G.O.; Zhitova, E.S.; Selivanova, E.A.; Pakhomovsky, Y.A.; Yakovenchuk, V.N.; Ivanyuk, G.Y.; Kasikov, A.G.; Drogobuzhskaya, S.V.; Elizarova, I.R.; Kiselev, Y.G.; et al. The new method for obtaining titanosilicate AM-4 and its decationated form: Crystal chemistry, properties and advanced areas of application. *Microporous Mesoporous Mater.* **2021**, *313*, 110787. [CrossRef]
4. Cruciani, G.; De Luca, P.; Nastro, A.; Pattison, P. Rietveld refinement of the zorite structure of ETS-4 molecular sieves. *Microporous Mesoporous Mater.* **1998**, *21*, 143–153. [CrossRef]
5. Anthony, R.G.; Dosch, R.G.; Gu, D.; Philip, C.V. Use of silicotitanates for removing cesium and strontium from defense waste. *Ind. Eng. Chem. Res.* **1994**, *33*, 2702–2705. [CrossRef]
6. Men'shikov, Y.P.; Sokolova, E.V.; Egorov-Tismenko, Y.K.; Khomyakov, A.P.; Polezhaeva, L.I. Sitinakite, Na$_2$KTi$_4$Si$_2$O$_{13}$(OH)·4H$_2$O— A new mineral. *Zap. RMO* **1992**, *121*, 94–99. (In Russian)
7. Panikorovskii, T.L.; Kalashnikova, G.O.; Nikolaev, A.I.; Perovskiy, I.A.; Bazai, A.V.; Yakovenchuk, V.N.; Bocharov, V.N.; Kabanova, N.A.; Krivovichev, S.V. Ion-Exchange-Induced Transformation and Mechanism of Cooperative Crystal Chemical Adaptation in Sitinakite: Theoretical and Experimental Study. *Minerals* **2022**, *12*, 248. [CrossRef]
8. Chapman, D.M.; Roe, A.L. Synthesis, characterization and crystal chemistry of microporous titanium-silicate materials. *Zeolites* **1990**, *10*, 730–737. [CrossRef]
9. Dadachov, M.; Rocha, J.; Ferreira, A.; Lin, Z.; Anderson, M. Ab initio structure determination of layered sodium titanium silicate containing edge-sharing titanate chains (AM-4) Na$_3$(Na,H)Ti$_2$O$_2$[Si$_2$O$_6$]$_2$·2H$_2$O. *Chem. Commun.* **1997**, *3*, 2371–2372. [CrossRef]
10. Britvin, S.N.; Gerasimova, L.G.; Ivanyuk, G.Y.; Kalashnikova, G.O.; Krzhizhanovskaya, M.G.; Krivovivhev, S.V.; Mararitsa, V.F.; Nikolaev, A.I.; Oginova, O.A.; Panteleev, V.N.; et al. Application of titanium-containing sorbents for treating liquid radioactive waste with the subsequent conservation of radionuclides in Synroc-type titanate ceramics. *Theor. Found. Chem. Eng.* **2016**, *50*, 598–606. [CrossRef]

11. Gerasimova, L.; Nikolaev, A.; Maslova, M.; Shchukina, E.; Samburov, G.; Yakovenchuk, V.; Ivanyuk, G. Titanite Ores of the Khibiny Apatite-Nepheline-Deposits: Selective Mining, Processing and Application for Titanosilicate Synthesis. *Minerals* **2018**, *8*, 446. [CrossRef]

12. Gerasimova, L.G.; Nikolaev, A.I.; Shchukina, E.S.; Maslova, M.V. Hydrothermal synthesis of framed titanosilicates with a structure of ivanyukite mineral. *Доклады Академии наук* **2019**, *487*, 289–292. [CrossRef]

13. Barcan, V.; Sylina, A. The appraisal of snow sampling for environmental pollution valuation. *Water. Air. Soil Pollut.* **1996**, *89*, 49–65. [CrossRef]

14. Yakovlev, E.; Druzhinina, A.; Druzhinin, S.; Zykov, S.; Ivanchenko, N. Assessment of physical and chemical properties, health risk of trace metals and quality indices of surface waters of the rivers and lakes of the Kola Peninsula (Murmansk Region, North-West Russia). *Environ. Geochem. Health* **2021**, 1–30. [CrossRef]

15. Barcan, V.; Kovnatsky, E. Soil Surface Geochemical Anomaly Around the Copper-Nickel Metallurgical Smelter. *Water. Air. Soil Pollut.* **1998**, *103*, 197–218. [CrossRef]

16. Koptsik, G.N.; Koptsik, S.V.; Smirnova, I.E.; Sinichkina, M.A. Remediation of Technogenic Barren Soils in the Kola Subarctic: Current State and Long-Term Dynamics. *Eurasian Soil Sci.* **2021**, *54*, 619–630. [CrossRef]

17. Nikolaev, A.I.; Samburov, G.O.; Kalashnikova, G.O. Method for the Extraction of Silver from Pyrometallurgical Waste. RU Patent Application No. 2021124566A, 16 February 2022.

18. Merlino, S.; Pasero, M.; Khomyakov, A.P. The crystal structure of lintisite, $Na_3LiTi_2[Si_2O_6]_2O_2 \cdot 2H_2O$, a new titanosilicate from Lovozero (USSR). *Z. Krist.* **1990**, *193*, 137–148. [CrossRef]

19. Timofeeva, M.N.; Kalashnikova, G.O.; Shefer, K.I.; Mel'gunova, E.A.; Panchenko, V.N.; Nikolaev, A.I.; Gil, A. Effect of the acid activation on a layered titanosilicate AM-4: The fine-tuning of structural and physicochemical properties. *Appl. Clay Sci.* **2020**, *186*, 105445. [CrossRef]

20. Gerasimova, L.G.; Shchukina, E.S.; Maslova, M.V.; Gladkikh, S.N.; Garaeva, G.R.; Kolobkova, V.M. Synthesis of rutile titanium dioxide from Russian raw materials. *Polym. Sci. Ser. D* **2017**, *10*, 23–27. [CrossRef]

21. Sheldrick, G.M. Crystal structure refinement with SHELXL. *Acta Crystallogr. Sect. C Struct. Chem.* **2015**, *71*, 3–8. [CrossRef]

22. Momma, K.; Izumi, F. VESTA 3 for three-dimensional visualization of crystal, volumetric and morphology data. *J. Appl. Crystallogr.* **2011**, *44*, 1272–1276. [CrossRef]

23. Frost, R.L.; Kloprogge, J.T. Raman spectroscopy of some complex arsenate minerals—implications for soil remediation. *Spectrochim. Acta Part A Mol. Biomol. Spectrosc.* **2003**, *59*, 2797–2804. [CrossRef]

24. Filippi, M. Oxidation of the arsenic-rich concentrate at the Přebuz abandoned mine (Erzgebirge Mts., CZ): Mineralogical evolution. *Sci. Total Environ.* **2004**, *322*, 271–282. [CrossRef]

25. Filippi, M.; Doušová, B.; Machovič, V. Mineralogical speciation of arsenic in soils above the Mokrsko-west gold deposit, Czech Republic. *Geoderma* **2007**, *139*, 154–170. [CrossRef]

26. Yakovenchuk, V.; Pakhomovsky, Y.; Panikorovskii, T.; Zolotarev, A.; Mikhailova, J.; Bocharov, V.; Krivovichev, S.; Ivanyuk, G. Chirvinskyite, $(Na,Ca)_{13}(Fe,Mn,\square)_2(Ti,Nb)_2(Zr,Ti)_3$-$(Si_2O_7)_4(OH,O,F)_{12}$, a New Mineral with a Modular Wallpaper Structure, from the Khibiny Alkaline Massif (Kola Peninsula, Russia). *Minerals* **2019**, *9*, 219. [CrossRef]

27. Pakhomovsky, Y.A.; Panikorovskii, T.L.; Yakovenchuk, V.N.; Ivanyuk, G.Y.; Mikhailova, J.A.; Krivovichev, S.V.; Bocharov, V.N.; Kalashnikov, A.O. Selivanovaite, $NaTi_3(Ti,Na,Fe,Mn)_4[(Si_2O_7)_2O_4(OH,H_2O)_4] \cdot nH_2O$, a new rock-forming mineral from the eudialyte-rich malignite of the Lovozero alkaline massif (Kola Peninsula, Russia). *Eur. J. Mineral.* **2018**, *30*, 525–535. [CrossRef]

28. Krivovichev, S. Topology of Microporous Structures. *Rev. Mineral. Geochem.* **2005**, *57*, 17–68. [CrossRef]

29. Panikorovskii, T.L.; Yakovenchuk, V.N.; Yanicheva, N.Y.; Pakhomovsky, Y.A.; Shilovskikh, V.V.; Bocharov, V.N.; Krivovichev, S.V. Crystal chemistry of ivanyukite-group minerals, $A_{3-x}H_{1+x}[Ti_4O_4(SiO_4)_3](H_2O)_n$ (A = Na, K, Cu), (n = 6–9, x = 0–2): Crystal structures, ion-exchange, chemical evolution. *Mineral. Mag.* **2021**, *85*, 607–619. [CrossRef]

30. Oleksiienko, O.; Wolkersdorfer, C.; Sillanpää, M. Titanosilicates in cation adsorption and cation exchange—A review. *Chem. Eng. J.* **2017**, *317*, 570–585. [CrossRef]

31. Rumsey, M.S.; Mills, S.J.; Spratt, J. Natropharmacoalumite, $NaAl_4[(OH)_4(AsO_4)_3] \cdot 4H_2O$, a new mineral of the pharmacosiderite supergroup and the renaming of aluminopharmacosiderite to pharmacoalumite. *Mineral. Mag.* **2010**, *74*, 929–936. [CrossRef]

32. Bedlivy, D.; Mereiter, K. Preisingerite, $Bi_3O(OH)(AsO_4)_2$, a new species from San Juan Province, Argentina: Its description and crystal structure. *Am. Mineral.* **1982**, *67*, 833–840.

33. Ridkosil, T.; Srein, V.; Fabry, J.; Hybler, J.; Maximov, B.A. Mrazekite, $Bi_2Cu_3(OH)_2O_2(PO_4)_2$. *Can. Mineral.* **1992**, *30*, 215–224.

34. Siidra, O.I.; Zinyakhina, D.O.; Zadoya, A.I.; Krivovichev, S.V.; Turner, R.W. Synthesis and Modular Structural Architectures of Mineralogically Inspired Novel Complex Pb Oxyhalides. *Inorg. Chem.* **2013**, *52*, 12799–12805. [CrossRef]

35. Voropanova, L.A.; Gagieva, Z.A.; Pukhova, V.P.; Vil'ner, N.A. Method of Extracting Lead Ions $Pb^{2+}$ from Acidic Solutions. RU Patent Application No. 2393244C1, 27 June 2010.

36. Lu, H.; Zhang, W.; Yang, Y.; Huang, X.; Wang, S.; Qiu, R. Relative distribution of $Pb^{2+}$ sorption mechanisms by sludge-derived biochar. *Water Res.* **2012**, *46*, 854–862. [CrossRef]

37. Wang, Q.; Zheng, C.; Cui, W.; He, F.; Zhang, J.; Zhang, T.C.; He, C. Adsorption of $Pb^{2+}$ and $Cu^{2+}$ ions on the CS2-modified alkaline lignin. *Chem. Eng. J.* **2020**, *391*, 123581. [CrossRef]

38. Pavlovic, I.; Perez, M.; Barriga, C.; Ulibarri, M. Adsorption of $Cu^{2+}$, $Cd^{2+}$ and $Pb^{2+}$ ions by layered double hydroxides intercalated with the chelating agents diethylenetriaminepentaacetate and meso-2,3-dimercaptosuccinate. *Appl. Clay Sci.* **2009**, *43*, 125–129. [CrossRef]

*Article*

# Preparation of NaA Zeolite from High Iron and Quartz Contents Coal Gangue by Acid Leaching—Alkali Melting Activation and Hydrothermal Synthesis

Deshun Kong [1,2] and Rongli Jiang [1,*]

[1] School of Chemical Engineering and Technology, China University of Mining and Technology, Xuzhou 221016, China; LB19040009@cumt.edu.cn
[2] Guizhou Provincial Key Laboratory of Coal Clean Utilization, School of Chemistry and Materials Engineering, Liupanshui Normal University, Liupanshui 553004, China
* Correspondence: ronglij@cumt.edu.cn

**Abstract:** In this study, NaA zeolite was successfully synthesized from coal gangue with high contents of iron and quartz as the main raw material. The results show that most iron ions can be removed from coal gangue after calcination at 700 °C for 2 h, leaching in hydrochloric acid with a mass fraction of 20% for 7 h and a liquid-solid ratio of 3.5:1. When $m$ (acid leached residue of calcined gangue):$m$ ($Na_2CO_3$) = 1.1 and melting at 750 °C for 2 h, the quartz and other aluminosilicates turn into nepheline, which dissolve in water. Finally, the optimum conditions of synthesis NaA zeolite are as follows: $n(SiO_2)/n(Al_2O_3) = 2.0$, $n(Na_2O)/n(SiO_2) = 2.1$, $n(H_2O)/n(Na_2O) = 55$, aging at 60 °C for 2 h, and crystallization at 94 °C for 4 h. This shows that the high iron and quartz contents coal gangue can be used for the synthesis of NaA zeolite.

**Keywords:** high iron and quartz contents coal gangue; acid leaching; alkali melting; hydrothermal reaction; NaA zeolite

**Citation:** Kong, D.; Jiang, R. Preparation of NaA Zeolite from High Iron and Quartz Contents Coal Gangue by Acid Leaching—Alkali Melting Activation and Hydrothermal Synthesis. *Crystals* **2021**, *11*, 1198. https://doi.org/10.3390/cryst11101198

Academic Editor: Sergio Brutti

Received: 12 September 2021
Accepted: 30 September 2021
Published: 3 October 2021

**Publisher's Note:** MDPI stays neutral with regard to jurisdictional claims in published maps and institutional affiliations.

## 1. Introduction

Coal gangue is the main discharged waste of the coal industry [1,2]. The arbitrary stacking of it seriously influences the safety of the ecological environment [3,4]. Recently, the development and utilization of coal gangue as a kind of new resource has gained interest [5–7]. Many products have been prepared from coal gangue, but the common utilization methods of coal gangue are as power plant fuel [8] and to prepare building materials [9], including brick [10], cement clinker products [11] and other products. The utilization technology level is lower and the industrial added value is not high. On the other hand, zeolite synthesis often uses cheap minerals or waste, which can save the cost of raw materials. Coal gangue can be used as raw material for preparing zeolites [12,13]. In recent years, researchers have noticed the great potential of coal gangue in zeolite synthesis [14–16].

Zeolites are micropore and mesoporous hydrated aluminosilicates containing alkali elements, alkaline earth metals, or other cations, whose structure is built up with a framework of tetrahedral molecules, which are linked with shared oxygen atoms [17]. Due to their unique properties, zeolites have been used in many fields, such as in agriculture [18], chemical technology [19], oil refining [20], and others for their porous characteristics, ion-exchange properties, and catalytic performance [21]. Many studies have succeeded in the conversion of low iron-bearing coal gangue into synthetic zeolites [22–24]. However, coal gangue with high iron content is a kind of material with poor quality. It is difficult to synthesize high quality zeolite because iron ions will affect the whiteness and performance of the products, therefore, there are only a few studies concerning what happens in the transformation process of the whole NaA zeolite preparation from high iron content coal gangue.

In this work, coal gangue is firstly calcined and then leached by acid to remove most of the iron ions; at the same time, there is very low chemical activity of kaolinite and quartz in the coal gangue, which are necessary to be activated before synthesizing zeolite. However, the quartz in gangue is difficult to activate by heating, which will directly enter the zeolite products [25]. In zeolite synthesis, the alkali melting method, as an activator for the formation of soluble aluminate and silicate [26–28], is adopted to activate the materials richly with inert silica or alumina in the presence of alkali. In the process, the acid-leached residue of coal gangue is activated absolutely by $Na_2CO_3$ powder at 750 °C for 2 h after calcination of the coal gangue powder at 700 °C for 2 h, and leaching in 20% hydrochloric acid at 90 °C for 7 h under stirring. The NaA zeolite is obtained following aging and hydrothermal synthesis. For further discussion of the mechanism of phase transformations, the gangue, acid-leached residue, alkali-melted intermediate products, and the NaA zeolite, which are prepared under different conditions, are detected by XRD; the phase transformation laws are obtained; and the suitable technological conditions for preparing NaA zeolite from high iron and quartz contents coal gangue are determined.

## 2. Experimental

### 2.1. Materials

The coal gangue comes from Wangjiazhai Coal Mine in Liupanshui, Guizhou Province, China, which is made into powder of less than 200 mesh. $NaAiO_2$ and $Na_2CO_3$ are analytically pure (AR), Tianjin city Beichen Founder Reagent Factory; Tianjin, China; NaOH is AR, Tianjin Zhiyuan Chemical Reagents Co., Ltd. Tianjin, China; and hydrochloric acid is AR, Chongqing Chuandong Chemical Co., Ltd. Chongqing, China.

### 2.2. Technological Process

The coal gangue is calcined at 700 °C for 2 h and then leached by 20% (in mass) hydrochloric acid; the liquid to solid ratio is 3.5:1 (volume/mL:mass/g) at 90 °C for 7 h; and the acid leached residue is obtained after filtering and drying, and then the residue is mixed with sodium carbonate at the ratio of $m$ (acid leaching residue):$m$ ($Na_2CO_3$) = 1.1, melting at 750 °C for 2 h in the muffle, by adjusting the reacting proportion and aging; the hydrothermal crystallization is completed under the stirring condition of 94 °C for several hours, and NaA powder zeolite is obtained after washing and drying.

The preparation process of NaA zeolites from high-iron content coal gangue is shown in Figure 1.

### 2.3. Characterization

TD-2500 type X-ray diffraction instrument (XRD, China) was used for $CuK_\alpha$ ($\lambda$ for $K_\alpha$ = 1.54059 Å), $2\theta = 3°$ (min)–65° (max), with a step width of 0.04°. The major chemical elemental compositions were detected by Thermo Electron ARL9900XP+ type X-ray fluorescence spectrometer (XRF, Massachusetts, USA). The morphology of the products was detected by Zeiss evo18 type scanning electron microscope (SEM, Jena, Germany).

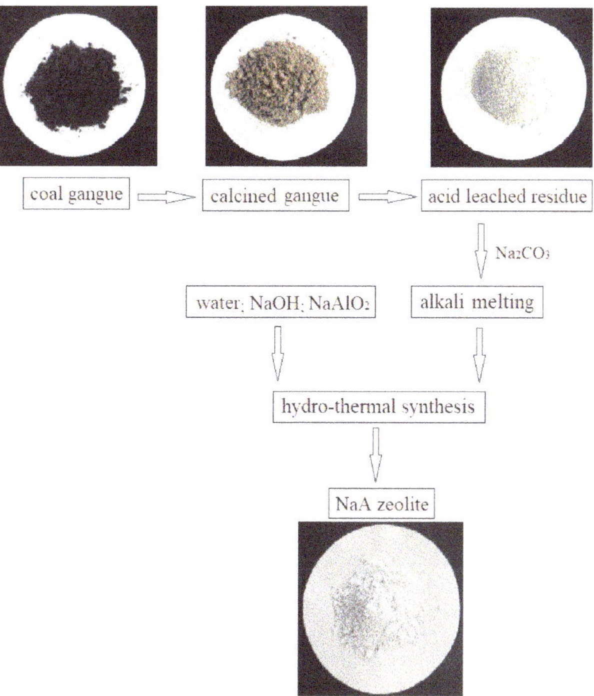



**Figure 1.** Process flow chart of preparing NaA zeolite from high iron coal gangue.

## 3. Results and Discussions

*3.1. Major Chemical Elements Analysis of Coal Gangue and the Acid Leached Residue*

The raw coal gangue and the acid leached residue are dried at 105 °C for 12 h [29], the main compositions are shown in Table 1.

**Table 1.** Main compositions of coal gangue and the acid-leaching residue/*wt* %.

| Compositions | Gangue | Residue |
|---|---|---|
| $SiO_2$ | 42.18 | 83.12 |
| $Al_2O_3$ | 20.43 | 10.50 |
| $Fe_2O_3$ | 15.36 | 0.99 |
| $K_2O$ | 1.24 | 1.04 |
| $Na_2O$ | 0.40 | 0.38 |
| CaO | 2.95 | 0.04 |
| MgO | 1.61 | 0.19 |
| MgO | 1.66 | 2.77 |
| $TiO_2$ | 0.25 | 0.04 |
| MnO | 0.54 | 0.03 |
| S | 0.30 | 0.06 |
| FC and other ignition loss | 13.08 | 0.84 |

We can see from Table 1 that the main compositions of the gangue are silicon, aluminum, and iron; obviously the iron element content is too high, which will affect the properties of the NaA product, so it must be removed. After acid leaching by hydrochloric acid, the major chemical elements of the residue are silicon and aluminum; the contents of other elements are low, $n(SiO_2)/n(Al_2O_3) = 13.46$, compared with NaA zeolite ($n(SiO_2)/n(Al_2O_3) = 2.0$), so it is necessary for $NaAlO_2$ to adjust the $n(SiO_2)/n(Al_2O_3)$ ratio of the system.

The morphology analyses of the gangue powder, calcined gangue, and residue are shown in Figures 2–5.

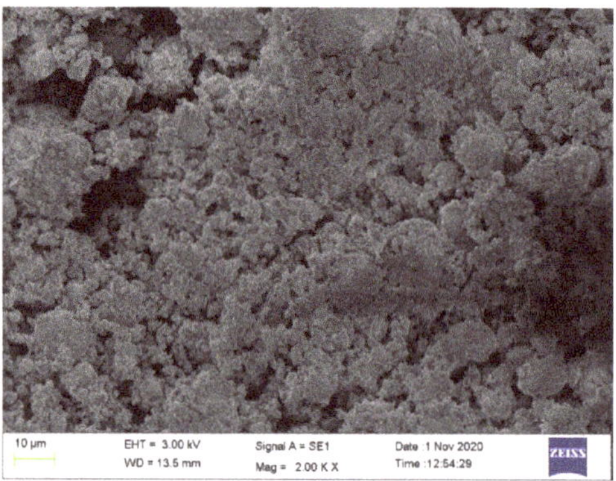

**Figure 2.** SEM image of raw gangue powder.

**Figure 3.** SEM image of 700 °C calcined gangue.

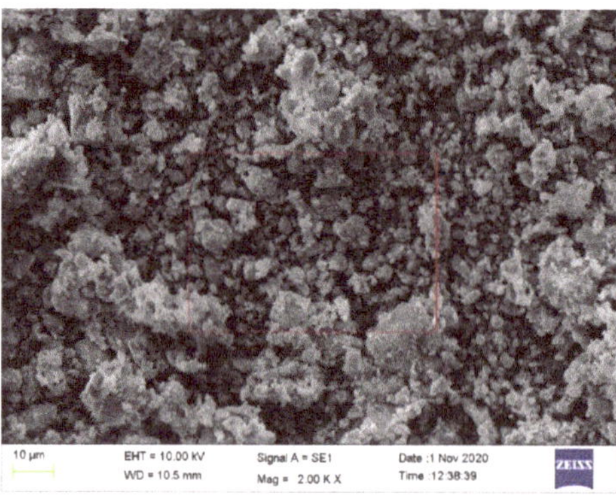

**Figure 4.** SEM image of acid-leached residue.

**Figure 5.** The SEM image of acid-leached residue (taken from Figure 4).

As seen, the particles of raw gangue powder are relatively uniform at 15–20 μm or larger in Figure 2; after calcination, the particle size decreases and fragments appear. After acid leaching, the number of flaky debris increases significantly. Most particle sizes are less than 5 μm, which is due to the overflow of water in coal gangue after calcination. Moreover, kaolinite decomposes into amorphous $Al_2O_3$ and $SiO_2$, at 700 °C; most of the iron content and part of $Al_2O_3$ are dissolved in the acid solution.

In order to observe the particle morphology of acid-leached residue more clearly, we enlarged a region in Figure 4 and marked it red, then got Figure 5. Figure 5 shows that many small particles become smoother, this is because the edges and corners of some particles are worn off by stirring, which makes the raw material more fragmented, so it is conducive to the alkali-melting reaction in the next step.

### 3.2. Phase Analysis of the Residue and Activation Product

Figure 6 shows that the crystal substances in the coal gangue mainly contain kaolinite and quartz, as well as a small amount of pyrite and siderite. Table 1 shows that the iron element content in the coal gangue is very high, but the diffraction peaks of iron-containing materials are very weak in Figure 6, which indicates that the iron materials are amorphous. After calcination, kaolinite turns into metakaolin, its crystal structure is destroyed, and the carbon in coal gangue is removed.

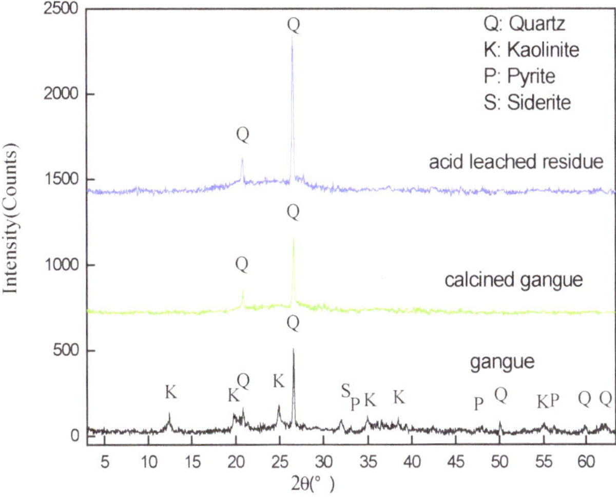



**Figure 6.** XRD patterns of coal gangue powder at 750 °C, calcined powder, and acid-leached residue.

The quartz is inert and it is difficult for it to participate in the crystallization reaction directly, so it must to be activated before the crystallization reaction. Therefore, according to the mass ratio of *m* (acid leached residue):*m* (sodium carbonate) = 1:1.1, after mixing the sodium carbonate evenly with the residue, the mixture is melted at 750 °C for 2 h, then the melted powder is analyzed by XRD and added to the water to be fully grinded then filtered; the XRD analyses are shown in Figure 7.

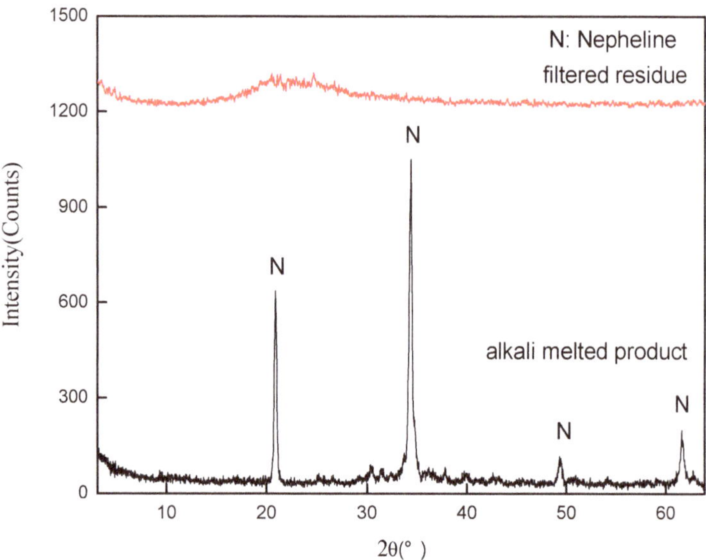


**Figure 7.** XRD spectra of alkali-melted product and its water washing filter residue.

Figure 7 shows that the alkali melted product is mainly nepheline (nepheline, PDF card number: 76-1733), its chemical formula is $NaAlSiO_4$, and the main reaction equations of acid-leached filter residue with sodium carbonate at 750 °C are as follows:

$$Na_2CO_3 + SiO_2 = Na_2SiO_3 + CO_2\uparrow \tag{1}$$

$$Al_2O_3 \cdot 2SiO_2 + Na_2CO_3 = 2NaAlSiO_4 + CO_2\uparrow \tag{2}$$

The metakaolin and quartz, which contain silicon and aluminum substances, can react with $Na_2CO_3$ at 750 °C, and nepheline can dissolve in water, so it can be used as silicon and aluminum sources to synthesize NaA zeolite. Therefore, this method can completely activate the quartz in the acid-leached residue; it not only takes full advantage of the silicon and aluminum sources in the residue but also avoids the quartz entering NaA zeolite products, which increases the quality of NaA products, so the soluble intermediate product $NaAlSiO_4$ with higher chemical activity is beneficial to the hydrothermal crystallization reaction.

The rest filter residue contains amorphous $SiO_2$ and $Al_2O_3$, they both have high chemical activity and can react with sodium aqueous solution in the hydrothermal synthesis period; the reactions are as follows:

$$SiO_2 + 2NaOH = Na_2SiO_3 + H_2O \tag{3}$$

$$Al_2O_3 + 2NaOH = 2NaAlO_2 + H_2O \tag{4}$$

$Na_2SiO_3$ and $NaAlO_2$ can be used as the active silicon source and aluminum sources to synthesis NaA zeolite. Because the synthesis of NaA zeolite is closely related to

$n(SiO_2)/n(Al_2O_3)$, $n(Na_2O)/n(SiO_2)$, $n(H_2O)/n(Na_2O)$, aging temperature, and crystallization time, so the experiments focused on these five factors.

### 3.3. Effects of n(SiO₂)/n(Al₂O₃) on the Product Phase

Setting the reaction system to $n(H_2O)/n(Na_2O) = 50$; $n(Na_2O)/n(SiO_2) = 1.2$; aging at 50 °C for 1 h; crystallization at 94 °C for 4 h,; and changing $n(SiO_2)/n(Al_2O_3) = 1.0, 1.5, 2.0$, and 2.5 respectively, after hydrothermal crystallization reaction, the XRD analyses are shown in Figure 8.

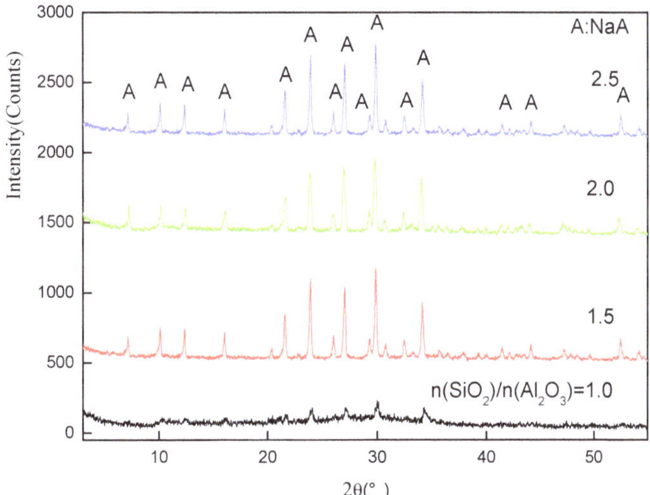


**Figure 8.** Effect of n(SiO₂)/n(Al₂O₃) on the phase.

Figure 8 shows when $n(SiO_2)/n(Al_2O_3) = 1.0$, the diffraction peak intensity of the product is very low, this is because the crystallinity and the zeolite type highly depend on the $n(SiO_2)/n(Al_2O_3)$ ratios [30–34]; when $n(SiO_2)/n(Al_2O_3) = 1.0$ or 1.5, they deviate from the target NaA zeolite product $n(SiO_2)/n(Al_2O_3) = 2.0$, which is not conducive to the nucleation and growth of zeolite; when $n(SiO_2)/n(Al_2O_3) = 2.0$, the diffraction intensity increases. The search result shows that the final product is NaA zeolite (PDF card number: 39-0223), there is a further increase of $n(SiO_2)/n(Al_2O_3) = 2.5$, and the diffraction intensity also increases. If the $n(SiO_2)/n(Al_2O_3)$ ratio is too high, the higher $n(SiO_2)/n(Al_2O_3)$ ratio products such as NaX and NaP zeolites may appear; in order to take full use of the silicon and aluminum components, $n(SiO_2)/n(Al_2O_3) = 2.0$ is selected.

### 3.4. Effects of n(Na₂O)/n(SiO₂) on the Phase

Setting the reaction system to $n(SiO_2)/n(Al_2O_3) = 2.0$; $n(H_2O)/n(Na_2O) = 60$; aging at 50 °C for 1 h; crystallization at 94 °C for 4 h; and changing $n(Na_2O)/n(SiO_2) = 1.2, 1.7, 1.9$, and 2.1, respectively, after finishing reactions, the XRD analyses of the products are shown in Figure 9.

In Figure 9, the intensity of these spectra is high and their shapes are similar. This is because $n(Na_2O)/n(SiO_2)$ works together with $n(H_2O)/n(Na_2O)$; they change the alkalinity of the solution and affect the growth rate of zeolite. When the $n(Na_2O)/n(SiO_2)$ ratio is larger, the system alkalinity is higher, which promotes the dissolution rate of the materials and the growth rate of NaA zeolite crystal; when $n(Na_2O)/n(SiO_2) = 2.1$, the intensity of diffraction peaks is high and there are no other miscellaneous crystals, so $n(Na_2O)/n(SiO_2) = 2.1$ is selected.

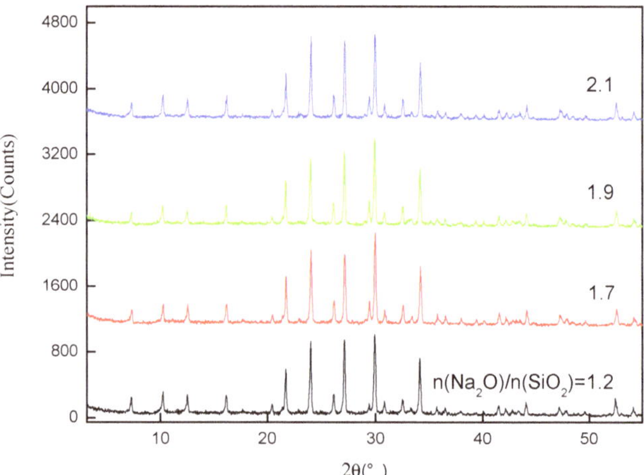



**Figure 9.** Effects of $n(Na_2O)/n(SiO_2)$ on the phase.

### 3.5. Effects of n(H₂O)/n(Na₂O) on the Phase

Setting the reaction system to $n(SiO_2)/n(Al_2O_3)$ = 2.0; $n(Na_2O)/n(SiO_2)$ = 1.7; the aging temperature as 50 °C; aging time as 2 h; crystallization at 94 °C for 4 h; and changing $n(H_2O)/n(Na_2O)$ = 50, 55, 60, and 65 respectively, after the reactions, the XRD analyses are shown in Figure 10.

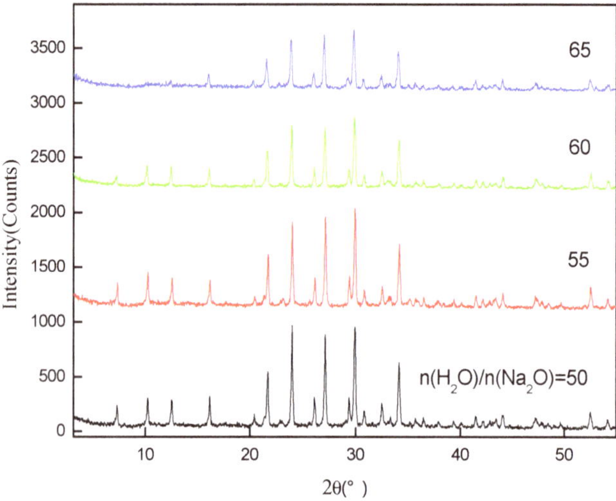



**Figure 10.** Effect of $n(H_2O)/n(Na_2O)$ on the phase.

We can see from Figure 10, as the value of $n(H_2O)/n(Na_2O)$ increases, the diffraction peak intensity decreases gradually. When $n(H_2O)/n(Na_2O)$ = 55, the intensity of the diffraction peaks are at maximum, and this is due to the high alkalinity concentration, which lead to large viscosity of the system and is not conducive to the mass transfer process, but is adverse to the dissolution of solid silicon and aluminum components in the system [35]. The reaction system cannot provide sufficient highly active raw material for crystal growth, which causes the intensity of the peaks to decrease, and so $n(H_2O)/n(Na_2O)$ = 55 is selected.

### 3.6. Effect of Aging Temperature on the Phase

Setting reaction system $n(SiO_2)/n(Al_2O_3) = 2.0$; $n(H_2O)/n(Na_2O) = 55$; $n(Na_2O)/n(SiO_2)$ = 2.1; the aging time as 2 h; crystallization at 94 °C for 4 h; and changing aging temperature to 30 °C, 40 °C, 50 °C, and 60 °C respectively, after the crystallization reaction, the XRD analyses of the products are shown in Figure 11.

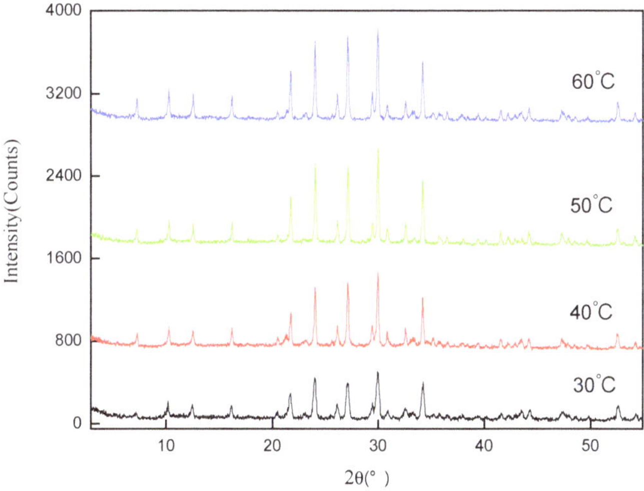


**Figure 11.** Effect of aging temperature on the phase.

The effect of aging temperature is reflected in the process of transforming the liquid sol into zeolite. It can be seen from Figure 11 that when the aging temperature is 30 °C or 40 °C, the intensity of diffraction peaks is low; when the aging temperature is 50 °C or 60 °C, the intensity is high, and this is because the low-temperature aging stage can improve the nucleation rate, reduce the grain size, and increase the number of crystals [36]. When the aging temperature is 50 °C or 60 °C, the raw materials can be fully dissolved and the nucleation will grow sooner, but if the aging temperature is too high, it will reduce the number of nucleation, so the proper aging temperature is 60 °C.

### 3.7. Effect of Crystallization Time on the Phase and Morphology

Setting reaction system $n(SiO_2)/n(Al_2O_3) = 2.0$; $n(H_2O)/n(Na_2O) = 65$; $n(Na_2O)/n(SiO_2)$ = 2.1; aging at 60 °C for 2 h; crystallization temperature as 94 °C, and changing the crystallization time to 1 h, 2 h, 3 h, and 4 h respectively, after the reactions, the XRD patterns of the products are shown in Figure 12, and the SEM images are shown in Figures 13–16.

The crystallization time is particularly important for synthesis zeolites; it mainly affects their crystallinity. We can see from Figure 12, when the crystallization time is 1 h or 2 h, the intensity of the peaks is low; when the crystallization time is 1 h, just a small number of little NaA crystals begin to appear (Figure 13); most of the crystals are about 1 μm in size. With the extension of reaction time to 2–4 h, a large number of crystals begin to appear, the crystals are a regular cubic shape and the particle size is about 2 μm. At the same time, when the crystallization time is 4 h, the intensity of the peaks is relatively higher, which is due to the longer crystallization time that gives zeolite enough time to grow, so the crystallization reaction time is determined to be 4 h.

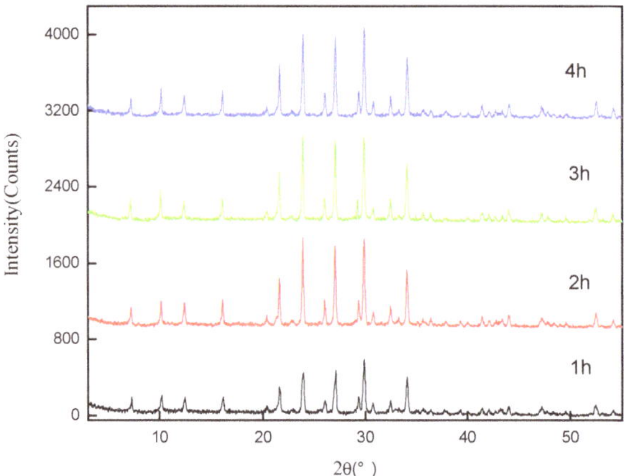

**Figure 12.** Effect of crystallization time on the phase.

**Figure 13.** The crystallization time is 1 h.

**Figure 14.** The crystallization time is 2 h.

**Figure 15.** The crystallization time is 3 h.

**Figure 16.** The crystallization time is 4 h.

*3.8. XRD and SEM Analyses of the Product under Optimized Conditions*

The product prepared under the above optimized conditions was detected by XRD and SEM; the results are shown in Figures 17 and 18.

After searching, the d values and 2 theta values are in good agreement with PDF card: 39-0223, so the product can be confirmed as NaA zeolite. As shown in Figure 17, the product is pure NaA zeolite, the diffraction peaks are sharp, and the intensity of the diffraction peaks are high, which indicates that the crystallinity of the product is high. Figure 18 shows that NaA zeolites prepared under the optimized conditions have regular cubic shapes and a uniform particle size of about 2.5 μm.

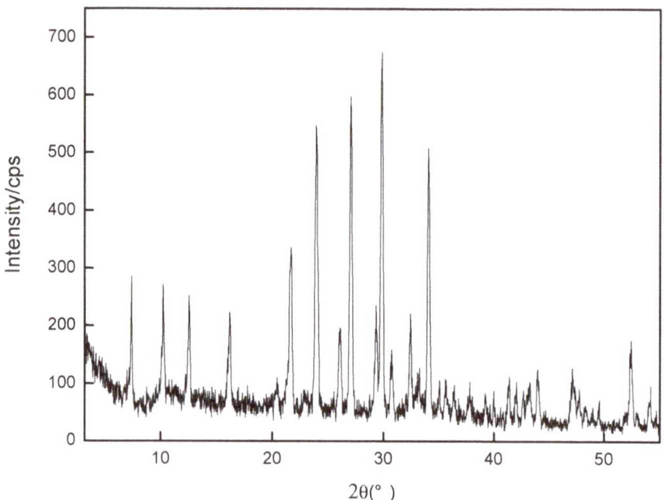

**Figure 17.** XRD spectra of the product under optimized conditions.

**Figure 18.** SEM photo of the product under optimized conditions.

### 3.9. Preparation Mechanism

The NaA zeolite is from high iron content coal gangue; many chemical reactions occurred in the process, in addition to the reactions mentioned above. Nepheline is produced by high temperature reaction and can be hydrolyzed under alkaline conditions. The equations are as follows:

$$2NaAlSiO_4 + 4NaOH = 2NaAlO_2 + 2Na_2SiO_3 + 2H_2O \tag{5}$$

$$SiO_2 + 2NaOH = Na_2SiO_3 + H_2O \tag{6}$$

$$Al_2O_3 + 2NaOH = 2NaAlO_2 + H_2O \tag{7}$$

The $NaAlSiO_4$ dissolves in NaOH solution to form $NaAlO_2$ and $Na_2SiO_3$, the amorphous $SiO_2$ in the residue reacts with NaOH solution to form $Na_2SiO_3$, and amorphous

Al$_2$O$_3$ in the residue reacts with NaOH solution to form NaAlO$_2$ [37–39]. In the silicate ions and aluminate ions system, the primary structural unit of silica–alumina zeolite skeleton is a silica–oxygen tetrahedron and alumina–oxygen tetrahedron, which are both called TO$_4$, as shown in Figure 19.

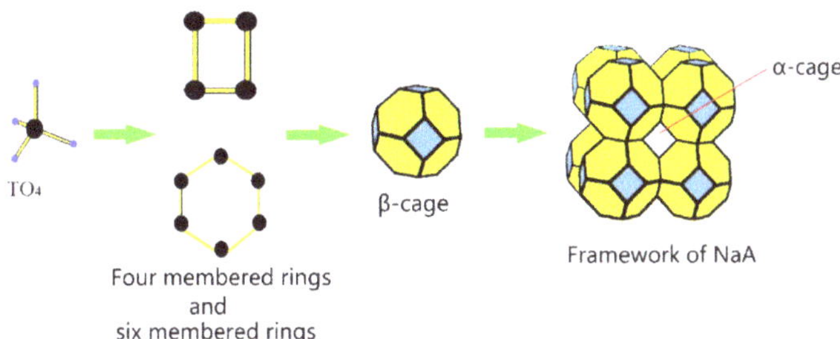


**Figure 19.** The formation process of NaA zeolite.

As the hydrothermal reaction continues, the TO$_4$ in the reaction system is connected through oxygen atoms, and the tetrahedrons formed are connected through an oxygen bridge to form a ring; four tetrahedrons make up a four-membered ring and six tetrahedrons make up a six-membered ring (the oxygen atoms are omitted, Figure 19) [40]. These rings are reduced to quadrangles and hexagons, and each of them have a Si or Al ion at the tip of the corner. Under the action bridging of cations, silicon and aluminum ions in the rings are further condensed around cations and continue to connect in three-dimensional space to form β cages (Figure 19). The β cage is a chamfered octahedron containing 6 four-membered rings; 8 six-membered rings; and 24 angular apex, β-cages arranged in the form of body-centered cubes, which are connected by a double quaternion ring, resulting in an α cage (Figure 19) and a three-dimensional skeleton structure at the center of the cell.

When the cages are formed, they will continue to form the zeolite cages; the zeolite cages will continue to expand in a three-dimensional direction according to the body-centered cubic structure. When there has been a certain amount of expansion, they will generate a certain geometric shape of the grain. The little grain seeds continue to grow in the hydrothermal system and finally form NaA zeolite crystals.

## 4. Conclusions

Through this study, these conclusions are drawn as follows:

(1) Calcination of the coal gangue at 700 °C for 2 h and then leaching the calcined powder by 20% hydrochloric acid; the liquid to solid ratio is 3:1 at 90 °C for 7 h and can remove most of the iron ions.

(2) Evenly mixing the sodium carbonate with the acid-leached filter residue according to the mass ratio of m(acid leached residue): m(sodium carbonate) = 1:1.1, the mixture is melted at 750 °C for 2 h, then acid-leached filter residue turns into NaAlSiO$_4$; the rests are amorphous SiO$_2$ and Al$_2$O$_3$, which both have high chemical activity and can participate in the crystallization reaction, so the acid-leached filter residue is activated absolutely.

(3) $n(SiO_2)/n(Al_2O_3) = 2.0$, $n(Na_2O)/n(SiO_2) = 2.1$, $n(H_2O)/n(Na_2O) = 55$, aging at 60 °C for 2 h, and crystallization at 94 °C for 4 h are the optimized conditions for NaA zeolite synthesis.

Compared with [41,42], the preparation conditions obtained in this study are more mild, the purity of the product is higher, and there are no impurity crystals in the product.

**Author Contributions:** D.K.: conceptualization, writing—original draft, investigation, methodology, data curation. R.J.: methodology, writing-reviewing, editing and supervision. All authors have read and agreed to the published version of the manuscript.

**Funding:** The study is financially supported by Guizhou Provincial Education Department's Scientific and Technological Innovation Team Project (NO. [2017] 054), Guizhou Science and Technology Foundation Project (NO. [2018] 1142), and Liupanshui City Science and Technology Foundation (NO. 52020-2019-05-17).

**Institutional Review Board Statement:** Not applicable.

**Informed Consent Statement:** Not applicable.

**Data Availability Statement:** Not applicable.

**Conflicts of Interest:** The authors declare no conflict of interest.

# References


1. Niu, X.; Guo, S.; Gao, L.; Cao, Y.; Wei, X. Mercury release during thermal treatment of two coal gangues and two coal slimes under $N_2$ and in air. *Energy Fuels* **2017**, *31*, 8648–8654. [CrossRef]
2. Xu, H.; Song, W.; Cao, W.; Gang, S.; Lu, H.; Yang, D.; Chen, D.; Zhang, R. Utilization of coal gangue for the production of brick. *J. Mater. Cycles Waste Manag.* **2016**, *19*, 1270–1278. [CrossRef]
3. Du, H.; Ma, L.; Liu, X.-Y.; Zhang, F.; Yang, X.; Wu, Y.; Zhang, J. A novel mesoporous $SiO_2$ material with MCM-41 structure from coal gangue: Preparation, ethylenediamine modification, and adsorption properties for $CO_2$ capture. *Energy Fuels* **2018**, *32*, 5374–5385. [CrossRef]
4. Liu, M.; Zhu, Z.; Zhang, Z.; Chu, Y.; Yuan, B.; Wei, Z. Development of highly porous mullite whisker ceramic membranes for oil-in-water separation and resource utilization of coal gangue. *Sep. Purif. Technol.* **2020**, *237*, 116483. [CrossRef]
5. Zhao, H.; Chen, Y.; Duan, X. Study on the factors affecting the deep reduction of coal gangue containing high contents of iron and sulfur. *Fuel* **2020**, *288*, 119571. [CrossRef]
6. Ma, H.; Zhu, H.; Wu, C.; Chen, H.; Sun, J.; Liu, J. Study on compressive strength and durability of alkali-activated coal gangue-slag concrete and its mechanism. *Powder Technol.* **2020**, *368*, 112–124. [CrossRef]
7. Xiao, M.; Ju, F.; He, Z. Research on shotcrete in mine using non-activated waste coal gangue aggregate. *J. Clean. Prod.* **2020**, *259*, 120810. [CrossRef]
8. Du, T.; Wang, D.; Bai, Y.; Zhang, Z. Optimizing the formulation of coal gangue planting substrate using wastes: The sustainability of coal mine ecological restoration. *Ecol. Eng.* **2020**, *143*, 105669. [CrossRef]
9. Jouni, R.; Katja, O.; Paivo, K.; Marcello, R.; Mirja, I. Milling of peat-wood fly ash: Effect on water demand of mortar and rheology of cement paste. *Constr. Build. Mater.* **2018**, *180*, 143–153.
10. Zhou, C.; Liu, G.; Wu, S.; Lam, P.K.S. The environmental characteristics of usage of coal gangue in bricking-making: A case study at Huainan, China. *Chemosphere* **2014**, *95*, 274–280. [CrossRef]
11. Yang, Z.; Zhang, Y.; Liu, L.; Seshadri, S.; Wang, X.; Zhang, Z. Integrated utilization of sewage sludge and coal gangue for cement clinker products: Promoting tricalcium silicate formation and trace elements immobilization. *Materials* **2016**, *9*, 275. [CrossRef]
12. Han, J.; Jin, X.; Song, C.; Bi, Y.; Liu, Q.; Liu, C.; Ji, N.; Lu, X.; Ma, D.; Li, Z. Rapid synthesis and NH3-SCR activity of SSZ-13 zeolite via coal gangue. *Green Chem.* **2019**, *22*, 219–229. [CrossRef]
13. Chen, J.; Lu, X. Synthesis and characterization of zeolites NaA and NaX from coal gangue. *J. Mater. Cycles Waste Manag.* **2017**, *20*, 489–495. [CrossRef]
14. Jin, Y.; Li, L.; Liu, Z.; Zhu, S.; Wang, D. Synthesis and characterization of low-cost zeolite NaA from coal gangue by hydrothermal method. *Adv. Powder Technol.* **2021**, *32*, 791–801. [CrossRef]
15. Wang, X.; Zhang, Y. Synthesis and characterization of zeolite A obtained from coal gangue for the adsorption of $F^-$ in wastewater. *Sci. Adv. Mater.* **2019**, *11*, 277–282. [CrossRef]
16. Ren, J.; Xie, C.; Guo, X.; Qin, Z.; Lin, J.; Li, Z. Combustion characteristics of coal gangue under an atmosphere of coal mine methane. *Energy Fuels* **2012**, *28*, 3688–3695. [CrossRef]
17. Merrikhpour, H.; Jalali, M. Comparative and competitive adsorption of cadmium, copper, nickel, and lead ions by iranian natural zeolite. *Clean. Technol. Environ. Policy* **2013**, *15*, 303–316. [CrossRef]
18. Polat, E.; Karaca, M.; Demir, H.; Onus, A.N. Use of natural zeolite (clinoptilolite) in agriculture. *J. Fruit Ornam. Plant Res.* **2004**, *12*, 183–189.
19. Muzic, M.; Sertic-Bionda, K.; Adzamic, T. Evaluation of commercial adsorbents and their application for desulfurization of model fuel. *Clean. Technol. Environ. Policy* **2012**, *14*, 283–290. [CrossRef]
20. Zhu, J.; Meng, X.; Xiao, F. Mesoporous zeolites as efficient catalysts for oil refining and natural gas conversion. *Front. Chem. Sci. Eng.* **2013**, *7*, 233–248. [CrossRef]
21. Yilmaz, B.; Müller, U. Catalytic applications of zeolites in chemical industry. *Top. Catal.* **2009**, *52*, 888–895. [CrossRef]
22. Qian, T.; Li, J. Synthesis of Na-A zeolite from coal gangue with the in-situ crystallization technique. *Adv. Powder Technol.* **2015**, *26*, 98–104. [CrossRef]
23. Lu, X.; Shi, D.; Chen, J. Sorption of $Cu^{2+}$ and $Co^{2+}$ using zeolite synthesized from coal gangue: Isotherm and kinetic studies. *Environ. Earth Sci.* **2017**, *76*, 591. [CrossRef]

24. Li, H.; Zheng, F.; Wang, J.; Zhou, J.; Huang, X.; Chen, L.; Hu, P.; Gao, J.; Zhen, Q.; Bashir, S.; et al. Facile preparation of zeolite-activated carbon composite from coal gangue with enhanced adsorption performance. *Chem. Eng. J.* **2020**, *390*, 124513. [CrossRef]

25. Jha, B.; Singh, D.N. A three step process for purification of fly ash zeolites by hydrothermal treatment. *Appl. Clay Sci.* **2014**, *90*, 122–129. [CrossRef]

26. Khaleque, A.; Alam, M.; Hoque, M.; Mondal, S.; Bin Haider, J.; Xu, B.; Johir, M.; Karmakar, A.K.; Zhou, J.; Ahmed, M.B.; et al. Zeolite synthesis from low-cost materials and environmental applications: A review. *Environ. Adv.* **2020**, *2*, 100019. [CrossRef]

27. Belviso, C.; Cavalcante, F.; Niceforo, G.; Lettino, A. Sodalite, faujasite and A-type zeolite from 2:1dioctahedral and 2:1:1 trioctahedral clay minerals. A singular review of synthesis methods through laboratory trials at a low incubation temperature. *Powder Technol.* **2017**, *320*, 483–497. [CrossRef]

28. Ayele, L.; Perez-Pariente, J.; Chebude, Y.; Diaz, I. Conventional versus alkali fusion synthesis of zeolite A from low grade kaolin. *Appl. Clay Sci.* **2016**, *132*, 485–490. [CrossRef]

29. Bi, H.; Wang, C.; Lin, Q.; Jiang, X.; Jiang, C.; Bao, L. Combustion behavior, kinetics, gas emission characteristics and artificial neural network modeling of coal gangue and biomass via TG-FTIR. *Energy* **2020**, *213*, 118790. [CrossRef]

30. Han, J.; Ha, Y.; Guo, M.; Zhao, P.; Liu, Q.; Liu, C.; Song, C.; Ji, N.; Lu, X.; Ma, D.; et al. Synthesis of zeolite SSZ-13 from coal gangue via ultrasonic pretreatment combined with hydrothermal growth method. *Ultrason. Sonochemistry* **2019**, *59*, 104703. [CrossRef] [PubMed]

31. Eilertsen, E.A.; Nilsen, M.H.; Wendelbo, R.; Olsbye, U.; Lillerud, K.P. Synthesis of high silica CHA zeolites with controlled Si/Al ratio. *Stud. Surf. Sci. Catal.* **2008**, *174*, 265–268.

32. Wang, Q.; Wang, L.; Wang, H.; Li, Z.; Zhang, X.; Zhang, S.; Zhou, K. Effect of $SiO_2/Al_2O_3$ ratio on the conversion of methanol to olefins over molecular sieve catalysts. *Front. Chem. Sci. Eng.* **2011**, *5*, 79–88. [CrossRef]

33. Kosinov, N.; Auffret, C.; Borghuis, G.J.; Sripathi, V.; Hensen, E. Influence of the si/al ratio on the separation properties of SSZ-13 zeolite membranes. *J. Membr. Sci.* **2015**, *484*, 140–145. [CrossRef]

34. Li, J.; Zhang, Z.; Li, X.; Bao, S.; Liu, L.; He, H.; Lai, X.; Zhang, L.; Zhang, P. Preparation of ZSM-5 molecular sieve by diatomaceous earth in-situ crystallization. *Integr. Ferroelectr.* **2018**, *189*, 165–174. [CrossRef]

35. Yao, G.; Lei, J.; Zhang, X.; Sun, Z.; Zheng, S. One-step hydrothermal synthesis of zeolite X powder from natural low-grade diatomite. *Materials* **2018**, *11*, 906. [CrossRef]

36. Maia, A.Á.B.; Dias, R.N.; Angélica, R.S.; Neves, R.F. Influence of an aging step on the synthesis of zeolite NaA from brazilian amazon kaolin waste. *J. Mater. Res. Technol.* **2019**, *8*, 2924–2929. [CrossRef]

37. Guo, Y.P.; Lee, N.H.; Oh, H.-J.; Yoon, C.R.; Rhee, C.K.; Lee, K.S.; Kim, S.J. Fabrication of oriented $TiO_2$ based nanotube array thin films. *Solid State Phenom.* **2008**, *135*, 19–22. [CrossRef]

38. Behin, J.; Bukhari, S.S.; Kazemian, H.; Rohan, S. Developing a zero liquid discharge process for zeolitization of coal fly ash to synthetic NaP zeolite. *Fuel* **2016**, *7*, 195–202. [CrossRef]

39. Bukhari, S.S.; Behin, J.; Kazemian, H.; Rohani, S. A comparative study using direct hydrothermal and indirect fusion methods to produce zeolites from coal fly ash utilizing single-mode microwave energy. *J. Mater. Sci. Lett.* **2011**, *19*, 8261–8271. [CrossRef]

40. Bu, N.; Liu, X.; Song, S.; Liu, J.; Yang, Q.; Li, R.; Zheng, F.; Yan, L.; Zhen, Q.; Zhang, J. Synthesis of NaY zeolite from coal gangue and its characterization for lead removal from aqueous solution. *Adv. Powder Technol.* **2020**, *31*, 2699–2710. [CrossRef]

41. Li, Y.; Peng, T.; Man, W.; Ju, L.; Zheng, F.; Zhang, M.; Guo, M. Hydrothermal synthesis of mixtures of NaA zeolite and sodalite from Ti-bearing electric arc furnace slag. *RSC Adv.* **2016**, *6*, 8358–8366. [CrossRef]

42. Su, Q.; He, Y.; Yang, S.; Wan, H.; Cui, X. Synthesis of NaA-zeolite microspheres by conversion of geopolymer and their performance of Pb (II) removal. *Appl. Clay Sci.* **2020**, *20*, 105914. [CrossRef]

*Article*

# Near-Infrared Spectroscopic Study of OH Stretching Modes in Kaolinite and Dickite

Shaokun Wu [1], Mingyue He [1,*], Mei Yang [2] and Bijie Peng [1]

[1] Gemological Institute, China University of Geosciences, Beijing 100083, China; 3009210005@cugb.com (S.W.); 300920003@cugb.com (B.P.)
[2] Sciences Institute, China University of Geosciences, Beijing 100083, China; yangmei@cugb.edu.cn
* Correspondence: hemy@cugb.edu.cn

**Abstract:** Kaolinite and dickite are differently ordered polytypes of kaolinite-group minerals, whose differences are in the stacking mode of layers and ion occupation. Fourier transform infrared spectroscopy was used to collect information about the differences between the two minerals. The common characteristics of kaolinite and dickite are bands near 4530 and 7068 $cm^{-1}$, which are attributed to the combination of the inner Al-OH stretching vibration and outer Al-OH bending vibration and the overtone of the inner Al-OH stretching vibration, respectively. The difference is that kaolinite has secondary peaks at 4610 and 7177 $cm^{-1}$, and the secondary peak of dickite is near 4588 $cm^{-1}$. The OH stretching vibration has the first fundamental overtone of the stretching vibration in the range of 7000–7250 $cm^{-1}$. In addition to the overtones generated by single OH stretching vibrations, overtones combining different OH stretching vibrations are also found, which are formed by adjacent peaks of OH stretching vibrations. The average factor of the first fundamental overtone with an OH-group stretching vibration is approximately 1.95. The near-infrared spectrum (NIR) of phyllosilicates is closely related to their structure and isomorphism. Therefore, the near-infrared region can distinguish between kaolinite and dickite and provide a basis for deposit research and geological remote sensing.

**Keywords:** near-infrared spectroscopy; kaolinite; dickite; OH group

**Citation:** Wu, S.; He, M.; Yang, M.; Peng, B. Near-Infrared Spectroscopic Study of OH Stretching Modes in Kaolinite and Dickite. *Crystals* **2022**, *12*, 907. https://doi.org/10.3390/cryst12070907

Academic Editor: Vladislav V. Gurzhiy

Received: 1 June 2022
Accepted: 23 June 2022
Published: 25 June 2022

**Publisher's Note:** MDPI stays neutral with regard to jurisdictional claims in published maps and institutional affiliations.

## 1. Introduction

Kaolinite minerals are the most common polytypes of aluminous phyllosilicates. They can be used in many types of materials, such as ceramics, glass, and refractory materials, and even as gemstones or handicrafts if the texture is pure and uniform. Kaolinite-group minerals include three ordered polytypes that are stable at normal temperature and pressure levels, namely, kaolinite, dickite, and nacrite, of which the first two are common. These minerals exist in a wide range of geological environments. Kaolinite exists in soil, sedimentary rocks, and hydrothermal deposits. Dickite is generally related to higher temperature and pressure conditions and mainly occurs in hydrothermal or sedimentary rocks; it can be used as a sign of exploration/zoning [1] and for detecting metallogenic environments [2] and geological activities [3].

Kaolinite and dickite display two different types of OH groups. The inner OH groups are located within the octahedral sheets, while the three nonequivalent inner surface OH groups are located on top of the octahedral sheets, sharing weak hydrogen bonds with the oxygen atoms of the next tetrahedral sheet. OH groups are very sensitive to the occurrence of defects, such as isomorphism and planar defects, and to the structural changes associated with phase transitions, which can be used to detect the structural order in hydrous minerals, especially in layered silicates [4]. Changes in the cationic environment will affect the vibrational properties of the OH groups, so the structural differences among clay minerals can be reflected by changes in hydroxyl bond vibrations.

Naturally occurring intergrowths of kaolinite-group minerals are very common, and their properties are very similar. Therefore, it is necessary to devise a simple characterization technique to identify and distinguish these minerals. Infrared spectroscopy is a powerful tool to characterize mineral species and study mineral structure. The mid-infrared region (MIR, 400–4000 cm$^{-1}$) is often used to detect the basic vibration of functional groups [5], and the near-infrared region (NIR, generally 4000–12,500 cm$^{-1}$) can reflect the vibration characteristics of the combination bands of hydroxyl and metal ions, as well as the overtone bands of water and some functional groups in minerals. These overtones will be sensitive to any change in the crystalline structure. Near-infrared spectroscopy has been widely used to identify kaolinite in mineral mixtures (e.g., [6–9]) and to study the stacking sequence of kaolinite-group minerals [10].

The purpose of this paper is to investigate the differences in the near-infrared spectra of kaolinite and dickite, which can thus realize the rapid identification of the two minerals. The results are applicable to ceramics, jade studies, and archaeology. In addition, the attribution of the OH group vibration peaks is discussed in detail, and attempts at rationalizing the origin of these bands are described. The careful measurement and interpretation of clay minerals may be very important for deposit research and geological remote sensing.

## 2. Materials and Methods

### 2.1. Materials

All samples were natural. The sample BL-DSL-4 is from Balin, Chifeng City, Inner Mongolia Autonomous Region, China; QJ-1 and QJ-8 are from the Shoushan area, Fuzhou City, Fujian Province, China. The dickite samples GS-1, GS-4, and GS-8 are from the Shoushan area, Fuzhou City, Fujian Province, China. Typical mineral samples were selected for this study. Each sample was moderately shattered, and the uniform part of each sample, without other phases, was used for the experiments.

Isomorphic substitution is very common in layered silicate minerals, and impurity elements will have varying degrees of influence on the results of spectral experiments. SEM-EDS analyses of the samples before this study showed that most samples contained the isomorphic elements potassium and iron, and a few samples contained other types of isomorphic elements (Table 1). Since the contents of impurity elements in the samples were very low, they were considered unlikely to affect the results of this study.

**Table 1.** Mineral composition of the samples.

| Sample | Color | Minerals | Refractive Index (RI) | Specific Gravity (SG) | Isomorphism (w%) |
|---|---|---|---|---|---|
| BL-DSL-4 | White | Kaolinite | 1.56 | 2.65 | Fe/0.1%, Ca/0.1%, Cu/<0.1% |
| QJ-1 | Gray | Kaolinite | 1.56 | 2.61 | K/0.2%, Fe/<0.1% |
| QJ-8 | White | Kaolinite | 1.56 | 2.62 | S/0.2%, K/0.1%, Ca/<0.1% |
| GS-1 | Gray | Dickite | 1.55 | 2.61 | Not detected |
| GS-4 | White | Dickite | 1.55 | 2.60 | K/0.1%, Fe/<0.1% |
| GS-8 | Light yellow | Dickite | 1.56 | 2.63 | K/0.4%, Fe/0.1% |

### 2.2. Methods

X-ray diffraction (XRD) data were acquired using the Smart lab X-ray powder diffractometer at the Institute of Earth Science, China University of Geosciences, Beijing (CUGB). The system was equipped with a conventional copper target X-ray tube (set to 45 kV and 200 mA) and a graphite monochromator, with a stepping scanning mode with a scanning speed of 4°/min and a step length of 0.02° in the range of 3–90°. The testing temperature was 15 °C, and the humidity was 22%. The samples were pulverized to 200-mesh powders using an agate mortar and stored immediately in a plastic bag to minimize contamination and oxidation. The results were analyzed using the MDI Jade 6.5 software and the International Center of Diffraction Database (ICDD). Baseline correction and normalization were performed before analysis.

Fourier transform infrared spectroscopy (FTIR) measurements were performed at room temperature using the Bruker Tensor II spectrometer at the Gem Research Center, the School of Gemology, China University of Geosciences, Beijing (CUGB). The spectra were collected in transmission mode in a tablet of KBr. The mid-infrared scanning range was 400–4000 cm$^{-1}$ with a resolution of 4 cm$^{-1}$, and the ratio of sample to KBr was 1:150. The near-infrared scanning range was 4000~8000 cm$^{-1}$, with a resolution of 8 cm$^{-1}$. Each spectrum was averaged from 64 scans to improve the signal-to-noise ratio. Baseline correction and normalization were performed.

Clay minerals often consist of aggregates of small particles of the same size as near-infrared waves, and if the size of the particles is larger than the IR wavelength, the IR spectrum is affected by multiple scattering within the particles, which can produce interference effects and reduce the signal strength [11,12]. The mineral shape also affects the signal [12]. In this study, because it was difficult to crush the sample to a size smaller than the wavelength of a near-infrared wave (more than 6000-mesh), the ratio of sample to KBr was increased to 2:150 in the near-infrared range to enhance the spectral signal.

Infrared hydroxyl vibration usually has small peak spacing, and a large number of spectral peaks coincide. Therefore, the original FTIR spectra were analyzed by taking the second derivative to find the spectral peaks [13–15]. The absorption maximum in the second-derivative spectrum was converted to the minimum. This greatly reduced the apparent spectral bandwidth and largely eliminated baseline differences between spectra [16]. Then, Peakfit v4.12 software was used to fit the peaks in order to find the exact locations and forms of their component peaks. All of the spectra were fitted using a combined Gauss–Lorentz area function, $r^2 > 0.95$.

Raman microprobe spectra were tested with the Horiba LabRAM HR-Evolution laser Raman spectrometer. The analytical conditions included a laser wavelength of 532 nm, a grating of 600 (500 nm), a range of 100–4000 cm$^{-1}$, a resolution of 1 cm$^{-1}$, a scanning time of 10 s, and an accumulation of 3 times, with the polished surface cleaned with alcohol before measurement.

## 3. Results

### 3.1. X-ray Diffraction

The XRD patterns of the six samples with PDF cards of kaolinite and dickite from the ICDD® database are shown in Figure 1. All of the samples are consistent with the characteristics of kaolinite-group minerals, showing common strong peaks of d(001) = 12.321–12.467° and d(002) = 24.839–24.970° and medium-intensity peaks of d(−110) = 20.281–20.460° and d(−133) = 38.460–38.823°. The classical method to distinguish kaolinite and dickite is to use the number and shape of peaks in the range of 34–40° (gray area in Figure 1). The samples BL-DSL-4, QJ-1, and QJ-8 have six peaks within this range, which is characteristic of kaolinite. The samples GS-1, GS-4, and GS-8 have four peaks within this range, which is characteristic of dickite. All of the samples were found to be pure minerals without any other phases.

### 3.2. Characteristics of MIR

There are four main spectral bands in the mid-infrared range: 400–600 cm$^{-1}$, 650–800 cm$^{-1}$, 850–1200 cm$^{-1}$, and 3500–3750 cm$^{-1}$. The spectral peaks are highly consistent with the standard spectra of kaolinite-group minerals (Figure 2), and no additional peaks appeared.

The number and intensity of peaks in the high-frequency region of 3600–3750 cm$^{-1}$ are often used to distinguish kaolinite-group minerals. Kaolinite shows three or four peaks in this range, with the strongest near 3700 cm$^{-1}$ and a weak peak in the middle. Dickite has three, which weaken in turn. We could observe more vibration modes after fitting the spectrum (Table 2 and Figure 3) [12,17]. It is worth noting that there was no change in the dipole moment at the positions near 3685 cm$^{-1}$ in kaolinite and 3645 cm$^{-1}$ in dickite, which should show no peak; however, weak peaks were fitted at the corresponding positions. This may have been caused by the structural ideality, which decreased due to the isomorphism, and then the dipole moment changed slightly.

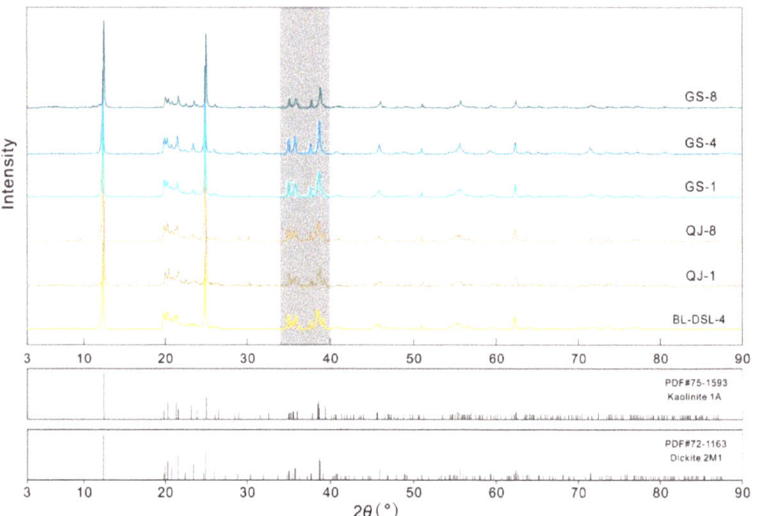

**Figure 1.** Powder XRD patterns of the samples.

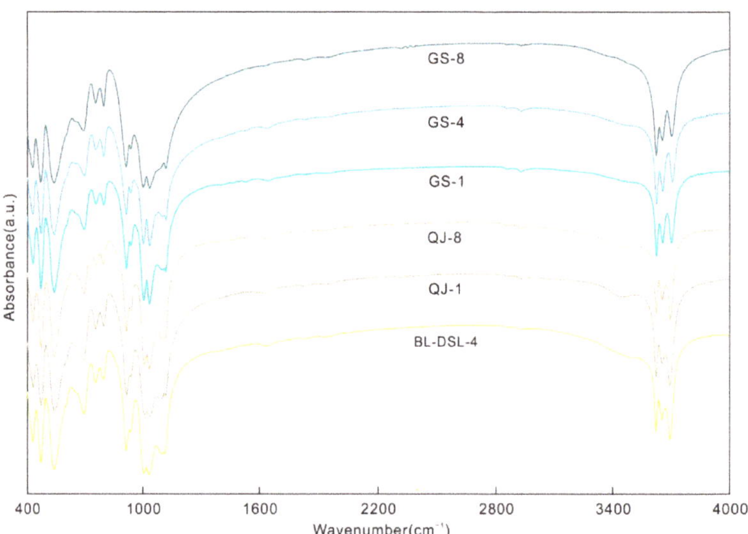

**Figure 2.** MIR spectra of the kaolinite and dickite samples.

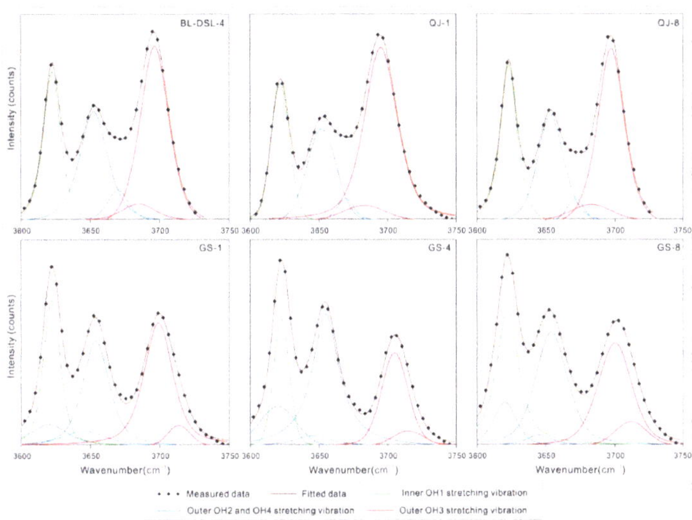



**Figure 3.** MIR spectral component analysis of the OH stretching region.

### 3.3. Raman

Raman spectroscopy has proven most useful for the elucidation of the kaolinite hydroxyl structure [18]. Raman spectra can detect the vibrations of nonpolar groups that cause a change in polarizability; these vibrations may be non-infrared-active. Thus, it can be a supplement to the infrared spectra. Kaolinite and dickite show 3685 and 3645 cm$^{-1}$ outer OH stretching vibration Raman peaks, respectively (Figures 4 and 5), which cannot be observed in normal infrared spectra. Although there is no infrared activity, it will still affect the combination bands and overtone.

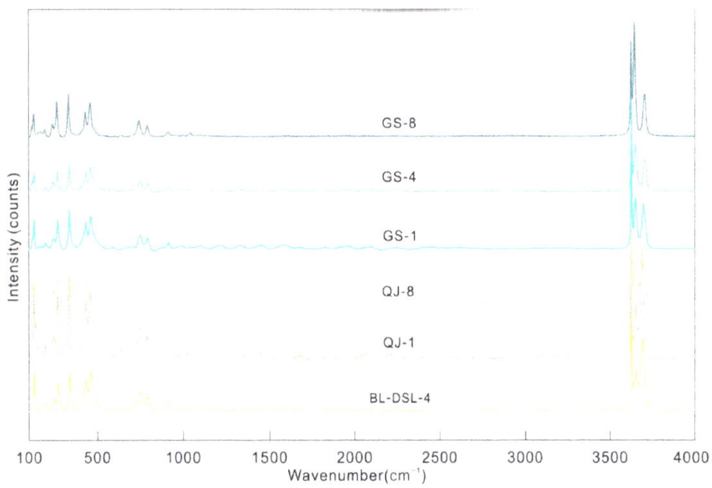



**Figure 4.** Raman spectra of the kaolinite and dickite samples.

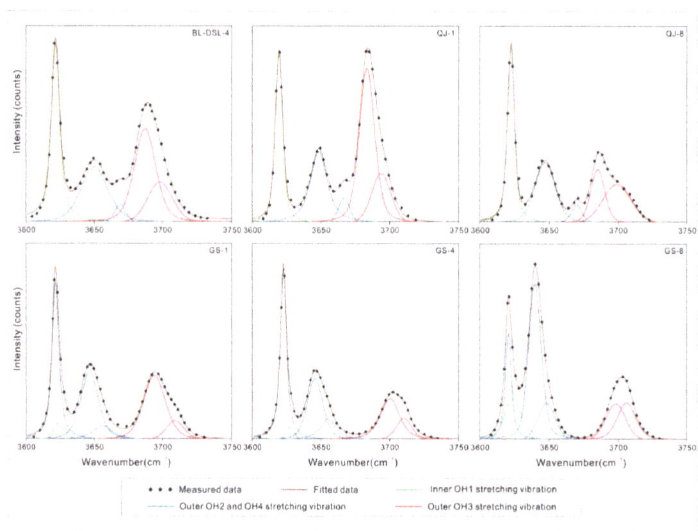



**Figure 5.** Raman spectral component analysis of the OH stretching region.

**Table 2.** The MIR and Raman bands related to kaolinite mineral samples and their assignments ($cm^{-1}$).

| Band Assignment [19–32] | Active | BL-DSL-4 | QJ-1 | QJ-8 | GS-1 | GS-4 | GS-8 |
|---|---|---|---|---|---|---|---|
| Lattice vibration | R | 122 | 118 | 122 | 122 | 120 | 118 |
| $Si_2O_5$ out of plane | R | 131 | 127 | 131 | 131 | 132 | 131 |
| $\nu_2$(e) of $AlO_6$ octahedron | R | | 140 | 142 | | | |
| $A_{1g}(\nu_1)$ of $AlO_6$ octahedron | R | 196 | 194 | 198 | 198 | 198 | 195 |
| O-H-O stretching vibration | R | 243 | 241 | 241 | 241 | 241 | 238 |
| O-H-O stretching vibration | R | 268 | 266 | 268 | 266 | 266 | 263 |
| Si-O stretching vibration | R | 334 | 332 | 334 | 334 | 334 | 331 |
| Si-O stretching vibration | IF | 412 | 412 | 412 | 411 | 412 | 412 |
| Si-O bending vibration | IF, R | 431 | 432 | 431 | 429 | 428 | 428 |
| Si-O bending vibration | IF, R | 472 | 473 | 472 | 471 | 471 | 471 |
| Si-O-Al vibration | IF | 541 | 542 | 541 | 539 | 541 | 540 |
| Si-O-Al vibration | IF | 694 | 692 | 695 | 694 | 695 | 692 |
| Si-O-Al vibration | IF, R | 754 | 754 | 754 | 754 | 754 | 752 |
| Si-O stretching vibration | IF, R | 794 | 795 | 794 | 794 | 794 | 795 |
| Outer Al-OH bending vibration | IF, R | 914 | 914 | 914 | 913 | 913 | 912 |
| Inner Al-OH bending vibration | IF | 937 | 937 | 938 | 937 | 937 | 937 |
| Si-O-Al vibration | IF | 1007 | 1009 | 1009 | 1004 | 1003 | 1001 |
| Si-O-Si vibration | IF | 1033 | 1034 | 1034 | 1033 | 1034 | 1036 |
| Si-O stretching vibration | IF | 1099 | 1094 | 1100 | 1097 | 1100 | 1092 |
| Si-O stretching vibration | IF | 1115 | 1117 | 1116 | 1116 | 1117 | 1118 |
| Outer OH2, OH4 stretching vibration | IF, R | | | | 3620 | 3620 | 3620 |
| Inner OH1 stretching vibration | IF, R | 3622 | 3622 | 3623 | 3622 | 3623 | 3622 |
| Outer OH2, OH4 stretching vibration | R | | | | 3645 | 3645 | 3645 |

**Table 2.** *Cont.*

| Band Assignment [19–32] | Active | BL-DSL-4 | QJ-1 | QJ-8 | GS-1 | GS-4 | GS-8 |
|---|---|---|---|---|---|---|---|
| Outer OH2, OH4 stretching vibration | IF, R | 3653 <br> 3670 | 3652 <br> 3669 | 3654 <br> 3672 | 3654 | 3655 | 3654 |
| Outer OH3 stretching vibration | R | 3685 | 3683 | 3683 | | | |
| Outer OH3 stretching vibration | IF, R | 3696 | 3695 | 3697 | 3699 <br> 3714 | 3705 <br> 3715 | 3700 <br> 3712 |

### 3.4. Characteristics of NIR

The main NIR spectral characteristics of kaolinite and dickite are located in the 4000–4800 cm$^{-1}$ and 7000–7600 cm$^{-1}$ regions (Table 3 and Figures 6–8).

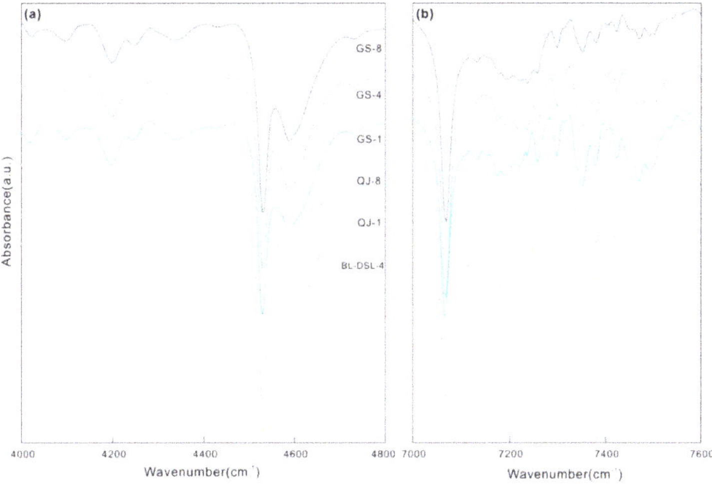


**Figure 6.** NIR spectra of the kaolinite and dickite samples. (**a**) The range of the OH combination bands (4000–4800 cm$^{-1}$); (**b**) the range of the OH stretching overtone bands (7000–7600 cm$^{-1}$).

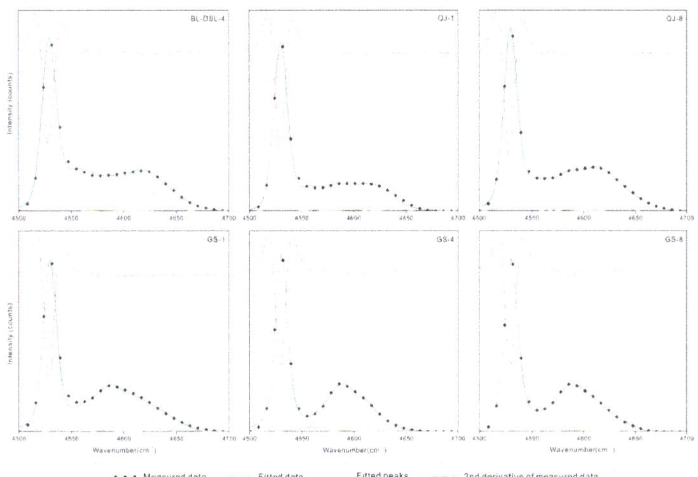


**Figure 7.** NIR spectral component analysis of the OH combination bands.

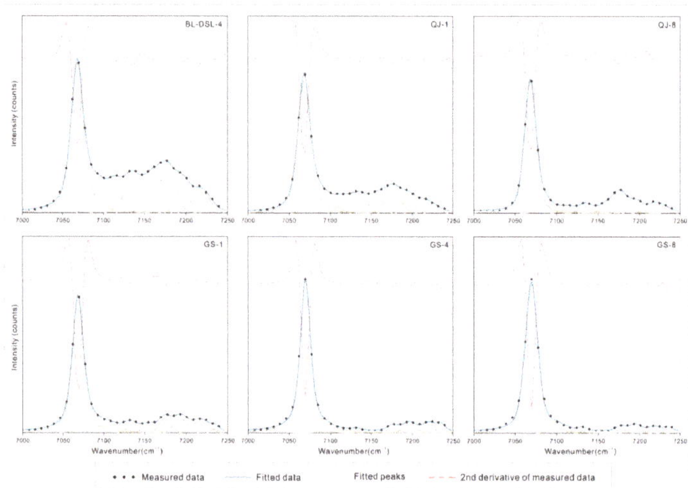



**Figure 8.** NIR spectral component analysis of the OH stretching overtone bands.

There are multiple weak peaks in the range of 4000–4400 cm$^{-1}$, assigned to the combination of lattice deformation vibrations and OH stretching vibrations. In the range of 4500–4700 cm$^{-1}$, kaolinite and dickite both fit five peaks with similar positions. The maximum peaks are located near 4530 cm$^{-1}$, which is a common characteristic of kaolinite-group minerals. A shoulder peak is present on the side of the higher wave number; the kaolinite shoulder is near 4613 cm$^{-1}$, and the dickite shoulder is near 4587 cm$^{-1}$, which can be used to distinguish the two minerals.

All samples have a large number of disordered but clear peaks in the range of 7000–8000 cm$^{-1}$. According to the theoretical calculation and the change in intensity before and after heating, a series of spectral peaks after 7250 cm$^{-1}$ is considered to be caused by $H_2O$ in the crystal structure of clay minerals. $H_2O$ causes many disordered peaks in this region, which are also common in serpentine, montmorillonite, and other phyllosilicates [6,33,34]. The first fundamental overtone of the OH stretching vibrations is in the range of 7000–7250 cm$^{-1}$, and a strong peak and multiple weak peaks were obtained after fitting. The strong peaks of both minerals are near 7068 cm$^{-1}$, and there is a secondary peak at 7177 cm$^{-1}$ for kaolinite, but not for dickite. GS-1 has a very weak spectral peak at 7155 cm$^{-1}$, while the peaks of GS-4 and GS-8 are too weak to be shown here.

**Table 3.** The bands in the range of 4000–8000 cm$^{-1}$ for related samples and their assignments (cm$^{-1}$).

| Assignment | Kaolinite [23] | Kaolinite [35] | BL-DSL-4 | QJ-1 | QJ-8 | GS-1 | GS-4 | GS-8 |
|---|---|---|---|---|---|---|---|---|
| | | | 4020 | 4022 | 4023 | 4019 | 4020 | 4023 |
| Combination of OH | | | 4093 | 4093 | 4093 | 4097 | 4096 | 4097 |
| stretching and lattice | | 4195 | 4197 | 4197 | 4197 | 4197 | 4197 | 4197 |
| vibration | | 4249 | 4247 | 4248 | 4247 | 4245 | 4246 | 4247 |
| | | 4312 | 4332 | 4330 | 4328 | 4335 | 4337 | 4336 |
| | 4527 | 4530 | 4530 | 4530 | 4530 | 4530 | 4531 | 4531 |
| Combination of OH | 4558 | | 4560 | 4565 | 4562 | 4567 | 4567 | 4564 |
| stretching and | 4590 | | 4587 | 4585 | 4587 | 4588 | 4589 | 4588 |
| bending vibration | 4610 | | 4613 | 4609 | 4610 | 4610 | 4614 | 4610 |
| | 4632 | 4633 | 4634 | 4631 | 4633 | 4636 | 4638 | 4635 |
| Combination of inner OH stretching and lattice vibration | | | 4729 | 4732 | 4733 | 4731 | 4734 | 4733 |

**Table 3.** *Cont.*

| Assignment | Kaolinite [23] | Kaolinite [35] | BL-DSL-4 | QJ-1 | QJ-8 | GS-1 | GS-4 | GS-8 |
|---|---|---|---|---|---|---|---|---|
| | 7067 | 7069 | 7068 | 7068 | 7068 | 7068 | 7070 | 7070 |
| | | 7099 | 7092 | 7089 | 7087 | 7098 | 7095 | 7110 |
| | 7118 | | 7113 | 7115 | 7113 | | | |
| The first fundamental | | 7124 | 7135 | 7135 | 7135 | 7133 | 7131 | 7131 |
| overtone of OH | 7156 | 7148 | 7158 | 7158 | 7156 | 7155 | — | — |
| stretching vibration | 7168 | 7175 | 7177 | 7177 | 7176 | 7175 | 7177 | 7178 |
| | 7182 | 7203 | 7196 | 7195 | 7196 | 7193 | 7197 | 7195 |
| | | 7222 | 7218 | 7217 | 7220 | 7217 | 7220 | 7217 |
| | | 7232 | | | | 7235 | 7234 | 7235 |

## 4. Discussion

*Assignment of OH Vibration in NIR Spectra*

The band frequency relationships between the MIR and NIR spectra were determined via trial-and-error summations. Combined with the data obtained in this experiment, we determined suitable peak assignments as far as possible.

The bands of kaolinite-group minerals in the region of 4000–5000 cm$^{-1}$ correspond to the combination of OH stretching vibration and lattice vibration or bending vibration ($\nu$OH + $\delta$OH). The band at 4530 cm$^{-1}$ is attributed to the combination of the band near 3622 cm$^{-1}$, assigned to inner OH stretching vibration, and the peak near 914 cm$^{-1}$, assigned to Al-OH bending vibration. The characteristic peaks near 4613 cm$^{-1}$ for kaolinite and 4587 cm$^{-1}$ for dickite correspond to the 3696 cm$^{-1}$ and 914 cm$^{-1}$ combination and the 3654 cm$^{-1}$ and 937 cm$^{-1}$ band combination, respectively. The specific corresponding values are shown in Table 4. All of the errors are within 10 cm$^{-1}$ [36], and the strength is also in agreement.

**Table 4.** The major NIR bands in 4500~4700 cm$^{-1}$ for related samples and their corresponding MIR peaks (cm$^{-1}$).

| | Measured Peak | Fundamental Peaks | Theoretical Peak | Δ | | Measured Peak | Fundamental Peaks | Theoretical Peak | Δ |
|---|---|---|---|---|---|---|---|---|---|
| BL-DSL-4 | 4530 | 3622 + 914 | 4536 | 6 | GS-1 | 4530 | 3622 + 913 | 4535 | 5 |
| | 4560 | 3653 + 914 | 4567 | 7 | | 4567 | 3654 + 913 | 4567 | 0 |
| | 4587 | 3670 + 914 | 4584 | −3 | | 4588 | 3654 + 937 | 4591 | 3 |
| | 4613 | 3696 + 914 | 4610 | −3 | | 4610 | 3699 + 913 | 4612 | 2 |
| | 4634 | 3696 + 937 | 4633 | −1 | | 4636 | 3699 + 937 | 4636 | 0 |
| QJ-1 | 4530 | 3622 + 914 | 4536 | 6 | GS-4 | 4531 | 3623 + 913 | 4536 | 5 |
| | 4565 | 3652 + 914 | 4566 | 1 | | 4567 | 3655 + 913 | 4568 | 1 |
| | 4585 | 3669 + 914 | 4583 | −2 | | 4589 | 3655 + 937 | 4592 | 3 |
| | 4609 | 3695 + 914 | 4609 | 0 | | 4614 | 3705 + 913 | 4618 | 4 |
| | 4631 | 3695 + 937 | 4632 | 1 | | 4638 | 3705 + 937 | 4642 | 4 |
| QJ-8 | 4530 | 3623 + 914 | 4537 | 7 | GS-8 | 4531 | 3622 + 912 | 4534 | 3 |
| | 4562 | 3654 + 914 | 4568 | 6 | | 4564 | 3654 + 912 | 4566 | 2 |
| | 4587 | 3672 + 914 | 4586 | −1 | | 4588 | 3654 + 937 | 4591 | 3 |
| | 4610 | 3697 + 914 | 4611 | 1 | | 4610 | 3700 + 912 | 4612 | 2 |
| | 4633 | 3697 + 938 | 4635 | 2 | | 4635 | 3700 + 937 | 4637 | 2 |

Both minerals have a weak peak near 4730 cm$^{-1}$, which may be the combination of the OH stretching vibration near 3620 cm$^{-1}$ and Si-O stretching vibration near 1100 cm$^{-1}$ [11].

The absorption in the range of 7000–8000 cm$^{-1}$ corresponds to the first fundamental overtone of OH stretching vibration (2$\nu$OH). The strongest peaks of both minerals are near 7068 cm$^{-1}$, assigned to inner OH stretching vibration. The secondary peak of kaolinite near 7177 cm$^{-1}$ corresponds to the outer OH stretching vibration, which is a non-infrared-active band shown in the Raman spectrum. The non-infrared-active band of dickite is located at 3645 cm$^{-1}$, corresponding to the weak peak near 7100 cm$^{-1}$. The specific corresponding values of the two minerals are presented in Table 5. Bishop [32] showed the overtone of the OH stretching vibration of kaolinite near 7232 cm$^{-1}$, but only the dickite samples showed a similar weak peak in this study. The average factor between the OH stretching vibration and its overtone is 1.95 ($\pm$0.003) (Figure 9), not simply two-fold but slightly less, as explained by quantum mechanics [35]. Petit [37] proposed a constant $k$ based on anharmonicity, and $k = -171$ cm$^{-1}$ for kaolinite minerals. He considered that the relationship between OH stretching vibration and its overtone should be two-fold minus this constant $k$. However, this method has an error in this study and cannot well explain the relationship.

**Table 5.** The major NIR bands in 7000~7250 cm$^{-1}$ for related samples and their corresponding MIR peaks (cm$^{-1}$).

| | Measured Peak | Fundamental Peaks | Factor | | Measured Peak | Fundamental Peaks | Factor |
|---|---|---|---|---|---|---|---|
| BL-DSL-4 | 7068 | 3622 | 1.9514 | GS-1 | 7068 | 3622 | 1.9514 |
| | 7092 | (3622 + 3653)/2 | 1.9497 | | 7098 | 3645 * | 1.9473 |
| | 7113 | 3653 | 1.9472 | | 7133 | 3654 | 1.9521 |
| | 7135 | 3670 | 1.9447 | | 7155 | (3645 * + 3699)/2 | 1.9485 |
| | 7158 | (3653 + 3685 *)/2 | 1.9504 | | 7175 | (3653 + 3699)/2 | 1.9518 |
| | 7177 | 3685 * | 1.9476 | | 7193 | (3653 + 3714)/2 | 1.9528 |
| | 7196 | (3685 * + 3696)/2 | 1.9499 | | 7217 | 3699 | 1.9511 |
| | 7218 | 3696 | 1.9529 | | 7235 | 3714 | 1.9480 |
| QJ-1 | 7068 | 3622 | 1.9514 | GS-4 | 7070 | 3623 | 1.9514 |
| | 7089 | (3622 + 3652)/2 | 1.9491 | | 7095 | 3645 * | 1.9465 |
| | 7115 | 3652 | 1.9482 | | 7131 | 3655 | 1.9510 |
| | 7135 | 3669 | 1.9460 | | — | — | — |
| | 7158 | (3652 + 3683 *)/2 | 1.9509 | | 7177 | (3655 + 3705)/2 | 1.9503 |
| | 7177 | 3683 * | 1.9487 | | 7197 | (3655 + 3715)/2 | 1.9531 |
| | 7195 | (3683 * + 3695)/2 | 1.9504 | | 7220 | 3705 | 1.9487 |
| | 7217 | 3695 | 1.9532 | | 7234 | 3715 | 1.9472 |
| QJ-8 | 7068 | 3623 | 1.9509 | GS-8 | 7070 | 3622 | 1.9520 |
| | 7087 | (3623 + 3654)/2 | 1.9478 | | 7110 | 3645 * | 1.9506 |
| | 7113 | 3654 | 1.9466 | | 7131 | 3654 | 1.9516 |
| | 7135 | 3672 | 1.9449 | | — | — | — |
| | 7156 | (3654 + 3683 *)/2 | 1.9488 | | 7178 | (3654 + 3700)/2 | 1.9521 |
| | 7176 | 3683 * | 1.9484 | | 7195 | (3654 + 3712)/2 | 1.9536 |
| | 7196 | (3683 * + 3697)/2 | 1.9501 | | 7217 | 3700 | 1.9505 |
| | 7220 | 3697 | 1.9529 | | 7235 | 3712 | 1.9491 |

* Non-infrared-active, only observed in Raman spectra.

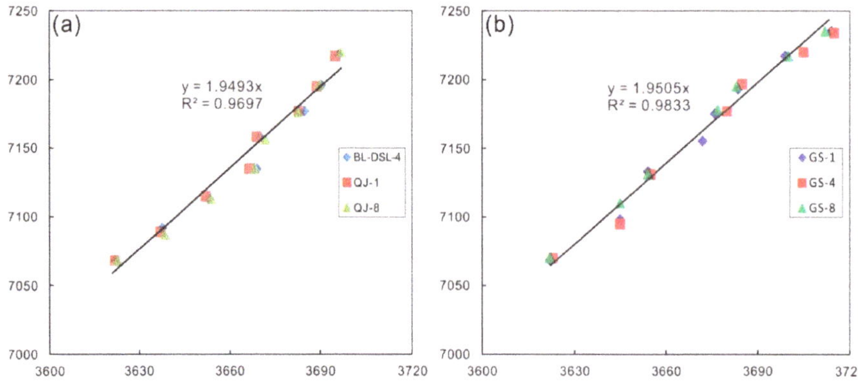


**Figure 9.** Relationship between the overtone vibrations and stretching vibrations of OH: (**a**) relationship for kaolinite; (**b**) relationship for dickite.

In addition to the fundamental overtone generated by OH stretching vibration, kaolinite and dickite have three bands each that cannot match a single OH stretching vibration. Comparing the NIR spectra of the samples before and after drying, it is found that there is no obvious change, so the influence of adsorbed water is excluded. The combination of double OH stretching vibration can lead to better matching results (Table 5: fitted peaks near 7092, 7158, and 7196 cm$^{-1}$ for kaolinite and fitted peaks near 7155, 7175, and 7193 cm$^{-1}$ for dickite), which is also proposed in the results of Petit [37]. This combination is related to the band position of OH stretching vibration, where adjacent peaks can produce combined absorption, but there is no intuitive correlation with band type or assignment. This has not been mentioned in previous studies.

The non-infrared-active bands at 3683 cm$^{-1}$ for kaolinite and 3645 cm$^{-1}$ for dickite do not have corresponding peaks in the range of OH combination but have their own overtone and combined overtone above 7000 cm$^{-1}$. This also provides a new method to identify the vibration of non-infrared-active bands, i.e., considering whether there is a corresponding fundamental overtone in the NIR region.

## 5. Conclusions

1.  The common characteristic peaks in the near-infrared spectra of kaolinite-group minerals are located near 4530 and 7068 cm$^{-1}$, corresponding to the combination of inner OH stretching vibration and outer OH bending vibration and the first fundamental overtone of inner OH. Next to the characteristic peak, kaolinite has two secondary peaks near 4610 and 7177 cm$^{-1}$, corresponding to the combination of outer OH stretching vibration and bending vibration and the overtone of outer OH stretching vibration. The secondary peak of dickite is near 4588 cm$^{-1}$, corresponding to the combination of outer OH stretching vibration and inner OH bending vibration. There is no secondary characteristic peak in the first fundamental overtone region of dickite.
2.  The OH combination vibration of kaolinite-group minerals is located at 4000–4800 cm$^{-1}$. The OH stretching vibration will be combined with the lattice deformation vibration, OH bending vibration, and Si-O vibration to form new vibration absorption. Moreover, 7000–7250 cm$^{-1}$ is the first fundamental overtone region for the OH stretching vibration, with a strong absorption peak and several weak peaks. Kaolinite and dickite each have three different combined overtones for the OH stretching vibration, which are composed of adjacent OH stretching vibrations.
3.  The non-infrared-active OH stretching vibration of kaolinite and dickite showed an infrared-active absorption peak in the overtone region. It can be used to identify vibration types that cannot be observed by MIR spectra.
4.  Due to the influence of non-ideal conditions, the actual peak position of the overtone peak is lower than the theoretical position. The factor of the first fundamental overtone of the OH group stretching vibration is approximately 1.95.

**Author Contributions:** Conceptualization, M.H.; data curation, S.W. and B.P.; formal analysis, S.W., M.Y. and B.P.; funding acquisition, M.H.; investigation, M.Y. and B.P.; methodology, S.W. and M.Y.; resources, M.H.; writing—original draft, S.W.; writing—review and editing, S.W., M.Y. and B.P. All authors have read and agreed to the published version of the manuscript.

**Funding:** This research was funded by the National Mineral Rock and Fossil Specimens Resource Center (http://www.nimrf.net.cn/, accessed on 1 June 2022) to Mingyue He.

**Data Availability Statement:** Not applicable.

**Acknowledgments:** I thank my classmates for their company and help during my graduate career. I hope that you all have a happy graduation and a bright future thanks to this paper.

**Conflicts of Interest:** The authors declare no conflict of interest.

# References


1. Chen, H.Y.; Zhang, S.T.; Chu, G.B.; Zhang, Y.; Cheng, J.M.; Tian, J.; Han, J.S. The short wave infrared (SWIR) spectral characteristics of alteration minerals and applications for ore exploration in the typical skarn-porphyry deposits, Edong ore district, eastern China. *Acta Petrol. Sin.* **2019**, *35*, 3629–3643.
2. Deng, Y.K.; Cao, J.J.; Dang, W.Q.; Wang, G.Q.; Liu, X.; Li, D.W. XRD and NIR Analysis of Oxidation Particles in Dabashan Polymetallic Deposit and Its Significance. *Spectrosc. Spectr. Anal.* **2019**, *39*, 2929–2934.
3. Zhang, C.; Ye, F.W.; Wu, D.; Wang, J.G.; Guo, B.J. Characteristics Recognition of Imaging Spectra for Uranium Mineralization Altered Mineral Assemblage in Xiangshan. *Uranium Geol.* **2021**, *37*, 69–77.
4. Farmer, V.C. (Ed.) *The Infrared Spectra of Minerals*; The Mineralogical Society of Great Britain & Ireland: Middlesex, UK, 1974; 539p.
5. Yu, X.Y. *Colored Gemmology*, 2nd ed.; Geological Publishing House: Beijing, China, 2016; pp. 86–87.
6. Bishop, J.L.; Lane, M.D.; Dyar, M.D.; Brown, A.J. Reflectance and emission spectroscopy study of four groups of phyllosilicates: Smectites, kaolinite-serpentines, chlorites and micas. *Clay Miner.* **2008**, *43*, 35–54. [CrossRef]
7. Post, J.L.; Crawford, S.M. Uses of near-infared spectra for the identification of clay minerals. *Appl. Clay Sci.* **2014**, *95*, 383–387. [CrossRef]
8. Liao, Y.P.; Cao, J.J.; Wu, Z.Q.; Luo, S.Y.; Wang, Z.Y. Near Infrared Spectroscopy of the Cretaceous Red Beds in Inner Mongolia Dongshengmiao. *Spectrosc. Spectr. Anal.* **2015**, *35*, 2521–2525.
9. Guo, X.F.; Zhu, X.; Zu, E.D. Near Infrared Spectroscopy Study for Different Types of Phyllosilicate Gemstones Minerals. *Bull. Chin. Ceram. Soc.* **2018**, *37*, 2270–2273.
10. Giese, R.F. Kaolin minerals: structures and stabilities. *Rev. Mineral. Geochem.* **1988**, *19*, 29–66.
11. Petit, S.; Madejová, J.; Decarreau, A. Characterization of Octahedral Substitutions in Kaolinites Using Near Infrared Spectroscopy. *Clays Clay Miner.* **1999**, *47*, 103–108. [CrossRef]
12. Balan, E.; Saitta, A.M.; Mauri, F.; Calas, G. First-principles modeling of the infrared spectrum of kaolinite. *Am. Mineral.* **2001**, *86*, 1321–1330. [CrossRef]
13. Medeghini, L.; Mignardi, S.; De Vito, C.; Conte, A.M. Evaluation of a FTIR data pretreatment method for Principal Component Analysis applied to archaeological ceramics. *Microchem. J.* **2016**, *125*, 224–229. [CrossRef]
14. Todorova, M.H.; Atanassova, S.L. Near infrared spectra and soft independent modelling of class analogy for discrimination of Chernozems, Luvisols and Vertisols. *J. Near Infrared Spectrosc.* **2016**, *24*, 271–280. [CrossRef]
15. Yin, Y.S.; Yin, J.; Zhang, W.; Tian, H.; Hu, Z.M.; Feng, L.H.; Chen, D.L. Characterization of Mineral Matter in Coal Ashes with Infrared and Raman Spectroscopy. *Spectrosc. Spectr. Anal.* **2018**, *38*, 789–793.
16. Rinnan, Å.; Berg, F.; Engelsen, S.B. Review of the most common pre-processing techniques for near-infrared spectra. *Trends Anal. Chem.* **2009**, *28*, 1201–1222. [CrossRef]
17. Balan, E.; Lazzeri, M.; Saitta, A.M.; Allard, T.; Fuchs, Y.; Mauri, F. First-principles study of OH-stretching modes in kaolinite, dickite, and nacrite. *Am. Mineral.* **2005**, *90*, 50–60. [CrossRef]
18. Johansson, U.; Frost, R.L.; Forsling, W.; Kloprogge, J.T. Raman Spectroscopy of the Kaolinite Hydroxyls at 77 K. *Appl. Spectrosc.* **1998**, *52*, 1277–1282. [CrossRef]
19. Han, X. Study on the Mineralogical Characteristics of Laos Stones. Master's Thesis, China University of Geosciences, Wuhan, China, 2019.
20. Jiang, J.P. Study on the Mineralogical Characteristics of Laos Stones. Master's Thesis, China University of Geosciences, Beijing, China, 2020.
21. Farmer, V.C. Differing effects of particle size and shape in the infrared and Raman spectra of kaolinite. *Clay Miner.* **1998**, *33*, 601–604. [CrossRef]
22. Balan, E.; Delattre, S.; Guillaumet, M.; Salje, E.K. Low-temperature infrared spectroscopic study of OH-stretching modes in kaolinite and dickite. *Am. Mineral.* **2010**, *95*, 1257–1266. [CrossRef]
23. Frost, R.L.; Johansson, U. Combination Bands in the Infrared Spectroscopy of Kaolins—A Drift Spectroscopic Study. *Clays Clay Miner.* **1998**, *46*, 466–477. [CrossRef]
24. Castellano, M.; Turturro, A.; Riani, P.; Montanari, T.; Finocchio, E.; Ramis, G.; Busca, G. Bulk and surface properties of commercial kaolins. *Appl. Clay Sci.* **2010**, *48*, 446–454. [CrossRef]
25. Frost, R.L.; Kloprogge, J.T. Towards a single crystal Raman spectrum of kaolinite at 77 K. *Spectrochim. Acta Part A: Mol. Biomol. Spectrosc.* **2001**, *57*, 163–175. [CrossRef]
26. Dong, J.K.; Du, Y.S. A Study of Mineralogical Characteristics of Larderite Deposits in Fujian Province. *Acta Geosci. Sin.* **2017**, *38*, 208–222.
27. Yuan, Y.; Shi, G.H.; Lou, F.S.; Wu, S.J.; Shi, M.; Huang, A.J. Mineralogical and Spectral Characteristics of "Gaozhou Stone" from Jiangxi Province. *Spectrosc. Spectr. Anal.* **2015**, *35*, 65–70.
28. Hu, X.C.; Yan, J.; Zhu, X.M. The Specific IR Spectra of Qingtian Stones from Zhejiang. *Phys. Test. Chem. Anal. (Part B Chem. Anal.)* **2016**, *52*, 24–28.
29. Han, W.; Ke, J.; Chen, H.; Lu, T.J.; Yin, K. Diffuse Reflectance Spectroscopy of Red Colored "Laowo Stone". *Spectrosc. Spectr. Anal.* **2016**, *36*, 2634–2638.
30. Liu, Y.G.; Chen, T. Infrared and Raman Spectra Study on Tianhuang. *Spectrosc. Spectr. Anal.* **2012**, *32*, 2143–2146.

31. Johnston, C.T.; Helsen, J.; Schoonheydt, R.A.; Bish, D.L.; Agnew, S.F. Single-crystal Raman spectroscopic study of dickite. *Am. Mineral.* **1998**, *83*, 75–84. [CrossRef]
32. Frost, R.L.; Tran, T.H.; Rintoul, L.; Kristof, J. Raman microscopy of dickite, kaolinite and their intercalates. *The Analyst* **1998**, *123*, 611–616. [CrossRef]
33. Bishop, J.L.; Pieters, C.M.; Edwards, J.O. Infrared spectroscopic analyses on the nature of water in montmorillonite. *Clays Clay Miner.* **1994**, *42*, 702–716. [CrossRef]
34. Wu, S.; He, M.Y.; Yang, M.; Zhang, B.Y.; Wang, F.; Li, Q.Z. Near-Infrared Spectroscopy Study of Serpentine Minerals and Assignment of the OH Group. *Crystals* **2021**, *11*, 1130. [CrossRef]
35. Bishop, J.; Murad, E.; Dyar, M.D. The influence of octahedral and tetrahedral cation substitution on the structure of smectites and serpentines as observed through infrared spectroscopy. *Clay Miner.* **2002**, *37*, 617–628. [CrossRef]
36. Baron, F.; Petit, S. Interpretation of the infrared spectra of the lizardite-nepouite series in the near- and mid-infrared range. *Am. Mineral.* **2016**, *101*, 423–430. [CrossRef]
37. Petit, S.; Decarreau, A.; Martin, F. Refined relationship between the position of the fundamental OH stretching and the first overtones for clays. *Phys. Chem. Miner.* **2005**, *31*, 585–592. [CrossRef]

*Article*

# The Behavior of Water in Orthoclase Crystal and Its Implications for Feldspar Alteration

Hongyan Zuo [1], Rui Liu [2] and Anhuai Lu [3,*]

[1] State Key Laboratory of Lunar and Planetary Sciences, Macau University of Science and Technology, Taipa, Macau 999078, China; hyzuo2014@gmail.com
[2] School of Prospecting and Survey, Jilin Oil Shale Drilling and Environmental Protecting Laboratory, Changchun Institute of Technology, Changchun 130021, China; gglr984@sohu.com
[3] Beijing Key Laboratory of Mineral Environmental Function, School of Earth and Space Sciences, Peking University, Beijing 100871, China
* Correspondence: ahlu@pku.edu.cn

**Abstract:** The phenomenon of feldspar alteration that occurs in the interior of feldspar crystals remains poorly understood. We observed experimentally that water can go into orthoclase crystals under pressures of up to 600 MPa at room temperature. With increasing pressure, the FTIR spectra of colorless orthoclase show a sharp increase in integral absorbance from $1.50 \ \text{cm}^{-1}$ to $14.54 \ \text{cm}^{-1}$ and normalized integral absorbance from $120 \ \text{cm}^{-2}$ to $1570 \ \text{cm}^{-2}$; the pink orthoclase saturates quickly with no significant change in either the integral absorbance or normalized integral absorbance. The different responses to the pressure between colorless orthoclase and pink orthoclase might be related to the K content in the structure. Moreover, FTIR spectra at atmospheric pressure collected in different crystallography directions show different absorbance intensities, which illustrates the characteristic of preferred crystallographic orientations. These results reveal that $H_2O$ molecules can occur as structural constituents entering the crystallographic channels of alkali feldspar crystals, preferentially along (001) orientation. These findings provide clues into the mechanism of feldspar alteration occurring in the interior of feldspar crystals, as well as the formation of micropores and microstructure in feldspar minerals. This study also provides important insights into the behavior of water molecules in nominally anhydrous minerals in the upper crust of the Earth.

**Keywords:** orthoclase crystal; ultra-microchannel; water; feldspar alteration; sericitization

**Citation:** Zuo, H.; Liu, R.; Lu, A. The Behavior of Water in Orthoclase Crystal and Its Implications for Feldspar Alteration. *Crystals* **2022**, *12*, 1042. https://doi.org/10.3390/cryst12081042

Academic Editor: Vladislav V. Gurzhiy

Received: 14 June 2022
Accepted: 24 July 2022
Published: 27 July 2022

**Publisher's Note:** MDPI stays neutral with regard to jurisdictional claims in published maps and institutional affiliations.



## 1. Introduction

Feldspars make up more than 50% of the Earth's crust. Thus, feldspar alteration is ubiquitous within the crust [1,2]. The ideal formula of a feldspar is $MT_4O_8$ [3]. Feldspar has a three-dimensional framework of corner-sharing $AlO_4$ and $SiO_4$ tetrahedra [4]. The degree of Al/Si order in the tetrahedral sites in feldspars is related to the formation conditions of the minerals [5,6]. The framework structure of feldspars is based on the linked $TO_4$ tetrahedra [4,7]. In the framework structure, the oxygen atoms lie at the corners of the tetrahedron centered by Al or Si atoms. All oxygen atoms are shared by two T (Al or Si) atoms to form a framework like a three-dimensional network [8]. The substitution of Si (+4) with Al (+3) in the tetrahedron makes the structure's charge imbalanced. Thus, the irregularly shaped cavities formed by the framework can be occupied by the large M cations to obtain a charge balance. Moreover, a special characteristic of the feldspar structure is the occurrence of chains of tetrahedra that are cross-linked in a special way to form crankshaft and elliptical rings of tetrahedra [9–11].

In the structure of feldspar, the eight-membered rings of tetrahedra overlap each other, forming ultra-microchannels of 0.1 nm × 0.1 nm width along (100), while the six-membered rings of tetrahedra form elliptical ultra-microchannels of 0.1 nm × 0.3 nm width along (001) [12]. However, it is more complex along (101). They are constructed with tetragonal

TO$_4$ sharing the apex, forming the double-crankshaft chain. The double chains connect to form two kinds of channels along (101), in which the aperture of six-membered rings is 0.1 nm × 0.1 nm. Another aperture formed by ten tetragonal TO$_4$ sharing the apex is 0.15 nm × 0.67 nm [12]. The 10-membered elliptical ring is the largest channel in feldspar, and alkali or alkaline-earth cations such as Na$^+$, K$^+$, Ca$^{2+}$, and Ba$^{2+}$ or even Pb$^{2+}$, Cd$^{2+}$, and Mg$^{2+}$ can occupy this type of channel [12,13]. The overall structural states of feldspars (including obliquity, triclinicity, and degree of disorder) are very complex and may correlate to the petrogenesis of their containing rocks [14–16].

Feldspar minerals can develop microcracks. These microcracks are mainly in three ranges: 1.0–2.0 nm, 10–15 nm, and >50 nm. Microcracks, including micropores and microstructures, are contributors to the processes of crystals coarsening and deuteric or hydrothermal alteration [17]. The microcracks, together with cleavage, twinning, and the grain-to-grain boundary, are thought to be responsible for weathering and alteration. Feldspar alteration is a consistent concern for geologists, mineralogists, and environment scientists [1,18–22]. Interestingly, fine-grained sericite was reported to occur as an alteration product in the interior of feldspar crystals [23–26]. However, the alteration that occurs in the interior of feldspar crystals is not fully understood yet. Given the fact that the diameter of water molecule is only 0.276 nm, water might be able to go through the ultra-microchannels.

In this work, we focus on the behavior of water going into the ultra-microchannels of feldspars. Two types of feldspars are selected for high-pressure experiments up to 600 MPa and oriented FTIR analysis. In situ Raman spectroscopic measurement in a cubic zirconia anvil cell was conducted in high-pressure experiments. FTIR analysis was used to detect the species and content of hydrogen (OH$^-$ and H$_2$O) in feldspars. The results of this study provide important insights into feldspar alteration and the behavior of water molecules in nominally anhydrous minerals in the upper crust of the Earth.

## 2. Materials and Methods

### 2.1. Materials

Colorless and pink feldspars were collected from granite in the Barkol area in eastern Junggar, Xinjiang. The compositions of these feldspars were measured with electron probe microanalysis (EPMA), and both the colorless and pink samples were determined to be orthoclase (hereafter called colorless orthoclase and pink orthoclase; contents in Table 1 refer to average analysis). Natural crystals of albite, rhodonite, plagioclase, orthoclase, pyrope, almandine, anhydride, and benitoite were used as EPMA standards. Multipoint measurements were carried out, and the total standard deviation is less than 0.3%. The mineralogical composition of the studied materials was further confirmed by transmission electron microscopy and selected area electron diffraction (Figure 1). The lattice image of the colorless orthoclase shows two d-spacings of 0.59 nm and 0.65 nm, which correspond to (110) and (11$\bar{1}$), respectively (Figure 1). Only microscopically inclusion-free, optically clear, and homogeneous parts of the two samples were selected for high-pressure experiments. Additionally, colorless orthoclase was prepared for FTIR spectroscopic measurements of oriented polished plane-parallel plates (3 orientations: (001), (010), and (100)).

**Table 1.** Electron-microprobe chemical compositions of orthoclase (wt%).

| Samples | SiO$_2$ | Al$_2$O$_3$ | Na$_2$O | K$_2$O | CaO | MgO | FeO | MnO | BaO | Or | Ab | An |
|---|---|---|---|---|---|---|---|---|---|---|---|---|
| colorless | 55.97 | 27.53 | 6.19 | 10.03 | 0.01 | 0.06 | 0.05 | 0.02 | 0.01 | 51.6 | 48.3 | 0.1 |
| pink | 54.67 | 28.33 | 6.43 | 9.26 | 0.61 | 0.06 | 0.25 | 0.05 | 0.02 | 47.7 | 49.6 | 2.7 |

In the high-pressure experiments, samples were subjected to the desired pressure in a cubic zirconia anvil cell for 12 h before being analyzed by in situ Raman spectroscopy. After Raman spectroscopy analysis, the pressure was released slowly. Meanwhile, samples that were maintained at pressures around 300 MPa and 600 MPa were taken out from the anvil for Fourier Transform Infrared (FTIR) spectroscopy analysis to evaluate the behavior of water in orthoclase.

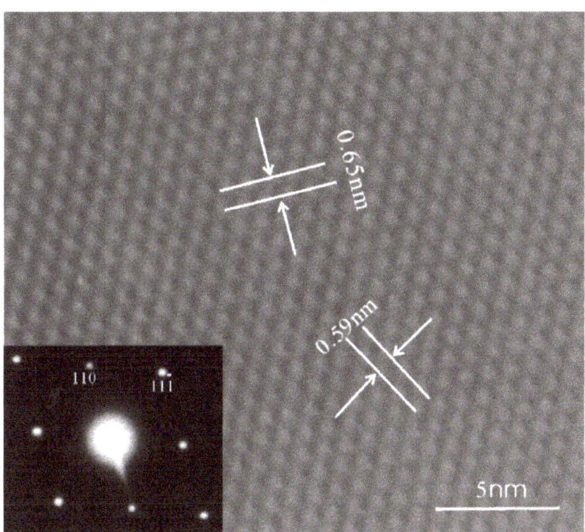

**Figure 1.** TEM image and electron diffraction pattern of colorless orthoclase.

### 2.2. Methods

**CZ anvil cell**. Metastable cubic zirconia (CZ) has been found to be inert, quite hard, and transparent to a large portion of the electromagnetic spectrum [27]. CZ anvil cell offers the advantage of a lower mid-IR cutoff point, enabling high-pressure experiments, and is a remarkable device for studying high-pressure crystallographic phenomena [27,28]. Cubic Zirconia (400 mg) for this study is a combination of $ZrO_2$ and $HfO_2$, containing 96.12 wt% and 3.88 wt%, respectively. High-pressure experiments in the CZ anvil cell were conducted using Cu gaskets with an initial thickness of 0.35 mm and a hole of 0.35 mm diameter. Pressure was measured using a quartz gauge as the pressure sensor. The force application was manual.

In the past several decades, 4:1 methanol–ethanol mixture has proven to be the most easily handled fluid used as a hydrostatic pressure-transmitting medium in connection with gasket cells [29–31]. Given the fact that the small molecular alcohols can enter the tunnels or structural cages of some microporous and mesoporous inorganic materials such as zeolites [32–34], it is reasonable that water can be used as a hydrostatic pressure fluid to study the tunnels of orthoclase. In this study, a 99.96% ions-free water was used as the pressure-transmitting medium.

**Raman spectroscopy.** In situ Raman spectroscopy together with a diamond cubic zirconia anvil cell was used to study the structural change at high pressures [35,36]. Raman spectra were collected on a Renishaw 1000 Raman microprobe spectrometer system coupled with a 20× microscope. To excite the samples, a continuous wave (CW) Ar ion laser with an incident laser power of 25 mW and a spot diameter of 1~2 μm was employed in the system. Raman scattering was detected in the range of 200–4000 $cm^{-1}$ with a spectral resolution of ~1 $cm^{-1}$. The data collection time was 60 s.

The occurring depth of feldspar minerals in the lithosphere ranges from 10 to 20 km; thus, the corresponding pressure range of 300–600 MPa has been chosen to conduct the high-pressure experiment. Quartz is used in the high-pressure experiment as the standard to determine the pressures that the laboratory instrument applied on samples. The following functions derived by Schmidt and Ziemann [37] are used:

$$P = \frac{(\triangle v_{\mathrm{p}})_{464} - 2.051 \times 10^{-11}T^4 - 1.465 \times 10^{-8}T^3 + 1.801 \times 10^{-5}T^2 + 0.01216T - 0.29}{0.009} + 0.1 \qquad (1)$$

$$p = 0.36079[(\Delta v_{\mathrm{p}})_{464}]^2 + 110.86(\Delta v_{\mathrm{p}})_{464}. \qquad (2)$$

These functions are reliable when the shift does not exceed 20 cm$^{-1}$. The static hydropressures were obtained based on function (2).

**FTIR spectroscopy.** Fourier transform infrared spectra were measured using a Nicolet 750 Fourier-transform spectrometer. Prior to FTIR performance, samples were kept in a temperature control box at 180 °C for 3 h to free them of surface absorbent water. FTIR spectra were collected in the range of 1000–4000 cm$^{-1}$, with a resolution of 16 cm$^{-1}$. The beam was focused to form a ~100 μm spot on samples. For each spectrum, 128 scans were accumulated. FTIR spectra were collected in different crystallographic directions for colorless orthoclase.

## 3. Results

### 3.1. Raman Spectra with Increasing Pressure

The tectosilicate structure with fully linked tetrahedra generates a Raman spectral pattern distinctly different from those of chain, ortho-, ring, and layer silicates, in which the TO$_4$ tetrahedra are not linked at all, or are only partially linked, with other TO$_4$ units [38]. The strongest Raman peak of feldspars falls within the narrow region of 505 to 515 cm$^{-1}$ (normally near ~510 cm$^{-1}$), which is distinctly different from those of other tectosilicates such as zeolites and quartz, whose strongest Raman peak is around 492 cm$^{-1}$ and 464 cm$^{-1}$, respectively [37,39–41].

The Raman spectra of orthoclase were obtained under different pressures, ranging from 0.1 MPa to about 600 MPa. The pressure response of the orthoclase samples is depicted in Figure 2. With increasing pressure, Raman bands shift to higher wavenumbers, and the three peaks become gradually broader due to the decreasing lattice constants [42]. The relative intensities of bands show no significant changes under pressure, which is also observed in previous high-pressure Raman spectroscopy studies of other minerals [43,44].

Peaks between 450 and 520 cm$^{-1}$ are attributed to the internal bending vibrational modes of the Si (Al$^{\mathrm{IV}}$)–O–Si bonds. The dominant feature of the orthoclase spectra is that the strongest peak in this region is near 510 cm$^{-1}$ [40,45]. This peak is the most characteristic feldspar peak and is most frequently used to identify feldspar in the multi-phase spectra from igneous materials. Peaks between 600 and 800 cm$^{-1}$ are assigned to the tetrahedral deformation modes [40,46]. The peak around 432 cm$^{-1}$ is assigned to M–O bending or stretching vibrational modes [47]. Positions of the above three peaks (near 432, 510, and 800 cm$^{-1}$) are listed in Table 2. The peak around 507.27 cm$^{-1}$ is shifted toward higher frequencies with increasing pressure (+2.19 cm$^{-1}$ from 0.1 MPa to 606.87 MPa, equal to 3.61 cm$^{-1}$/GPa), as is the peak around 796.69 cm$^{-1}$ (3.61 cm$^{-1}$/GPa). The peak around 432.33 cm$^{-1}$ totally shifts 2.5 cm$^{-1}$ (4.13 cm$^{-1}$/GPa) towards the higher wavenumber. The shifting to higher wavenumber in all three peaks means the chemical bonds become shorter under high pressures.

**Table 2.** Raman spectra data of colorless orthoclase at different pressures.

| Wavenumber of quartz (cm$^{-1}$) | 464 | 465.10 | 466.73 | 467.56 | 468.60 | 469.38 |
|---|---|---|---|---|---|---|
| Pressure (MPa) | 0.1 | 122.38 | 305.34 | 399.23 | 517.59 | 606.87 |
| M–O vibrational peak (cm$^{-1}$) | 432.33 | 433.02 | 433.20 | 433.50 | 433.50 | 434.83 |
| Al$^{\mathrm{IV}}$–O–Si vibrational peak (cm$^{-1}$) | 507.27 | 507.43 | 507.80 | 508.46 | 508.13 | 509.46 |
| [SiO$_4$] vibrational peak (cm$^{-1}$) | 796.69 | 797.07 | 795.47 | 792.80 | 797.07 | 798.40 |

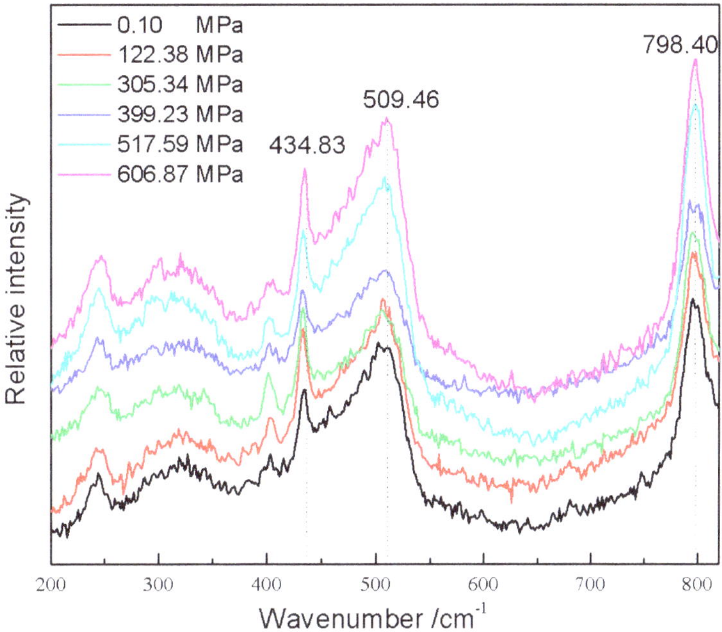


**Figure 2.** Raman spectra of colorless orthoclase at different pressures. The labelled numbers refer to the peak position at around 600 MPa pressure. The spectra are stacked for comparison.

## 3.2. FTIR Spectra with Increasing Pressure

Polarized spectra of orthoclase in the 1000–4000 cm$^{-1}$ range are shown in Figure 3, and band positions are listed in Table 3. Although Galeener and Mikkelsen [48] proposed to quantify the $H_2O$ content of silica glass at the ppm level with micro-Raman analytical procedures four decades ago, Freeman et al. [40] explained that the Raman peak positions of the felspars cannot be used to extract quantitative information due to the complex cation composition of the feldspar phases in comparison with olivine, pyroxene, and some Fe-oxides. Instead, micro-Fourier transform infrared spectroscopy is more efficient and the most widely used method to measure the hydrous absorbing species (including hydroxyl $OH^-$ and molecular $H_2O$) and content, especially nominally anhydrous minerals with trace to minor amounts of hydrous components.

**Table 3.** FTIR spectra and analysis of two types of orthoclase at 300 and 600 MPa pressure.

| Samples | Polarized Directions | Pressure | Wavenumber (cm$^{-1}$) | Integral Absorbance * (cm$^{-1}$) | Normalized Integral Absorbance * (cm$^{-2}$) |
|---|---|---|---|---|---|
| Pink orthoclase | Perpendicular to (001) | 1 atm | 3420 | 17.98 | 1383 |
| | | 300 MPa | 3420 | 19.89 | 1904 |
| | | 600 MPa | 3410 | 20.92 | 2001 |
| Colorless orthoclase | Perpendicular to (001) | 1 atm | 3420 | 1.50 | 120 |
| | | 300 MPa | 3310 | 11.32 | 1124 |
| | | 600 MPa | 3290 | 14.54 | 1570 |

* Note: Integral absorbance and Normalized integral absorbance are calculated based on 1 cm thickness.

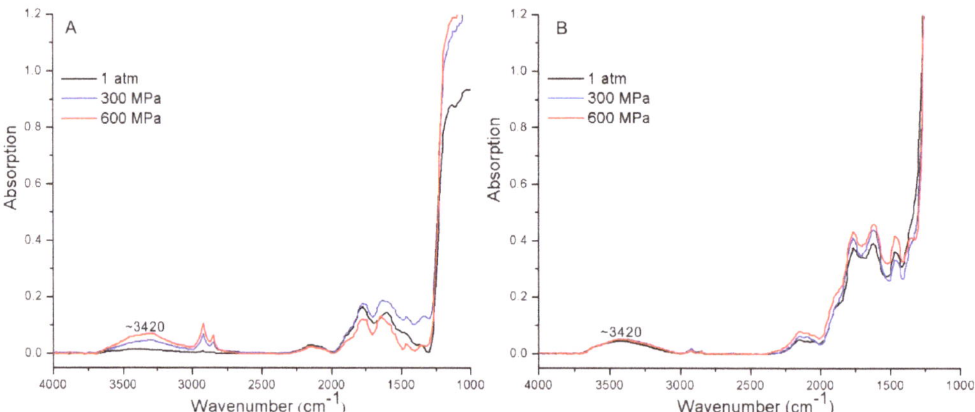



**Figure 3.** FTIR spectra of 300 and 600 MPa pressure: (**A**) colorless feldspar; (**B**) pink feldspar. The spectra are stacked for comparison.

The stretching bands of $OH^-$ groups and the symmetric/asymmetric $H_2O$ molecules occur between 4000 cm$^{-1}$ and 2000 cm$^{-1}$. The most prominent absorption band produced by orthoclase crystals in this study is the broad symmetric peak centered at approximately 3420 cm$^{-1}$ under 1 atmospheric pressure (1 atm) condition (Figure 3). For colorless orthoclase, the 3420 cm$^{-1}$ peak shift to lower wavenumber, 3310 cm$^{-1}$ at ~300 MPa and 3290 cm$^{-1}$ at ~600 MPa, indicates that $H_2O$ molecules are increased [49]. The remarkable observation is an intensity increase in $H_2O$ broad band with increasing pressure of colorless orthoclase (Figure 3A). Statistically, integral absorbance increases sharply with pressure from 1 atm to 300 MPa, with only 1.50 cm$^{-1}$ at 1 atm but 11.32 cm$^{-1}$ at 300 MPa (Table 3). With pressure up to 600 MPa, the integral absorbance still rises slightly, to 14.54 cm$^{-1}$ for colorless orthoclase. The normalized integral absorbance goes up to 1124 cm$^{-2}$ and 1570 cm$^{-2}$ at ~300 MPa and ~600 MPa, respectively, compared to only 120 cm$^{-2}$ at 1 atm. The normalized integral absorbance is about 2.40 cm$^{-2}$/MPa, indicating that water goes into the channel of orthoclase under high pressures.

For pink orthoclase (Figure 3B), the integral absorbance increases gradually with no significant change in the position of peak (Table 3). The small increase in intensity is interpreted to reflect the existence of water in orthoclase. Little pressure dependence was observed between 300 MPa and 600 MPa. Doubling the pressure made a trace increase, which probably means the hydrous component is already saturated under 300 MPa pressure.

### 3.3. FTIR Spectra in Different Crystallography Direction at 1 atm

The FTIR spectra of colorless orthoclase collected from three different crystallography directions show differences in either position or intensity of peaks (Figure 4). Two groups of bands are observed in the spectra from 3000 cm$^{-1}$ to 4000 cm$^{-1}$. The three different orientations, (001), (010), and (100), have a prominent broad band near 3418 cm$^{-1}$, which is in good agreement with those determined previously by Seaman et al. [50]. This peak is assigned to molecular $H_2O$ [51]. The 3418 cm$^{-1}$ band in the direction of (001) slightly shifts to 3420 cm$^{-1}$ in (100) and (010) directions, indicating more molecular $H_2O$ in the direction of (001) (Table 4). The intensity of this peak is very sensitive to the orientation of the orthoclase crystals. The 3418 cm$^{-1}$ band has a stronger intensity in the direction of (001), compared to the other two directions (Figure 4). The normalized (to 1 cm thickness) integral absorbance in the direction of (001) is 1717.5 cm$^{-2}$, but the other two directions are much lower (Table 4). The differences in position and intensity of this peak in three different directions illustrate that the entrance of water into orthoclase structure has a preferred crystallographic orientation. A small but sharp asymmetric band around 3619 cm$^{-1}$ is obvious in the direction of (010), which is assigned to the OH$^-$ stretching vibrational band [49].

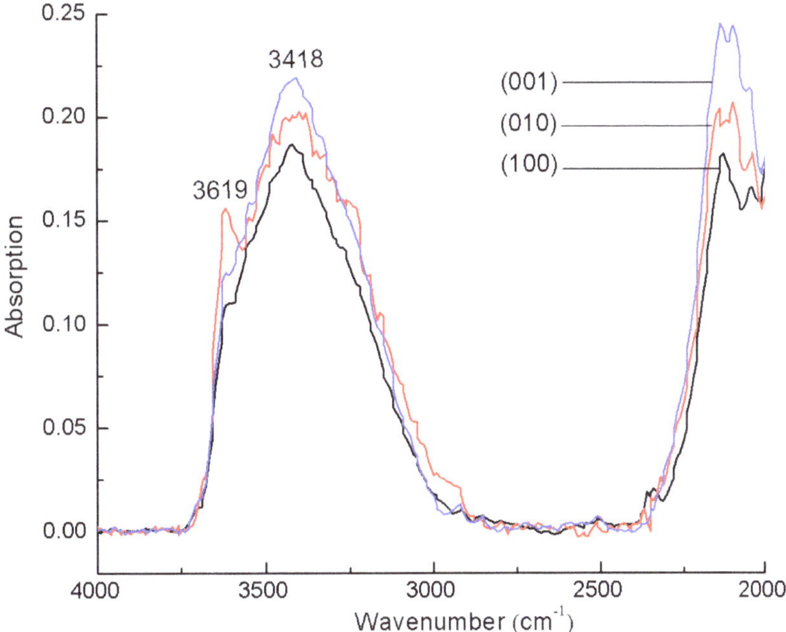


**Figure 4.** FTIR spectra of colorless orthoclase with the polarization direction of the IR radiation oriented perpendicular to (001), (010), and (100) vibration directions. The spectra are stacked for comparison.

**Table 4.** FTIR spectra and analysis of colorless orthoclase at 1 atm pressure.

| Sample | Polarized Directions | Conditions | Wavenumber (cm$^{-1}$) | Integral Absorbance * (cm$^{-1}$) | Normalized Integral Absorbance * (cm$^{-2}$) |
|---|---|---|---|---|---|
| Colorless Orthoclase | Nearly perpendicular to (100) | 1 atm, room temperature | 3420 | 18.19 | 1365.3 |
| | Perpendicular to (010) | | 3420, 3619 | 19.40 | 1434 |
| | Perpendicular to (001) | | 3418 | 20.86 | 1717.5 |

* Note: Integral absorbance and Normalized integral absorbance are calculated based on 1 cm thick.

## 4. Discussion

### 4.1. The Mechanism of Hydrogen in Feldspar

The idea that the hydrous species in alkali feldspar is mainly molecular $H_2O$ whereas the plagioclase is OH$^-$ has been proposed by several investigators. The hydrogen speciation in previously studied alkali feldspars, such as adularia [52,53], sanidine [53–55], microcline [55], anorthoclase [51,56] and orthoclase [51], are all mostly molecular $H_2O$, except for sanidine from Eifel in Germany, researched by Hofmeister and Rossman [55], deemed as OH$^-$. However, hydrogen in plagioclase mainly presents as OH$^-$ [49,57]. Yang [58] found a good correlation between H solubility and K content, and he suggested a rough trend that H solubility increases with increasing K content from albite to labradorite to anorthoclase. Smith [59] and Beran [60] tried to explain $H_2O$ molecules in alkali feldspars as entering the M site of the structure. However, the M site is usually occupied by alkaline ions to balance the charge of aluminum (III) replacing silica (IV) and alkaline ions stay in fixed site. To facilitate the discussion, Figure 5A presents the crystallographic structure with $K^+$ in the M site in the channel along (101).

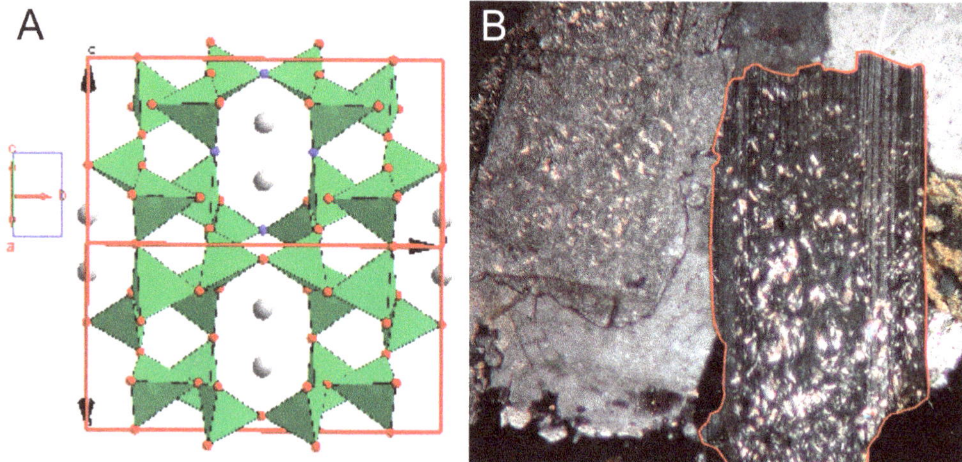

**Figure 5.** (**A**) Projection of the K-feldspar structure along the (101) direction with K (grey) in the channel; (**B**) sericitization alteration happened in the interior of K-feldspar crystal (e.g., red line).

A water molecule has a diameter of 0.276 nm, and its radius is 0.138 nm [61]. The order of ionic radius occupied in framework structure is $K^+$ (0.155 nm, coordination 9) > $Na^+$ (0.102 nm, coordination 6) > $Ca^{2+}$ (0.100 nm, coordination 6) [62]. Considering the non-collapsible framework structure, large radius ions with high coordination can expand the aperture larger than those with relatively small ionic radius within limits. Baur and Joswig [63] explained the flexible but non-collapsible feldspar framework with individual T-O-T angles vary up to 27° without any significant effect on the overall mean T-O-T angle. Large ions prop up the apertures and result in crystallographic channels. Combing these properties together, it is reasonable to infer that K-feldspars have bigger channels than those of Na- or Ca-feldspar, which may allow molecular $H_2O$ to enter.

*4.2. The Speciation of Hydrogen in Orthoclase*

This work presents detailed evidence of the behavior of hydrogen in orthoclase via various experimental techniques, especially the high-pressure method. With increasing pressure, Raman spectra present a shift toward higher wavenumber. Corner-sharing $SiO_4$ tetrahedra show positive relationships between a shorter $SiO_4$ bonding and a bigger force constant, which causes the shift toward the higher wavenumber [51,64]. In the present study, the chemical bonds of Si ($Al^{IV}$)–O–Si become shorter with increasing pressures. Unfortunately, whether the presence of $H_2O$ can lead to a higher wavenumber shift is not well studied in the literature.

The FTIR spectra at atmospheric pressure show orientation-dependent absorbance (Figure 4). As stated in the introduction, alkali feldspar minerals possess good ultra-micro channels, and the apertures of the channels also have crystallographic orientation. In this study, it is shown that there is a correlation between the intensity of FTIR spectra at atmospheric pressure and the aperture size of different orientations. Two peaks on the FTIR spectra in this study (3418 $cm^{-1}$ and 3619 $cm^{-1}$, Table 4) are in good agreement with those previously studied bands (3400 $cm^{-1}$ and 3630 $cm^{-1}$). Johnson and Rossman [49] contributed these two peaks to molecular $H_2O$. Since it is difficult to prepare samples along the (101) direction, we could not collect the corresponding FTIR spectra. However, we expect higher intensity of molecular $H_2O$ peak along the (101) direction based on the results of (001) and (100) (Figure 3).

In high-pressure experiments, the FTIR spectra of colorless orthoclase show good pressure-dependence. At atmospheric pressure, the peak still exists with very low intensity. With increasing pressure, more water in orthoclase raises the intensities which proves the possibility to make "$H_2O$" enter structural channels of orthoclase under certain hydrostatic pressure. The FTIR spectra intensities of pink orthoclase stay relatively stable with increasing pressure, illustrating a limit to the amount of water that can enter their structure via apertures or channels, especially if they are already saturated. The different responses to the pressure between colorless orthoclase and pink orthoclase might be related to the chemical compositions, with $K_2O$ 10.03% and 9.26%, respectively (Table 1). Colorless orthoclase has a higher K content than that of pink orthoclase. The large radius of K can expand the aperture to become large enough to allow more water to enter. To point it out, this is just one possibility to explain the different responses to the pressure between colorless orthoclase and pink orthoclase based on our study.

### 4.3. Implications for Feldspar Alteration, Micropores/Microstructure, and Crustal Process

Sericitization is one of the most common types of hydrothermal alteration found in felsic rocks [65,66]. Sericitization of feldspars (including plagioclase) is one of the most ubiquitous characteristics of many granite-intrusive complexes [65,67]. An interesting phenomenon is that sericitization always occurs in an impregnated state in the interior of potassium feldspar crystals [23]. However, other alterations such as chloritization always occur along the crystal fringe of minerals such as pyroxene, hornblende, and biotite because these minerals have no channel structure to allow hydrothermal alteration to take place in the interior of the crystals. We propose that ultra-microchannels in potassium feldspar facilitate the entrance of $H_2O$ molecules into their structure, leading to sericitization in the interior of potassium feldspar crystals. The reaction transpires as follows: $3KalSi_3O_8$ (potassium feldspar) + $H_2O$ = $Kal_2AlSi_3O_{10}$ $(OH)_2$ (sericite) + $6SiO_2$ + $K_2O$. As shown in Figure 5B, the sericitization takes place in the interior of alkali feldspar. Moreover, sericitization also shows crystallographic orientation (Figure 5B)

Alkali feldspars with large-radius ions in the M sites might bring micropores and microstructures. Walker et al. [17] presented the size of micropores, from ~1 μm to a few nm in length, typically 0.1 μm. Micropores or microtextures are contributed to the processes of crystals coarsening and deuteric or hydrothermal alteration. A process called 'unzipping', i.e., dissolution and reprecipitation around dislocation cores, is believed to result in turbid, microporous, and micropermeable feldspars (Parsons et al., 2000). Baur et al. [63] compared all feldspars with alkaline ions from the least open structure of $LiAlSi_3O_8$ to the most open structure of $RbAlSi_3O_8$. They found the non-collapsible nature of the feldspar-type aluminosilicate framework is because the T-O-T angles anti-rotate when a tetrahedron rotates upon compression or expansion in the framework in order to maintain open pores without expansion by inserted cations. Considering the non-collapsible frameworks achieving their non-collapsibility by being extremely flexible within certain limits, the replacement of $K^+$ with large ionic radius by those ions with small ionic radius such as $Na^+$ or $Ca^{2+}$ is also mechanically allowed, resulting in compression of the framework. Consequently, the compression of the framework may lead to micropores and microstructures; however, these kinds of micropores and microstructures vary on a sub-micrometer scale due to the heterogeneous replacement. However, the TEM analysis shows that the studied feldspars are homogeneous and no microstructures are observed.

Feldspar is the most abundant constituent of the crust, mainly in the upper crust at the depth of 10–20 km (alkali feldspars can also form in more superficial conditions). Pressures ranging from 300 MPa to 600 MPa in this study correspond to the lithostatic pressures at those depths. On the surface of the crust, zeolite minerals contain water molecules in the channels and cages of its framework structure. In the upper crust, alkali feldspars maintain trace amounts of molecular $H_2O$ in its crystallographic channels. The apparent dependence of molecular $H_2O$ on pressure in this study displays the importance of water in potassium feldspar in the upper crust. Although it saturates at a certain concentration and the overall water abundance in feldspar is small, the abundant feldspars content means the total mass of molecular water in the upper crust is potentially significant.

## 5. Conclusions

A high-pressure experiment coupled with FTIR spectroscopy of orthoclase provides evidence of water entering ultra-microchannels in feldspar structure. With increasing pressure, colorless orthoclase exhibits higher peak intensities of water in the FTIR spectra. The normalized integral absorbance increases up to 1124 cm$^{-2}$ and 1570 cm$^{-2}$ at ~300 MPa and ~600 MPa, respectively, compared to only 120 cm$^{-2}$ at 1 atm. An orientation-dependent FTIR reveals the preference of orientation when water enters the structure of orthoclase. Nearly perpendicular to (001) is preferred, with the highest normalized integral absorbance in this study (1717.5 cm$^{-2}$). Water in the ultra-microchannels excellently explains the phenomenon that sericitization always occurs in an impregnated state in the interior of potassium feldspar crystals. This study provides a mechanism for feldspar alteration and important insights into the behavior of water molecules in nominally anhydrous minerals in the upper crust of the Earth.

**Author Contributions:** Writing—original draft preparation: H.Z.; investigation: R.L. and H.Z.; writing—review and editing: H.Z. and A.L. All authors have read and agreed to the published version of the manuscript.

**Funding:** This work was funded by the Science and Technology Development Fund, Macau SAR (File no. SKL-LPS(MUST)-2021-2023) and the Natural Science Foundation of China (no. 41820104003 & 41172043).

**Institutional Review Board Statement:** Not applicable.

**Informed Consent Statement:** Not applicable.

**Data Availability Statement:** Data are contained within the article.

**Acknowledgments:** Technical comments and English review by Sky P. Beard are appreciated. Constructive reviews by anonymous reviewers were very helpful in revising the manuscript.

**Conflicts of Interest:** The authors declare no conflict of interest.

## References

1. Yuan, G.; Cao, Y.; Schulz, H.-M.; Hao, F.; Gluyas, J.; Liu, K.; Yang, T.; Wang, Y.; Xi, K.; Li, F. A review of feldspar alteration and its geological significance in sedimentary basins: From shallow aquifers to deep hydrocarbon reservoirs. *Earth-Sci. Rev.* **2019**, *191*, 114–140. [CrossRef]
2. Parsons, I. *Feldspars and Their Reactions*; Springer Science & Business Media: Berlin/Heidelberg, Germany, 2012.
3. Solé, J.; Pi-Puig, T.; Ortega-Rivera, A. A Mineralogical, Geochemical, and Geochronological Study of 'Valencianite' from La Valenciana Mine, Guanajuato, Mexico. *Minerals* **2021**, *11*, 741. [CrossRef]
4. Machatschki, F. The structure and constitution of feldspars. *Centr. Min. Abt. A* **1928**, *8*, 97–104.
5. Jones, J. Order in alkali feldspars. *Nature* **1966**, *210*, 1352–1353. [CrossRef]
6. Stewart, D.; Ribbe, P. Structural explanation for variations in cell parameters of alkali feldspar with Al/Si ordering. *Am. J. Sci.* **1969**, *267*, 444–462.
7. Taylor, W.H.; Darbyshire, J.A.; Strunz, H. An X-ray investigation of the feldspars. *ZK* **1934**, *87*, 35.
8. Deer, W.A.; Howie, R.; Zussman, J. *Rock-Forming Minerals, Volume 4A: Framework Silicates–Feldspars*; The Geological Society: London, UK, 2001; Volume viii, p. 972.
9. Smith, J.V. *Feldspar Minerals: Crystal Structure and Physical Properties 1*; Springer Science & Business Media: Berlin/Heidelberg, Germany, 2013.
10. Smith, J.V. *Feldspar Minerals: In Three Volumes. 2. Chemical and Textural Properties*; Springer: Berlin/Heidelberg, Germany, 1974.
11. Brown, W.L. *Feldspars and Feldspathoids: Structures, Properties and Occurrences*; Springer Science & Business Media: Berlin/Heidelberg, Germany, 2013.

12. Lu, A.; Huang, S.; Liu, R.; Zhao, D.; Qin, S. Environmental Effects of Micro- and Ultra-microchannel Structures of Natural Minerals. *Acta Geol. Sin.-Engl. Ed.* **2006**, *80*, 161–169.
13. Shamloo, H.I.; Till, C.B.; Hervig, R.L. Multi-mode magnesium diffusion in sanidine: Applications for geospeedometry in magmatic systems. *Geochim. Cosmochim. Acta* **2021**, *298*, 55–69. [CrossRef]
14. Dietrich, R. K-feldspar structural states as petrogenetic indicators. *Nor. Geol. Tidsskr. Bind.* **1962**, *42*, 394–414.
15. Ragland, P.C. Composition and structural state of the potassic phase in perthites as related to petrogenesis of a granitic pluton. *Lithos* **1970**, *3*, 167–189. [CrossRef]
16. Parsons, I.; Gerald, J.D.F.; Lee, M.R. Routine characterization and interpretation of complex alkali feldspar intergrowths. *Am. Mineral.* **2015**, *100*, 1277–1303. [CrossRef]
17. Walker, F.D.L.; Lee, M.R.; Parsons, I. Micropores and micropermeable texture in alkali feldspars: Geochemical and geophysical implications. *Mineral. Mag.* **1995**, *59*, 505–534. [CrossRef]
18. Mamonov, A.; Puntervold, T.; Strand, S.; Hetland, B.; Andersen, Y.; Wealth, A.; Nadeau, P.H. Contribution of feldspar minerals to pH during Smart Water EOR processes in sandstones. *Energy Fuels* **2019**, *34*, 55–64. [CrossRef]
19. Yun, J.; Kumar, A.; Removski, N.; Shchukarev, A.; Link, N.; Boily, J.-F.; Bertram, A.K. Effects of Inorganic Acids and Organic Solutes on the Ice Nucleating Ability and Surface Properties of Potassium-Rich Feldspar. *ACS Earth Space Chem.* **2021**, *5*, 1212–1222. [CrossRef]
20. Choudhary, A.; Khandelwal, N.; Singh, N.; Tiwari, E.; Ganie, Z.A.; Darbha, G.K. Nanoplastics interaction with feldspar and weathering originated secondary minerals (kaolinite and gibbsite) in the riverine environment. *Sci. Total Environ.* **2022**, *818*, 151831. [CrossRef] [PubMed]
21. Wollast, R. Kinetics of the alteration of K-feldspar in buffered solutions at low temperature. *Geochim. Cosmochim. Acta* **1967**, *31*, 635–648. [CrossRef]
22. Duan, G.; Ram, R.; Xing, Y.; Etschmann, B.; Brugger, J. Kinetically driven successive sodic and potassic alteration of feldspar. *Nat. Commun.* **2021**, *12*, 4435. [CrossRef]
23. Engelhardt, W.V.; Matthäi, S.K.; Walzebuck, J. Araguainha impact crater, Brazil. I. The interior part of the uplift. *Meteoritics* **1992**, *27*, 442–457. [CrossRef]
24. Wibberley, C. Are feldspar-to-mica reactions necessarily reaction-softening processes in fault zones? *JSG* **1999**, *21*, 1219–1227. [CrossRef]
25. Kawano, M.; Tomita, K. Formation of mica during experimental alteration of K-feldspar. *Clays Clay Miner.* **1995**, *43*, 397–405. [CrossRef]
26. Garrels, R.; Howard, P. Reactions of feldspar and mica with water at low temperature and pressure. *Clays Clay Miner.* **1957**, *6*, 68–88. [CrossRef]
27. Patterson, D.E.; Margrave, J.L. The use of gem-cut cubic zirconia in the diamond anvil cell. *J. Phys. Chem.* **1990**, *94*, 1094–1096. [CrossRef]
28. Chen, J.; Zheng, H.; Xiao, W.; Zeng, Y. High-temperature and high-pressure cubic zirconia anvil cell for Raman spectroscopy. *Appl. Spectrosc.* **2003**, *57*, 1295–1299. [CrossRef] [PubMed]
29. Grocholski, B.; Jeanloz, R. High-pressure and-temperature viscosity measurements of methanol and 4: 1 methanol: Ethanol solution. *J. Chem. Phys.* **2005**, *123*, 204503. [CrossRef]
30. Eggert, J.H.; Xu, L.; Che, R.; Chen, L.; Wang, J. High pressure refractive index measurements of 4: 1 methanol: Ethanol. *J. Appl. Phys.* **1992**, *72*, 2453–2461. [CrossRef]
31. Lemos, V.; Camargo, F. Effects of pressure on the Raman spectra of a 4: 1 methanol–ethanol mixture. *J. Raman Spectrosc.* **1990**, *21*, 123–126. [CrossRef]
32. Hazen, R. Zeolite molecular sieve 4A: Anomalous compressibility and volume discontinuities at high pressure. *Science* **1983**, *219*, 1065–1067. [CrossRef]
33. Lee, Y.; Vogt, T.; Hriljac, J.A.; Parise, J.B.; Hanson, J.C.; Kim, S.J. Non-framework cation migration and irreversible pressure-induced hydration in a zeolite. *Nature* **2002**, *420*, 485–489. [CrossRef]
34. Angel, R.J.; Bujak, M.; Zhao, J.; Gatta, G.D.; Jacobsen, S.D. Effective hydrostatic limits of pressure media for high-pressure crystallographic studies. *J. Appl. Crystallogr.* **2007**, *40*, 26–32. [CrossRef]
35. Goncharov, A.F. Raman spectroscopy at high pressures. *Int. J. Spectrosc.* **2012**, *2012*, 617528. [CrossRef]
36. Sharma, S.K. Raman spectroscopy at very high pressures. *Raman Scatt. Lumin. Spectrosc. Instrum. Technol.* **1989**, *1055*, 105–116.
37. Schmidt, C.; Ziemann, M.A. In-situ Raman spectroscopy of quartz: A pressure sensor for hydrothermal diamond-anvil cell experiments at elevated temperatures. *Am. Mineral.* **2000**, *85*, 1725–1734. [CrossRef]
38. Deer, W.A. *Rock-Forming Minerals.* 2011. Available online: https://books.google.co.uk/books?hl=zh-CN&lr=lang_en&id=2GLGx6A0d0UC&oi=fnd&pg=PR5&dq=Rock-Forming+Minerals&ots=o-Nx3qJx35&sig=yh_WekveZ2goe7JZvFCvlhnOiKE#v=onepage&q&f=false (accessed on 6 June 2022).
39. Mernagh, T. Use of the laser Raman microprobe for discrimination amongst feldspar minerals. *J. Raman Spectrosc.* **1991**, *22*, 453–457. [CrossRef]
40. Freeman, J.J.; Wang, A.; Kuebler, K.E.; Jolliff, B.L.; Haskin, L.A. Characterization of natural feldspars by Raman spectroscopy for future planetary exploration. *Can. Mineral.* **2008**, *46*, 1477–1500. [CrossRef]

41. Sharma, S.K.; Simons, B.; Yoder, H. Raman study of anorthite, calcium Tschermak's pyroxene, and gehlenite in crystalline and glassy states. *Am. Mineral.* **1983**, *68*, 1113–1125.
42. Ostertag, R. Shock experiments on feldspar crystals. *J. Geophys. Res. Solid Earth* **1983**, *88*, B364–B376. [CrossRef]
43. Kyono, A.; Ahart, M.; Yamanaka, T.; Gramsch, S.; Mao, H.-k.; Hemley, R.J. High-pressure Raman spectroscopic studies of ulvospinel $Fe_2TiO_4$. *Am. Mineral.* **2011**, *96*, 1193–1198. [CrossRef]
44. Ruiz-Fuertes, J.; Errandonea, D.; López-Moreno, S.; González, J.; Gomis, O.; Vilaplana, R.; Manjón, F.; Muñoz, A.; Rodríguez-Hernández, P.; Friedrich, A. High-pressure Raman spectroscopy and lattice-dynamics calculations on scintillating $MgWO_4$: Comparison with isomorphic compounds. *Phys. Rev. B* **2011**, *83*, 214112. [CrossRef]
45. Tan, J.; Zhao, S.; Huang, Q.; Davies, G.; Mo, X. The microstructure of silicate varying with crystal and melt properties under the same cooling condition. *Mater. Res. Bull.* **2004**, *39*, 939–948. [CrossRef]
46. Holtz, F.; Bény, J.-M.; Mysen, B.O.; Pichavant, M. High-temperature Raman spectroscopy of silicate and aluminosilicate hydrous glasses: Implications for water speciation. *Chem. Geol.* **1996**, *128*, 25–39. [CrossRef]
47. Makreski, P.; Jovanovski, G.; Kaitner, B. Minerals from Macedonia. XXIV. Spectra-structure characterization of tectosilicates. *J. Mol. Struct.* **2009**, *924*, 413–419. [CrossRef]
48. Galeener, F.L.; Mikkelsen, J., Jr. Vibrational dynamics in $^{18}O$-substituted vitreous $SiO_2$. *Phys. Rev. B* **1981**, *23*, 5527. [CrossRef]
49. Johnson, E.A.; Rossman, G.R. A survey of hydrous species and concentrations in igneous feldspars. *Am. Mineral.* **2004**, *89*, 586–600. [CrossRef]
50. Seaman, S.J.; Dyar, M.D.; Marinkovic, N.; Dunbar, N.W. An FTIR study of hydrogen in anorthoclase and associated melt inclusions. *Am. Mineral.* **2006**, *91*, 12–20. [CrossRef]
51. Wilkins, R.; Sabine, W. Water content of some nominally anhydrous silicates. *Am. Mineral. J. Earth Planet. Mater.* **1973**, *58*, 508–516.
52. Kronenberg, A.K.; Yund, R.A.; Rossman, G.R. Stationary and mobile hydrogen defects in potassium feldspar. *Geochim. Cosmochim. Acta* **1996**, *60*, 4075–4094. [CrossRef]
53. Lehmann, G. *Spectroscopy of Feldspars. Feldspars and Feldspathoids*; Springer: Berlin/Heidelberg, Germany, 1984; pp. 121–162.
54. Hofmeister, A.M.; Rossman, G.R. A spectroscopic study of irradiation coloring of amazonite: Structurally hydrous, Pb-bearing feldspar. *Am. Mineral.* **1985**, *70*, 794–804.
55. Hofmeister, A.M.; Rossman, G.R. A model for the irradiative coloration of smoky feldspar and the inhibiting influence of water. *Phys. Chem. Miner.* **1985**, *12*, 324–332. [CrossRef]
56. Xia, Q.; Pan, Y.; Chen, D.; Kohn, S.; Zhi, X.; Guo, L.; Cheng, H.; Wu, Y. Structural water in anorthoclase megacrysts from alkalic basalts: FTIR and NMR study. *Acta Petrol. Sin.* **2000**, *16*, 485–491.
57. Rossman, G. Studies of OH in nominally anhydrous minerals. *Phys. Chem. Miner.* **1996**, *23*, 299–304. [CrossRef]
58. Yang, X. An experimental study of H solubility in feldspars: Effect of composition, oxygen fugacity, temperature and pressure and implications for crustal processes. *Geochim. Cosmochim. Acta* **2012**, *97*, 46–57. [CrossRef]
59. Smith, J.V. *Feldspar Minerals: 2 Chemical and Textural Properties*; Springer Science & Business Media. Available online: https://books.google.co.uk/books?hl=zh-CN&lr=lang_en&id=IObtCAAAQBAJ&oi=fnd&pg=PA3&dq=Feldspar+Minerals:+2+Chemical+and+Textural+Properties%3B+Springer+Science+%26+Business+Media&ots=RTL4w_rdrJ&sig=OqHau4WuboNuzeH2ooV_zwtKlOc#v=onepage&q=Feldspar%20Minerals%3A%202%20Chemical%20and%20Textural%20Properties%3B%20Springer%20Science%20%26%20Business%20Media&f=false (accessed on 5 June 2022).
60. Beran, A. A model of water allocation in alkali feldspar, derived from infrared-spectroscopic investigations. *Phys. Chem. Miner.* **1986**, *13*, 306–310. [CrossRef]
61. Marcus, Y. On Water Structure in Concentrated Salt Solutions. *J. Solut. Chem.* **2009**, *38*, 513–516. [CrossRef]
62. Shannon, R.D. Revised effective ionic radii and systematic studies of interatomic distances in halides and chalcogenides. *Acta Crystallogr. Sect. A Cryst. Phys. Diffr. Theor. Gen. Crystallogr.* **1976**, *32*, 751–767. [CrossRef]
63. Baur, W.H.; Joswig, W.; Müller, G. Mechanics of the feldspar framework; Crystal structure of Li-feldspar. *J. Solid State Chem.* **1996**, *121*, 12–23. [CrossRef]
64. Hemley, R.; Mao, H.; Bell, P.; Mysen, B. Raman spectroscopy of $SiO_2$ glass at high pressure. *Phys. Rev. Lett.* **1986**, *57*, 747. [CrossRef]
65. Que, M.; Allen, A.R. Sericitization of plagioclase in the Rosses granite complex, Co. Donegal, Ireland. *Mineral. Mag.* **1996**, *60*, 927–936. [CrossRef]
66. Zuo, H.; Lu, A.; Gu, X.; Ma, W.; Cui, Y.; Yi, L.; Lei, H.; Wang, Z.; Zhang, D.; Liu, J. Typomorphic feature of chromium sericite in granite hosted gold deposits in Jiaodong Peninsula, China. *Appl. Clay Sci.* **2016**, *119*, 49–58. [CrossRef]
67. Verati, C.; Jourdan, F. Modelling effect of sericitization of plagioclase on the 40K/40Ar and 40Ar/39Ar chronometers: Implication for dating basaltic rocks and mineral deposits. *Geol. Soc. Lond. Spec. Publ.* **2014**, *378*, 155–174. [CrossRef]

*Article*

# Characteristics of Channel-Water in Blue-Green Beryl and Its Influence on Colour

Hui Wang, Tong Shu, Jingyi Chen and Ying Guo *

School of Gemmology, China University of Geosciences, Beijing 100083, China; 2009190007@cugb.edu.cn (H.W.);
1009191105@cugb.edu.cn (T.S.); 1009191104@cugb.edu.cn (J.C.)
* Correspondence: guoying@cugb.edu.cn

**Abstract:** This study reports the characteristics of water in channels of blue-green beryl and its effect on color. An industrial camera was used to measure color in the CIELAB color space. X-ray fluorescence (XRF), X-ray diffraction (XRD), infrared spectroscopy (IR), ultraviolet-visible (UV–vis) spectroscopy, and silicate rock chemical analysis method were used for analysis. The peaks at $5105\ cm^{-1}$ and $5269\ cm^{-1}$ were the combination tone of type II water, which were negatively correlated with b*, and positively correlated with the peak area at $3162\ cm^{-1}$ (Na–H) and cell parameter $a_0$. The peaks at $7097\ cm^{-1}$ and $7142\ cm^{-1}$ were related to the metal ions types in the channels. Part of the water in the channel combined with $Fe^{3+}$ to form $[Fe_2(OH)_4]^{2+}$ and cause a yellow tone, and when the yellow tone combined with the blue tone caused by $Fe^{2+}$, the beryl has a blue-green colour.

**Keywords:** channel-water; X-ray fluorescence; X-ray diffraction; infrared spectroscopy; UV–vis spectroscopy; transition metal ions

**Citation:** Wang, H.; Shu, T.; Chen, J.;
Guo, Y. Characteristics of
Channel-Water in Blue-Green Beryl
and Its Influence on Colour. *Crystals*
**2022**, *12*, 435. https://doi.org/
10.3390/cryst12030435

Academic Editor: Vladislav V.
Gurzhiy

Received: 20 February 2022
Accepted: 17 March 2022
Published: 21 March 2022

**Publisher's Note:** MDPI stays neutral
with regard to jurisdictional claims in
published maps and institutional affiliations.



## 1. Introduction

Beryl ($Be_3Al_2Si_6O_{18}$) is a ring silicate mineral, mainly occurring in pegmatites. Pure beryl is colorless and transparent, but when it contains chromophores (Figure 1), it will show different colors and form different varieties. Many varieties of beryl, such as emerald, aquamarine, morganite, and heliodor occupy a large market share. However, high quality beryl is still in short supply, in recent years, it has been the hot issues to optimize the poor colour of beryl and achieve the desired colour. Therefore, it has become an urgent problem to study the chromaticity mechanism of beryl and guide the chromaticity experiments.

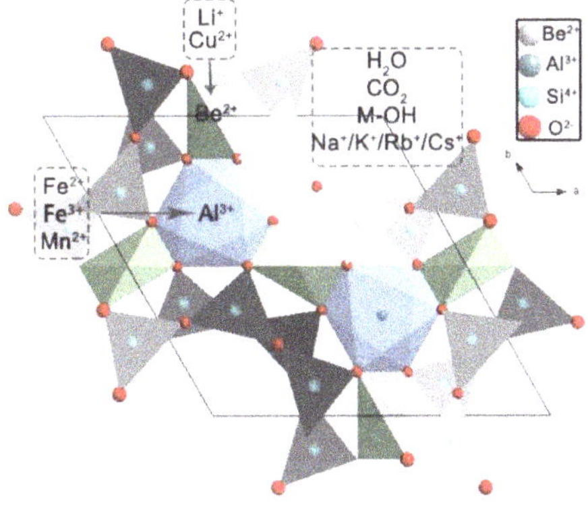


**Figure 1.** The crystal structure of beryl. $Al^{3+}$ can be replaced by $Fe^{2+}$, $Fe^{3+}$, and $Mn^{2+}$; $Be^{2+}$ can be replaced by $Li^+$ and $Cu^{2+}$; $H_2O$, $CO_2$, alkali ions, and M–OH are present in channels.

Blue-green beryl is mainly colored by iron ions. By heating and irradiation, high-quality aquamarine can be obtained [1–3], so it has an important research value. The chromaticity mechanism has also been studied by many scholars. Goldman et al. [4] concluded that the $Fe^{2+}$ channel occupancy in beryl caused a light-blue color, the $Fe^{2+}$–$Fe^{3+}$ charge transfer caused a blue-green color, the coexistence of the $Fe^{3+}$ charge transfer and the $Fe^{2+}$ channel occupancy caused a blue-green color, and the $Fe^{3+}$ in channel caused a yellow-orange color [4,5]. Hu [6] speculated that $[Fe_2(OH)_4]^{2+}$ in the channel was the cause of the yellow tone of beryl. Zhong et al. [7] concluded that the blue-green color of the paraiba-color beryl was influenced by $Cu^{2+}$, $Fe^{2+}$, $Fe^{3+}$, and $Ni^{2+}$ in the crystal structure. Wang et al. [8] concluded that the d–d electron transition of $Fe^{3+}$ and the charge transition of the binuclear metal M–M complexes formed by $[Fe_2(OH)_4]^{2+}$ in the channel would produce a yellow tone, and when combined with the blue color caused by the $Fe^{2+}/Fe^{3+}$ charge transfer, the beryl would produce a yellow-green color. The lightness value would decrease with the total amount of transition metal ions increasing, and the chroma would decrease when $Cs^+$ and $Mn^{2+}$ exist.

In addition to the influence of transition metal ions on the color of beryl, some scholars believed that the channel-water also had an effect. It is well known that water exists in minerals in the following five main forms: hydroscopic water, crystallization water, constitution water, interlayer water, and zeolitic water. Beryl mainly contains constitution water [9], which can exist in the form of $H_2O$ molecules, or be polymerized with cations in channels, such as $[Fe_2(OH)_4]^{2+}$. Zou [10] found that the color of beryl was closely related to the content ratio of type I and type II water. From golden beryl–cat's-eye beryl–green-beryl–aquamarine, the content of type I water gradually decreased. Especially in the bright color aquamarine, the value of type I water/type II water was close to 1:1, which was speculated as the best condition to promote the combination of $Fe^{2+}$ and channel $H_2O$ to form hydration ions. However, most scholars [9,11–16] concluded that the substitution of isomorphism in beryl was the root cause. When $Al^{3+}$ and $Be^{2+}$ were replaced by different valence ions, alkaline ions entered the channel to compensate for the charge difference and caused the transition from type I water to type II water in the channel. Therefore, the beryl color showed a correlation with the type and content of channel-water; although the quantitative relationship has not been obtained so far. The purpose of this study was to use rigorous data to explore the effect of channel-water on the color of blue-green beryl, so that in the future, the characteristics of channel-water can be obtained through simple spectral testing to help predict the chromaticity mechanism of blue-green beryl. This will greatly simplify the experimental process and make a great contribution to the study of chromaticity mechanism and the color optimization of beryl.

A CIE standard color system by the International Commission on Illumination was introduced. Scholars have used this system to quantitatively describe color gems. For example, Jiang et al. [17] classified Australian chalcedony in a CIE 1976 LAB uniform color space. Tang et al. [18] studied the influence of illumination on the color of magnesium olivine in a CIE 1976 LAB uniform color space. Wang [8] explored the influence of transition metal ions on the color of blue-green beryl in a CIE 1976 LAB uniform color space.

## 2. Materials and Methods

### 2.1. Materials

Twenty-three pieces of hexagonal prism beryl with diameters of 6–12 mm were selected as the experimental samples, which were cut and ground into plate shapes perpendicular to the c-axis and polished on both sides. The samples were blue-green, glassy, and translucent to nontranslucent (Figure 2).

**Figure 2.** Top view photo of 23 samples, hexagonal short columns.

*2.2. Colour Collection*

The CIE 1976 LAB uniform color space was adopted, which was the latest and most widely used uniform color space.

L* is the lightness component and ranged from 0 to 100 (from black to white). A* and B* are the chromaticity components. A* ranges from $-128$ to $+127$ (from negative green to positive red) and B* ranges from $-128$ to $+127$ (from negative blue to positive yellow). In addition, the $C^*_{ab}$ (chroma value) and $h_{ab}$ (hue angle) can be calculated according to L*, a*, and b* values:

$$C^*_{ab} = [(a^*)^2 + (b^*)^2]^{1/2} \tag{1}$$

$$h_{ab} = \arctan\left(\frac{b*}{a*}\right) \tag{2}$$

The color measuring instrument used was the industrial camera Mako G-507C (manufactured by Allied Vision Technologies, Stadtroda, Germany), with the following specifications: resolution: $2464 \times 2056$; frame rate: 23.7; sensitive chip: Sony IMX264; chip type: CMOS; target surface: $2/3''$; pixel: $3.45 \times 3.45$; exposure mode: global, color; and digital interface: GigE.

The light source was standard D65 and the test background was a PANTONE Cool Gray 1C colorless background.

*2.3. Structure and Composition Testing*

Absorption spectra in the ultraviolet to visible (UV–vis) range were recorded with a UV-3600 UV–vis spectrophotometer (SHIMADZU, Kyoto, Japan). The samples were ground into transparent slices with a thickness of 1 mm. The transmission method was used to test the absorption value. The wavelength range for the test was 300–900 nm, the scanning speed was set to 'high', the sampling interval was 1.0s and the scanning mode was 'single'.

Absorption spectra in the infrared (IR) range were recorded with a Brooke Tensor 27 Fourier transform infrared spectrometer. The test method was transmission for 2000–8000 $cm^{-1}$. The resolution was 4 $cm^{-1}$ and the scans were set to '32'.

Composition semiquantitative results were obtained with an EDX 700 X-ray fluorescence (XRF) spectrometer (SHIMADZU, Kyoto, Japan). The samples were cut and ground into plate shapes perpendicular to the c-axis and polished, and one of the planes perpendicular to the c-axis were selected as the test plane. The collimator diameter was 5 mm. The testing atmosphere was air. The element test conditions in the Al–U range were 50 kV and

272 µA, the analysis was set to '0.00–40.00', the time was set to 'live-100', and the DT% was set to '29'. The element test conditions in the Na–Sc range were 15 kV and 1000 µA, the analysis was set to '0.00–4.40', the time was set to 'live-100', and the DT% was set to '17'.

Samples were grounded to 400 mesh powder and placed in a silicon glass recess, and measured using an Ultima IV polycrystalline powder X-ray diffractometer (XRD). A nickel filter was used and the copper target was used as the detector. The wavelength was 0.15406 nm, the current was 50 mA, and the voltage was 40 kV. The continuous scanning was set to '5°/min'. The scanning range was 5–90°, and the step was 0.02. X-ray scan samples were at a different θ angle and detector was at 2θ angle position. The standard PDF card '09-4310' for beryl was used as a reference.

Ferrous oxide was determined for 10 samples, using the silicate rock chemical analysis method. The samples were grounded to 400 mesh powder and decomposed with HF ($\rho$ = 1.15 g/mL) and $H_2SO_4$ ($\rho$ = 1.84 g/mL), and the remaining $F^-$ in the solution was added to saturated $H_3BO_3$ for complexation. The $C_{12}H_{10}NNaO_3S$ ($\rho$ = 5 g/L) was used as indicator, and the solution was titrated with $K_2Cr_2O_7$ (c = 0.002319 mol/L) to calculate the amount of ferrous oxide. The electronic analytical balance BSA 124S-CW was used. The ambient temperature was 21 °C and the humidity was 42%.

The K-means clustering method was adopted to simplify the data processing. When using this method, samples were classified to several clusters according to a certain variable within the cluster of similarity, and the difference between clusters. SPSS software was used to classfy. Other variables of different samples were also divided into the corresponding clusters. To make the figure clearer, box diagrams were used to represent different clusters, and the change rule of these box diagrams could represent the whole data.

## 3. Results

### 3.1. Colour Parameters

According to the test results, the color parameters of the experimental samples L* ∈ (41.7, 68.05), a* ∈ (−10.8, −5.3), b* ∈ (−0.85, 7.84), C* ∈ (7.55, 12.78), and h ∈ (134.2°, 185.2°), belonged to a blue-green tone interval, medium lightness, and low chroma (Figure 3). The test data can be found in Table S1 (see Supplementary Materials).

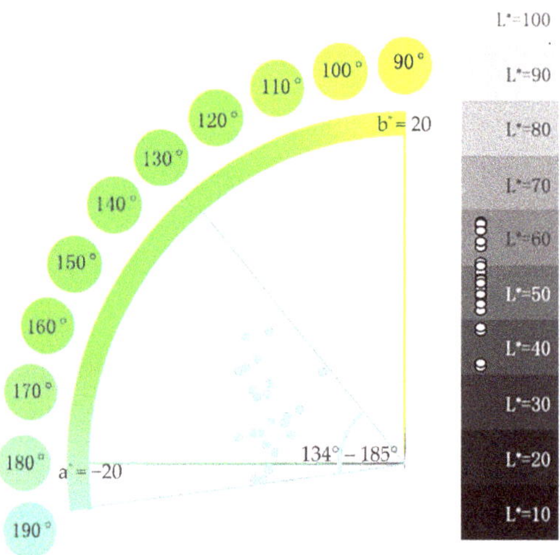


**Figure 3.** Colour distribution of samples in the CIELAB space. $H_{ab}$ ranges from 134° to 185°. L* ranges from 41 to 68 (from Wang H. [8]).

### 3.2. Characteristics of IR Spectroscopy

The IR spectra of 23 samples in the range of 2000–8000 cm$^{-1}$ were obtained as follows (Figure 4), and the test data can be found in Table S2 (see Supplementary Materials).

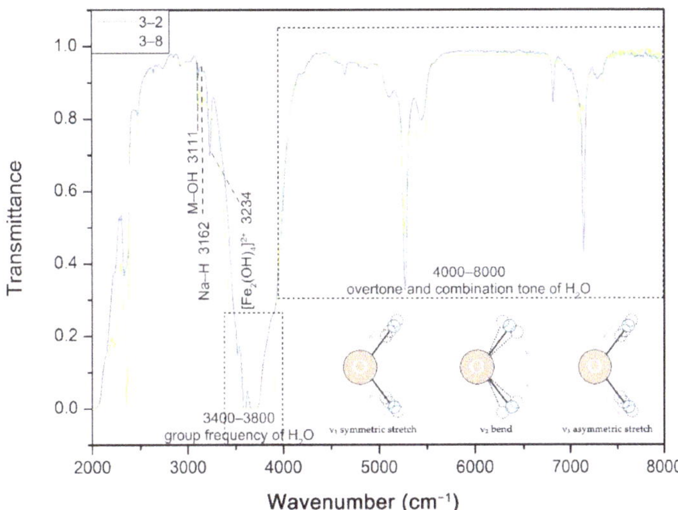



**Figure 4.** IR spectrum of 2000–8000 $cm^{-1}$ and characteristic peaks. Schematic diagram of the vibration model of the $H_2O$ molecule is in the lower right.

The 3111 $cm^{-1}$ was the absorption peak of M–OH, 3162 $cm^{-1}$ was the absorption peak of Na–H, and 3234 $cm^{-1}$ was the absorption peak of $[Fe_2(OH)_4]^{2+}$. Some peaks for type I and type II water (Box 1) were valuable for this study. The fundamental frequency vibration zone of $H_2O$ was 3400–3800 $cm^{-1}$, which including symmetric stretching vibration ($v_1$) and asymmetric stretching vibration ($v_3$) (the bending vibration ($v_2$) was below 2000 $cm^{-1}$, and it was not tested and would not be discussed this time (Figure 4). 4000–8000 $cm^{-1}$ was the combination and overtone zone of $H_2O$.

Among the samples, absorption peaks of No. 2-3, 3-2, and 3-6 were slightly different from others, showing strong absorption at 3234 $cm^{-1}$, 3520 $cm^{-1}$, 4645 $cm^{-1}$, 5269 $cm^{-1}$, 7142 $cm^{-1}$, and 3550–3650 $cm^{-1}$, as shown by the blue curve in Figure 4. Other samples of spectra were similar with No. 3-8, shown by the green curve in Figure 4. There were several wavenumbers shift of absorption peaks, which were caused by different types of alkali ions in the channels [19].

**Box 1.** Type I and type II water in beryl channel.

There are two types of $H_2O$ molecules in the beryl channel: (1) type I water, whose symmetry axis is parallel to the c-axis; (2) type II water, whose symmetry axis is perpendicular to the c-axis. Standard beryl contains only type I water. When there are alkali ions in the channel, the type I water will be rotated 90° to type II water due to the charge attraction between the alkali cation and the $O^{2-}$ [1–3,13,14,20,21]. The H–H axis of type I water is bound to Be and Al, respectively, to form chemical bonds such as Be–HOH–Al. The O–H vibration is mainly controlled by Be, Al, and the longitudinal crystal field. The $O^{2-}$ of type II water is bound to the alkali ion, and the O–H vibration is mainly controlled by the alkali ion and the transverse crystal field. So, the vibration frequencies of the two types of water are different [14].

Furthermore, D. G. et al. and K. B. et al. [20,22,23] found that in most beryl, each alkali ion combined with two type II water (Figure 5, left), but in the water-poor samples, each alkali ion only combined with one type II water (Figure 5, middle). This resulted in a vibration frequency shift of type II water. D. G. et al. [20] also explained this in terms of bond length and bond angle (Figure 5, right): in the latter case, the M–O bond length was shortened because of the inductive effect, the O–H bond length increased, and bond angle decreased, so the bending vibration ($v2$) increased and the symmetric stretching vibration ($v1$) and asymmetric stretching vibration ($v3$) decreased.

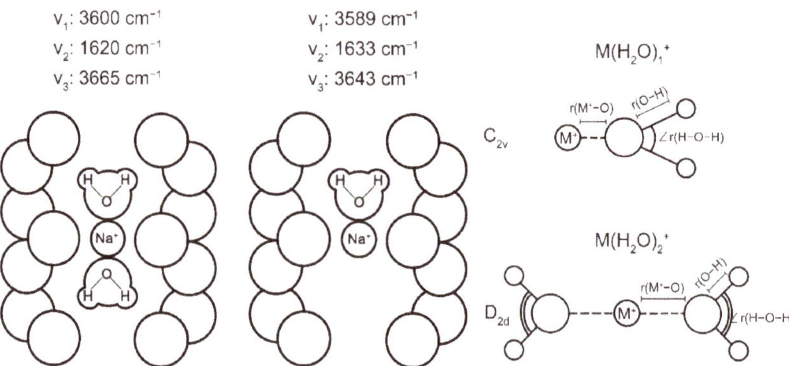


**Figure 5.** Schematic diagram of type II water in water-rich and water-poor condition (from G. Della Ventura [20], redrawn).

After a comprehensive analysis of previous research, the assignment of characteristic spectrum peaks is shown in Table 1.

**Table 1.** Characteristic absorption peaks and assignment within 2000–8000 $cm^{-1}$.

| Wavenumber ($cm^{-1}$) | Assignment | Reference |
|---|---|---|
| 2291 | $CO_2$ | [24] |
| 2358 | $CO_2$ | [11,13,15] |
| 3111 | M–OH | [14] |
| 3162 | Na–H | [14] |
| 3234 | $[Fe_2(OH)_4]^{2+}$ | [1,14] |
| 3558 | $\nu_1^{I}$ | [13] |
| 3590 | $\nu_1^{IIs}$ | [15,19,20] |
| 3605 | $\nu_1^{IId}$ | [15,19,20] |
| 3661 | $\nu_3^{IId}$ | [13,15,19,20] |
| 3694 | $\nu_3^{I}$ | [13] |
| 5105 | Need to be discussed | |
| 5269 | $\nu_2^{II} + \nu_3^{II}$ | [14,25] |
| 7098 | Need to be discussed | |
| 7142 | Need to be discussed | |

### 3.3. Total Iron and $Fe^{2+}$ Content

The relative total Fe content of 23 samples was tested by XRF and the test data was shown in Table S3 (see Supplementary Materials). The relative content of $Fe^{2+}$ in 10 samples was obtained by a chemical analysis of silicate rocks and the test data was shown in Table S4 (see Supplementary Materials). The relative content of $Fe^{3+}$ is the difference between total Fe and $Fe^{2+}$. The relative content of $Al^{3+}$ was also determined by XRF. The above data are listed in Table 2.

**Table 2.** Different ion content and cell parameters in 23 samples.

| No. | Al$^{3+}$ | Total Fe Ions | Fe$^{2+}$ | a0 | c0 | No. | Al$^{3+}$ | Total Fe Ions | Fe$^{2+}$ | a0 | c0 |
|---|---|---|---|---|---|---|---|---|---|---|---|
| 1-3 | 17.04 | 1.81 | 0.60 | 9.21 | 9.20 | 2-10 | 15.24 | 3.32 | | 9.22 | 9.19 |
| 1-4 | 17.17 | 3.02 | | 9.22 | 9.20 | 3-1 | 17.35 | 3.05 | | 9.22 | 9.21 |
| 1-6 | 17.95 | 2.15 | 0.42 | 9.22 | 9.20 | 3-2 | 16.16 | 2.77 | 0.48 | 9.21 | 9.19 |
| 2-1 | 16.54 | 3.41 | | 9.23 | 9.20 | 3-3 | 16.57 | 3.41 | 0.71 | 9.23 | 9.21 |
| 2-2 | 16.87 | 3.22 | | 9.23 | 9.20 | 3-4 | 14.18 | 2.14 | | 9.22 | 9.21 |
| 2-3 | 14.40 | 3.67 | | 9.23 | 9.20 | 3-5 | 16.35 | 3.40 | 0.86 | 9.22 | 9.21 |
| 2-4 | 15.82 | 3.44 | 0.92 | 9.22 | 9.18 | 3-6 | 15.37 | 3.53 | 0.96 | 9.23 | 9.20 |
| 2-5 | 15.58 | 3.12 | | 9.23 | 9.20 | 3-7 | 16.66 | 2.71 | | 9.22 | 9.21 |
| 2-6 | 15.04 | 4.00 | | 9.22 | 9.19 | 3-8 | 16.36 | 2.25 | 0.57 | 9.22 | 9.20 |
| 2-7 | 15.48 | 3.08 | | 9.21 | 9.19 | 3-9 | 16.66 | 2.75 | | 9.22 | 9.19 |
| 2-8 | 15.46 | 3.36 | | 9.23 | 9.20 | 3-10 | 15.93 | 3.58 | 0.79 | 9.23 | 9.21 |
| 2-9 | 15.86 | 3.16 | | 9.23 | 9.20 | | | | | | |

## 3.4. Characteristics of XRD

The samples were multiphase beryl and other diffraction lines could be detected besides the standard beryl diffraction lines (Figure 6).

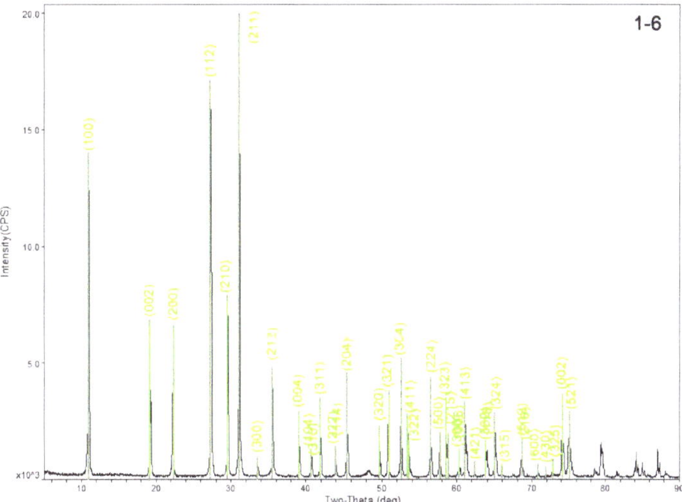


**Figure 6.** XRD pattern of no.1–6. Beryl peaks were marked by green lines.

Beryl's symmetry is P6/mcc (D$_{6h}^{2}$). The crystal cell parameters can be obtained by XRD and refinement. In standard beryl samples (PDF card: 09-4310), a$_0$ = 9.217 Å and c$_0$ = 9.192 Å. When the octahedral Al$^{3+}$ is replaced by transition metal ions, a$_0$ increases due to the increase of the ion radius, whereas when the tetrahedral Be$^{2+}$ is replaced by Li$^{+}$ or Cu$^{2+}$, c$_0$ increases. Beryl can be divided into two types according to the c/a value [26]:
Octahedral substitution of Al$^{3+}$ is dominant: c/a = 0.991–0.998;
Tetrahedral substitution of Be$^{2+}$ is dominant: c/a = 0.999–1.003.

XRD was carried out on the samples, and JADE 6.5 was used to refine the diffraction lines. With reference to the standard beryl PDF card 09-4310, the crystal cell parameters were obtained (Table 2). Both tetrahedral and octahedral substitutions can be found because the a$_0$ and c$_0$ of the samples were higher than the standard beryl, and the octahedral substitutions were the main one.

## 4. Discussion

### 4.1. The Relative Content of Water in the Channel Deduced by IR

Although Li [21] concluded that the peak at about 5105 cm$^{-1}$ was the combination peak of type I water, through the study of IR spectrum, it was found that the peak at 5105 cm$^{-1}$ was positively correlated with the sum of the peak areas of 3111 cm$^{-1}$ (M–OH) and 3162 cm$^{-1}$ (Na–H), and with the peak at 5269 cm$^{-1}$ (type II water, Figure 7). Furthermore, $\nu_1^{I} + \nu_2^{I} = 1542$ cm$^{-1}$ (this data from D.L. [13]) + 3558 cm$^{-1}$ = 5100 cm$^{-1}$ was slightly less than 5105 cm$^{-1}$, whereas $\nu_1^{II} + \nu_2^{II} = 1628$ cm$^{-1}$ (this data from D.L. [13]) + 3605 cm$^{-1}$ = 5233 cm$^{-1}$, as according to Weng [27], the combined frequency must be less than the sum of fundamental frequency, so the peak here was concluded to be $\nu_1^{II} + \nu_2^{II}$ more likely.

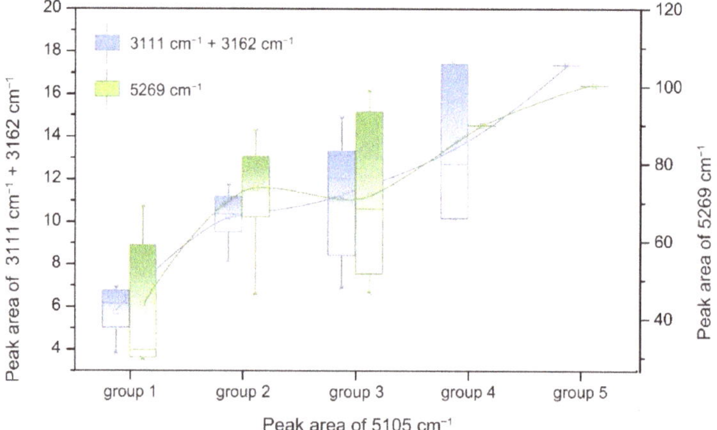


**Figure 7.** (1) The peak areas at 5105 cm$^{-1}$ were positively correlated with the sum of peak areas at 3111 cm$^{-1}$ and 3162 cm$^{-1}$, which are represented by blue boxes. The samples were divided into five groups by K-means clustering. (2) The peak areas at 5105 cm$^{-1}$ were positively correlated with the peak areas at 5269 cm$^{-1}$, which are represented by green boxes. The samples were divided into five groups by K-means clustering.

In addition, Li [21] assigned 7098 cm$^{-1}$ as the first-order overtone peak of type I water, and Mashkovtsev et al. [19] assigned 7142 cm$^{-1}$ as the combined peak of type I water. Theoretically, with the change of type I water, the two peaks in different samples should show the same trend of change. However, as can be seen from Figure 8, the two peaks in different samples differed greatly. Furthermore, it was found that the peaks at 7097 cm$^{-1}$ and 7142 cm$^{-1}$ were controlled by the types of metal ions in the channel. When the sum of peak areas at 3111 cm$^{-1}$ + 3234 cm$^{-1}$ was large (representing large radius cation content), the peak area at 7097 cm$^{-1}$ was large. When the peak area at 3162 cm$^{-1}$ (representing Na$^{+}$, small radius cation content) was large, the peak area at 7142 cm$^{-1}$ was large. Therefore, it was speculated that these two peaks were not the overtone and combination peaks of type I water, but probably related to the overtone or combination peaks of metal cation in the channel.

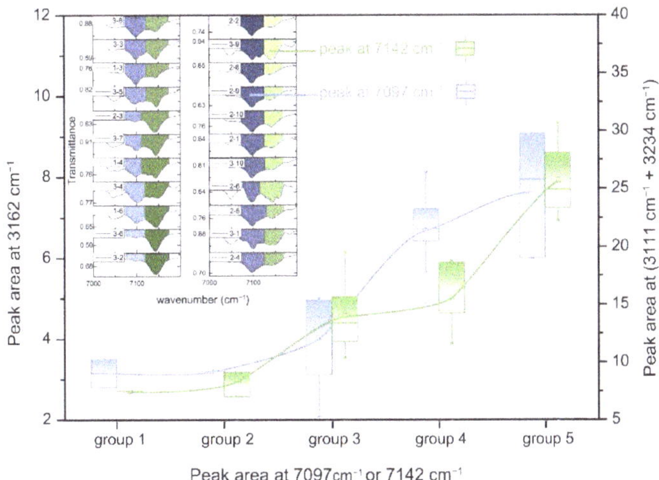



**Figure 8.** The peak area at 7097 cm$^{-1}$ was positively correlated with the peak area at 3162 cm$^{-1}$, as shown by the blue boxes. The peak area at 7142 cm$^{-1}$ was positively correlated with the peak areas at 3111 cm$^{-1}$ + 3234 cm$^{-1}$, as shown by the green boxes. The upper left corner was the IR absorption spectrum within the range of 7000–7200 cm$^{-1}$. The blue shaded part represented the absorption peak at 7097 cm$^{-1}$, and the green shaded part represented the absorption peak at 7142 cm$^{-1}$.

### 4.2. Relationship between Type II Water Content and Colour

The strong absorption peak at 5269 cm$^{-1}$ was selected as the characteristic peak of type II water.

With the increase of the peak area at 5269 cm$^{-1}$, the value of the cell parameter $a_0$ and the peak area at 3162 cm$^{-1}$ increased, whereas the color parameter b* decreased; that is, with the increase of type II water, the beryl color transitioned from yellow to blue (Figure 9). The increase of the cell parameter $a_0$ value indicated that the different charge substitution in the octahedral position (mainly $Fe^{2+}$ replacing $Al^{3+}$) increased, which led to the increase of $Na^+$ in the channel to compensate for the charge difference. Thus, the peak area at 3162 cm$^{-1}$ increased, and the amount of type II water increased with the increase of alkali ions in the channel.

### 4.3. Characteristics of Channel-Water in the Form of Hydration Ions and Its Influence on Colour

Wang [8] found that iron ions mainly replaced the $Al^{3+}$ in the octahedral position, but the specific valence state could not be determined. Through the analysis of IR spectra and UV–vis spectra, it was speculated that the charge transfer of $Fe^{2+}$ and $Fe^{3+}$ in the octahedral position would produce absorption in the red region of UV–vis spectra. The d–d transition of the octahedral $Fe^{3+}$ and the charge transfer between it and $O^{2-}$ and the $[Fe_2(OH)_4]^{2+}$ in the channel would produce absorption in the blue–violet region. The absorptions in the two regions produced blue-green color. This conclusion could also be proven in the two illustrations at the bottom of Figure 10. With the increase of $Fe^{2+}/Fe^{3+}$, the absorption area ratio of the red region to the purple region in the UV–vis spectrum increased, as well as the hue angle (h). The test data of UV-vis can be found in Table S5 (see Supplementary Materials).

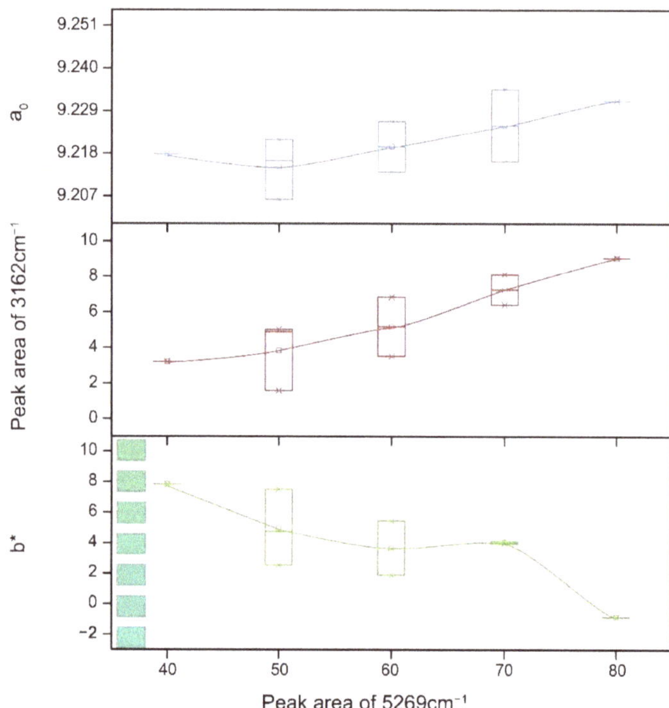

**Figure 9.** The IR peak area at 5269 cm$^{-1}$ was negatively correlated with the color parameter b*, and positively correlated with the Na–H absorption peak at 3162 cm$^{-1}$, and positively correlated with cell parameter a$_0$.

In this study, the total iron content tested by XRF and the Fe$^{2+}$ tested by titration method were combined to further study the valence state of iron ions at different positions and their influence on color. Qi et al. [2] concluded that Fe$^{3+}$ would enter the hexagonal ring channel in the form of filler impurity ions and hydrolyses with water molecules. Under the polymerization of alkali ions, [Fe$_2$(OH)$_4$]$^{2+}$ copolymer and ion are generated:

$$2Fe^{3+} + 4H_2O = [Fe_2(OH)_4]^{2+} + 4H^+ \quad k = 10^{-291} \tag{3}$$

It can be seen from Figure 11 that with the increase of Fe$^{3+}$ content, the peak area at 3234 cm$^{-1}$ (representing the content of [Fe$_2$(OH)$_4$]$^{2+}$) also increased, proving that Fe$^{3+}$ mainly exists in the channel in the form of hydration ions. However, Fe$^{2+}$ didn't completely exist in the octahedral position. It can be seen from Figure 10 that a$_0$ showed an upward trend at the beginning with the increase of Fe$^{2+}$, but when the Fe$^{2+}$ increased to a certain extent, a$_0$ did not increase, indicating that Fe$^{2+}$ existed in other positions at this time, such as the channel and tetrahedron [28]. No matter what position it existed in, it would produce absorption in the red region of the UV–vis spectrum and produce blue tone. As shown in Figure 10, the three samples had the highest Fe$^{2+}$ content but not the maximum cell parameter a$_0$ value. In addition, in their UV–vis spectrum, the ratio of absorption area of the red region to that of the purple region is the largest, and the hue angle (h) is the largest. It indicated that the presence of Fe$^{2+}$ in other positions except the octahedral position was also responsible for the blue tone. Absorptions in blue–violet region caused by [Fe$_2$(OH)$_4$]$^{2+}$ and in red region caused by Fe$^{2+}$ combined and produced the blue-green color of beryl.

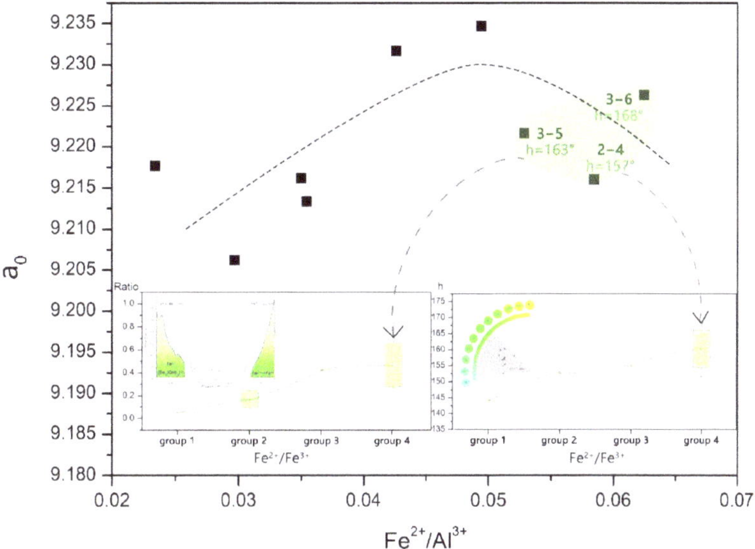

**Figure 10.** $Fe^{2+}/Al^{3+}$ content was positively correlated with the cell parameter $a_0$ at first, but no longer correlated when $Fe^{2+}$ reached a certain content. $Fe^{2+}$ content was the highest in the 2–4, 3–2, and 3–6 samples, the red area/purple area was the largest in UV–vis spectrum, and the hue angle (h) was the largest. The illustration on the left shows a positive correlation between the $Fe^{2+}/Fe^{3+}$ content and the red region absorption area/purple region absorption area in the UV–vis spectrum. On the right illustration, $Fe^{2+}/Fe^{3+}$ content was positively correlated with h.

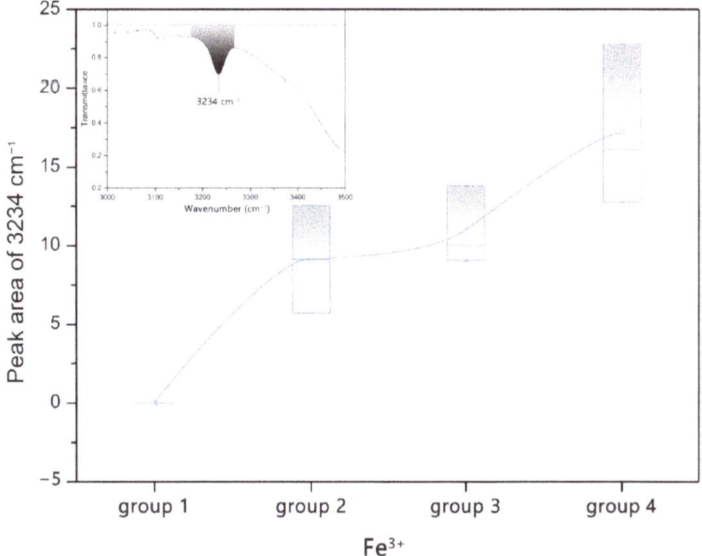

**Figure 11.** $Fe^{3+}$ was positively correlated with the area of the IR absorption peak area at 3234 cm$^{-1}$. The samples were divided into four groups by K-means clustering method, and each group of data was represented by a box.

## 5. Conclusions

Through the analysis of infrared spectrum, $5105$ cm$^{-1}$ and $5269$ cm$^{-1}$ were determined as the combined peaks of type II water, and the areas of these two peaks were used as the basis to infer the relative content of type II water. However, the absorption peaks at $7097$ cm$^{-1}$ and $7142$ cm$^{-1}$ were unlikely to be the combination and the overtone peak of type I water. It was found that the absorption peak at $7097$ cm$^{-1}$ was positively correlated with the absorption peak at $3162$ cm$^{-1}$ caused by Na–H, the absorption peak at $7142$ cm$^{-1}$ was positively correlated with the absorption peak at $3111$ cm$^{-1}$ caused by M–OH, and the absorption peak at $3234$ cm$^{-1}$ was caused by $[Fe_2(OH)_4]^{2+}$.

With the different charge substitution of $Fe^{2+} \rightarrow Al^{3+}$ in the octahedral position, the blue-green beryl transition from yellow to blue and the increase of $Na^+$ in the channel led to the increase of type II water content. In the IR spectrum, the absorption peak area at $5269$ cm$^{-1}$ was negatively correlated with b*, and positively correlated with the area at $3162$ cm$^{-1}$ and $a_0$.

$Fe^{3+}$ mainly existed in the channel as hydrated ions, but $Fe^{2+}$ did not only exist in the octahedral position. $[Fe_2(OH)_4]^{2+}$ produced a yellow tone, which, when combined with the blue tone produced by $Fe^{2+}$, made beryl blue-green.

**Supplementary Materials:** The following supporting information can be downloaded at: https://www.mdpi.com/article/10.3390/cryst12030435/s1. Table S1: The test data of colour parameters of 23 samples; Table S2: The test data of IR; Table S3: The main elements content of 23 samples tested by XRF; Table S4: The $Fe^{2+}$ content of 10 samples; Table S5: The test data of UV-vis.

**Author Contributions:** Conceptualization, H.W.; methodology, H.W. and T.S.; validation, Y.G.; formal analysis, J.C.; investigation, H.W.; resources, T.S. and J.C.; data curation, T.S. and J.C.; writing—original draft preparation and editing, H.W.; supervision, Y.G.; project administration, T.S.; funding acquisition, J.C. All authors have read and agreed to the published version of the manuscript.

**Funding:** This research received no external funding.

**Institutional Review Board Statement:** Not applicable.

**Informed Consent Statement:** Not applicable.

**Data Availability Statement:** Not applicable.

**Acknowledgments:** The experiments in this research were conducted in the laboratories of the Gemological Institute, China University of Geosciences, Beijing.

**Conflicts of Interest:** The authors declare no conflict of interest.

## References

1. Yang, S. Aquamarine Optimization Process and Mechanism. Master's Thesis, Tong Ji University, Shanghai, China, 2014.
2. Qi, L.; Ye, S.; Xiang, C. Vibration Spectrum and Irradiation Splitting of Mixture in Beryl Channels. *Geol. Sci. Technol. Inf.* **2001**, *20*, 59–64.
3. Qi, L.; Zhao, B.; Zhou, Z. Constitution Water Irradiation Cracking and F-NIR Spectra Analysis of Yellow Beryl from Xinjiang. *Acta Mineral. Sin.* **2012**, *10*, 103–105.
4. Goldman, D.S.; Rossman, G.R.; Parkin, K.M. Channel Constituents in Beryl. *Phys. Chem. Miner.* **1978**, *31*, 225–235. [CrossRef]
5. Loeffer, B.M.; Burns, R.G. Shedding Light Other Colour of Gems and Minerals. *Am. Sci.* **1976**, *64*, 636–647.
6. Hu, D. Colouration Mechanism and Positions of Impurities in Yellow Beryl from Wulateqianqi, Inner Mongolia. Master's Thesis, China University of Geosciences, Beijing, China, 2010.
7. Zhong, Q.; Liao, Z.; Zhou, Z. Gemmological Characteristic of Hydrothermal Synthetic Paraiba-Colour Beryl. *J. Gems Gemmol.* **2016**, *18*, 1–7.
8. Wang, H.; Guan, Q.; Liu, Y.; Guo, Y. Effects of Transition Metal Ions on the Colour of Blue-Green Beryl. *Minerals* **2022**, *12*, 86. [CrossRef]
9. Qiao, X.; Zhou, Z.; Nong, P. Study on the Infrared Spectral Characteristics of $H_2O$ I-type Emerald and the Controlling Factors. *Rock Miner. Anal.* **2019**, *38*, 169–178.
10. Zou, T. Colouration Mechanism and Controlling Factor of Aquamarine from China. *Miner. Depos.* **1996**, *10*, 55–61.
11. Aurisicchio, C.; Grubessi, O.; Zecchini, P. Infrared spectroscopy and crystal chemistry of the beryl group. *Can. Mineral.* **1994**, *32*, 55–68.

12. Charoy, B.; De Donato, P.; Barres, O.; Pinto-Coelho, C. Channel Occupancy in an Alkali-Poor Beryl from Serra Branca (Goias, Brazil): Spectroscopic Characterization. *Am. Mineral.* **1996**, *81*, 395–403. [CrossRef]
13. Zhang, H.; Liu, C.; Ma, Y. Advances in Mineralogy and Hydrogen Isotope of $H_2O$ in Channel of Pegmatitic Beryl. *Acta Mineral. Sin.* **1999**, *19*, 370–378.
14. Wood, D.L.; Nassau, K. The Characterization of Beryl and Emerald by Visible and Infrared Absorption Spectroscopy. *Am. Mineral.* **1968**, *53*, 777–800.
15. Fridrichová, J.; Bacík, P.; Bizovská, V.; Libowitzky, E.; Škoda, R.; Uher, P.; Ozdín, D.; Števko, M. Spectroscopic and bond-topological investigation of interstitial volatiles in beryl from Slovakia. *Phys. Chem. Miner.* **2016**, *43*, 419–437. [CrossRef]
16. Qi, L.; Xia, Y.; Yuan, X. Channel-Water Molecular Pattern and [1]H, [23]Na NMR Spectra Representation in Synthetic Red Beryl. *Gems Gemmol.* **2002**, *4*, 8–14.
17. Jiang, Y.; Guo, Y.; Zhou, Y.; Li, X.; Liu, S. The Effects of Munsell Neutral Grey Backgrounds on the Colour of Chrysoprase and the Application of AP Clustering to Chrysoprase Colour Grading. *Minerals* **2021**, *11*, 1092. [CrossRef]
18. Tang, J.; Guo, Y.; Xu, C. Colour effect of light sources on peridot based on CIE1976L*a*b*colour system and round RGB diagram system. *Colour Res. Appl.* **2019**, *44*, 932–940. [CrossRef]
19. Mashkovtsev, R.I.; Thoms, V.G.; Fursenko, D.A.; Zhukova, E.S.; Uskov, V.V.; Gorshunov, B.P. FTIR Spectroscopy of $D_2O$ and HDO molecules in the C-Axis Channels of Synthetic Beryl. *Am. Mineral.* **2016**, *101*, 175–180. [CrossRef]
20. Della Ventura, G.; Radica, F.; Bellatreccia, F.; Freda, C.; Guidi, M.C. Speciation and Diffusion Profiles of $H_2O$ in Water-Poor Beryl: Comparison with Cordierite. *Phys. Chem. Miner.* **2015**, *42*, 735–745. [CrossRef]
21. Peng, M.; Wang, H. A Study on the Vibrational Spectra of Water in Tourmaline. *Acta Mineral. Sin.* **1995**, *15*, 372–377.
22. Kolesov, B. Vibrational States of $H_2O$ in Beryl: Physical aspects. *Phys. Chem. Min.* **2008**, *34*, 727–731. [CrossRef]
23. Kolesov, B.; Geiger, C.A. The Orientation and Vibrational States of $H_2O$ in Synthetic Alkali-Free Beryl. *Phys. Chem. Min.* **2000**, *27*, 557–564. [CrossRef]
24. Le Breton, N. Infrared Investigation of $CO_2$-Bearing Cordierites: Some Implications for the Study of Metapelitic Granulites. *Contrib. Mineral. Petrol.* **1989**, *103*, 387–396. [CrossRef]
25. Li, X.; Zu, E. Near-Infrared Spectrum Analysis of Cyclosilicates Gem Minerals. *Bull. Chin. Ceram. Soc.* **2016**, *35*, 1318–1321.
26. Guo, Y.; Wang, R.; Xu, S. A Study of the Structure of a Rare Tabular Crystal of Beryl. *Geol. Rev.* **2000**, *46*, 312–317.
27. Weng, S.; Xu, Y. *Fourier Transform Infrared Spectroscopy Analysis*; Chemical Industry Press: Beijing, China, 2016; pp. 30–32.
28. Qi, L.; Xiang, C.; Liu, G.; Pei, J.; Luo, Y. ESR Behaviour of Paramagnetic Mixture in Irradiated Beryl. *Geol. Sci. Technol. Inf.* **2001**, *20*, 59–64.

Article

# Gemological and Mineralogical Studies of Greenish Blue Apatite in Madagascar

Zhi-Yi Zhang, Bo Xu *, Peng-Yu Yuan and Zi-Xuan Wang

School of Gemmology, China University of Geosciences Beijing, 29 Xueyuan Road, Haidian District, Beijing 100083, China; 1009191124@cugb.edu.cn (Z.-Y.Z.); yuanpengyu@cugb.edu.cn (P.-Y.Y.); wangzx@cugb.edu.cn (Z.-X.W.)
* Correspondence: bo.xu@cugb.edu.cn

**Abstract:** Madagascar is known as the 'Island of Gemstones' because it is full of gemstone resources. Apatite from Madagascar is widely popular because of its greenish blue Paraiba-like color. This study analyzes apatite from Madagascar through standard gemological characteristic methods, spectroscopic tests and chemical analyses (i.e., electron probe and laser ablation inductively coupled plasma mass spectrometry). This work explores the gemological and the diagenesis information recorded on Madagascar apatite by comparing them with apatite from other sources and establishes the origin information of Madagascar apatite. The origin characteristics are as follows: Apatite from Madagascar is fluorapatite, with excellent diaphaneity, greenish–blue color caused by Ce and Nd and crystal structure distortion indicated by spectroscopic tests. The F/Cl ratio (16.47 to 21.89) suggests its magmatic origin Cl loss during the weathering processes forming the source rocks, and lg $fO_2$ ($-10.30$ to $-10.35$) reflects the high oxidation degree of magma.

**Keywords:** gemological; mineralogical; greenish blue apatite; fluorapatite

**Citation:** Zhang, Z.-Y.; Xu, B.; Yuan, P.-Y.; Wang, Z.-X. Gemological and Mineralogical Studies of Greenish Blue Apatite in Madagascar. *Crystals* **2022**, *12*, 1067. https://doi.org/10.3390/cryst12081067

Academic Editors: Vladislav V. Gurzhiy and Jolanta Prywer

Received: 22 June 2022
Accepted: 23 July 2022
Published: 30 July 2022

**Publisher's Note:** MDPI stays neutral with regard to jurisdictional claims in published maps and institutional affiliations.

## 1. Introduction

Apatite is a chain phosphate mineral ($CaPO_4$); the high-quality ones are considered semi-precious gemstones [1]. Apatite has no preferential precipitation temperature; hence, it is a common accessory mineral widely distributed in Earth's crust and found in magmatic, sedimentary, metamorphic and hydrothermal systems [2–4]. Mafic magmatic rocks contain up to 20–30% of apatite. Apatite is of indicative significance for magma evolution and diagenesis because it is not affected by metamorphism and hydrothermal alteration [5–7]. It is enriched with trace elements and halogen elements, such as iron, manganese, rare earth elements (REE), fluorine and chlorine. The REE content generally ranges from 5% up to 11.4% [8]. These elements not only record and preserve information about the parent magma and indicate the magma's redox state but also assist in identifying and determining the rock type.

Myanmar, Sri Lanka, India, Madagascar and China are among the main sources of apatite. Madagascar mainly produces blue, green and sky-blue gem-quality apatite [1].

Apatite belongs to the hexagonal crystal system, specifically the symmetrical type 6/m. Its chemical formula is $X_5(ZO_4)_3Y$, where X is an ion represented by $Ca^{2+}$, which can be isomorphically replaced by $Mg^{2+}$, $Fe^{2+}$, $Sr^{2+}$, $Mn^{2+}$, $Pb^{2+}$, $Zn^{2+}$, $Ba^{2+}$, $Ag^+$ and $REE^{3+}$. Z = P, As, V, Si, S, C, etc. Y is an additional anion [9]. The most common phase in rocks is fluorapatite ($Ca_5[PO_4]_3F$) [10]. $Ca^{2+}$ occupies two kinds of positions in the apatite lattice represented by Ca1 (coordination 9 with nine oxygen ions) and Ca2 (coordination 7 with six oxygen ions and an additional anion). The number ratio of Ca1 and Ca2 is 4:6, and their ion radii are 1.18 Å and 1.06 Å, respectively [8]. Apatite structure can tolerate relatively large structural distortions and allow different substitutions [11]. Light REE preferably occupy the Ca2 site [12]. $Na^+$, $Eu^{2+}$, $Fe^{2+}$, $Mg^{2+}$, $Pb^{2+}$, $Mn^{2+}$, $Sr^{2+}$, $REE^{3+}$, $U^{4+}$ and $Th^{4+}$ occupy the X site, $Si^{4+}$, $S^{6+}$ and $C^{4+}$ tend to occupy the Z site. $Na^+$ is usually situated at the Ca1 site.

It can access fluorapatite through three substitutions (below, where V = vacancy) [12–14]: $REE^{3+} + Na^+ = 2Ca^{2+}$; $Na^+ + S^{6+} = Ca^{2+} + P^{5+}$; and $2Na^+ = Ca^{2+} + [V]$. Due to the law of charge conversation, the monovalent ion addition is indicated in the existence of alternative ionic groups, such as $REE^{3+}$ and $SO_4^{2-}$. $REE^{3+}$ and $Y^{3+}$ can occur in higher concentrations and occupy the Ca2 site. The possible alternative reactions for $REE^{3+}$ and $Y^{3+}$ are as follows [11,12,15–17]: $REE^{3+} + Mn^{2+} + Na^+ = 3Ca^{2+}$; $REE^{3+} + SiO_4^{4-} = Ca^{2+} + PO_4^{3-}$ and $REE^{3+} + O^{2-} = Ca^{2+} + F^-$. Divalent ions can be doped into apatite at any concentration because they have a common charge and a similar cation size as $Ca^{2+}$. In natural apatite, the most abundant tetravalent ions (i.e., $Th^{4+}$ and $U^{4+}$) can enter apatite through the following complex substitution [11,18–20]: $Th^{4+}(U^{4+}) + [V] = 2Ca^{2+}$. The $PO_4^{3-}$ ion cluster can be substituted by the $AsO_4^{3-}$, $SO_4^{2-}$, $CO_3^{2-}$ and $SiO_4^{4-}$ ion clusters [21–24]. The following substitutions can occur for an instance of $SO_4^{2-}$ [22]: $SO_4^{2-} + SiO_4^{4-} = 2PO_4^{3-}$ and $SO_4^{2-} + Na^+ = PO_4^{3-} + Ca^{2+}$.

As accessory minerals, apatites could provide an extremely reliable record of magma conditions and preserve in the sedimentary record in order to document the formation and evolution of magmas now lost in the geological record. Elements supporting varying states in accessory minerals can reflect the redox conditions through the valence state. Fe, Mn, Ce and Eu offer great potential in obtaining robust estimates of redox conditions. For instance, Ce in zircon has been studied for its relationship to $fO_2$, an indicator of the availability and ability of oxygen to participate in mineral and liquid reactions [25].

In this study, gem-quality apatite crystal samples from Madagascar were selected, and gemological and spectroscopy tests were performed to determine specific types. The results obtained herein are useful for subsequent studies on and developments of apatite. Given the gaps in the gemological and spectroscopic studies of Madagascar apatite, in this study, the gemological and spectroscopic analyses will be more detailed, providing a theoretical basis for the forthcoming research of blue to green−blue Paraiba-like apatite and establishing the origin information of Madagascar apatite.

## 2. Geological Setting

Located in southeastern Africa, Madagascar is the fourth largest island in the world. It is formed by the cleavage of Gondwanaland, part of the African craton, and it is mainly composed of Precambrian tectonic-magmatic-metamorphic heterogeneous rocks, Late Paleozoic–Mesozoic silty clastic carbonate rocks and Mesozoic volcanoes–sedimentary rock systems [26]. Among them, the Precambrian hybrid rocks are widely distributed, accounting for approximately 75% of the total area (Figure 1). Recent studies have shown that the Precambrian here mainly experienced tectonic–thermal events, such as Archaic-Paleoproterozoic substrate formation and cover deposition, Middle–Neoproterozoic Continent of Rhodesia's disintegration, oceanic crust subduction, late Neoproterozoic-Early Paleozoic East-West Gondwana continent convergence, collage, collision orogeny and post−orogenic lithospheric demolition [27]. The origin and composition of the plot are very complex [28].

According to previous research, apatite can be produced as an accessory mineral in granite in the Ambatondrazaka region and as a vein mineral in the same region of magma-type vanadium–titanium magnetite [28,29]. The magmatic rocks in this area are developed in the form of bands and rock plants, spreading in the north-west and -east directions, mainly in the north-west direction. The granite outcrop area is 50 km$^2$, making it possible for apatite to be exploited. Magma-type V-Ti magnetite is mainly produced in Precambrian super-base hybrid rocks [29]. Further gemological and mineralogical studies on apatite from Madagascar are presented herein (Figure 2).

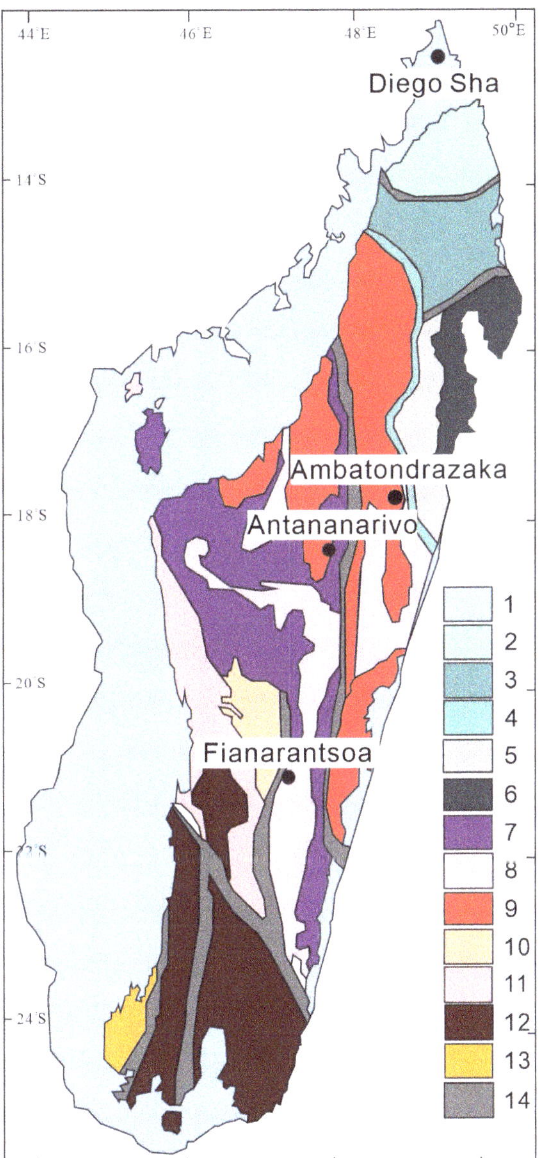



**Figure 1.** Madagascar map: outline of the structural–lithological units during the Precambrian Period. 1. Sedimentary strata. 2. Daraina Supergroup, Manambata Rock Set. 3. Sahantaha Group, North Antsirabe Rock Formation (Bemarivo units 2 and 3). 4. Betsimisaraka Snake Green Mixed Rock Belt. 5. Ambodiriana Group, Nosy Boraha Rock Formation. 6. Mananara Group, Masoala Rock Sets (Antongil units 5 and 6). 7. Granite mixed gneiss. 8. Graphite metamorphic rocks (Antananarivo units 7 and 8). 9. Tsaratanana Greenstone Belt. 10. Itermo group. 11. Amborompotsy Group (10 and 11 for Itremo units). 12. Vohibory group. 13. Androyen Anosyen blocks (12 and 13 for the Tsivory unit). 14. Cutting tapes: Ant: Antsaba; San: Sandratota; Agv: Angove; Bet: Betsileo; Ran: Ranotsara; Apm: Ampanihy; and Ber: Beraketa. From [26].

**Figure 2.** (a–d) Apatite crystals MADB-1 to MADB-4 from Madagascar. One millimeter per line.

## 3. Materials and Methods

Four apatite samples (i.e., MADB-1–MADB-4, Figure 2) from Madagascar were subjected to standard gemological tests.

The classic gemological analyses were carried out at the Gemological Research Laboratory of China University of Geosciences, Beijing, including observation, RI, ultraviolet fluorescence detection and specific gravity testing. The optical observations were carried out with a tenfold magnification gem microscope. RI was determined using the spot measurement method. The UV fluorescence detection of samples was performed using UV fluorescent lamps with long and short wavelengths. The specific gravity was tested by the hydrostatic method, and each sample was tested thrice.

The IR analyses were carried out with a Tensor 27 Fourier infrared spectrometer (Bruker, Karlsruhe, Germany) using the reflection method. The test conditions were as follows: 18–25 °C scanning temperature; <70% humidity; 85–265 V scanning voltage; 4 cm$^{-1}$ resolution; 6 mm grating; 400–2000 cm$^{-1}$ test range; and 32 times of scanning signal accumulation.

Raman spectroscopy was performed with the HR Evolution Raman microspectrometer (HORIBA, Kyoto, Japan) with the following analytical conditions: 532 nm laser wavelength; 200–2000 cm$^{-1}$ scanning range; 50 mW laser power; 4 cm$^{-1}$ resolution; 100 μm slit width; 600 gr/mm grating; 4 s scanning time; and 3 s integration time.

A UV-3600 UV-VIS spectrophotometer (Shimadzu Corporation, Kyoto, Japan) with a reflection method was used to perform the UV-VIS spectroscopy tests. The method conditions were as follows: 200–900 nm wavelength; 20 nm slit width; 1.0 s time constant; medium scanning speed; and 0.5 s sampling interval.

A homogeneous portion of the apatite samples without inclusions was selected for testing to obtain a more representative elemental content.

We selected six spots at MADB-1 and -2 and three spots for each sample. The apatite crystals were first mechanically crushed. The pure inner parts of the sample were then selected under a binocular microscope and placed in an epoxy block to be polished to maximum surface. Before the analyses, the polishing portion was carbon-blasted on the surface. The elements were measured using a JXA-8230 electron probe microanalyser (JEOL, Tokyo, Japan) at Macquarie University, Australia. The following test conditions were applied: $2 \times 10^{-8}$ A electron beam current; 15.0 kV accelerating voltage; and 5 μm beam spot. Each element was tested with an accuracy greater than 0.001% and corrected by ZAF calibration.

The trace element compositions were measured using laser ablation inductively coupled plasma mass spectrometer in the Institute of Geomechanics, Chinese Academy of Geological Sciences. Inline testing was performed using a 193 nm excimer laser ablation system (GeoLas HD; Coherent, Santa Clara, CA, USA) and a four-stage rod mass spectrometer (Agilent 7900, Agilent, Palo Alto, Santa Clara, CA, USA) with the carrier gases of Ar and He. NIST SRM 610 and 612 were used as the external standard, whilst 43Ca was employed as an internal marker for the trace element content.

## 4. Results

### 4.1. Visual Appearance and Gemological Properties of Apatite

Apatite samples analyzed in this work were greenish-blue uniform color with a glassy luster. The samples were plate-like or irregular in shape; the diaphaneity was transparent. The crystals show internal fissures, black short-columnar and orange-red inclusions. Moreover, greasy and shiny shell-like and irregular fractures were found. The crystal size ranged from 9 to 12 mm, with 6–10 mm thickness.

The RI of the samples was 1.63–1.64, and all samples were inert at the UV long and short wavelengths. The specific gravity values varied from 3.172 to 3.195.

### 4.2. Spectral Characteristics

### 4.2.1. Infrared Spectrum

The infrared spectrum is used for determining the phase of apatite and the presence of crystal distortion in order to establish origin information. Previous literature on apatites showed that infrared spectrum vibration was mainly manifested in the $[PO_4]^{3-}$ ion vibration. Free $[PO_4]^{3-}$ ions had four fundamental frequencies, namely symmetrical telescopic vibration ($v_1$), bending vibration ($v_2$), anti–telescopic vibration ($v_3$) and bending vibration ($v_4$) [11,13]. The $v_2$ band came from two strong absorption peaks at 320 and 270 cm$^{-1}$ [9]. The test wavelength was between 2000 and 400 cm$^{-1}$. The $v_2$ bend was invisible and is not discussed below.

The following characteristic peaks were found (Figure 3): a weak, single absorption band at 962 cm$^{-1}$ assigned to the symmetric stretching vibration ($v_1$); a wide, strong absorption shoulder near 1057 cm$^{-1}$; a strong absorption band at 1101 cm$^{-1}$ assigned to the anti-symmetric stretching vibration ($v_3$) of $[PO_4]^{3-}$. The moderately strong absorption bands at 575 and 608 cm$^{-1}$ were assigned to the bending vibration ($v_4$) of the $[PO_4]^{3-}$ ion; a slightly weak absorption band at 591 cm$^{-1}$ was observed in MADB−3 and MADB−4 spectra. This absorption band was also assigned to the bending vibration ($v_4$) of $[PO_4]^{3-}$.

The R050274 fluorapatite from the RRUFF database was selected for comparison with the Madagascar samples. This sample from Minas Gerais, Brazil, was blue, with two moderately strong $v_4$ bands at 567 and 600 cm$^{-1}$, a sharp $v_1$ band found at 962 cm$^{-1}$ and two $v_3$ bands near 1022 cm$^{-1}$ and 1090 cm$^{-1}$. Comparing the MADB samples with R050274, we found an extra 591 cm$^{-1}$ weak analyzed band because of the orientation. It was also assigned to $v_4$ vibration of $[PO_4]^{3-}$. The $v_3$ vibration of $[PO_4]^{3-}$ showed an obvious right shift, and the $v_3$ bands widened. The splitting of the P−O bond and the interaction of the $[PO_4]^{3-}$ tetrahedron vibration pattern with the crystal lattice of fluorapatite have been responsible for the $v_3$ band widening [30]. In addition, the infrared spectra results of the MADB samples were consistent with fluorapatite. Therefore, apatite of Malagasy origin could preliminarily be judged as fluorapatite, and there is splitting of P−O bond in fluorapatite crystal lattice indicated by infrared spectrum.

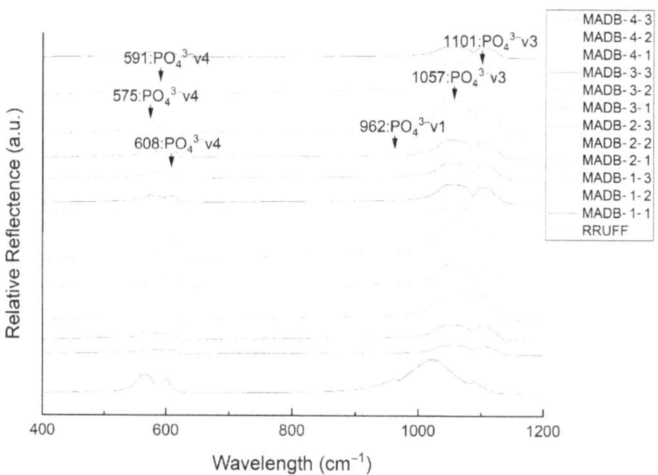



**Figure 3.** Infrared spectra of the Madagascar apatite samples.

### 4.2.2. Raman Spectra

The Raman spectra of the Madagascar apatite appeared at $400-1500$ cm$^{-1}$ [8]. The Raman feature bands of the samples were chiefly reflected in the vibration of the $[PO_4]^{3-}$ ion cluster [31]. The symmetric stretching vibration $(v_1)$ bands of the ion cluster were seen at 962–965 cm$^{-1}$. The bending vibration $(v_2)$ bands were observed at 419–431 cm$^{-1}$. Asymmetric stretching vibration $(v_3)$ bands were detected at 1040–1049 cm$^{-1}$, whilst asymmetric bending vibration$(v_4)$ bands were detected at 575–593 cm$^{-1}$.

Figure 4 shows the Raman characteristic absorption peaks of the apatite samples. The strongest band at 962 cm$^{-1}$ was assigned to the symmetrical telescopic vibration$(v_1)$ band of $[PO_4]^{3-}$. The symmetrical bending vibration band $(v_2)$ at 430 cm$^{-1}$ was a weak band. Bands were found near 1052 cm$^{-1}$ and 1086 cm$^{-1}$, which were assigned to the asymmetric stretching vibration of $[PO_4]^{3-}$. In addition, a medium-strong absorption band at 591 cm$^{-1}$ assigned to the asymmetric bending vibration $(v_4)$ was also observed. Since its Raman spectrum matches the standard Raman spectrum of fluorapatite well, the crystals from Madagascar resulted in fluorapatites, which confirms the results of infrared spectra.

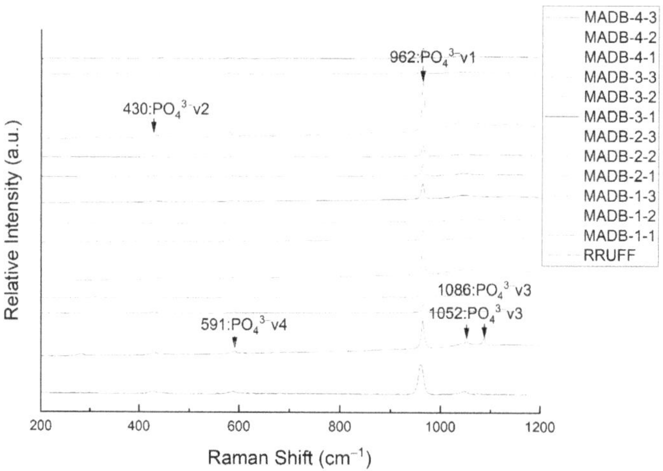



**Figure 4.** Raman spectra of the Madagascar apatite samples.

### 4.2.3. UV-VIS Spectrum

The types and contents of REE directly affect the gemological properties of apatite, such as color and luminescence. We can determine the causes of the different colors of apatite through the UV-VIS spectrum and indirectly determine REE content in Madagascar apatite. Apatite with a high REE content absorbs more strongly in the ultraviolet region. This is mainly related to the high content of light REE (LREE) ions [8]. As the concentration of REE, the apatite absorption of visible increases. Figure 5 shows the MADB apatite UV-VIS spectra.

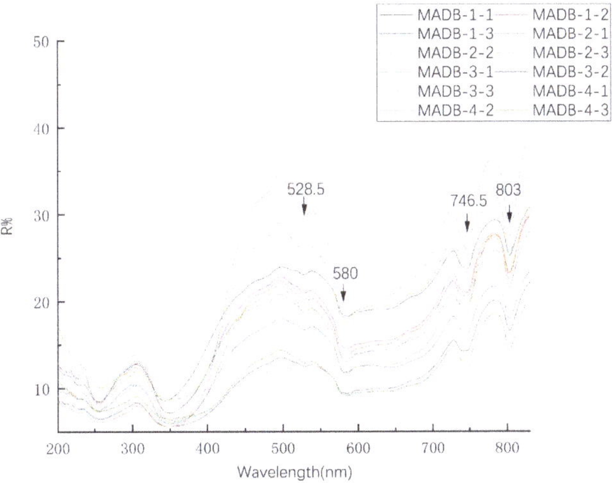


**Figure 5.** UV-VIS spectra of the Madagascar apatite samples.

The UV-VIS spectrum revealed that the samples have significant absorptions between 580 nm and 700 nm. Strong reflections were found in 400–550 nm and 700–850 nm, that is, the green, blue–violet and near-infrared regions. The reflectivity in the ultraviolet region was generally low due to the strong absorption of Ce in the ultraviolet region. Moreover, compared with other rare earth elements, Ce has a more significant effect on the weakening of absorption in the blue and orange-red regions [8]. The visible region had moderately pronounced reflectance troughs near 580 nm and 528.5 nm, whilst the infrared region showed sharp, strong hollow near 746.5 nm and 803 nm related to the $Nd^{3+}$ ion absorption [13]. In a word, the color mechanism of MADB apatite was selective absorption of Ce and Nd in the orange-red areas and the ultraviolet regions. The appearance color and the UV-VIS spectrum can corroborate each other. Additionally, the absorptions at 528.5 nm, 580 nm, 746.5 nm and 803 nm can be one of the origin characteristics of Madagascar apatite.

### 4.3. Major and Trace Elements

Tables A1 and A2 present the major and trace elements of the analyzed apatite crystals. On the basis of the analytical results, the main chemical components of Madagascar apatite were CaO (54.16–54.92 wt%) and $P_2O_5$ (39.10–40.79 wt.%). The results were consistent with the compositional range of pyrogenic apatite (CaO = 54–57 wt.% and $P_2O_5$ = 39–44 wt.%) described by Belousova [4].

According to the EPMA results, the chemical formula of MADB−1 was calculated to be $(Ca_{4.499}, Na_{0.015}, Mn_{0.002}, Sr_{0.013}, Ce_{0.012})[P_{2.605}Si_{0.047}S_{0.009}O_{12}](F_{0.804}, Cl_{0.022})$, whilst that of MADB−2 was $(Ca_{4.513}, Na_{0.006}, Sr_{0.009}, Ce_{0.012})[P_{2.607}Si_{0.052}S_{0.010}O_{12}](F_{0.780}, Cl_{0.025})$, which is that of fluorapatite.

The trace elements in the Madagascar apatite included REE, Mn, Sr, Fe and Th. The samples contained 250–270 ppm of Mn, 2900–3095 ppm of Sr, 350–370 ppm of Fe and

665–900 ppm of Th. The total REE content of samples was 9000–10225 ppm. The LREE content was 8690–9860 ppm. The analytical results showed that in the chondrite standardization model (Figure 6), the LREE were slightly enriched, and the HREE were relatively deficient. The δEu range was 0.62–0.66, showing a weak negative Eu anomaly. δCe ranged from 0.94 to 1.00 with no obvious Ce anomaly [32].

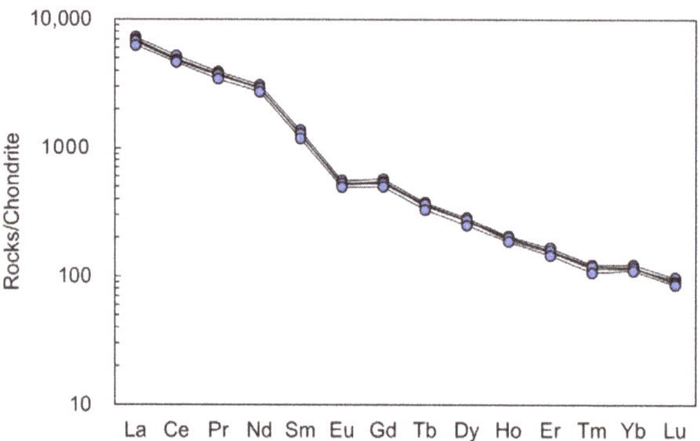


**Figure 6.** Chondrite-normalized REE distribution patterns of Madagascar apatite. The normalization values for chondrite were from Herman (1971) [33].

## 5. Discussion

### 5.1. Gemological Characteristics

In this part, apatites from Madagascar, Morocco, Mexico and China (Shanxi, Shaanxi, Anhui) were in comparison (Table A3). The color of the Madagascar apatite analyzed in this study was greenish blue, which is considered optimal in apatite gem evaluations. Apatites from other origins are yellow or green. Madagascar apatite diaphaneity is better than any other origin's in comparison. The SG of Madagascar apatite was slightly lower than that of the apatite from other origins. Apatites from Madagascar and China (Shanxi and Shaanxi) were inert at the UV, while those from Mexico, Morocco and China (Anhui) had varying degrees of fluorescence. The shape of Madagascar apatite was plate-like or anhedral instead of hexagonal columnar from Shanxi, China.

### 5.2. Major and Trace Elements

### 5.2.1. Major Elements Characteristics of Apatite

The previous studies [31,34–36] tested the following apatite samples: Durango in an iron mine from Mexico; Moro in igneous phosphate block rock from Morocco; SDG in an alkaline mafic complex, XZ in mafic sill and M1 in phosphorite-type rare earth deposit from China; ASP-I in Quaternary ignimbrites from Japan. The relationship between $SiO_2$, FeO and MnO contents in apatite can reflect the apatite origin [31,37,38]; therefore, the MADB apatite were plotted against those from the other sources (Figure 7). The plots confirm the magmatic origin of the blue–green apatite analyzed in this work.

The F content of apatite in Madagascar ranged from 3.10 wt.% to 3.90 wt.%, and the Cl content ranged from 0.17 wt.% to 0.19 wt.% (Appendix A Table A1), showing that the MADB apatite is fluorapatite. The Madagascar apatite had a high F/Cl ratio (16.47 to 21.89) compared with apatite from Mexico, Morocco and Japan (Figure 8, Table A4). The Cl content in the apatite was directly related to the Cl content in the parent magma, suggesting that the Madagascar magma had a lower Cl content than magma from the other regions. F was difficult to remove due to its low solubility in water. Therefore, the rock mass remelted in the crust often exhibited F-rich and Cl-poor characteristics reflected in the F and Cl compositions of apatite [39]. The MADB apatite was verified to be of magmatic origin [40,41]. Sha et al. (2018) verified that the high F/Cl ratio is also related to the Cl loss during the weathering processes forming the source rocks [13].

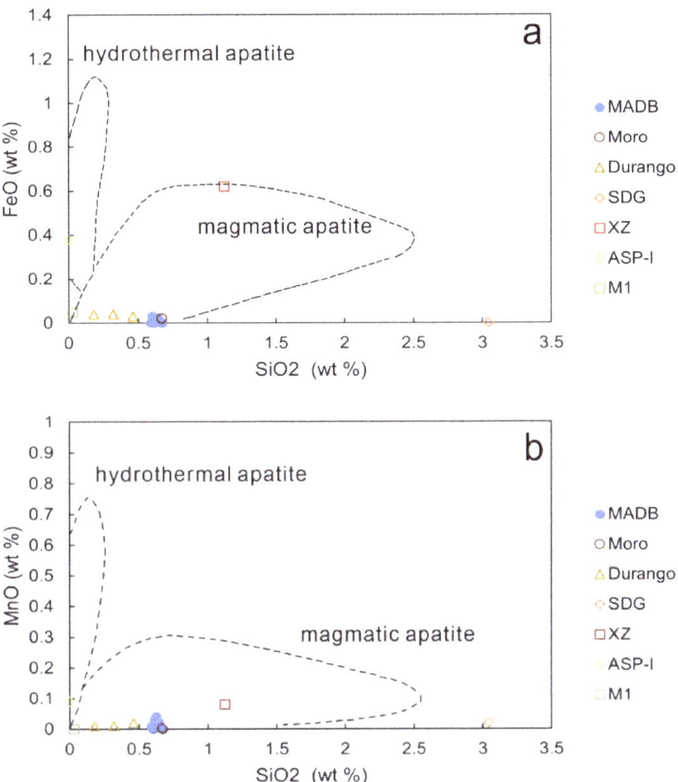


**Figure 7.** Illustration of the origin of the Madagascar apatite and those from other sources [31,34]: (**a**) FeO–SiO$_2$ and (**b**) MnO–SiO$_2$.

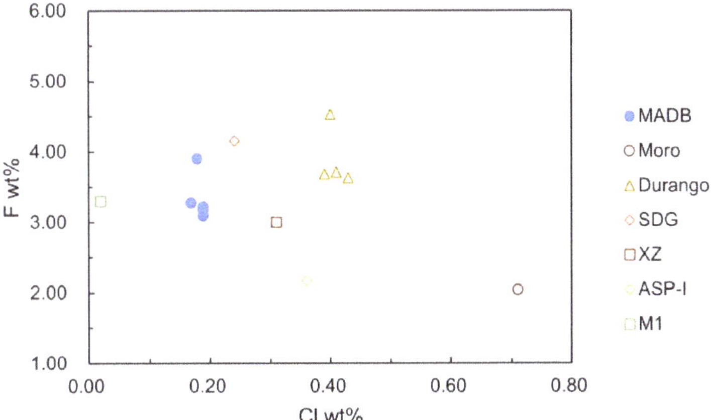


**Figure 8.** F versus Cl wt% in apatite from different origins.

## 5.2.2. Rare Earth Element Characteristics of Apatite

The trace element data for apatite showed that apatite can accommodate a wide range of structural distortion due to chemical substitutions [25] (Tables A2 and A5).

The chondrite-normalized REE distribution patterns of apatite usually show negative slopes (high $(Ce/Yb)_N$), indicating a relative enrichment in the LREE (Figure 9) [4]. The Madagascar apatite had $(Ce/Yb)_N$ equaling 41.07–43.36, showing its LREE enrichment (Figure 10). The Moro and Durango samples had $(Ce/Yb)_N$ equaling 24.51 and 18.52–27.21, respectively. M1 apatite only had that equaling 2.68. In general, the $\delta Ce$ of igneous apatite was almost equal to 1, indicating the absence of an obvious Ce anomaly. However, microcrystalline apatite M1 from sedimentary carbonate rocks showed a negative Ce anomaly, which could be one of the differences between the igneous and sedimentary apatite. In contrast, a negative Eu anomaly was found in the magmatic apatite samples (i.e., Durango, Moro and MADB), whilst the M1 apatite had a weak negative Eu anomaly, and the SDG apatite had no obvious Eu anomaly. Conversely, the $\Sigma REE$ in magmatic apatite were much more abundant compared to the sedimentary apatite. As shown in the slope in Figure 9, the distribution of the REE content of the sedimentary apatite is more uniform.

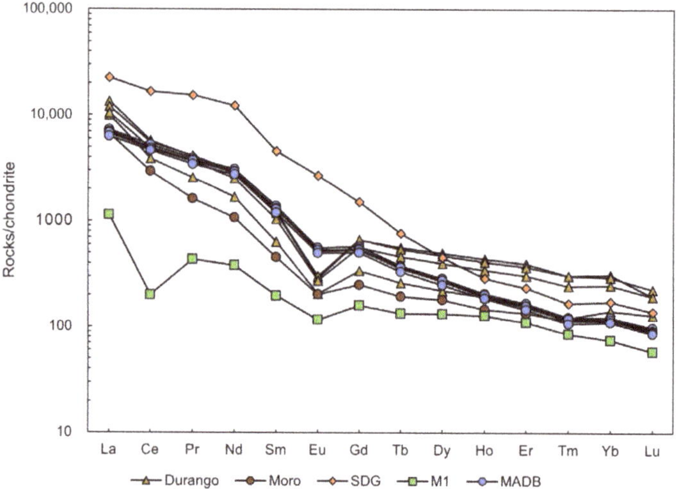


**Figure 9.** Chondrite-normalized REE distribution patterns of the apatite from Madagascar analyzed in this work compared to apatite from literature. Data from [31,34,42]. Normalization values for chondrite are from Herman (1971) [33].

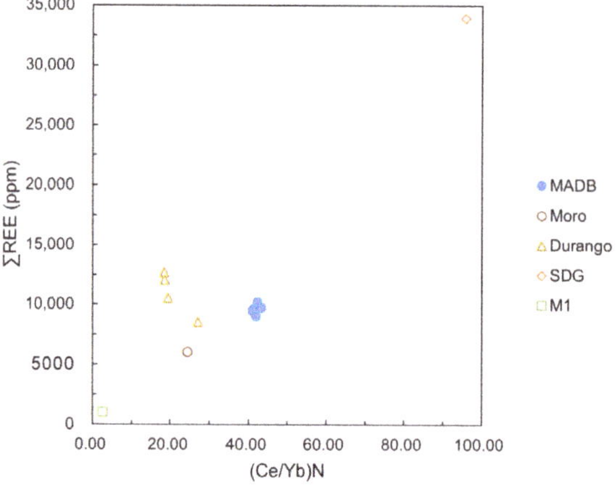


**Figure 10.** $\Sigma REE$ versus $(Ce/Yb)_N$ of apatite from different origins.

### 5.2.3. Redox Characteristics of Magma

The Mn content in the apatite was negatively correlated with the magma oxygen fugacity. Miles et al. [25] proposed an empirical formula for calculating the magma oxygen fugacity with the Mn content in apatite: $\lg fO_2 = -0.0022(\pm 0.0003)\ Mn\ (ppm) - 9.75(\pm 0.46)$. Introducing the experimental data from the MADB apatite into the formula, we obtained the results of $-10.30$ to $-10.35$. Overall, the magma in the Madagascar area had a high oxygen fugacity. Sha et al. [13] also pointed out that apatite with <900 ppm Mn and <2100 ppm Fe had a high oxygen fugacity, which was consistent with the above-mentioned description.

The study of Prowatke et al. [43] indicated that the Mg and Fe contents of apatite were positively related to those of the source magma. The Mg and Fe contents of Madagascar apatite were extremely low, with 21.65–23.44 ppm of Mg and 349.36–372.50 ppm of Fe. It can be rolled out that the magma had very few Mg and Fe elements, which is an iconic feature of felsic magmatic rocks. The XZ and ASP−I apatite were rich in Mg and Fe, corresponding to the characteristics of the parent rock (i.e., XZ was from mafic rocks) [44].

## 6. Conclusions

After the analyses above, the origin information of Madagascar apatite has been established.

Madagascar apatite crystals have a rare greenish–blue color and excellent diaphaneity, with a low degree of self-formation and slightly lower SG (3.172–3.195). The color mechanism of Madagascar apatite is selective absorption of Ce and Nd in the orange−red areas and the ultraviolet regions indicated by UV–VIS spectrum. In the crystal lattice, there exists splitting of the P–O bond indicated by the infrared spectrum.

According to the spectroscopic and major element analyses, Madagascar apatite is determined as igneous fluorapatite, with an extremely high F/Cl ratio (16.47–21.89). Due to the Cl loss during the weathering processes forming the source rocks, the F/Cl ratio of Madagascar apatite was the highest among the igneous apatite from different origins in comparison, and it also indicates the F-rich and Cl-poor characteristics in the parent rock. In addition, the FeO, MgO and $SiO_2$ content of Madagascar apatite can also reflect its magmatic origin.

Madagascar apatite has a relatively high LREE enrichment compared with other origins apatite, except SDG. The high $(Ce/Yb)_N$ (41.07–43.36) makes Madagascar apatite different from others. Additionally, the Mn, Mg and Fe concentrations of Madagascar apatite indicated the high oxygen fugacity and the felsic magmatic parent rocks (such as granite in the Ambatondrazaka region).

In summary, the igneous greenish blue apatite from Madagascar has a good scientific value and can be further studied as a standard sample.

**Author Contributions:** Writing—original draft, Z.-Y.Z.; writing—review and editing, Z.-Y.Z., B.X., P.-Y.Y. and Z.-X.W.; investigation, B.X.; data curation, Z.-Y.Z.; software, Z.-Y.Z.; methodology, B.X.; resources, B.X. All authors have read and agreed to the published version of the manuscript.

**Funding:** This research was funded by the National Key Technologies R&D Program 2019YFC0605201, 2019YFA0708602, 2020YFA0714802, National Natural Science Foundation of China (42073038, 41803045), Young Talent Support Project of CAST (IGCP-662), Fundamental Research Funds for the Central Universities (265QZ2021012) and Innovation and Entrepreneurship Training Program for College Students of China University of Geosciences (Beijing) (202211415060).

**Data Availability Statement:** The data presented in this study are available within the article.

**Acknowledgments:** This is the 9th contribution of BX to the National Mineral Rock and Fossil Specimens Resource Center. Thanks to the two reviewers and the editors for their comprehensive and professional suggestions.

**Conflicts of Interest:** The authors declare no conflict of interest.

# Appendix A

**Table A1.** Chemical composition and structural formula of Madagascar apatite, determined by EMPA (in wt%).

| | F | Na$_2$O | CaO | P$_2$O$_5$ | SO$_3$ | SiO$_2$ | FeO | MnO | Cl | SrO | Ce$_2$O$_3$ | Total |
|---|---|---|---|---|---|---|---|---|---|---|---|---|
| MADB−1−1 | 3.28 | 0.10 | 54.16 | 39.68 | 0.45 | 0.61 | 0.00 | 0.03 | 0.17 | 0.21 | 0.42 | 99.10 |
| MADB−1−2 | 3.90 | 0.11 | 54.68 | 39.10 | 0.41 | 0.62 | 0.01 | 0.04 | 0.18 | 0.13 | 0.42 | 99.59 |
| MADB−1−3 | 3.10 | 0.08 | 54.92 | 39.56 | 0.50 | 0.64 | 0.02 | 0.02 | 0.19 | 0.25 | 0.30 | 99.57 |
| MADB−2−1 | 3.18 | 0.04 | 54.31 | 39.71 | 0.50 | 0.67 | 0.00 | 0.00 | 0.19 | 0.15 | 0.42 | 99.14 |
| MADB−2−2 | 3.22 | 0.06 | 54.54 | 40.10 | 0.50 | 0.60 | 0.03 | 0.00 | 0.19 | 0.24 | 0.38 | 99.86 |
| MADB−2−3 | 3.22 | 0.12 | 54.26 | 40.79 | 0.50 | 0.59 | 0.00 | 0.01 | 0.19 | 0.21 | 0.34 | 100.22 |
| | Si$^{4+}$ | Fe$^{2+}$ | Mn$^{2+}$ | Ca$^{2+}$ | Na$^+$ | F$^-$ | Cl$^-$ | S$^{6+}$ | Sr$^{2+}$ | P$^{5+}$ | Ce$^{2+}$ | |
| MADB−1−1 | 0.047 | 0.000 | 0.002 | 4.503 | 0.015 | 0.805 | 0.022 | 0.009 | 0.004 | 2.607 | 0.012 | |
| MADB−1−2 | 0.048 | 0.001 | 0.003 | 4.508 | 0.016 | 0.949 | 0.023 | 0.008 | 0.003 | 2.547 | 0.012 | |
| MADB−1−3 | 0.050 | 0.001 | 0.001 | 4.564 | 0.012 | 0.760 | 0.025 | 0.010 | 0.005 | 2.598 | 0.009 | |
| MADB−2−1 | 0.047 | 0.000 | 0.002 | 4.503 | 0.015 | 0.805 | 0.022 | 0.009 | 0.004 | 2.607 | 0.012 | |
| MADB−2−2 | 0.045 | 0.000 | 0.000 | 4.515 | 0.006 | 0.780 | 0.025 | 0.010 | 0.003 | 2.609 | 0.012 | |
| MADB−2−3 | 0.047 | 0.000 | 0.001 | 4.446 | 0.018 | 0.779 | 0.025 | 0.010 | 0.004 | 2.641 | 0.010 | |

**Table A2.** Chemical composition and structural formula of Madagascar apatite, analyzed by LA−ICP−MS.

| Element | | MADB−1−1 | MADB−1−2 | MADB−1−3 | MADB−1−4 | MADB−1−5 | MADB−2−1 |
|---|---|---|---|---|---|---|---|
| Mg | ppm | 22.50 | 22.26 | 22.31 | 23.44 | 22.00 | 21.65 |
| Si | ppm | 1.13 | 0.00 | 1.13 | 0.98 | 0.00 | 0.00 |
| P | % | 28.762672 | 29.974847 | 28.265906 | 28.425884 | 27.471936 | 27.462787 |
| Cl | ppm | 1530.50 | 1830.53 | 1450.26 | 1229.59 | 1162.63 | 962.15 |
| Ca | % | 57.194298 | 57.452847 | 55.776887 | 54.186530 | 54.832854 | 52.793121 |
| Ti | ppm | 0.02 | 0.04 | 0.10 | 0.00 | 0.33 | 0.44 |
| Mn | ppm | 270.14 | 271.57 | 265.52 | 266.64 | 252.03 | 254.65 |

**Table A2.** *Cont.*

| Element | | MADB−1−1 | MADB−1−2 | MADB−1−3 | MADB−1−4 | MADB−1−5 | MADB−2−1 |
|---|---|---|---|---|---|---|---|
| Fe | ppm | 369.65 | 372.50 | 350.25 | 353.04 | 349.36 | 353.09 |
| Rb | ppm | 0.00 | 0.00 | 0.09 | 0.00 | 0.00 | 0.00 |
| Sr | ppm | 3093.09 | 3132.23 | 3044.83 | 2944.21 | 2902.41 | 2941.15 |
| Zr | ppm | 0.00 | 0.00 | 0.00 | 0.00 | 0.00 | 0.00 |
| Nb | ppm | 0.74 | 0.92 | 1.12 | 1.29 | 0.69 | 1.19 |
| Ba | ppm | 0.29 | 0.81 | 0.80 | 0.52 | 0.17 | 0.52 |
| La | ppm | 2231.74 | 2319.83 | 2216.59 | 2145.67 | 2161.80 | 2008.82 |
| Ce | ppm | 4651.92 | 4903.57 | 4529.55 | 4498.58 | 4509.20 | 4353.56 |
| Pr | ppm | 447.20 | 467.68 | 452.40 | 437.65 | 445.81 | 411.92 |
| Nd | ppm | 1769.42 | 1852.47 | 1755.73 | 1735.78 | 1745.71 | 1641.37 |
| Sm | ppm | 254.67 | 275.52 | 258.72 | 260.48 | 257.39 | 237.86 |
| Eu | ppm | 39.20 | 40.59 | 38.63 | 37.56 | 38.42 | 36.17 |
| Gd | ppm | 163.59 | 177.31 | 168.50 | 165.87 | 164.65 | 154.73 |
| Tb | ppm | 17.93 | 18.82 | 18.23 | 17.75 | 18.29 | 16.53 |
| Dy | ppm | 83.52 | 88.63 | 84.73 | 84.05 | 85.11 | 77.26 |
| Ho | ppm | 14.29 | 14.98 | 14.68 | 13.98 | 14.63 | 13.55 |
| Er | ppm | 33.09 | 35.00 | 33.17 | 32.34 | 32.93 | 30.49 |
| Tm | ppm | 3.79 | 4.06 | 3.96 | 3.81 | 3.94 | 3.52 |
| Yb | ppm | 21.68 | 23.40 | 22.21 | 22.14 | 22.19 | 20.98 |
| Lu | ppm | 2.92 | 3.04 | 2.82 | 2.78 | 2.75 | 2.65 |
| Hf | ppm | 0.04 | 0.02 | 0.03 | 0.03 | 0.03 | 0.03 |
| Ta | ppm | 0.01 | 0.02 | 0.01 | 0.02 | 0.02 | 0.02 |
| Hg | ppm | 0.00 | 0.00 | 0.00 | 0.00 | 0.00 | 0.00 |
| Pb | ppm | 14.53 | 18.24 | 17.54 | 17.56 | 17.64 | 16.70 |

**Table A2.** *Cont.*

| Element | | MADB−1−1 | MADB−1−2 | MADB−1−3 | MADB−1−4 | MADB−1−5 | MADB−2−1 |
|---|---|---|---|---|---|---|---|
| Th | ppm | 665.42 | 835.65 | 825.05 | 819.24 | 854.73 | 800.74 |
| U | ppm | 22.98 | 25.80 | 23.07 | 22.63 | 22.93 | 21.45 |
| ΣLREE | ppm | 9394.14 | 9859.66 | 9251.60 | 9115.72 | 9158.34 | 8689.70 |
| ΣHREE | ppm | 340.80 | 365.24 | 348.29 | 342.71 | 344.48 | 319.71 |
| ΣREE | ppm | 9734.94 | 10224.90 | 9599.89 | 9458.43 | 9502.81 | 9009.41 |
| δEu | | 0.66 | 0.63 | 0.63 | 0.62 | 0.64 | 0.64 |
| δCe | | 0.97 | 0.98 | 0.94 | 0.97 | 0.96 | 1.00 |

**Table A3.** Characteristics of apatite from different production areas. Data derived from [31,38].

| Origin | Color | Lustre | Diaphaneity | RI | DR | SG | UV | Size | Feature |
|---|---|---|---|---|---|---|---|---|---|
| Madagascar | Greenish blue | Glassy luster | Transparent | 1.63−1.64 | | 3.17−3.20 | Inert | The largest of which measured 12*10*9 mm | Plate-like or anhedral, with internal fissures, flat surface black short−columnar and orange−red inclusions |
| Durango, Mexico | Uniform yellow−green | Glassy luster | Transparent | 1.631−1.636 | 0.005 | 3.21 | Inert to long−wave; weak yellow to short−wave | The largest of which measured 14.97*9.47*7.97 mm | Emerald- and cushion-cut. The cushion-cut stone revealed straight growth zoning. The emerald-cut contained a small liquid feather |
| Anemzi, Morocco | Yellow−green | Glassy to weak glassy luster | Translucent | 1.635−1.640 | | 3.18−3.20 | Inert to long−wave; weak purple to short−wave | The largest of which measured 14*8*7.97 mm | Hexagonal columnar crystal shape, with many fissures, shell-like or irregular fractures, visible crystal face longitudinal pattern |
| Shanxi, China | Yellow, dark green to dark blue−green | Glassy luster | Translucent | 1.632−1.639 | 0.004 | 3.19−3.23 | Inert | The largest of which measured 30*10*5 mm | Hexagonal columnar or subhexagonal columnar crystal form; high degree of self-formation showing massive or columnar |

**Table A3.** *Cont.*

| Origin | Color | Lustre | Diaphaneity | RI | DR | SG | UV | Size | Feature |
|---|---|---|---|---|---|---|---|---|---|
| Shaanxi, China | Gray−green | Glassy luster | Sub−transparent | 1.533−1.637 | 0.004 | 3.21−3.23 | Inert | The largest of which measured 6.5*4*4 mm | Massive, with hexagonal columnar crystal form, high degree of self−formation, smooth surface, and poor cleavage |
| Anhui, China | Light yellow | Glassy luster to weak glassy luster | Transparent to translucent | 1.533−1.637 | 0.004 | 3.18−3.20 | Medium yellow−green | The largest of which measured 15*4*5 mm | Sheet-like and plate-like, poorly self-formation, no obvious crystalline shape, longitudinal lines between crystal faces |

**Table A4.** Chemical composition and structural formula of apatite from other origins, determined by EMPA (in wt%). Data derived from [31,34−36,42].

| | DurangoChew | DurangoFishier | DurangoGriffin | DurangoHou | Moro | SDG | XZ | ASP−I | M1 |
|---|---|---|---|---|---|---|---|---|---|
| F | 3.71 | 3.63 | 4.53 | 3.68 | 2.04 | 4.15 | 3.00 | 2.17 | 3.30 |
| CaO | 53.90 | 53.99 | 53.85 | 53.85 | 54.10 | 51.58 | 54.58 | 53.89 | 55.57 |
| $P_2O_5$ | 41.88 | 42.16 | 41.91 | 41.91 | 39.63 | 34.48 | 41.57 | 43.56 | 38.17 |
| $SiO_2$ | 0.46 | 0.18 | 0.32 | 0.32 | 0.67 | 3.04 | 1.12 | − | 0.03 |
| FeO | 0.03 | 0.04 | 0.04 | 0.04 | 0.02 | − | 0.62 | 0.38 | 0.05 |
| MnO | 0.02 | 0.01 | 0.01 | 0.01 | 0.00 | 0.02 | 0.08 | 0.09 | 0.00 |
| Cl | 0.41 | 0.43 | 0.40 | 0.39 | 0.71 | 0.24 | 0.31 | 0.36 | 0.02 |
| SrO | 0.05 | 0.04 | 0.05 | 0.05 | 0.03 | 1.44 | − | − | |
| MgO | 0.04 | 0.02 | 0.02 | 0.02 | − | 0.01 | 0.20 | 0.20 | 0.06 |
| $Ce_2O_3$ | − | − | − | − | − | − | − | 0.28 | |
| $Y_2O_3$ | − | − | − | − | − | − | − | 0.32 | |
| $Al_2O_3$ | − | − | − | − | − | − | 0.17 | − | 0.01 |
| ions | | | | | | | | | |
| $Ca^{2+}$ | 4.667 | 4.651 | 4.665 | 4.648 | 4.736 | 4.946 | 4.389 | 4.371 | 4.731 |

**Table A4.** *Cont.*

| | DurangoChew | DurangoFishier | DurangoGriffin | DurangoHou | Moro | SDG | XZ | ASP–I | M1 |
|---|---|---|---|---|---|---|---|---|---|
| $Fe^{2+}$ | 0.002 | 0.003 | 0.003 | 0.002 | 0.001 | 0.001 | 0.039 | 0.024 | 0.024 |
| $Mn^{2+}$ | 0.001 | 0.001 | 0.001 | 0.001 | 0.000 | 0.001 | 0.005 | 0.006 | 0.006 |
| $P^{5+}$ | 2.865 | 2.870 | 2.869 | 2.877 | 2.741 | 2.812 | 2.641 | 2.792 | 2.792 |
| $Mg^{2+}$ | 0.005 | 0.002 | 0.002 | 0.004 | – | 0.000 | 0.022 | 0.023 | 0.007 |
| $F^-$ | 0.105 | 0.109 | 0.102 | 0.099 | 0.184 | 0.014 | 0.712 | 0.520 | 0.520 |
| $Cl^-$ | 0.056 | 0.059 | 0.055 | 0.053 | 0.099 | 0.007 | 0.039 | 0.046 | 0.046 |

**Table A5.** Chemical composition and structural formula of Moroccan apatite, analyzed by LA–ICP–MS. Data derived from [31,34,42].

| | | DurangoChew | DurangoFishier | DurangoGriffin | DurangoHou | Moro | SDG | M1 |
|---|---|---|---|---|---|---|---|---|
| Rb | ppm | 0.12 | 0.11 | 0.13 | 0.12 | 0.12 | 0.2 | – |
| Sr | ppm | 482 | 456 | 491 | 476 | 618.13 | 11368 | – |
| Ba | ppm | 1.7 | 1.4 | 1.8 | 1.5 | 1.29 | 1.3 | – |
| Nb | ppm | 1 | 0.02 | 0.03 | 0.02 | 0.01 | 2.4 | – |
| Ta | ppm | 0 | 0 | 0 | 0 | 0 | 0.03 | – |
| Zr | ppm | 1.4 | 0.6 | 1.1 | 0.8 | 0 | 48 | – |
| Hf | ppm | 0.23 | 0.19 | 0.23 | 0.26 | 0.01 | 0.39 | – |
| Pb | ppm | 0.9 | 0.4 | 0.7 | 0.6 | 1.62 | 50 | – |
| Th | ppm | 320 | 151 | 270 | 231 | 184.37 | 705 | – |
| U | ppm | 20 | 7 | 11 | 11 | 16.48 | 47 | – |
| La | ppm | 4285 | 3194 | 3819 | 3334 | 2117.01 | 7209 | 364 |
| Ce | ppm | 5405 | 3635 | 5178 | 4561 | 2763.91 | 15668 | 188 |
| Pr | ppm | 488 | 307 | 496 | 436 | 193.71 | 1843 | 52 |
| Nd | ppm | 1677 | 1009 | 1745 | 1514 | 639.77 | 7344 | 227 |
| Sm | ppm | 237 | 127 | 244 | 207 | 90.72 | 911 | 39.4 |
| Eu | ppm | 21 | 15 | 22 | 20 | 14.78 | 196 | 8.5 |

**Table A5.** *Cont.*

| | | DurangoChew | DurangoFishier | DurangoGriffin | DurangoHou | Moro | SDG | M1 |
|---|---|---|---|---|---|---|---|---|
| Gd | ppm | 204 | 105 | 206 | 174 | 77.45 | 468 | 49.4 |
| Tb | ppm | 28 | 13 | 27 | 23 | 9.68 | 38 | 6.7 |
| Dy | ppm | 154 | 68 | 146 | 123 | 55.94 | 140 | 41 |
| Ho | ppm | 32 | 14 | 30 | 25 | 10.67 | 21 | 9.3 |
| Er | ppm | 83 | 34 | 77 | 64 | 28.34 | 49 | 23.2 |
| Tm | ppm | 10 | 4 | 10 | 8 | 3.78 | 5.5 | 2.8 |
| Yb | ppm | 59 | 27 | 56 | 47 | 22.79 | 33 | 14.2 |
| Lu | ppm | 6 | 4 | 7 | 6 | 2.99 | 4.3 | 1.8 |
| Y | ppm | 911 | 427 | 886 | 762 | — | 605 | 472 |
| ΣLREE | ppm | 12113 | 8269 | 11504 | 10072 | 5819.89 | 33171 | 878.9 |
| ΣHREE | ppm | 576 | 269 | 559 | 470 | 211.65 | 758.8 | 148.4 |
| ΣREE | ppm | 12689 | 8538 | 12063 | 10542 | 6031.54 | 33929.8 | 1027.3 |
| δEu | | 0.29 | 0.4 | 0.3 | 0.32 | 0.54 | 0.92 | 0.91 |
| δCe | | 0.92 | 0.9 | 0.92 | 0.93 | 1.06 | 1.05 | 0.31 |

## References


1. Zhang, B.-L. *Systematic Gemology*; Geological Publishing House: Beijing, China, 2006; pp. 324–327.
2. Bruand, E.; Fowler, M.; Storey, C.; Darling, J. Apatite trace element and isotope applications to petrogenesis and provenance. *Am. Mineral.* **2017**, *102*, 75–84. [CrossRef]
3. Boyce, J.W.; Hervig, R.L. Apatite as a Monitor of Late-Stage Magmatic Processes at Volcán Irazú, Costa Rica. *Contrib. Mineral. Petrol.* **2009**, *57*, 135–145. [CrossRef]
4. Belousova, E.A.; Griffin, W.L.; O'Reilly, S.Y.; Fisher, N.I. Apatite as an indicator mineral for mineral exploration: Trace-element compositions and their relationship to host rock type. *Geochem. Explor.* **2002**, *76*, 45–69. [CrossRef]
5. Xu, B.; Hou, Z.-Q.; Griffin, W.L.; O'Reilly, S.Y. Apatite halogens and Sr–O and zircon Hf–O isotopes: Recycled volatiles in Jurassic porphyry ore systems in southern Tibet. *Chem. Geol.* **2022**, *605*, 10. [CrossRef]
6. Zhang, F.; Li, W.; White, N.; Zhang, L.; Qiao, X.; Yao, Z. Geochemical and isotopic study of metasomatic apatite: Implications for gold mineralization in Xindigou, northern China. *Ore Geol. Rev.* **2020**, *127*, 103853. [CrossRef]
7. Ren, Z.; Cui, J.; Liu, C.; Li, T.; Chen, G.; Dou, S.; Tian, T.; Luo, Y. Apatite Fission Track Evidence of Uplift Cooling in the Qiangtang Basin and Constraints on the Tibetan Plateau Uplift. *Acta Geol. Sin.-Engl. Ed.* **2015**, *89*, 467–484.
8. Li, W. Study on Gemological and Chromatic Characteristics of Blue-Green Apatite. Master's Thesis, China University of Geosciences, Beijing, China, 2021.
9. Yang, Y.-F. Gem Mineralogy of Different Colors of Apatite. Master's Thesis, China University of Geosciences, Beijing, China, 2019.
10. Rossi, M.; Ghiara, M.R.; Chita, G.; Capitelli, F. Crystal-Chemical and Structural Characterization of Fluorapatites in Ejecta from Somma−Vesuvius Volcanic Complex. *Am. Mineral.* **2011**, *96*, 1828–1837. [CrossRef]
11. Pan, Y.; Fleet, M.E. Compositions of the Apatite-Group Minerals: Substitution Mechanisms and Controlling Factors. *Rev. Mineral. Geochem.* **2002**, *48*, 13–49. [CrossRef]
12. Zhu, X.Q.; Wang, Z.G.; Huang, Y. Rare earth composition of apatite and its tracing significance. *Rare Earths* **2004**, *25*, 41–45+63. (In Chinese) [CrossRef]
13. Sha, L.-K.; Chappell, B.W. Apatite Chemical Composition, Determined by Electron Microprobe and Laser-Ablation Inductively Coupled Plasma Mass Spectrometry, as a Probe into Granite Petrogenesis. *Geochim. Cosmochim. Acta* **1999**, *63*, 3861–3881. [CrossRef]
14. Xu, B.; Hou, Z.-Q.; Griffin, W.L.; Lu, Y.; Belousova, E.; Xu, J.-F.; O'Reilly, S.Y. Recycled Volatiles Determine Fertility of Porphyry Deposits in Collisional Settings. *Am. Mineral.* **2021**, *106*, 656–661. [CrossRef]
15. Chen, N.; Pan, Y.; Weil, J.A. Electron Paramagnetic Resonance Spectroscopic Study of Synthetic Fluorapatite: Part I. Local Structural Environment and Substitution Mechanism of $Gd^{3+}$ at the Ca2 Site. *Am. Mineral.* **2002**, *87*, 37–46. [CrossRef]
16. Fleet, M.E.; Pan, Y. Site Preference of Rare Earth Elements in Fluorapatite. *Am. Mineral.* **1995**, *80*, 329–335. [CrossRef]
17. Mao, M.; Rukhlov, A.S.; Rowins, S.M.; Spence, J.; Coogan, L.A. Apatite Trace Element Compositions: A Robust New Tool for Mineral Exploration. *Econ. Geol.* **2016**, *111*, 1187–1222. [CrossRef]
18. Rakovan, J.F.; Hughes, J.M. Strontium in the apatite structure: Strontium fluorapatite and belovite-(Ce). *Can. Mineral.* **2000**, *38*, 839–845. [CrossRef]
19. Piccoli, P.M.; Candela, P.A. Apatite in Igneous Systems. *Rev. Mineral. Geochem.* **2002**, *48*, 255–292. [CrossRef]
20. Hughes, J.M.; Ertl, A.; Bernhardt, H.-J.; Rossman, G.R.; Rakovan, J. Mn−Rich Fluorapatite from Austria: Crystal Structure, Chemical Analysis, and Spectroscopic Investigations. *Am. Mineral.* **2004**, *89*, 629–632. [CrossRef]
21. Sudarsanan, K.; Young, R.A.; Wilson, A.J.C. The Structures of Some Cadmium 'apatites' $Cd_5(MO_4)_3X$. I. Determination of the Structures of $Cd_5(VO_4)_3I$, $Cd_5(PO_4)_3Br$, $Cd_3(AsO_4)_3Br$ and $Cd_5(VO_4)_3Br$. *Acta Crystallogr. Sect. B Struct. Crystallogr. Cryst. Chem.* **1977**, *33*, 3136–3142. [CrossRef]
22. Peng, G.; Luhr, J.F.; McGee, J.J. Factors Controlling Sulfur Concentrations in Volcanic Apatite. *Am. Mineral.* **1997**, *82*, 1210–1224. [CrossRef]
23. Perseil, E.-A.; Blanc, P.; Ohnenstetter, D. As-Bearing Fluorapatite in Manganiferous Deposits from St. Marcel-praborna, val d'aosta, Italy. *Can. Mineral.* **2000**, *38*, 101–117. [CrossRef]
24. Xu, B.; Hou, Z.-Q.; Griffin, W.L.; Zheng, Y.-C.; Wang, T.; Guo, Z.; Hou, J.; Santosh, M.; O'Reilly, S.Y. Cenozoic Lithospheric Architecture and Metallogenesis in Southeastern Tibet. *Earth-Sci. Rev.* **2021**, *214*, 103472. [CrossRef]
25. Miles, A.J.; Graham, C.M.; Hawkesworth, C.J.; Gillespie, M.R.; Hinton, R.W.; Bromiley, G.D. Apatite: A New Redox Proxy for Silicic Magmas? *Geochim. Cosmochim. Acta* **2014**, *132*, 101–119. [CrossRef]
26. Che, J.Y.; Zhao, Y.D. A review of the basal characteristics of Precambrian metamorphic substrates in Madagascar. *Geol. Resour.* **2013**, *4*, 341.
27. Xu, B.; Griffin, W.L.; Xiong, Q.; Hou, Z.-Q.; O'Reilly, S.Y.; Guo, Z.; Pearson, N.; Greau, Y.; Zheng, Y.-C. Ultrapotassic rocks and xenoliths from South Tibet: Contrasting styles of interaction between lithospheric mantle and asthenosphere during continental collision. *Geology* **2017**, *45*, 51–54. [CrossRef]
28. Huang, G.P.; Hu, Q.L.; Chen, D.M.; Li, L.; Zhang, Z.; Zhu, A.A.; Xu, H.B. Overview of Geology and Mineral Resources of Madagascar. *Resour. Environ. Eng.* **2014**, *28*, 626–632.
29. Wang, H.B.; Xia, F.F.; Wu, H.X. A brief analysis of geological characteristics and minerals in Ambatondrazaka area. *Mod. Min.* **2012**, *27*, 33−34, 37.

30. Zolotarev, V.M. Optical Constants of an Apatite Single Crystal in the IR Range of 6–28 Mm. *Opt. Spectrosc.* **2018**, *124*, 262–272. [CrossRef]
31. Yuan, P.Y.; Xu, B.; Wang, Z.X.; Liu, D.Y. A Study on Apatite from Mesozoic Alkaline Intrusive Complexes, Central High Atlas, Morocco. *Crystals* **2022**, *12*, 461. [CrossRef]
32. Zhao, Z.-G.; Gao, L.-M. Standardization of δEu, δCe calculation methods. *Stand. Cover.* **1998**, *5*, 24–26.
33. Wang, R.J. Application of rare earth elements in petrology. *Geol. Sci. Technol. Inf.* **1983**, *3*, 32–39.
34. Yang, Y.-H.; Wu, F.-Y.; Yang, J.-H.; Chew, D.M.; Xie, L.-W.; Chu, Z.-Y.; Zhang, Y.-B.; Huang, C. Sr and Nd Isotopic Compositions of Apatite Reference Materials Used in U–Th–Pb Geochronology. *Chem. Geol.* **2014**, *385*, 35–55. [CrossRef]
35. Su, X.D.; Peng, P.; Wang, C.; Sun, F.B.; Zhang, Z.Y.; Zhou, X.T. Whole-rock and mineral chemical data from a profile of the ~900 Ma Niutishan Fe-Ti-rich sill in XuZhou, North China. *Data Brief* **2018**, *21*, 727–735. [CrossRef] [PubMed]
36. Takashima, R.; Kuwabara, S.; Sato, T.; Takemura, K.; Nishi, H. Utility of trace elements in apatite for discrimination and correlation of Quaternary ignimbrites and co−ignimbrite ashes, Japan. *Quat. Geochronol.* **2017**, *41*, 151–162. [CrossRef]
37. Liu, J.-W. Characterization of Granitic Apatite Speciation in Jiu Dong and Xiao Qin Ling and Its Geological Significance. Master's Thesis, China University of Geosciences, Beijing, China, 2019.
38. Liu, D.-Y. Gemological Characteristics of Apatite in Three Origins. Master's Thesis, China University of Geosciences, Beijing, China, 2021.
39. Xu, B.; Hou, Z.-Q.; Griffin, W.L.; Zhou, Y.; Zhang, Y.F.; Lu, Y.J.; Belousova., E.A.; Xu, J.F.; O'Reilly, S.Y. Elevated Magmatic Chlorine and Sulfur Concentrations in the Eocene–Oligocene Machangqing Cu–Mo Porphyry System. *SEG Spec. Publ.* **2021**, *24*, 257–276.
40. O'Reilly, S.Y.; Griffin, W.L. Apatite in the mantle: Implications for metasomatic processes and high heat production in Phanerozoic mantle. *Lithos* **2000**, *53*, 217–232. [CrossRef]
41. Chen, W.; Simonetti, A. In-situ determination of major and trace elements in calcite and apatite, and U-Pb ages of apatite from the Oka carbonatite complex: Insights into a complex crystallization history. *Chem. Geol.* **2013**, *353*, 151–172. [CrossRef]
42. Liu, X.Q.; Zhang, H.; Tang, Y.; Liu, Y.L. REE Geochemical Characteristic of Apatite: Implications for Ore Genesis of the Zhijin Phosphorite. *Minerals* **2020**, *10*, 1012. [CrossRef]
43. Prowatke, S.; Klemme, S. Trace element partitioning between apatite and silicate melts. *Geochim. Cosmochim. Acta* **2006**, *70*, 4513–4527. [CrossRef]
44. Xu, B.; Hou, Z.-Q.; Griffin, W.L.; O'Reilly, S.Y.; Zheng, Y.-C.; Wang, T.; Fu, B.; Xu, J.F. In-situ mineralogical interpretation of the mantle geophysical signature of the Gangdese Cu-porphyry mineral system. *Gondwana Res.* **2022**, *111*, 53–63. [CrossRef]

*Article*

# Natural Forsterite Strongly Enriched in Boron: Crystal Structure and Spectroscopy

Bijie Peng [1], Mingyue He [1,*], Mei Yang [2], Shaokun Wu [1] and Jingxin Fan [1]

[1] School of Gemmology, China University of Geosciences, Beijing 100083, China; pengbijie@163.com (B.P.); 3009210005@email.cugb.edu.cn (S.W.); 2009210011@email.cugb.edu.cn (J.F.)
[2] Sciences Institute, China University of Geosciences, Beijing 100083, China; yangmei@cugb.edu.cn
* Correspondence: hemy@cugb.edu.cn

**Abstract:** Boron is a typical crustal element and largely incompatible in olivine. Most natural olivine samples have very low concentrations of boron. Recently, forsterite with high boron content (up to 60.53 wt% MgO and 1795.91 ppm B) has been discovered in the Jian forsterite jade in the Jian area of northeast China. In this study, B-rich forsterite was examined by electron microprobes, Laser Ablation-Inductively Coupled Plasma Mass Spectrometry, Single crystal X-ray diffraction, Raman spectroscopy, and infrared spectroscopy. The B-rich forsterite is orthorhombic, existing in space group *Pnma*, and its unit-cell parameters are: a = 10.1918(7) Å, b = 5.9689(4) Å, c = 4.7484(3) Å, $\alpha = 90°$, $\beta = 90°$, $\gamma = 90°$, and V = 288.86(3) $Å^3$. The results of single crystal X-ray diffraction analysis indicate that the unit-cell parameters (a, b, and c) and unit-cell volume of forsterite in Jian forsterite jade are much smaller than those of known olivine. An equivalent set of Raman and infrared spectra were measured for the natural B-rich forsterite and compared to the results for mantle forsterite with a Fo value of ~91. The Raman spectrum of B-rich forsterite is similar to that of mantle olivine. We conclude that the systematic peak position shifts towards higher Raman shift with increasing Fo content. The infrared spectrum of B-rich forsterite crystals is characterized by strong absorption bands at 761, 1168, 1259, and 1303 $cm^{-1}$, which are assigned to stretching vibrations of $BO_3$ groups. Our data further confirm the existence of the $B(F, OH)Si_{-1}O_{-1}$ coupled substitution in natural B-rich forsterite.

**Keywords:** forsterite; boron; crystal structure; spectroscopy

**Citation:** Peng, B.; He, M.; Yang, M.; Wu, S.; Fan, J. Natural Forsterite Strongly Enriched in Boron: Crystal Structure and Spectroscopy. *Crystals* **2022**, *12*, 975. https://doi.org/10.3390/cryst12070975

Academic Editor: Carlos Rodriguez-Navarro

Received: 21 June 2022
Accepted: 11 July 2022
Published: 12 July 2022

**Publisher's Note:** MDPI stays neutral with regard to jurisdictional claims in published maps and institutional affiliations.

## 1. Introduction

Olivine is one of the simplest silicate minerals, and the general crystal chemical formula of which is $(A)_2SiO_4$, where A—$Mg^{2+}$, $Fe^{2+}$, $Mn^{2+}$, etc. The olivine crystal structure is orthorhombic with a slightly distorted hexagonal close packing array of oxygen atoms. Si is on tetrahedral interstices and Mg and Fe ions are on octahedral sites (labeled M1 and M2). The M1 site is a distorted octahedra at the center of symmetry whereas M2 is a regular octahedra on the mirror plane [1–4]. Olivine is a general name of forsterite-fayalite sold solution [5]. In addition to the complete isomorphism of Mg and Fe, olivine also contains some petrogenetically significant minor components, such as B, Li, Co, P, and As [4,6,7].

A growing body of crystal structure and spectroscopy investigations have been carried out on both synthetic and natural olivine over the past few years. The majority of the previous studies have primarily focused on crystal structure refinement [3,4,8,9], the effect of temperature and pressure on the crystal structure [10–14], the chemical composition of the forsterite-fayalite series determination [15,16], residual pressure calculation [17], and water content estimation [18]. Although extensive studies have been conducted on olivine over a range of compositions, temperatures, and pressures using a variety of methods, more research is needed on several aspects of the properties of B-rich olivine.

Boron is a typical crustal element with high concentrations in rocks closely related to continents and rocks interacting with the hydrosphere, but low concentrations in mantle peridotite [19,20]. B-rich olivine, associated with clinohumite K-rich pargasite, uvite, spinel, magnetite, ludwigite, and sinhalite, was first described from the Tayozhnoye Fe deposit of Russia [21]. Sykes et al. [22] studied the infrared spectrum and transmission electron microscopy of B-rich olivine and presented the first direct evidence of coupled substitution of boron and hydrogen for silicon. More precise and complex coupled substitution mechanisms were presented by Kent and Rossman [6] and Gose et al. [23]. More recently, Ingrin et al. [24] identified the position of the OH bands associated with the boron substitution through the infrared spectrum. Unfortunately, due to the incompatibility of boron in olivine, the connection between boron and olivine is often neglected. Hence, we addressed the key question: Whether the incorporation of B in olivine will affect the structure of olivine? Whether the Raman spectrum and infrared spectrum of boron-containing olivine be different from those of boron-free olivine? Additionally, most previous studies on the B-contained olivine were performed on synthetic olivine. Therefore, it is inevitable to investigate the properties of natural B-rich olivine.

In this paper, we report a new occurrence of natural end-member forsterite in Jian forsterite jade from Jilin province, China. Besides the high content of Mg, this forsterite has another unusual feature, namely strong enrichment by B. We present chemical composition, single crystal X-ray diffraction, Raman spectrum, and infrared spectrum studies for B-rich forsterite. Our results refine the crystal structure of B-rich forsterite and provide strong evidence for the coupled substitution of H and B for Si in natural B-rich forsterite. This study presents a basis for understanding the formation of natural B-rich forsterite.

## 2. Materials and Methods

The natural B-rich forsterite crystals, directly selected from the Jian forsterite jade, were used for detailed data analyses. The Jian forsterite jade is a new type of jade that has been recently found in the Jian area of northeast China [25]. It is mainly composed of forsterite, serpentine, and brucite (Figure 1a,b). We crushed the jade to 60–80 mesh, and hand-picked crack-free and inter-transparent olivine particles under binoculars for analysis. The size of the selected forsterite is 0.3–1 mm (Figure 1c,d). Forsterite with Fo~91 in mantle peridotite xenolith from Jilin province, China was chosen for comparison (Sample O-1 and O-2).

**Figure 1.** Sample photographs. (**a**) the Jian forsterite jade. (**b**) 40× magnification observation of (**a**) under gem microscope. (**c**) and (**d**) hand-picked B-rich forsterite.

The chemical compositions of forsterite were obtained on polished thin sections using an EPMA-1720 electron microprobe at the Electron Probe Laboratory, China University of Geosciences (Beijing, China). The primary analyzing settings were a 5 μm beam spot diameter, 10 nA beam, and 15 kV acceleration voltage. The standards used to calibrate the electron microprobe were 52 standard minerals from the SPI Company of the United States, Washington, DC, USA.

In situ trace element analyses were carried out at the Mineral Laser Microprobe Analysis Laboratory (Milma Lab, Beijing, China), China University of Geosciences (Beijing) using Laser Ablation-Inductively Coupled Plasma Mass Spectrometry (LA-ICP-MS). An Agilent 7900 ICP–MS fitted with a NewWave 193UC excimer laser ablation system was used. A laser repetition rate of 8 Hz at 3 J/cm$^3$ and a spot diameter of 50 μm were used in the analyses. NIST 610 was used as the external standard, Si was used as the internal standard, and ARM-1 and BCR-2G were used as the monitoring standards.

Single crystal X-ray diffraction measurements were obtained at ambient conditions with a Rigaku Xtalab PRO diffractometer system and HyPix-6000HE detector at the Laboratory of X-ray single crystal diffraction, China University of Geosciences (Beijing). The experiment was conducted with φ and Ω scanning mode, and the scanning step size was 0.5. Monocrystalline silicon array was used to conduct monochromatic processing on the wavelength, and X-ray (λ = 0.71073 Å) was used as the diffraction source to collect diffraction data. A 1.2 Kw water cooled microfocus source with a Mo rotor target and multilayer mirrors were used to collect intensity data. The atomic coordinates of samples are provided as CIF files in the Supplementary Materials. COD (entries 3000401) contains the supplementary crystallographic data for this paper.

The Raman spectrum was obtained using a Horiba HR-Evolution Raman microspectrometer with an Ar-ion laser operating at 532 nm excitation at the School of Gemmology, China University of Geosciences (Beijing). A 100 μm entrance slit and a grating with 1200 grooves per mm were used to collect the scattered light. Every Raman spectrum was acquired by 10 scans with 1 cm$^{-1}$ resolution in the 4000–100 cm$^{-1}$ range.

The infrared spectrum was obtained using the Tensor 27 Fourier infrared spectrometer at the School of Gemmology, China University of Geosciences (Beijing). All sample analyses adopted the transmission method. The experimental test conditions were as follows: test voltage was 220 V, the resolution was 4 cm$^{-1}$ with 64 scans, the scanning range was 4000–400 cm$^{-1}$, and the scanning speed was 10 kHz.

## 3. Results

### 3.1. EPMA

The chemical composition of olivine is end-member forsterite. All tested samples are compositionally homogeneous. The characteristics of zonation or exsolution lamellae are not observed. The Forsterite is Mg-rich (MgO > 57 wt%), and Fe-poor (FeOtot < 1 wt%). The Fo (100 × Mg/[Mg + Fe], mol%) of forsterite varies from 99.62 to nearly 99.74. The NiO content of this olivine is very low (<0.08 wt%), and the CaO content varies between 0.01 and 0.04 wt%. In addition, forsterite contains a certain amount of fluorine (0.13–1.06 wt%). Olivine in mantle peridotite xenolith is relatively iron-rich with Fo values of ~91. The content of NiO and CaO is higher than that of end-member forsterite. The chemical compositions are listed in Table 1.

**Table 1.** Chemical composition of B-rich forsterite and mantle olivine.

| Sample | S-1 | S-2 | S-3 | S-4 | O-1 | O-2 |
|---|---|---|---|---|---|---|
| $SiO_2$ | 39.41 | 41.98 | 41.84 | 41.71 | 41.31 | 41.33 |
| $TiO_2$ | 0.12 | 0.05 | 0.02 | 0.03 | 0.00 | 0.00 |
| $Al_2O_3$ | 0.03 | 0.01 | 0.02 | 0.01 | 0.00 | 0.00 |
| $Cr_2O_3$ | 0.00 | 0.00 | 0.00 | 0.00 | 0.02 | 0.00 |
| FeO | 0.28 | 0.37 | 0.35 | 0.39 | 8.17 | 7.86 |

**Table 1.** *Cont.*

| Sample | S-1 | S-2 | S-3 | S-4 | O-1 | O-2 |
|---|---|---|---|---|---|---|
| MnO | 0.04 | 0.01 | 0.08 | 0.00 | 0.06 | 0.15 |
| NiO | 0.08 | 0.00 | 0.05 | 0.00 | 0.44 | 0.36 |
| MgO | 60.53 | 58.21 | 57.12 | 57.20 | 49.54 | 49.74 |
| CaO | 0.04 | 0.00 | 0.01 | 0.01 | 0.021 | 0.03 |
| $Na_2O$ | 0.00 | 0.00 | 0.01 | 0.03 | 0.00 | 0.00 |
| $K_2O$ | 0.01 | 0.00 | 0.02 | 0.03 | 0.00 | 0.00 |
| F | 1.06 | 0.13 | 0.17 | 0.18 | 0.00 | 0.00 |
| Total | 101.59 | 100.75 | 99.69 | 99.59 | 99.55 | 99.47 |
| Fo | 99.74 | 99.64 | 99.66 | 99.62 | 91.53 | 91.86 |

### 3.2. LA-ICP-MS

The main trace element feature of the end-member forsterite is the extremely high content of B (1773.4–1795.91 ppm) (Figure 2). The content of compatible elements Cr, Co, and Ni is depleted. The concentrations of other trace elements are relatively low. Li and Ti concentrations range from 1.93 to 2.62 and 4.17 to 9.61 ppm, respectively. Other elements do not show any notable characteristics. The distribution of trace elements in olivine is very uniform [25]. Detailed trace elements contents are listed in Table 2. The trace element characteristics of B-rich forsterite are different from those of common mantle peridotite [26].

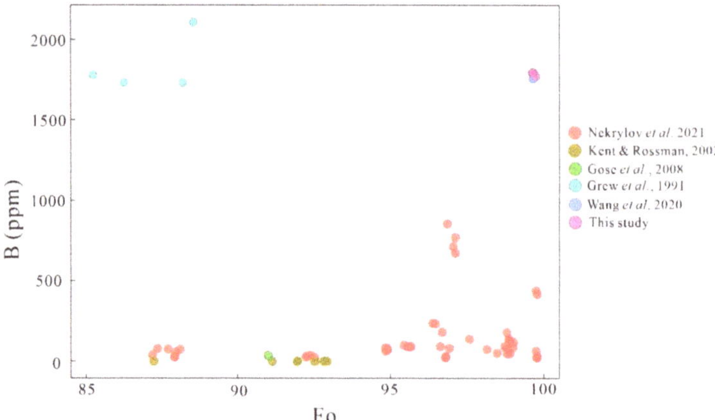


**Figure 2.** The B concentration of forsterite from the Jian forsterite jade compared with olivine from other origins.

**Table 2.** Trace element composition (ppm) of B-rich forsterite.

| Element | S-1 | S-2 | S-3 | S-4 | O-1 | O-2 |
|---|---|---|---|---|---|---|
| Li | 1.93 | 2.17 | 2.51 | 2.62 | 1.18 | 1.05 |
| B | 1773.4 | 1792.54 | 1795.56 | 1795.91 | b.d.l. | b.d.l. |
| Na | b.d.l. [1] | b.d.l. | b.d.l. | b.d.l. | 92 | 90 |
| Al | 37.82 | 43.16 | 45.21 | 48.94 | 27 | 23 |
| P | 118.2 | 120.18 | 135.78 | 142.24 | 38 | 43 |
| Ca | b.d.l. | b.d.l. | b.d.l. | b.d.l. | 74 | 58 |
| Sc | 1.22 | 1.247 | 1.249 | 1.25 | 0.58 | 0.76 |
| Ti | 9.58 | 6.51 | 9.61 | 4.17 | 165 | 35 |
| V | 0.587 | 0.609 | 0.614 | 0.654 | 3.8 | 2.9 |
| Cr | b.d.l. | b.d.l. | b.d.l. | b.d.l. | 88 | 102 |
| Mn | 200.71 | 201.72 | 203.69 | 210.69 | 610 | 636 |
| Co | 0.592 | 0.594 | 0.604 | 0.613 | 127 | 121 |

**Table 2.** *Cont.*

| Element | S-1 | S-2 | S-3 | S-4 | O-1 | O-2 |
|---|---|---|---|---|---|---|
| Ni | 0.914 | 0.928 | 0.944 | 0.974 | 2890 | 2868 |
| Cu | b.d.l. | b.d.l. | b.d.l. | b.d.l. | 1.52 | 1.10 |
| Zn | 43.92 | 43.96 | 45.49 | 46.34 | 56 | 42 |
| Ga | 0.271 | 0.29 | 0.3 | 0.301 | b.d.l. | b.d.l. |
| Sr | b.d.l. | b.d.l. | b.d.l. | b.d.l. | b.d.l. | b.d.l. |
| Y | 0.0675 | 0.07 | 0.0736 | 0.0738 | 0.02 | b.d.l. |
| Zr | 1.002 | 1.134 | 1.148 | 1.95 | 0.21 | 0.19 |
| Nb | 0.119 | 0.1232 | 0.1392 | 0.146 | 0.26 | 0.015 |
| Ce | b.d.l. | b.d.l. | b.d.l. | b.d.l. | b.d.l. | b.d.l. |

[1] b.d.l., Below the detected line.

### 3.3. Single Crystal X-ray Diffraction

Crystal data for B-rich forsterite: orthorhombic, space group *Pnma*, a = 10.1918(7) Å, b = 5.9689(4) Å, c = 4.7484(3) Å, $\alpha = 90°$, $\beta = 90°$, $\gamma = 90°$, and V = 288.86(3) Å$^3$, Z = 4, T = 293(2)K, $\mu$(MoK$\alpha$) = 1.16 mm$^{-1}$, Dcalc = 3.25 g/cm$^3$, 3161 reflections measured ($7.998° \leq 2\Theta \leq 60.592°$), 439 unique (Rint = 0.0824, Rsigma = 0.0500) which were used in all calculations. The final R1 was 0.0350 (I > 2$\sigma$(I)) and wR2 was 0.0948 (all data). In the crystal structure of forsterite, each structure unit contains one Si site, two cation sites, and three O sites (Figure 3). All the silicon atoms are coordinated to four oxygen atoms to form [SiO$_4$] tetrahedrons and account for one-eighth of tetrahedral voids. Isolated [SiO$_4$] tetrahedron (T) is surrounded by [MgO$_6$] octahedra (M1 and M2). The Si–O bond distances vary from 1.613(3) to 1.6312(14) Å (Table 3). The average O–Si–O bond angle is 109.15°. As shown in Figure 3, magnesium atoms are coordinated with six oxygen atoms to form [MgO$_6$] octahedra. Two cation M sites M(1) and M(2) are occupied by magnesium atoms. The M(1) site is located at the center of symmetry, while the M(2) site is located on the mirror plane. The Mg1–O bond distances vary from 2.0622(14) to 2.1283(14) Å and the Mg2–O bond distances from 2.049(3) to 2.2068(16) Å. The O–Mg–O angles vary from 71.65(8)° to 109.93(9)° (Table 3). Table 3 shows bond distances and angles parameters for M1, M2, and T.

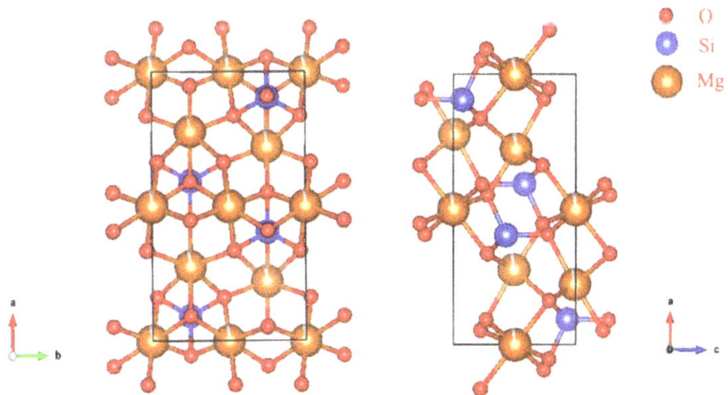


**Figure 3.** Unit-cell of B-rich forsterite structure viewed along c and b axes.

Compared with published olivine cell parameters [3,8,13,27–30], B-rich forsterite tested in this study shows the smallest unit cell parameters (Figure 4). As shown in Figure 4a–c, a negative and significant correlation between unit-cell parameters and the Fo value, which is in agreement with the results of Princivalle and Secco [3]. The parameter a is most affected by compositional changes ($1/k_a > 1/k_b > 1/k_c$). Figure 4d–f show that cell volume and coordination polyhedron volume of tested forsterite are both inversely proportional to magnesium contents. Figure 4g,h show the shortening of M1–O, and M2–O bond distances with increasing magnesium contents.

**Table 3.** Bond distances (Å) and bond angles of B-rich forsterite.

| | Bond Distance/Å | Bond Angle |
|---|---|---|
| octahedra M1 | (Mg1–O1)[2] = 2.0622(14) Å<br>(Mg1–O2)[2] = 2.0816(16) Å<br>(Mg1–O3)[2] = 2.1283(14) Å<br><br>Average 2.0907 | (O3–Mg1–O2) = 85.02(7)<br>(O2–Mg1–O1) = 86.53(7)<br>(O2–Mg1–O3) = 94.98(7)<br>(O2–Mg1–O1) = 93.47(7)<br>(O3–Mg1–O2) = 94.98(7)<br>(O1–Mg1–O2) = 93.47(7)<br>(O3–Mg1–O2) = 85.02(7)<br>(O1–Mg1–O2) = 86.53(7)<br>(O1–Mg1–O3) = 105.06(7)<br>(O3–Mg1–O1) = 105.06(7)<br>(O3–Mg1–O1) = 74.94(7)<br>(O3–Mg1–O1) = 74.94(7)<br>Average 90 |
| octahedra M2 | (Mg2–O1)[1] = 2.049(3) Å<br>(Mg2–O2)[1] = 2.175(3) Å<br>(Mg2–O3)[2] = 2.2068(16) Å<br>(Mg2–O3′)[2] = 2.0674(14) Å<br><br>Average 2.1286 | (O1–Mg2–O3) = 96.76(7)<br>(O3–Mg2–O3) = 88.83(4)<br>(O3–Mg2–O2) = 80.97(7)<br>(O3–Mg2–O3) = 71.65(8)<br>(O1–Mg2–O3) = 90.79(6)<br>(O3–Mg2–O3) = 109.93(9)<br>(O2–Mg2–O3) = 90.82(6)<br>(O3–Mg2–O3) = 88.83(4)<br>(O1–Mg2–O3) = 90.79(6)<br>(O3–Mg2–O2) = 90.82(6)<br>(O2–Mg2–O3) = 80.97(7)<br>(O1–Mg2–O3) = 96.76(7)<br>Average 89.83 |
| tetrahedron T | (Si–O3) = 1.6312(14) Å<br>(Si–O1) = 1.656(3) Å<br>(Si–O2) = 1.613(3) Å<br>(Si–O3) = 1.6312(14) Å<br><br>Average 1.6327 | (O3–Si–O2) = 116.09(7)<br>(O2–Si–O1) = 114.46(11)<br>(O1–Si–O3) = 101.75(7)<br>(O3–Si–O3) = 104.72(11)<br>(O2–Si–O3) = 116.09(7)<br>(O1–Si–O3) = 101.75(7)<br>Average 109.15 |

Numbers in square brackets represent the number of repeats of bonds and bond angles in a polyhedron.

### 3.4. Raman Spectrum

Laser Raman spectroscopy is a powerful method for structural and compositional characterization of minerals. The factor group analysis indicates that forsterite has 36 Raman-active vibration modes: $11A_g + 11B_{1g} + 7B_{2g} + 7B_{3g}$ [31,32]. Symmetry and assignment for the Raman modes of B-rich forsterite are listed in Table 4. A Raman spectrum of B-rich forsterite can be divided into three spectrum regions: (1) 700–1100 cm$^{-1}$, (2) 400–700 cm$^{-1}$, and (3) < 400cm$^{-1}$ [15]. The bands of the region (1) (at approximately 824, 857, 882, 919, and 965 cm$^{-1}$ for B-rich forsterite) are attributed to the internal symmetric and asymmetric stretching vibrational modes of the SiO$_4$ ionic group [15]. Bands between 700 and 1100 cm$^{-1}$ are the most characteristic peak of the olivine Raman spectrum, which can be used to identify olivine in the multi-phase spectrum [15,16,31]. Low intense bands of the region (2) (at 437, 544, 586, and 609 cm$^{-1}$ for B-rich forsterite) are related to the internal bending vibrational modes of the SiO$_4$ ionic groups [15]. The bands of the region (3) (at around 227, and 305 cm$^{-1}$ for B-rich forsterite) are assigned to the lattice vibration modes, including rotational and translational vibrations of SiO$_4$ tetrahedra, and translational vibrations of magnesium and iron cations [31]. In addition, bands associated with the vibration of B-O are not detected, which is attributed to the concentration of boron. Generally speaking, boron has no effect on the Raman spectra of olivine.

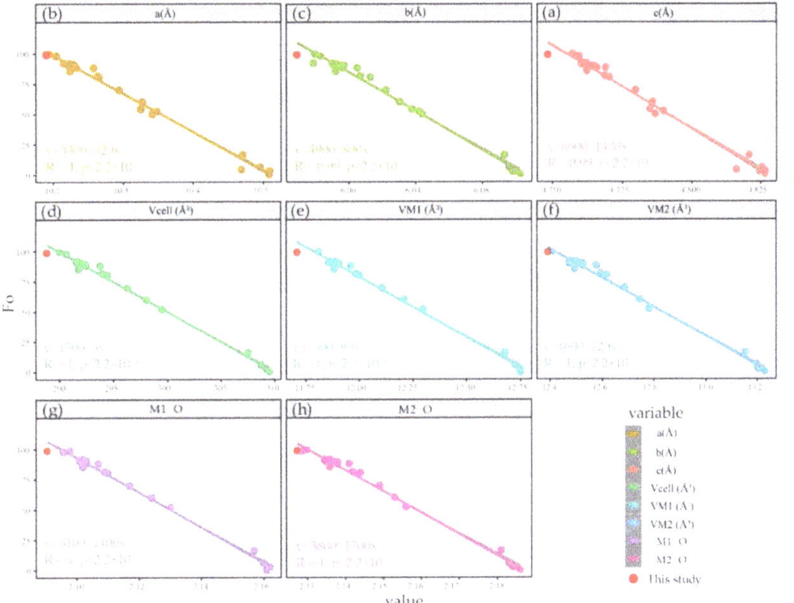

**Figure 4.** (**a–c**) Unit-cell parameters **a**, **b**, **c** vs. Fo content of olivine. (**d**) Unit-cell volume vs. Fo content of olivine. (**e,f**) The volume of coordination polyhedron M1, and M2 vs. Fo content of olivine. (**g,h**) Average M1–O, and M2–O bond distances vs. Fo content of olivine.

**Table 4.** Assignment of bands in the Raman spectrum of the forsterite.

| Band (cm$^{-1}$) | | Symmetry | Assignment |
|---|---|---|---|
| **B-Rich Olivine (Fo = 99)** | **B-Free Olivine (Fo = 91)** | | |
| 227 | 220 | $A_g$ | $SiO_4$ translation |
| 305 | 300 | $A_g$ | $M_2$ translation |
| 437 | 433 | $B_{1g}$ | $v_2$ |
| 544 | 540 | $A_g$ | $v_4$ |
| 586 | 581 | $B_{2g}$ | $v_4$ |
| 609 | 603 | $A_g$ | $v_4$ |
| 824 | 820 | $A_g$ | $v_1 + v_3$ |
| 857 | 852 | $A_g$ | $v_1 + v_3$ |
| 882 | 878 | $B_{2g}$ | $v_3$ |
| 919 | 916 | $B_{3g}$ | $v_3$ |
| 965 | 958 | $A_g$ | $v_3$ |

In our analysis, another feature of the Raman spectrum is that the systematic peak position shift towards higher Raman shift compared with Fo~91 olivine (820, 852 for Fo~91 olivine and 824, 857 for Fo~99 olivine, respectively) (Figure 5b). The behavior of atomics at the M2 octahedral site is the main factor affecting the Raman spectrum of olivine [33]. Our results indicate that the relative intensities and position of the Raman peak in olivine can be correlated with the type of atomic substitutions involved. Kuebler et al. [15] explained that systematic peak-position is related to the decrease in atomic mass and polyhedral volume in octahedral sites, and to the degree of coupling of the symmetric and asymmetric stretching vibrational modes of $SiO_4$ groups.

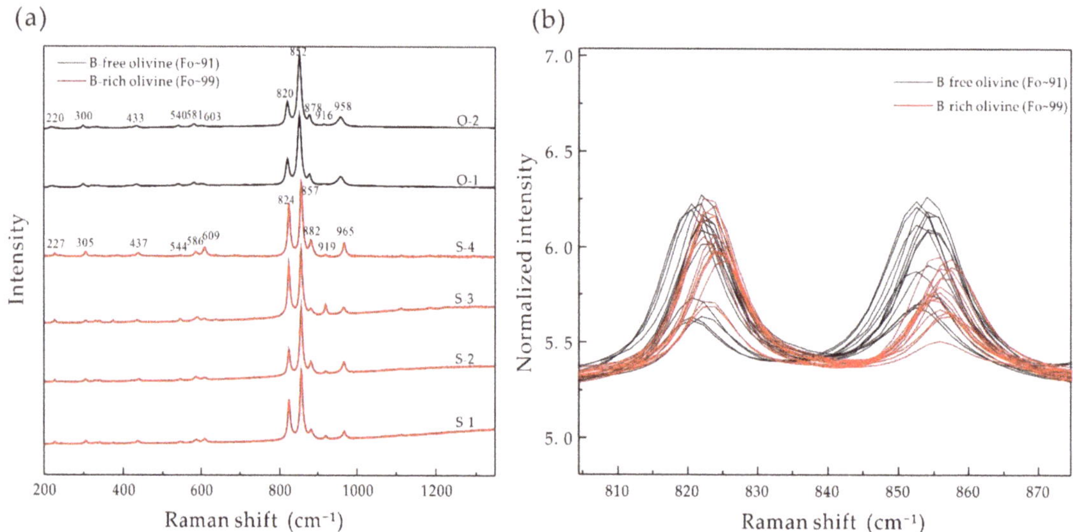

**Figure 5.** Raman spectrum of B-rich forsterite compared with mantle olivine. (**a**) The Raman spectrum of samples. (**b**) The magnification of (**a**).

### 3.5. Infrared Spectrum

The characteristic peaks of B-rich forsterite appear mainly at approximately 1303, 1259, 1168, 990, 958, 883, 839, 761, 610, 508, 468, and 424 cm$^{-1}$. The bands around 990, 958, and 883 cm$^{-1}$ are related to the symmetrical stretching vibration of the Si−O−Si group. The bands near 468, 508, and 610 cm$^{-1}$ represent the bending vibration of the Si−O group. Internal vibrations and lattice vibrations appear at 468 and 424 cm$^{-1}$. The IR spectrum of B-rich forsterite displays strong OH band at 3696 cm$^{-1}$ and weaker band at 3593 cm$^{-1}$. Both bands are caused by OH stretching vibration, which indicates the existence of constitutional water in forsterite.

Figure 6 compares the IR spectra of B-rich forsterite from Jian forsterite jade and B-free olivine (Fo~91) from mantle peridotite. The spectra are similar, except for five bands at 761, 1168, 1259, 1303, and 3593 cm$^{-1}$ in the spectrum of the B-rich forsterite. Similar bands have been reported in B-rich olivine samples [22,24]. The band at 761 cm$^{-1}$ is close to the 758 cm$^{-1}$ band observed in the infrared spectrum of B-rich olivine from Tayozhnoye, Russia. Sykes et al. [22] assigned the band at 758 cm$^{-1}$ to the $v_1$ symmetric stretching mode of the BO$_3$ groups. The bands at 1168 and 1259 cm$^{-1}$ in the B-rich forsterite spectra are assigned to $v_3$ asymmetric stretching vibration modes of BO$_3$ groups [22,24]. The bands at 1301 cm$^{-1}$ likely correspond to the isotopic shift of the bands at 1168 and 1256 cm$^{-1}$, respectively, due to $^{10}$B [24].

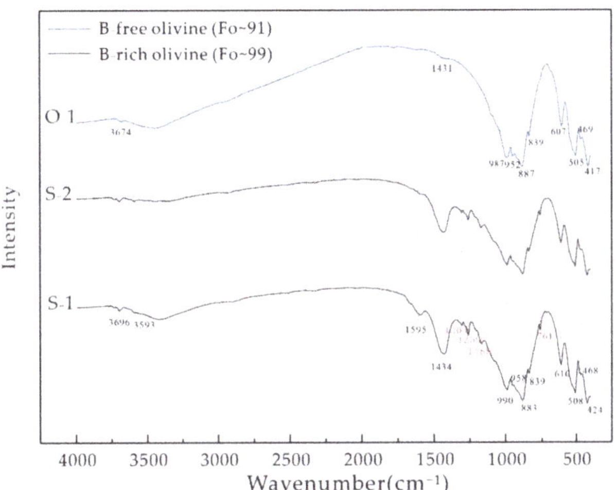



**Figure 6.** The infrared spectrum of B-rich forsterite compared with mantle olivine.

## 4. Discussion

### 4.1. The Coupled Substitution B(F, OH)Si$_{-1}$, O$_{-1}$

The incorporation of boron in olivine is mainly related to the coupled substitution of boron and H for silicon [6,21–24]. The BO$_3$ group lies on the (O3−O1−O3) face of the tetrahedral site, inclined at 17° from the (x,y) plane, and the H atom is bonded to the O2 atom [24]. The O2−H group points either to the O1 oxygen or out of the tetrahedral site. The former configuration is more stable than the latter [24]. The introduction of B promotes the formation of defects in the crystal structure of forsterite and may result in a distortion of the tetrahedral site [22,34], which is interpreted as the formation of terminal B−(OH) or B−F bonds or the perturbation of the B−O−M linkage. Guo et al. [35] pointed out that the substitution of B for Si may result in the reduction in cell parameters and cell volume of forsterite.

The results of single crystal X-ray diffraction analysis indicate that the unit-cell parameters (a, b, and c) and unit-cell volume of forsterite in Jian forsterite jade are much smaller than those of known olivine. The small crystal parameters are rather due to an increased content of Mg. It's well known that lattice parameters linearly decrease with increasing forsterite [3]. However, we compared the lattice parameters of B-rich forsterite (Fo = 99.7) with pure forsterite (Fo = 100) (Figure 4) [1,13]. The lattice parameters of the former are smaller than those of the latter. Similar phenomena were described in B-rich diopside [34]. Halenius et al. [34] presented that a B−Si substitution in B-rich diopside decreases the T site volume and leads to disruption of one of the T−O bonds compared with end-member diopside. In our study, V$_T$ (Å$^3$) in B-rich forsterite (2.196 Å$^3$) is smaller than that of pure forsterite (2.21 Å$^3$). Thus, we speculated that the incorporation of B in olivine slightly reduces the lattice parameters, especially decreases the T site volume and affects the neighboring M1 and M2 sites. Due to the low concentration of B, the X-ray diffraction techniques can't resolve the true coordination of B in forsterite. Nevertheless, the occurrence of the BO$_3$ group is further demonstrated by our spectra data.

B incorporation into forsterite has no effect on the Raman spectrum. In our study, the difference between B-rich and B-free forsterite Raman spectrum is caused by the different magnesian contents. No evident bands of B−O vibration appear in the Raman spectrum. While significant differences are observed between the B-rich and B-free forsterite IR spectrum. B-rich forsterite displays extra five bands at 761, 1168, 1259, 1303, and 3593 cm$^{-1}$ compared with B-free forsterite. The vibrational bands at 1200–1400 cm$^{-1}$ are related to B−O vibrations [22,24,34]. Halenius et al. [34] reported the existence of the replacement of $SiO_4$ by $BO_3$ in synthetic minerals (i.e., diopside and forsterite). The bands at 761, 1168, 1259, and 1303 cm$^{-1}$ are consistent with the bands of vibration modes of the $BO_3$ group [22,24,34]. In addition, strong OH band at 3696 cm$^{-1}$ and weaker band at 3593 cm$^{-1}$ are displayed in B-rich forsterite. OH groups commonly exist in nominally anhydrous minerals through a variety of pathways (commonly known as "water") [23,36,37]. Due to the variety and complexity of the infrared spectra of olivine associated with O-H stretching modes, uncertainties remain in the determination of the OH bands associated with the $B(OH)Si_{-1}O_{-1}$ substitution in B-rich olivine. Ingrin et al. [24] presented that the OH bands at 3704 cm$^{-1}$ (//z), 3598 cm$^{-1}$ (//x,y), and 3525 cm$^{-1}$ (//x) are associated with the $B(OH)Si_{-1}O_{-1}$ substitution in synthetic forsterite and natural olivine. Matsyuk and Langer [38] noted that two bands at 3672 and 3535 cm$^{-1}$ are assigned to boron-related defects. Gose et al. [37] speculated that the OH defect at 3597 cm$^{-1}$ may be linked to the coupled $B(OH)Si_{-1}O_{-1}$ substitution. Thus, the OH band at 3593 cm$^{-1}$ is most likely associated with the $B(F, OH)Si_{-1}O_{-1}$ substitution in our B-rich forsterite. The OH band at 3696 cm$^{-1}$ is assigned to OH stretching vibrations of OH defects. Therefore, the IR spectra of B-rich forsterite provided evidence that the incorporation of B into forsterite is mainly caused by the coupled substitution of boron and H for silicon. Theoretically, the ratio of the contents of B to H is close to the 1:1 trend expected from the $B(F, OH)Si_{-1}O_{-1}$ substitution. However, the contents of B (1773.4–1795.91 ppm) in our sample are higher than that of water (1450(300)) wt ppm estimated by Wang et al. [25]. The deviation from the 1:1 trend can be explained by additional fluorine (usually 0.13–0.18 wt% in our sample) [6].

### 4.2. The Rarity of Mg- and B-Rich Forsterite

The composition of olivine in the mantle and magmatic rocks is typically Fo85~96 [5,39]. Olivine with nearly forsterite end composition is rare and can only be formed in special geological environments. For example, rare forsterite (Fo = 97–99) inclusion in magnesiochromite was reported by Xiong et al. [40], and Majumdar et al. [41] found a pseudomorph rim of Mg-rich olivine (Fo = 98) in serpentinized dunite. Blondes et al. [42] found several olivine grains with Fo at up to 99.8 in multiple primitive basaltic lava flows from Big Pine Volcanic Field (Inyo, CA, USA). Nekrylov et al. [43] reported high-Mg olivine (Fo value up to 99.8) from magnesian skarns and silicate marbles from different locations. In addition, boron-rich olivines are less well-known. Most natural olivine samples have very low concentrations of boron [19,20]. Sykes et al. [22] noted that olivine from the Tayozhnoye iron deposit, Siberia, Russia, contains substantial $B_2O_3$ (1.11–1.35 wt%). Majumdar et al. [41] reported a boron content up to 10.4 ppm in high-magnesium olivine. Nekrylov et al. [43] presented that B concentrations of olivine from magnesian skarns and silicate marbles vary from 23 to 856 ppm. However, the natural forsterite tested in this study surpasses all olivines known in geological objects in magnesium and boron content (Figure 2). The formation of B-rich forsterite remains disputed.

Olivine with high boron content is generally considered to be associated with the metasomatism of boron-rich fluids [22,43]. The B-rich forsterite may be derived from mantle peridotite metasomatized by B-rich fluids. However, the composition of olivine in mantle peridotite is typical Fo$_{86-92}$ [5]. Moreover, mantle olivine has high Ni, Mn, and Co concentrations (Ni, 2040–3310 ppm; Mn, 460–850 ppm; Co, 87–137 ppm) [26]. However, the trace element compositions of B-rich forsterite are evidently inconsistent with the mantle olivine. The above speculation about the origin seems to be incorrect. In addition, Nekrylov et al. [43] presented that olivine from magnesian skarns and silicate marbles are enriched in B and depleted in Ni, Co, and Cr concentrations, which are consistent with our trace element characteristics. According to the description of the geology setting in Wang et al. [25], the formation in which B-rich forsterite forms contains various altered felsic and granitic rocks, dolomitic marble, and serpentinized olivine marble. Moreover, previous study confirms that the formation is a boron-bearing sequence [25]. Thus, we suggest that B-rich forsterite is derived from the metamorphism of magnesian marbles. The origin and formation process of B-rich forsterite needs further study.

## 5. Conclusions

(1)  The chemical composition of forsterite from Jian forsterite jade of Jilin province, China, is B-rich end-member forsterite. This forsterite is the most magnesium- and boron-rich natural olivine.

(2)  The B-rich forsterite is orthorhombic, in the space group Pnma, and its unit-cell dimensions are: a = 10.1918(7) Å, b = 5.9689(4) Å, c = 4.7484(3) Å, $\alpha = 90°$, $\beta = 90°$, $\gamma = 90°$, and V = 288.86(3) $Å^3$. The unit-cell parameters (a, b, and c) and unit-cell volume of forsterite in Jian forsterite jade are smaller than common natural olivine. There is a negative and significant correlation between unit-cell parameters and Fo values.

(3)  The Raman of B-rich forsterite is consistent with that of mantle olivine (Fo = 91). The Raman shift of B-rich forsterite is mainly in 227, 305, 437, 544, 586, 609, 824, 857, 882, 919, and 967 $cm^{-1}$. With the substitution of magnesium for iron increasing in the forsterite-fayalite series, the bands of the Raman spectrum of B-rich forsterite will systematically shift to high Raman shift. The infrared spectrum characteristic peaks of B-rich forsterite appear mainly at approximately 1303, 1259, 1168, 990, 958, 883, 839, 761, 610, 508, 468, and 424 $cm^{-1}$. In particular, the stretching vibrations of BO3 groups occur at 761, 1168, 1259, and 1303 $cm^{-1}$.

**Supplementary Materials:** The following supporting information can be downloaded at: https://www.mdpi.com/article/10.3390/cryst12070975/s1, crystallographic information files (CIF).

**Author Contributions:** Conceptualization, M.H.; methodology, M.H. and B.P.; software, B.P. and S.W.; formal analysis, B.P., S.W. and J.F.; investigation, B.P. and S.W.; data curation, B.P.; writing—original draft preparation, B.P., S.W. and J.F.; writing—review and editing, M.H. and M.Y.; supervision, M.H. and M.Y.; All authors have read and agreed to the published version of the manuscript.

**Funding:** National Mineral Rock and Fossil Specimens Resource Center.

**Institutional Review Board Statement:** Not applicable.

**Informed Consent Statement:** Not applicable.

**Data Availability Statement:** Not applicable.

**Acknowledgments:** We thank Yingliang Jiang and Haiyang Yu from Ji'an Xinli Mining Co, LTD (Jilin, China) for providing samples for this study. We thank Xi Liu and Qiangwei Su for fruitful discussions and suggestions.

**Conflicts of Interest:** The authors declare no conflict of interest.

## References

1.  Kadziolka-Gawel, M.; Dulski, M.; Kalinowski, L.; Wojtyniak, M. The effect of gamma irradiation on the structural properties of olivine. *J. Radioanal Nucl. Chem.* **2018**, *317*, 261–268. [CrossRef] [PubMed]

2.  Ottonello, G.; Princivalle, F.; Della Giusta, A. Temperature, composition, and f $O_2$ effects on intersite distribution of Mg and $Fe^{2+}$ in olivines. *Phys. Chem. Miner.* **1990**, *17*, 301–312. [CrossRef]

3.  Princivalle, F.; Secco, L. Crystal structure refinement of 13 olivines in the forsterite-fayalite series from volcanic rocks and ultramafic nodules. *Tschermaks Mineral. Petrogr. Mitt.* **1985**, *34*, 105–115. [CrossRef]

4.  Shchipalkina, N.V.; Pekov, I.V.; Zubkova, N.V.; Koshlyakova, N.N.; Sidorov, E.G. Natural forsterite strongly enriched by arsenic and phosphorus: Chemistry, crystal structure, crystal morphology and zonation. *Phys. Chem. Miner.* **2019**, *46*, 889–898. [CrossRef]

5.  Plechov, P.Y.; Shcherbakov, V.D.; Nekrylov, N.A. Extremely magnesian olivine in igneous rocks. *Russ. Geol. Geophys.* **2018**, *59*, 1702–1717. [CrossRef]

6.  Kent, A.J.R.; Rossman, G.R. Hydrogen, lithium, and boron in mantle-derived olivine: The role of coupled substitutions. *Am. Mineral.* **2002**, *87*, 1432–1436. [CrossRef]

7.  Müller-Sommer, M.; Hock, R.; Kirfel, A. Rietveld refinement study of the cation distribution in (Co, Mg)-olivine solid solution. *Phys. Chem. Miner.* **1997**, *24*, 17–23. [CrossRef]

8.  Nestola, F.; Nimis, P.; Ziberna, L.; Longo, M.; Marzoli, A.; Harris, J.W.; Manghnani, M.H.; Fedortchouk, Y. First crystal-structure determination of olivine in diamond: Composition and implications for provenance in the Earth's mantle. *Earth Planet. Sci. Lett.* **2011**, *305*, 249–255. [CrossRef]

9.  Taran, M.N.; Matsyuk, S.S. $Fe^{2+}$, Mg-distribution among non-equivalent structural sites M1 and M2 in natural olivines: An optical spectroscopy study. *Phys. Chem. Miner.* **2013**, *40*, 309–318. [CrossRef]

10. Hazen, R.M. Effects of temperature and pressure on the crystal structure of forsterite. *Am. Mineral.* **1976**, *61*, 1280–1293.

11. Smyth, J.R.; Hazen, R.M. The crystal structure and horhonolite at seeral temperatures up to 900 °C. *Am. Mineral.* **1973**, *58*, 588–593.

12. Xu, J.; Fan, D.; Zhang, D.; Li, B.; Zhou, W.; Dera, P.K. Investigation of the crystal structure of a low water content hydrous olivine to 29.9 GPa: A high-pressure single-crystal X-ray diffraction study. *Am. Mineral.* **2020**, *105*, 1857–1865. [CrossRef]

13. Pamato, M.G.; Nestola, F.; Novella, D.; Smyth, J.R.; Pasqual, D.; Gatta, G.D.; Alvaro, M.; Secco, L. The High-Pressure Structural Evolution of Olivine along the Forsterite—Fayalite Join. *Minerals* **2019**, *9*, 790. [CrossRef]

14. Artioli, G.; Rinaldi, R.; Wilson, C.; Zanazzi, P. High-temperature Fe-Mg cation partitioning in olivine: In-situ single-crystal neutron diffraction study. *Am. Mineral.* **1995**, *80*, 197–200. [CrossRef]

15. Kuebler, K.E.; Jolliff, B.L.; Wang, A.; Haskin, L.A. Extracting olivine (Fo-Fa) compositions from Raman spectral peak positions. *Geochim. Cosmochim. Acta* **2006**, *70*, 6201–6222. [CrossRef]

16. Breitenfeld, L.B.; Dyar, M.D.; Carey, C.J.; Tague, T.J.; Wang, P.; Mullen, T.; Parente, M. Predicting olivine composition using Raman spectroscopy through band shift and multivariate analyses. *Am. Mineral.* **2018**, *103*, 1827–1836. [CrossRef]

17. Yasuzuka, T.; Ishibashi, H.; Arakawa, M.; Yamamoto, J.; Kagi, H. Simultaneous determination of Mg# and residual pressure in olivine using micro-Raman spectroscopy. *J. Mineral. Petrol. Sci.* **2009**, *104*, 395–400. [CrossRef]

18. Bolfan-Casanova, N.; Montagnac, G.; Reynard, B. Measurement of water contents in olivine using Raman spectroscopy. *Am. Mineral.* **2014**, *99*, 149–156. [CrossRef]

19. Marschall, H.R.; Wanless, V.D.; Shimizu, N.; Pogge von Strandmann, P.A.E.; Elliott, T.; Monteleone, B.D. The boron and lithium isotopic composition of mid-ocean ridge basalts and the mantle. *Geochim. Cosmochim. Acta* **2017**, *207*, 102–138. [CrossRef]

20. Ottolini, L.; Le Fèvre, B.; Vannucci, R. Direct assessment of mantle boron and lithium contents and distribution by SIMS analyses of peridotite minerals. *Earth Planet. Sci. Lett.* **2004**, *228*, 19–36. [CrossRef]

21. Grew, E.S.; Pertsev, N.N.; Boronikhin, V.A.; Borisovskiy, S.Y.; Yates, M.G.; Marquez, N. Serendibite in the Tayozhnoye deposit of the Aldan Shield, eastern Siberia, U.S.S.R. *Am. Mineral.* **1991**, *76*, 1061–1080.

22. Sykes, D.; Rossman, G.R.; Veblen, D.R.; Grew, E.S. Enhanced H and F incorporation in borian olivine. *Am. Mineral.* **1994**, *79*, 904–908.

23. Gose, J.; Reichart, P.; Dollinger, G.; Schmadicke, E. Water in natural olivine-determined by proton-proton scattering analysis. *Am. Mineral.* **2008**, *93*, 1613–1619. [CrossRef]

24. Ingrin, J.; Kovacs, I.; Deloule, E.; Balan, E.; Blanchard, M.; Kohn, S.C.; Hermann, J. Identification of hydrogen defects linked to boron substitution in synthetic forsterite and natural olivine. *Am. Mineral.* **2014**, *99*, 2138–2141. [CrossRef]

25. Wang, Y.; He, M.; Yan, W.; Yang, M.; Liu, X. Jianite: Massive Dunite Solely Made of Virtually Pure Forsterite from Ji'an County, Jilin Province, Northeast China. *Minerals* **2020**, *10*, 220. [CrossRef]

26. De Hoog, J.C.M.; Gall, L.; Cornell, D.H. Trace-element geochemistry of mantle olivine and application to mantle petrogenesis and geothermobarometry. *Chem. Geol.* **2010**, *270*, 196–215. [CrossRef]

27. Princivalle, F.; De Min, A.; Lenaz, D.; Scarbolo, M.; Zanetti, A. Ultramafic xenoliths from Damaping (Hannuoba region, NE-China): Petrogenetic implications from crystal chemistry of pyroxenes, olivine and Cr-spinel and trace element content of clinopyroxene. *Lithos* **2014**, *188*, 3–14. [CrossRef]

28. Birle, J.D.; Gibbs, G.V.; Moore, P.B.; ASmith, J.V. Crystal structure of natural olivines. *Am. Mineral.* **1968**, *53*, 807–824.

29. Princivalle, F. Influence of Temperature and Composition on Mg–$Fe^{2+}$ Intracrystalline Distribution in Olivines. *Mineral. Petrol.* **1990**, *43*, 121–129. [CrossRef]

30. Bai, W.; Fang, Q.; Zhang, Z.; Yan, B. Crystal structure and significance of magnesium olivine from pod-shaped chromite from The Lubsha ophiolite in Tibet. *J. Rock Mineral.* **2001**, *20*, 1–10.

31. Chopelas, A. Single crystal Raman spectra of forsterite, fayalite, and monticellite. *Am. Mineral.* **1991**, *76*, 1101–1109.

32. Kolesov, B.A.; Geiger, C.A. A Raman spectroscopic study of Fe–Mg olivines. *Phys. Chem. Miner.* **2004**, *31*, 142–154. [CrossRef]

33. Mouri, T.; Enami, M. Raman spectroscopic study of olivine-group minerals. *J. Mineral. Petrol. Sci.* **2008**, *103*, 100–104. [CrossRef]

34. Halenius, U.; Skogby, H.; Eden, M.; Nazzareni, S.; Kristiansson, P.; Resmark, J. Coordination of boron in nominally boron-free rock forming silicates: Evidence for incorporation of $BO_3$ groups in clinopyroxene. *Geochim. Cosmochim. Acta* **2010**, *74*, 5672–5679. [CrossRef]

35. Guo, Y.X.; Qu, D.l.; Li, Z.; Yang, C.G. Effect of boron oxide on crystal structure and properties of magnesium olivine synthesized from magnesite tailings. *Bull. Chin. Ceram. Soc.* **2016**, *35*, 865–869. [CrossRef]

36. Kovacs, I.; O'Neill, H.S.C.; Hermann, J.; Hauri, E.H. Site-specific infrared O–H absorption coefficients for water substitution into olivine. *Am. Mineral.* **2010**, *95*, 292–299. [CrossRef]

37. Gose, J.; Schmädicke, E.; Markowitz, M.; Beran, A. OH point defects in olivine from Pakistan. *Miner. Petrol.* **2009**, *99*, 105–111. [CrossRef]

38. Matsyuk, S.S.; Langer, K. Hydroxyl in olivines from mantle xenoliths in kimberlites of the Siberian platform. *Contrib. Miner. Petrol.* **2004**, *147*, 413–437. [CrossRef]

39. Sobolev, A.V.; Hofmann, A.W.; Kuzmin, D.V.; Yaxley, G.M.; Arndt, N.T.; Chung, S.L.; Danyushevsky, L.V.; Elliott, T.; Frey, F.A.; Garcia, M.O.; et al. The amount of recycled crust in sources of mantle-derived melts. *Science* **2007**, *316*, 412–417. [CrossRef]

40. Xiong, F.; Yang, J.; Robinson, P.T.; Xu, X.; Liu, Z.; Li, Y.; Li, J.; Chen, S. Origin of podiform chromitite, a new model based on the Luobusa ophiolite, Tibet. *Gondwana Res.* **2015**, *27*, 525–542. [CrossRef]
41. Majumdar, A.S.; Hövelmann, J.; Vollmer, C.; Berndt, J.; Mondal, S.K.; Putnis, A. Formation of Mg-rich Olivine Pseudomorphs in Serpentinized Dunite from the Mesoarchean Nuasahi Massif, Eastern India: Insights into the Evolution of Fluid Composition at the Mineral–Fluid Interface. *J. Petrol.* **2016**, *57*, 3–26. [CrossRef]
42. Blondes, M.S.; Brandon, M.T.; Reiners, P.W.; Page, F.Z.; Kita, N.T. Generation of Forsteritic Olivine (Fo99·8) by Subsolidus Oxidation in Basaltic Flows. *J. Petrol.* **2012**, *53*, 971–984. [CrossRef]
43. Nekrylov, N.; Plechov, P.Y.; Gritsenko, Y.D.; Portnyagin, M.V.; Shcherbakov, V.D.; Aydov, V.A.; Garbe-Schönberg, D. Major and trace element composition of olivine from magnesian skarns and silicate marbles. *Am. Mineral.* **2021**, *106*, 206–215. [CrossRef]

# Bentonites in Southern Spain. Characterization and Applications

Jorge Luis Costafreda * and Domingo Alfonso Martín

Escuela Técnica Superior de Ingenieros de Minas y Energía, Universidad Politécnica de Madrid, C/Ríos Rosas, 21, 28003 Madrid, Spain; domingoalfonso.martin@upm.es
* Correspondence: jorgeluis.costafreda@upm.es

**Abstract:** The objective of this work was to investigate and demonstrate the pozzolanic properties of the bentonites found at the San José–Los Escullos deposit, located in the southeast of the Iberian Peninsula, to be used in the manufacturing of more durable and environmentally compatible pozzolanic cements, mortars and concretes. These bentonites are mainly composed of smectites, with montmorillonite as the main clay mineral. They were formed by the hydrothermal alteration of tuffs, volcanic glasses, dacites, rhyolites and andesites. For this research, samples were taken from outcrops on the south, north and west side of the San José–Los Escullos deposit, and in the Los Trancos deposit located 19.3 km to the northeast. All samples consisted of bentonites, except for a zeolite sample taken from the northern flank of the San José–Los Escullos deposit, which was used to contrast and compare the behaviour of bentonite in some of the analyses that were done. An investigation of the mineralogical, petrological, chemical and thermogravimetric characteristics of the samples was carried out using various methods, such as XRD, OA (Oriented aggregates), TGA, XRF, SEM and thin section petrography (TSP). In addition, a chemical analysis of pozzolanicity (CAP) was done at 8 and 15 days to determine the pozzolanic capacity of the samples. XRD, XRF, SEM and TSP studies showed that these bentonites have a complex mineralogical constitution, composed mainly of smectites of the montmorillonite variety, as well as halloysite, illite, vermiculite, biotite, muscovite, kaolinite, chlorite, mordenite, feldspar, pyroxene, amphibole, calcite, volcanic glass and quartz. Thermogravimetric analysis (TGA) established the thermal stability of the bentonites studied at above 800 °C. Chemical analysis of pozzolanicity (CAP) confirmed the pozzolanic character of the bentonites, exhibited in their reactive behaviour with $Ca(OH)_2$. The pozzolanic reactivity increased significantly from 8 to 15 days. These results show that the materials studied can be used as quality pozzolans for the manufacture of pozzolanic cements, mortars and concretes.

**Keywords:** bentonite; mordenite; smectite; pozzolanicity test; cement; concrete

**Citation:** Costafreda, J.L.; Martín, D.A. Bentonites in Southern Spain. Characterization and Applications. *Crystals* **2021**, *11*, 706. https://doi.org/10.3390/cryst11060706

Academic Editor: Zhaohui Li

Received: 23 May 2021
Accepted: 17 June 2021
Published: 20 June 2021



## 1. Introduction

This research focused on the characterisation and uses of bentonites from the San José–Los Escullos deposit as pozzolans for the manufacture of pozzolanic cements, mortars and concretes. There are no previous reports on the use of bentonites from this deposit for these specific purposes, so the results presented here are considered new. However, the use of bentonites is broad and diverse, as described below.

Bentonites have been used in many fields for several decades. Due to their chemical, mineral and technical properties, the bentonites from Sardinia (Italy) were used to prepare bentonite pastes to be used in pelotherapy [1], thus substituting traditional peloids. Nowadays, the use of bentonites is so diverse that they are used in many sectors of the economy, science and technology [2]. There is a current trend to thermally activate the bentonites to facilitate the production of geo-polymer binding agents [3]. The heated bentonites quickly acquire pozzolanic activity, as has been proven by Habert et al. [4]. They proved that by heating the bentonites between 20 and 900 °C it is possible to raise the mechanical resistance of the mortar up to 60 MPa in 28 days. Kaci et al. [5] proved the influence of bentonite on the rheologic behaviour of the mortar, especially regarding the fluency and thixotropy, and arrived at the conclusion that bentonite improves the resistance of the mortar. To mitigate the effects of copper mining, Bertagnolli et al. [6] used thermically activated bentonitic clay in the porous beds to absorb the copper. Slamova et al. [7] proved the role of clay in animal food and formulated procedures of application to avoid gastrointestinal disorders in these animals; they highlighted that the clay minerals with a fine granulometry and a high capacity for absorption are the most effective for veterinarian use. Pelayo [8] proved the behaviour of bentonite as an efficient insulator in storing radioactive residue in his research on bentonites from the Morrón de Mateo mineral deposit, in the south east of the Iberian Peninsula. Hassan and Abdel [9] made a mixture out of bentonite–zeolite with a 1:10 ratio to improve the productivity of the crops in sandy and arid soils. Recent research shows great appreciation for products that are absorbent, catalysts and biomaterials [10]. Park et al. [11] list a series of uses for the clay bentonites in drilling mud, whitening products, stabilising emulsions, in the smelting industry, drying products, catalysts, adhesive seals, cosmetics and pharmaceutical products. Panday and Ramontja [12] list in detail the most recent progress in the use of bentonite and its compounds to eliminate contaminating synthetic dyes in water. Luqman and Enobong [13] used bentonite as a base to prepare sol-gel to obtain bioactive glass, significantly lowering the costs generated using old precursors to alkoxysilane. Aravindhraj and Sapna [14] were able to raise the resistance to compression and flection of concrete, mortars and cements, substituting the cement for bentonite in 15%, which also made it exceptionally durable against the effects of sulphate and chlorides. In the process of de-fluoridisation of the waters, Masindi [15] patented a simple, new and innovative method in the synthesis of cryptocrystalline magnesite from bentonites, which notably surpasses the absorption capacity of fluoride in conventional products. Masindi [15] also proved that bentonite that is ground up to the correct size can be used efficiently in the management, neutralisation and remediation of acid waters from mines. The effectiveness of this fact is because the thermic activation of the pores and micropores increases, thus facilitating an ionic exchange between the copper and the interlaminar cations. Kim et al. [16] have recently proposed a numerical model based on the relationship between density and thermal–hydraulic properties of compacted bentonite for sealing high-activity waste tanks. Masood et al. [17] made compound concretes with low calcium and arid bentonite compounds that came from recycled concrete. The bentonite substituted the cement in 5%, 10%, 15% and 20%. The concretes were significantly efficient regarding durability and mechanical resistances; on the other hand, it was highlighted as a good response to the environmental problem. Wu et al. [18] made a mixed compound based on reactive MgO, blast furnace grinded granulated slag, bentonite and clay sand to improve soils. Bentonites are currently being used successfully to make high activity radioactive waste inert, in the form of a paste fabricated from grains, particulates and bentonite dust [19].

## 2. Geological Settings

The study area is in the province of Almería (Spain) between the municipalities of San José de Níjar and Los Escullos (Figure 1). The research took place inside of the Los Frailes caldera, a volcanic structure that was formed 14.4 million years ago in the southeast of the Iberian Peninsula, before the other surrounding calderas were formed, as Rodalquilar and La Lomilla. The Los Frailes caldera is a circular megastructure slightly larger than 5 km in diameter [20,21].

The geology is made up of rocks that were formed during the volcanic processes of the Neogene and are mainly andesites and dacites from the Unidad Frailes 1 (UFR-1), dacites and rhyolites from the Rodalquilar Complex, as well as basaltic andesites and post-caldera pyroxene andesites from the Unidad Frailes 2 (UFR-2) [21–23]. The rocks of a half acid composition, as dacites and rhyolites, make dykes and domes, whereas the most basic lithologies like the andesites, form great lava flows and pyroclastic breccia (Figure 2) [24].

**Figure 1.** Geographic location map.

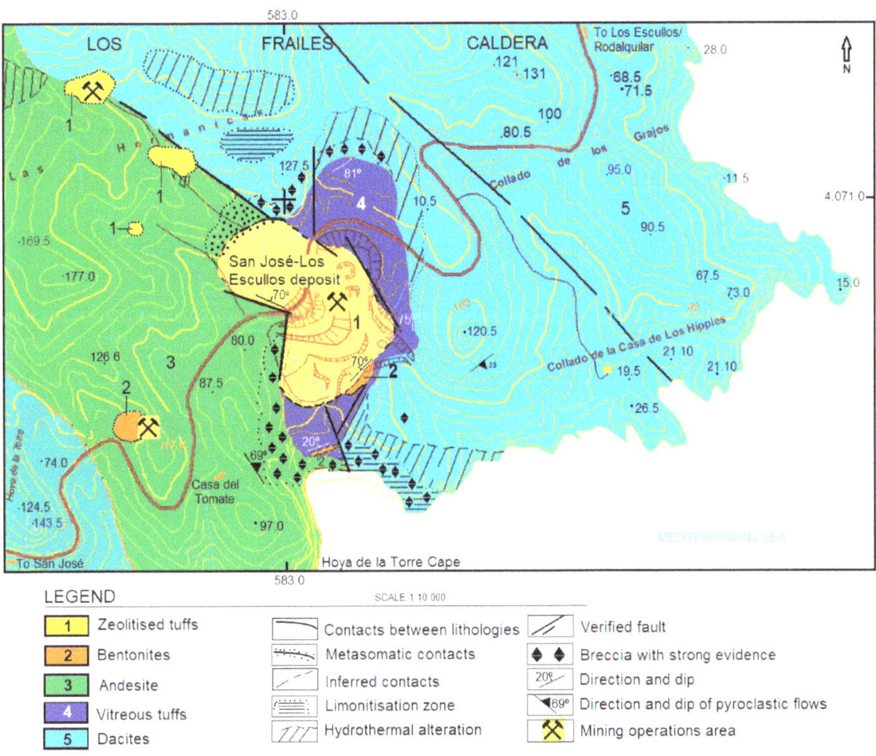



**Figure 2.** Geological map of the research area. Modified from Costafreda [24].

Generally, bentonites are stratified, banded and changed by spheroidal weathering. They are highly friable and mylonitised. Their colour is a characteristic beige (Figure 3). Occasionally they include dacite pyroclastic and tuff materials. The outcrops are cut by vertical and subvertical diaclases with a dip between 70° and 80° towards the west [24].

**Figure 3.** View of two bentonite outcrops in the south (**a**) and west (**b**) of the research area.

The bentonite deposits are well represented in the study area. In UFR-1, the bentonite has been formed by fine lapilli and bentonised ash; these are in Cerro de la Palma and in Barranco de Cala Higuera. In Rodalquilar, the dacite tuffs are bentonised and form industrial deposits in the Rambla del Plomo [8,25–27].

The most representative outcrops of the area of research are in a small area of the southern flank of the San José–Los Escullos deposit, as well as in the western part of the same area (Figure 3a,b).

## 3. Materials and Methods

### 3.1. Materials

The research area is in the east of San José and to the west of Los Escullos. In this place the bentonites form a representative outcrop on the southern side of the San José–Los Escullos deposit. Five samples of bentonite were taken directly from the outcrops for this research and were designated with the codes B-01 through B-05 (Figure 4a).

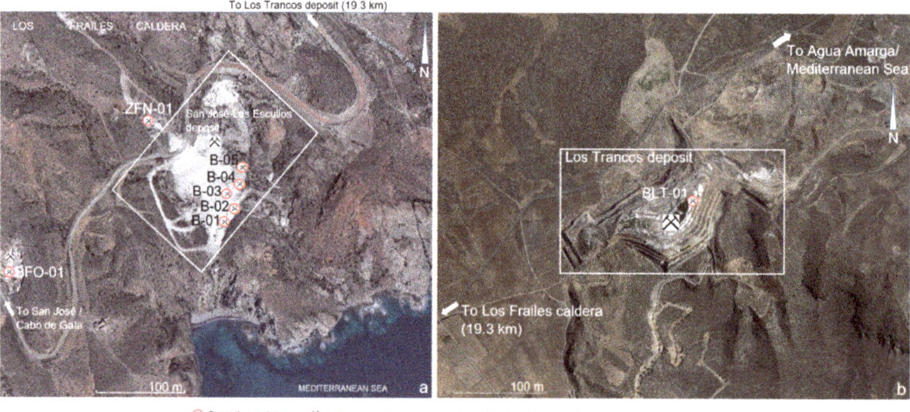


**Figure 4.** Location of sampling points: (**a**) samples taken in the San José–Los Escullos deposit; (**b**) sample taken in the Los Trancos deposit [28].

Another three samples were taken outside of the deposit limits to make comparative studies. These are: BFO-01, ZFN-01 and BLT-01 (Figure 4a,b).

The dacite has a massive, uniform, vesicular and brecciated structure. It has a medium-grained, porphyritic, phaneritic texture. Its colour varies from dark grey to green. It is generally altered by limonite, sericite and chlorite. The dacite is composed of phenocrystals of amphibole, pyroxene and plagioclase, as well as sericite, glass, iron oxide and opaques.

Andesite is dark grey to green in colour, with a brecciated and prismatic-columnar structure. It is composed of plagioclase, pyroxene, glass, sericite and opaques. Its texture is porphyritic. It is strongly altered to zeolite near the fault systems. Plagioclases are sericitised, while pyroxenes are chloritised. The pyroxenes and plagioclases are replaced by zeolites.

Bentonite varies in colour from light grey, green, beige to yellow. It has a stratified structure. The texture is cinereous and devitrified. It is formed mainly by smectite, and montmorillonite the main mineral, as well as altered glass, relict crystals and fragments of rocks altered to smectite, zeolite and serpentine. The bentonite is weathered and fractured in the outcrops.

Zeolite is white, light grey and green in colour. Its structure is massive and compact, becoming brecciated. It is mainly composed of the mordenite, accompanied by smectite, plagioclase, potassium feldspar, quartz, hematite, amorphous materials and traces of gypsum. They are altered in outcrops, forming shallow, residual and porous clay layers. Outcrops are marked by abundant diaclases and fissures. Mordenite appears as elongated, fibrous and tabular crystals associated with montmorillonite, halloysite, kaolinite and micas.

### 3.2. Methods

#### 3.2.1. Petrography Characterisation by Thin Section (PTS)

Petrographic analysis was carried out to describe the properties of the samples analysed: mineral type, texture, colour, grain size, alterations (zeolitisation and bentonisation) as well as petrogenesis. The equipment used was a Leica DM600M microscope (Leica, Weztlar, Germany). It has an adapted system of 13 filters (DTA-RPFMAX) for visible and infrared light at 13 wavelengths, from 350 nm to 1000 nm, in 50 nm intervals. It has a LAS control and a Märzhäuser motorised plate. The image analysis software used was APHELION.

#### 3.2.2. X-ray Diffraction (XRD)

In the sample analyses the Phillips diffractometer 1710 BASED (Amsterdam, Netherlands) was used with the following characteristics: tube anode of Cu, generator tenson (45 Kv), generator current 30 mA, wavelength Alpha 1 [Å] 1.54060, wavelength Alpha 2 [Å] 1.54439, intensity ratio (alpha1/alpha2) 0.500, scan range of software 2theta (4–60°), monochromator, and PC-APD diffraction software.

#### 3.2.3. Oriented Aggregates (OA)

This method was used to determine the clay minerals associated with more crystalline species, as feldspars, quartz and zeolites, present in the researched samples. Some types of clay, such as smectites, chlorites and vermiculites have a d = 14 Å. The test procedure consisted of taking 500 mg of each bentonite sample previously crushed, ground and sieved to 2 μm. Next, the samples were placed in test tubes and mixed with 4–5 drops of sodium hexametaphosphate. They were then allowed to settle. A part of the saturated clay suspension was placed in a glass sample holder for 24 h at room temperature. After evaporation of the water, a thin layer of clay remained on the sample holder, which was treated with ethylene glycol and heated in an oven at 70–80 °C for 24 h. During this time, the vapours acted directly on the sample. The samples were then analysed by XRD and the corresponding patterns were obtained.

#### 3.2.4. Thermogravimetric Analysis (TGA)

This method consisted in heating the samples and a pattern in a controlled manner. During this process, the temperature difference (ΔT) was measured based on time. The thermogravimetric analysis (TGA) measured the variation and behaviour of the mass of the samples under the influence of temperature in a controlled atmosphere. This variation manifested in mass loss. The recording and monitoring of the changes gave information regarding the decomposition of the sample or of its reaction to other components. This test used Universal equipment V4.1DTA Instruments/2960 SDT V3.OF. A total of 13.1971 g was analysed for each sample using the 10C-1100N2, N2 10 mL/min method, and the TGA curves were obtained.

It was important to apply this method to the samples because it was important to know the sorbed phases (adsorbed, absorbed and constituent water), porosity, reactions that occur with increasing temperature, weight loss, variations in cation exchange capacity (CEC), increase in active surface area, thermal stability and pozzolanic capacity. These properties are decisive in the manufacture of pozzolanic cements mixed with natural and activated bentonites, which is highly important for the preservation of the environment and the durability of sustainable building materials.

### 3.2.5. X-ray Fluorescence (XRF)

This research used a Philips PW 1404 (Amsterdam, Netherlands) with a collimator, with a radiation intensity between 10 and 100 kV, and monochrome was used to isolate the radiation. The preparation of the samples consisted of a previous grinding to 200 mesh. Next, 8 g of dust with a 1.5 mL of elbaite were mixed in to make a 5 cm in diameter paste to be analysed by XRF.

### 3.2.6. Scanning Electron Microscopy (SEM)

The following equipment was used to prepare and analyse the samples: Hitachi S-570 Scanning Electron Microscope (Tokyo, Japan) with a Kevex 1728 analyser, the BIORAD Polaron Division Carbon Evaporation Power Supply and the Polaron SEM Coating System. The Winshell and Printerface software were also used. All the samples were reduced to between 0.2 and 0.5 cm; and were then covered by a layer of graphite and placed in the sample carrier for their analysis.

### 3.2.7. Chemical Pozzolanicity Analysis (CPA)

Pozzolanicity is the property of rocks and minerals to actively react with calcium hydroxide at 40 degrees, once crushed, ground up and micronised to around 63 µm. Under these conditions they can acquire cementitious properties and develop mechanical strength over time [29]. The aim of this work is to determine the pozzolanic behaviour of the analysed bentonites and to establish their uses in the manufacture of cements, mortars and concretes with excellent durability and mechanical strength.

Pozzolanicity was determined by comparing the amount of calcium hydroxide that exists in a watery dissolution which contains hydrated cement and pozzolan, with enough calcium hydroxide necessary to obtain a saturated dissolution with the same alkalinity as the other sample. The chemical trial for pozzolanicity was done following standard UNE-EN 196-5:2011 [29] to 8 to 15 days. An amount of 100 mL of distilled water was heated to 40 °C, for 60 s. Twenty grams of sample (bentonite) and cement made up of 75:25 ratio was added. After a week, the solution was filtered. The concentration of the hydroxyl ions [OH⁻] was calculated with the following equation:

$$[OH^-] = \frac{1000 \times 0.1 \times V_3 \times f_2}{50} = 2 \times V_3 \times f_2$$

where:

- $[OH^-]$: is the concentration in hydroxyl ions (mmol/L).
- $V_3$: is the volume of the hydrochloric acid solution (0.1 mol/L).
- $f_2$: is the factor of the hydrochloric acid solution (0.1 mol/L).

The concentration of calcium oxide (CaO) was calculated using the following equation:

$$[CaO] = \frac{1000 \times 0.025 \times V_4 \times f_1}{50} = 2 \times V_4 \times f_1$$

where:

- [CaO]: is the concentration in calcium oxide (mmol/L).
- $V_4$: is the volume of EDTA solution used in the titration.
- $f_1$: is the factor of the EDTA solution.

The pozzolanicity test is positive when the hydroxide concentration of calcium in dissolution is lower than the saturation concentration [29]. The results are seen in the graph, where all samples below the solubility isotherm curve are considered as pozzolans.

## 4. Results and Discussion

### 4.1. Petrographic Study by Thin Section

The analysed bentonites have a texture that is porphydic, relict, hemivitreous, crystal-lolithoclastic, pyroclastic and oriented. The predominant minerals are clays (montmorillonite-type smectite), mordenite, plagioclase, biotite, muscovite, hematite, apart from the glass. Subordinately there is quartz, kaolinite and polygenic lithic fragments of a small diameter. Figure 5a,b (N//) represents a process of strong pyroxene and plagioclase alterations being substituted partially or totally by smectite and mordenite.

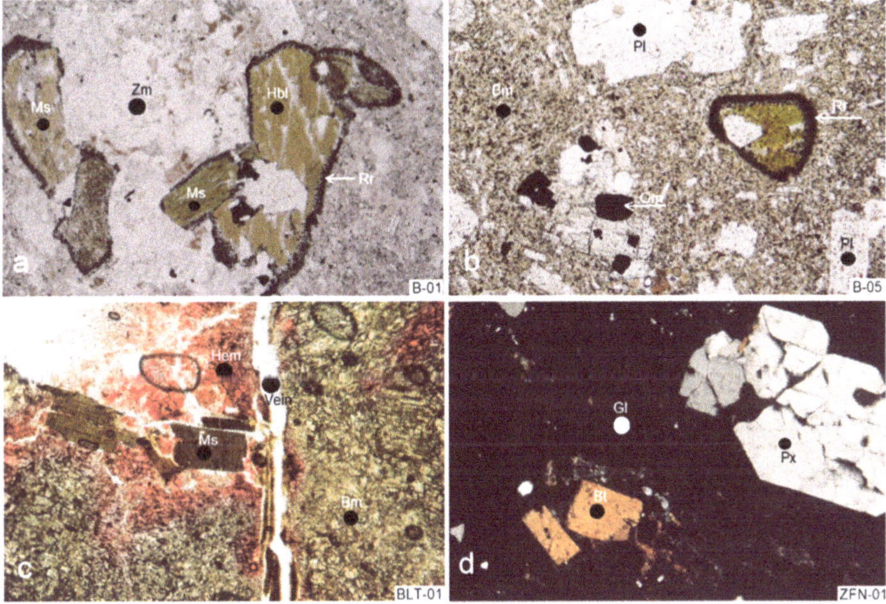


Ms: Muscovite / Hbl: Hornblende / Pl: Plagioclase / Bt: Biotite / Px: Pyroxene / Om: Opaque minerals / Gl: Glass / Bm: Bentonitised matrix / Vein: Carbonated-filled vein
Zm: Matrix altered by zeolitisation processes / Rr: Reation rim between glassy matrix and relict crystals (Hornblende, pyroxene, muscovite, biotite and others)

**Figure 5.** Petrographic thin section micrograph of bentonite samples from San José–Los Escullos deposit. N// (parallel nicols, (**a**,**b**)) and Nx (crossed nicols (**c**,**d**)).

The micrograph of the sample BLT-01 (Figure 5c) (Nx) shows a predominance of smectite and muscovite amid transforming into smectite. The hematite is in small fissures and covers the smectite and the mica. Figure 5d (Nx) shows fractured pyroxene phenocrystal, completely altered to zeolite, floating on a glass fabric. The biotite crystals show a reaction with the fabric and are partially surrounded by halos of hematite in the paste. There is a biotite crystal in process of matrix alteration (Figure 5a,c,d). Three types of matrices can be distinguished: in sample B-01, zeolitised; in B-05, combined (zeolitised–bentonised); in BLT-01, bentonised; and in ZFN-01, glassy, with zeolite formation. The comparison among the samples analysed shows that the bentonisation process was more effective in sample

BLT-01 (Figure 5c), with total development of the clay minerals. In the case of samples B-01, B-05 and ZFN-01, bentonisation did not totally alter the pyroxenes, amphiboles, plagioclase, biotite and muscovite, as these minerals still persist. This fact is observed in Figure 5a–d.

### 4.2. X-ray Diffraction

The phase study showed that the samples analysed have a complex and heterogeneous mineralogy, with the exception of sample BLT-01. The following minerals were detected: montmorillonite-type smectite, illite, vermiculite, halloysite, biotite, muscovite, mordenite, orthoclase, plagioclase, hematite, kaolinite, chlorite, calcite and quartz (Figure 6). According to Figure 6, each sample behaves according to its mineral content. Comparison of X-ray diffraction patterns reveals a clear similarity between samples BLT-01, BFO-01 and B-05, suggesting the presence of similar mineralogical phases. Something very different is observed in samples B-01 to B-03, which shows a greater complexity than the other samples. This is due to the presence of mordenite, plagioclase, muscovite, quartz and kaolinite in greater quantities. The presence of smectite in sample BLT-01 is more evident than in samples BFO-01 and B-05, while in B-01 to B-03, there is a greater abundance of mordenite.

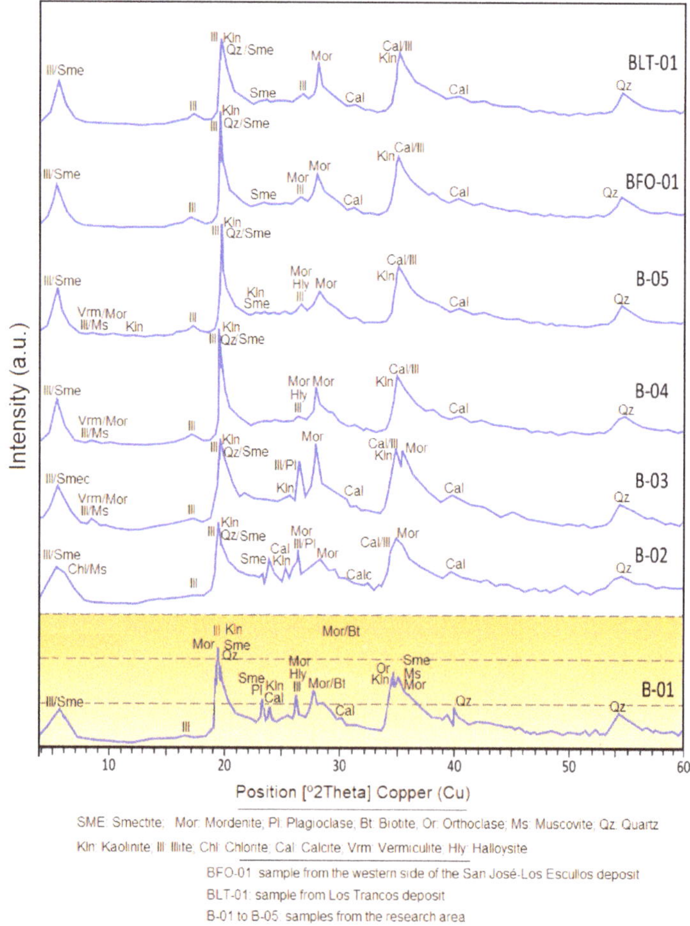


**Figure 6.** X-ray diffraction patterns of bentonite from the research area.

According to the abovementioned and to the position of the samples in the deposit (Figure 4), it can be deduced that the bentonisation process was less intense towards the centre of the San José–Los Escullos deposit. Samples B-01 to B-03 represent another case of typical zonality in this deposit, as they indicate that the zeolitisation process, unlike bentonisation, decreases from the centre towards the flanks.

The presence of orthoclase, kaolinite and quartz could indicate a previous deposition of acidic volcanic material which was altered to form mordenite and smectites. The presence of hematite suggests that hydrothermal activity took place in this region. This interpretation agrees with Costafreda [24].

The bentonisation and zeolitisation process in the south and southeast of Spain was practically ubiquitous, as has been stated by Martín-Vivaldi [30]; Leone et al. [26]; Linares [31]; Delgado and Reyes [32]; García-Romero et al. [33]; and Martínez et al. [34]. Samples BFO-01 and BLT-01 lie to the west and south of the San José–Los Escullos deposit, respectively. However, the chemical composition (Table 1), as well as the mineralogy, indicate that both bentonisation and zeolitisation processes were caused by large-scale hydrothermal alteration of materials of calc-alkaline composition.

**Table 1.** Chemical composition of bentonites from the San José–Los Escullos deposit and surrounding areas (% weight).

| Sample | $S_iO_2$ | $Al_2O_3$ | CaO | $Na_2O$ | $K_2O$ | MgO | $Fe_2O_3$ | $TiO_2$ | LOI * | $Na_2O$/CaO | MgO/$Al_2O_3$ | Si/Al |
|---|---|---|---|---|---|---|---|---|---|---|---|---|
| B-01 | 64.30 | 13.71 | 1.03 | 3.3 | 2.3 | 2.5 | 1.3 | 0.123 | 11.5 | 0.58 | 0.18 | 4.1 |
| B-02 | 51.29 | 13.13 | 1.20 | 1.40 | 0.53 | 7.61 | 2.75 | 0.122 | 22.0 | 1.17 | 0.58 | 3.4 |
| B-03 | 52.53 | 13.69 | 1.11 | 0.99 | 0.70 | 7.67 | 2.43 | 0.133 | 20.5 | 0.89 | 0.56 | 3.3 |
| B-04 | 61.3 | 14.76 | 1.24 | 1.56 | 1.92 | 5.0 | 2.46 | 0.116 | 11.5 | 1.26 | 0.34 | 3.6 |
| B-05 | 62.46 | 14.57 | 0.94 | 2.64 | 2.21 | 4.0 | 1.83 | 0.133 | 11.5 | 2.8 | 0.27 | 3.7 |
| BFO-01 | 52.13 | 17.05 | 1.12 | 1.96 | 0.35 | 5.91 | 1.61 | 0.123 | 19.2 | 1.75 | 0.35 | 2.7 |
| BLT-01 | 47.26 | 19.02 | 1.22 | 0.30 | 0.87 | 4.5 | 2.45 | 0.173 | 24.4 | 0.25 | 0.24 | 2.1 |
| ZFN-01 | 67.04 | 12.55 | 1.54 | 2.64 | 1.79 | 1.40 | 1.17 | 0.085 | 11.7 | 1.75 | 0.11 | 4.7 |

* LOI: Loss on ignition.

### 4.3. Oriented Aggregates

The analysis of the oriented aggregates attenuated most of the more crystalline phases present in the samples, like mordenite, plagioclase, quartz, orthoclase and others; thus, the following clay minerals were clearly observed in the X-ray diffraction patterns: smectite, vermiculite, illite, kaolinite and halloysite (Figure 7).

Figure 7 represents a case where there is an appreciable amount of mordenite in the sample (ZFN-01), taken at the north of the deposit. Initially the peaks of mordenite of greater intensity masked the clay minerals, whereas, once the sample was treated with ethylene glycol, the vermiculite peaks became visible, as well as illite, kaolinite and halloysite. Another detail worth highlighting is that the main peak of smectite in sample (ZFN-01) is initially in position $2\theta = 6.5$ with interplanar spacing equal to 14 Å; once the sample was treated with ethylene glycol it produced a strong peak reflection of the smectite and a displacement from $2\theta = 6.5$ to $2\theta = 5.4$ (Figure 7). In this case, the peak reflection of the smectite has a maximum intensity of 100% and a maximum spacing of 17 Å. This fact proved that the bentonite of the San José–Los Escullos deposit is expansive.

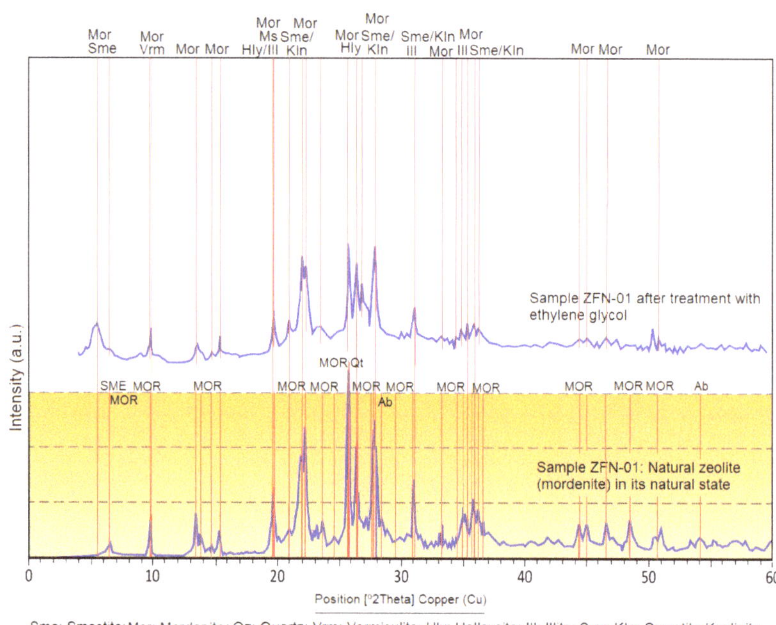



**Figure 7.** X-ray diffraction patterns of mordenite accompanied by bentonite clay minerals before and after ethylene glycol treatment.

### 4.4. Thermogravimetric Analysis

The thermal behaviour of the samples analysed shows similarities and differences, according to the variables in the curves (Figure 8). The curve of sample ZFN-01 shows a higher mass loss compared to the other samples. Like Costafreda [24], the authors have linked this fact to the complex composition of the zeolite, due to its Si/Al ratio, its capacity for cationic exchange and the proportion of smectite–mordenite. The BLT-01 sample is noted because of its mass loss, which is coherent with the data of LOI represented in Table 1. The BFO-01, B-02 and B-03 show similar behaviour which may be related to its low ratio of Si/Al, its high values of LOI and the appreciable content of MgO. The samples B-01 and B-04 behave similarly with regards to their chemical composition, high contents of $SiO_2$ and $Na_2O$ and the lowest values of LOI.

In addition, it is noted that the TGA curves show a simple decomposition in all samples analysed (Figure 8). In all cases there are three significant inflection points that signal different thermogravimetric processes. These changes occur at intervals of temperature 39.2–86.1 °C, 86.1–198.1 °C and 198.1–502.4 °C, which indicate phases of mass loss during heating. Between 39.2 and 86.1 °C the mass loss is 2.362%, and it indicates a loss of humidity and superficial dehydration in the smectite–mordenite phase. Between 86.1 and 198.1 °C, the mass loss is of 3.972%, which points to the continuation of the process of dehydration, loss of superficial humidity and gasses removal. Between 198.1 and 502.4 °C, the curve shows a loss of mass equivalent to 2.291% due to the loss of intrareticular water. At 750 °C the curve stabilises, which could be due to the high Si/Al ratio of the mordenite. At this temperature range, the mordenite collapses and the sample structure is reordered.

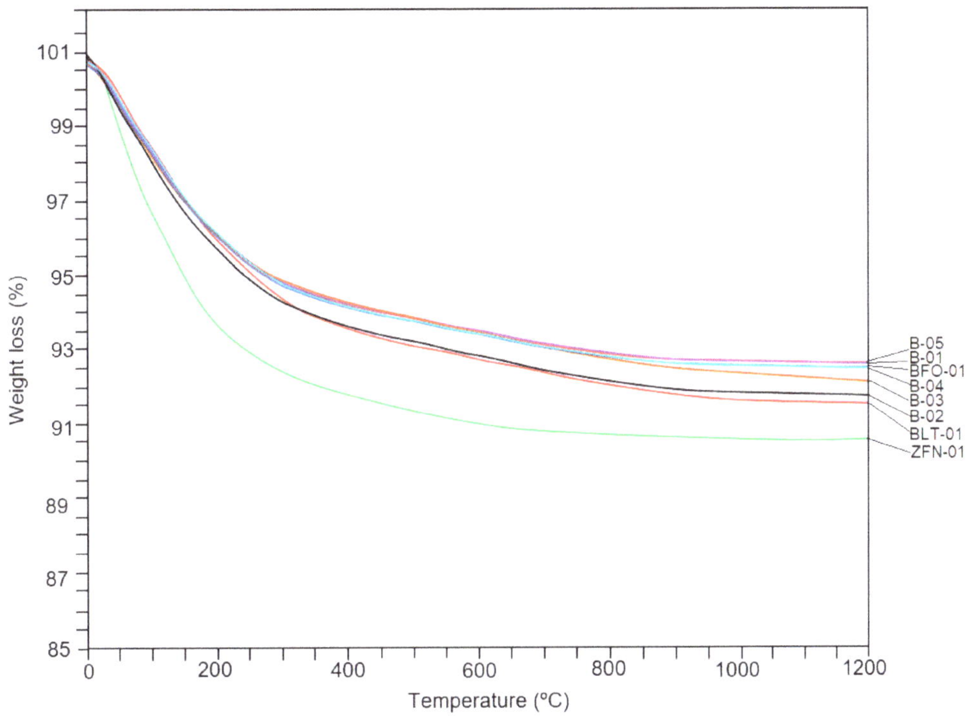


**Figure 8.** TGA thermograms of bentonites from the research and surrounding areas.

### 4.5. X-ray Fluorescence

The results that stem from the analyses indicate a variable composition in the analysed samples (Table 1). Some bentonites are more siliceous (61.3–64.31% $SiO_2$) (e.g., B-01, B-04 and B-05) regarding the less siliceous (51.29–52.53% $SiO_2$) (e.g., B-02, B-03). The high $SiO_2$ contents indicate a marked presence of mordenite in paragenesis with smectite which concurs with what was established by Costafreda [24]; this fact has been proven by comparing the ZFN-01 sample (67.04% $SiO_2$), from the San José–Los Escullos deposit, with BLT-01 (47.26% $SiO_2$), from the Los Trancos deposit. García-Romero et al. [33] calculated the $SiO_2$ (51.25–53.47%) content in the Los Trancos bentonites, and found it to be slightly higher than the values obtained in this research; however, their calculations are similar for the $Al_2O_3$ (17.01–17.75%), the MgO (4.56–4.87%), the $Na_2O$ (0.51–1.07%), the $K_2O$ (0.14–0.43%), the CaO (1.05–1.32%) and the $Fe_2O_3$ (2.21–2.41%) (Table 2). The CaO content in the samples from the San José–Los Escullos deposit is discrete and homogeneous with respect to $Na_2O$ and $K_2O$ (Figure 9a), emphasising its alkaline character. This conclusion coincides with what was established by predecessor researchers [32,35–37].

**Table 2.** Chemical composition (wt% oxides) of bentonites from the Los Trancos deposit, by García-Romero et al. [33].

| Sample | $S_iO_2$ | $Al_2O_3$ | CaO | $Na_2O$ | $K_2O$ | MgO | $Fe_2O_3$ | $TiO_2$ | LOI * |
|---|---|---|---|---|---|---|---|---|---|
| **LTBB** | 52.12 | 17.75 | 1.29 | 0.51 | 0.14 | 4.56 | 2.41 | 0.20 | 20.77 |
| **LTBV** | 51.25 | 17.01 | 1.32 | 0.51 | 0.43 | 4.78 | 2.31 | 0.17 | 20.61 |
| **LTBN** | 53.47 | 17.21 | 1.05 | 1.07 | 0.27 | 4.87 | 2.21 | 0.16 | 19.42 |

* LOI: Loss on ignition.

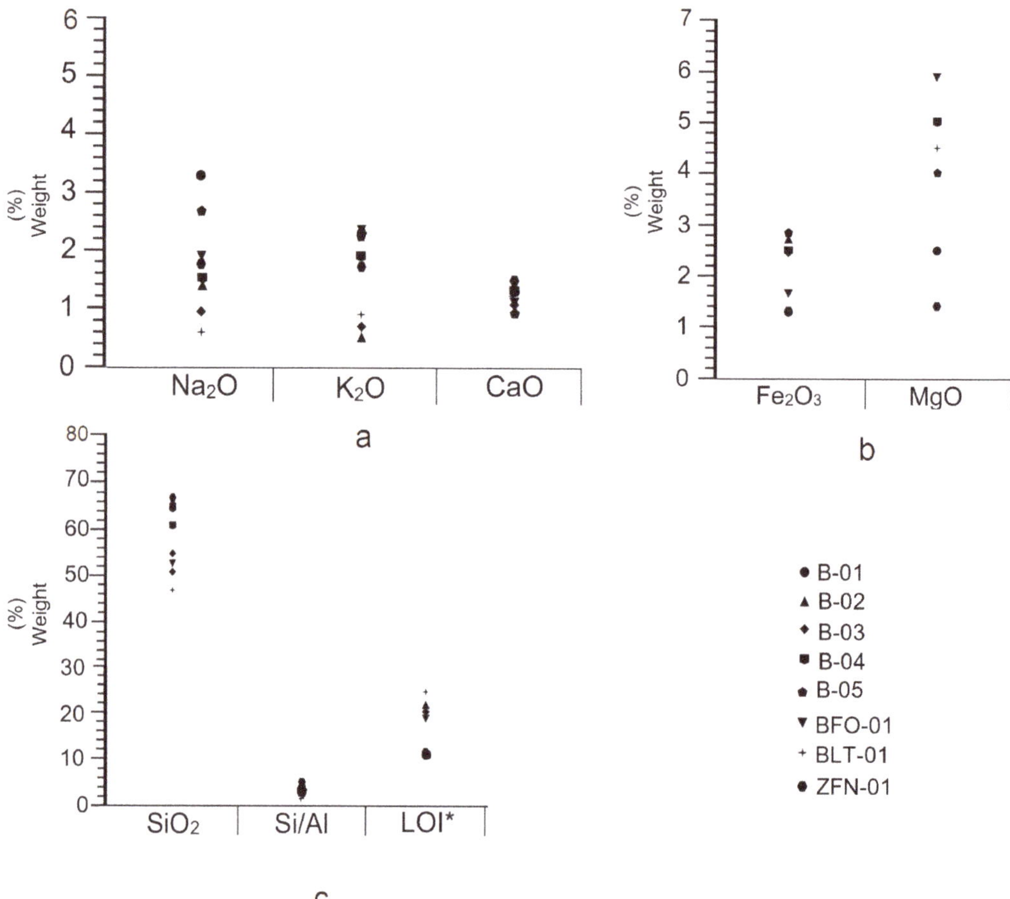



**Figure 9.** (**a–c**). Behaviour of alkaline, alkaline earth and other compounds in bentonites from the San José–Los Escullos deposit and surrounding areas. * Loss on ignition.

The hydrothermal processes displaced the iron ions and concentrated them in the bentonites, producing a marked alteration in the bentonites with typical beige and yellow colours of samples B-01 to 05 and BFO-1 (Figure 3), which differentiates them notably from sample BLT-01. The contents of $Fe_2O_3$ (1.3–2.75%), as seen in Table 1, focus of this type of variation in the southern and western sides of the San José–Los Escullos deposit. The mobilisation of the iron ions could have been helped by the placement of subvolcanic bodies of dacitic composition in the peripheries of the Los Frailes caldera [24]; this reasoning fits well with the description made by Martinez et al. [34] and Perez del Villar et al. [38] regarding a dome sited in the Cala del Tomate. These researchers applied the name rubefaction to refer to the ferruginous alteration that some bentonites possess in the south of the Iberian Peninsula.

The MgO contents are even more of an anomaly (4.0–7.67%) in samples B-01 to B-05 (Table 1 and Figure 9b), possibly provoked by a combined process of removal–precipitation of the $Mg^{2+}$ ions from the amphiboles and pyroxene of the dacites and andesites altered by hydrothermal processes.

There is an inverse relationship among $SiO_2$, loss on ignition (LOI) and Si/Al ratio in the analysed samples (Figure 9c). It was observed that the bentonites richer in $SiO_2$ have lower LOI values and high Si/Al ratios (e.g., B-01, B-04, B-05 and ZFN-01). A different case occurred with the other samples analysed (B-02, B-03, BFO-01, BLT-01) where low $SiO_2$ contents are related to high LOI values and a low Si/Al ratio. The first case characterises bentonites that are not at all expansive or only slightly, while the second case indicates that the bentonites are expansive. All of the above can be summarised as follows: more siliceous bentonites, with higher Si/Al ratios, have lower loss on ignition (LOI) caused by excess silica (Figure 9c); however, when the Si/Al ratio is low, the samples have a high loss on ignition. Therefore, an inverse ratio was observed, as shown in Figure 9c.

### 4.6. Scanning Electron Microscopy

Bentonite sample analysis by way of scanning electron microscopy outlined the presence of various types of minerals. These were: smectite, mordenite, halloysite and pyroxene (Figure 10a–d).

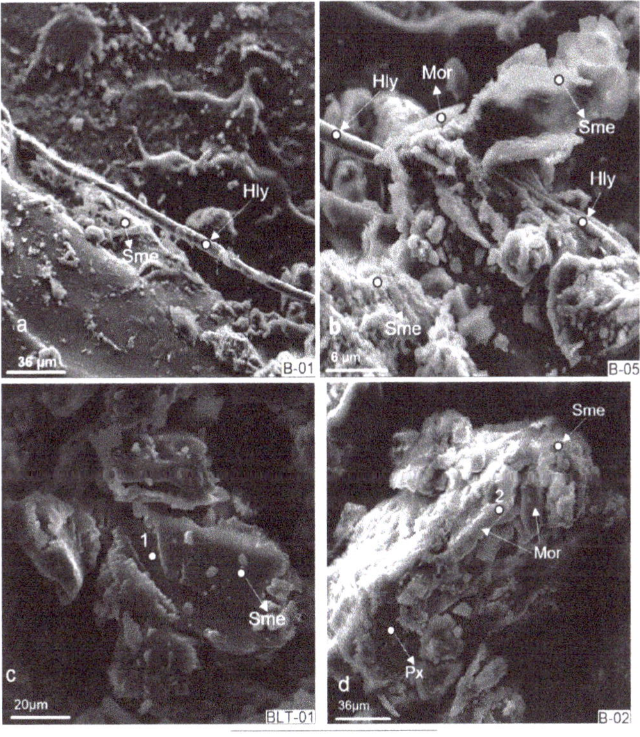


**Figure 10.** Scanning electron microscopy (SEM) micrographs of the bentonite samples (**a–d**).

Smectite in its montmorillonite variety is the most common mineral, as observed in Figure 10a–d. A comparison between the different samples analysed allowed to establish different morphological properties of the mineral phases. For example, in sample B-01 smectite forms small irregular white to light grey masses (Figure 10a). In sample B-05 it forms light grey and white and can be translucent with compact efflorescence (Figure 10b). In sample BLT-01 it appears to form large compact and irregular masses (Figure 10c), and is practically monomineral, which indicates a high purity. In sample B-02 (Figure 10d) it forms compact aggregates together with mordenite. Both smectite and mordenite developed at the expense of relict minerals such as pyroxene and plagioclase, as well as volcanic glass.

The mordenite found in samples B-05 and B-02 forms elongated crystals >36 μm in size, as well as compact crystalline aggregates (Figure 10b,d). Mordenite is syngenetic with smectite. In Figure 10d both mordenite and smectite grow at the expense of an altered pyroxene crystal.

### 4.7. Pozzolanicity Test

According to Figure 11a, all the samples analysed have pozzolanic behaviour at 8 days of testing, as their position is below the solubility isotherm (40 °C), as indicated by UNE-EN 196-5:2011 [29]. Specifically, the zeolite sample (ZFN-01) is the most reactive, followed by B-01 and B-05. The other samples, according to their reactivity with $Ca(OH)_2$, are in the following order: B-04, BFO-01, B-02, B-03 and BLT-01. These data reflect the capacity of the samples to trap the free lime in the solution and the formation of stable reaction products, such as portlandite and tobermorite. The pozzolanic activity is reflected in the decrease of $Ca(OH)_2$ concentration values from 7.9 mmol/L to 4.1 mmol/L, where sample ZFN-01 is the most pozzolanic. Taking this into account, it should be noted that the bentonite samples contribute effectively to the reaction process. The concentration of the bentonite samples under the curve indicates that B-02, B-03, B-04 and BFO-01 behave similarly, which differentiates them from BLT-01, B-01 and B-05. On the other hand, B-01 and B-05 show behaviour that is quite similar.

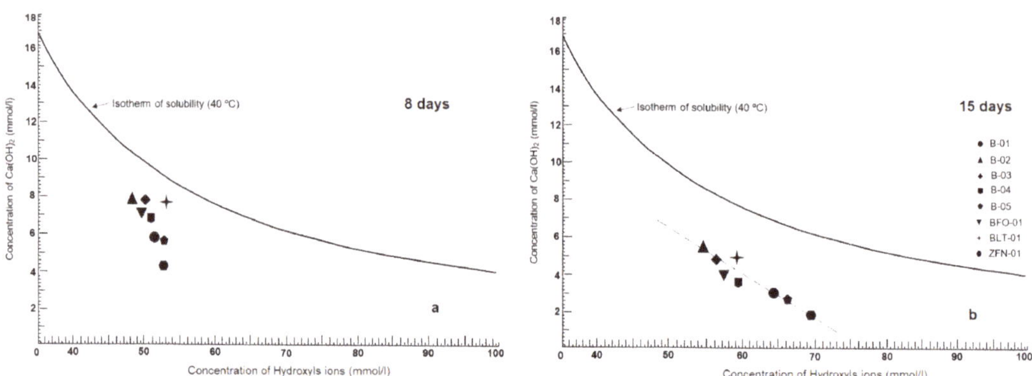


**Figure 11.** Graphs of $Ca(OH)_2$ concentration vs. hydroxyl ions (mmol/L) showing the pozzolanic reactivity at 8 and 15 days (**a,b**).

According to the data provided in Table 1, B-01 and B-05 are the samples with the highest $SiO_2$ and $Al_2O_3$ contents, which is an indispensable condition for a sample to be considered a pozzolan [24]; this is because both $SiO_2$ and $Al_2O_3$ represent the acid phase that causes the pozzolanic reaction in a very alkaline solution [29]. Additionally, the high $Na_2O$ and $K_2O$ contents of these samples also confirm the outstanding pozzolanic properties of both samples. Because of this fact, it shows why ZFN-01 is the most pozzolanic reference sample, from which the hierarchy of the quality of each sample studied has been traced.

In addition, it can be deduced that the bentonites from the San José–Los Escullos deposit show a more efficient pozzolanic behaviour than the sample from the Los Trancos deposit (BLT-01) in the 8 days of testing (Figure 11a).

According to the graph in Figure 11b, for 15 days, the pozzolanic reaction of all samples visibly increased. Sample ZFN-01 is still more pozzolanic, followed by B-05, B-01, B-04, BFO-01, B-03, BLT-01 and B-02. The pozzolanic behaviour of BLT-01 stands out in relation to B-02 and B-03 from 8 to 15 days of testing. The set of samples analysed has caused a decrease in $Ca(OH)_2$ concentration from 4.25 mmol/L to <2 mmol/L at 15 days of analysis.

Another important aspect to highlight in Figure 11b is the linear trend in the behaviour of the samples analysed, which is different from that observed in Figure 11a. This is interpreted as a consequence of the complex mineralogical, petrological and chemical constitution of the samples discussed in Section 4.1 to Section 4.6, which causes different behaviours in the first stages of reaction with $Ca(OH)_2$. One of the consequences of this is the slow initial reactivity. However, the reactivity becomes stronger with time (15 days), when all components of the samples start to react with $Ca(OH)_2$ (Figure 11b). This aspect is important as these are the same mechanisms that provide mechanical strength to cements, mortars and concretes [24].

The remarkable inclination of the trend line towards the abscissa axis (Figure 11b) is highly relevant in this investigation, because it permits the prediction of an uninterrupted development of the pozzolanic reaction over time. This is interpreted as a decisive factor to use the researched bentonites as pozzolans.

The presence of zeolite in almost all bentonite samples, according to the XRD and SEM tests (Figures 6, 7 and 10), is a very important factor contributing to the reinforcement of pozzolanic properties; this is due to several intrinsic factors: porosity, cation exchange capacity, high active surface area and high adsorption capacity.

According to the thermograms in Figure 8, the bentonites studied lose water and cations with increasing temperature, which activates their ion-exchange capacities and their pozzolanic reactivity.

## 5. Conclusions

The bentonites from the San José–Los Escullos deposit have pozzolanic behaviour. The complex mineralogical, petrological and chemical constitution reinforces its pozzolanic capacity. These bentonites are able to react with $Ca(OH)_2$ for a long time, which allows the effective neutralisation of free lime in the reaction system and the formation of secondary reaction products, such as tobermorite, which prevents further reaction with sulphates and chlorides. The use of these bentonites as pozzolans could have several advantages, for example, sustainable manufacture of pozzolanic cements, mortars and concretes with high durability and no negative impact on the environment. On the other hand, the specific use of these bentonites in the manufacture of clinker could prevent the emission of $CO_2$ into the atmosphere, thus avoiding the greenhouse effect. Furthermore, concretes and mortars manufactured with bentonite could prevent the degradation of structures built in coastal environments and control the formation of pollutant flows. Bentonite has good pozzolanic capacity compared to zeolite. It is therefore concluded that bentonite and zeolite from the San José–Los Escullos deposit could be mined together, which would increase the productivity and sustainability of the mining activity.

**Author Contributions:** Conceptualisation, J.L.C. and D.A.M.; methodology, J.L.C. and D.A.M.; software, J.L.C. and D.A.M.; validation, J.L.C. and D.A.M.; formal analysis, J.L.C. and D.A.M.; investigation, J.L.C. and D.A.M.; resources, J.L.C. and D.A.M.; data curation, J.L.C. and D.A.M.; writing—original draft preparation, J.L.C. and D.A.M.; writing—review and editing, J.L.C. and D.A.M.; visualisation, J.L.C. and D.A.M.; supervision, J.L.C. and D.A.M.; project administration, J.L.C. and D.A.M.; funding acquisition, J.L.C. and D.A.M. All authors have read and agreed to the published version of the manuscript.

**Funding:** This work has been partially funded for the Universidad Politécnica de Madrid, through the Projects AL15-PDI-32 and AL16-PDI-11.

**Institutional Review Board Statement:** Not applicable.

**Informed Consent Statement:** Not applicable.

**Data Availability Statement:** Data is contained within the article.

**Acknowledgments:** The authors wish to thank Laboratorio LOEMCO for preparing the samples and conducting chemical pozzolanicity and petrographic tests. J.L.C. and D.A.M. would like to thank the centralised laboratory of the Escuela Técnica Superior de Ingenieros de Minas y Energía (Universidad Politécnica de Madrid) for the OA analysis. The authors would like to thank A.L.S., of the Department of Geological and Mining Engineering, for processing the images. J.L.C. thanks the Universidad Federal de Pernambuco (Brazil) for thermic analysis (TGA).



## References


1.  Cara, S.; Carcangiu, G.; Padalino, G.; Palomba, M.; Tamanini, M. The bentonites in pelotherapy: Thermal properties of clay pastes from Sardinia, Italy. *Appl. Clay Sci.* **2000**, *16*, 125–132. [CrossRef]
2.  Bergaya, F.; Lagaly, G. General introduction: Clays, clay minerals, and clay science. In *Handbook of Clay Science. Developments in Clay Science*; Elsevier: Amsterdam, The Netherlands, 2006; Volume 1, pp. 1–18. [CrossRef]
3.  Buchwald, A.; Hohmann, M.; Posern, K.; Brendler, E. The suitability of thermally activated illite/smectite clay as raw material for geopolymer binders. *Appl. Clay Sci.* **2009**, *46*, 300–304. [CrossRef]
4.  Habert, G.; Choupay, N.; Escadeillas, G.; Guillaume, D.; Montel, J.M. Clay content of argillites: Influence on cement-based mortars. *Appl. Clay Sci.* **2009**, *43*, 322–330. [CrossRef]
5.  Kaci, A.; Chaouche, M.; Andréani, P.-A. Influence of bentonite clay on the rheological behaviour of fresh mortars. *Cem. Concr. Res.* **2011**, *41*, 373–379. [CrossRef]
6.  Bertagnolli, C.; Kleinübing, S.J.; Carlos da Silva, G.M. Preparation and characterization of a Brazilian bentonite clay for removal of copper in porous beds. *Appl. Clay Sci.* **2011**, *53*, 73–79. [CrossRef]
7.  Slamova, R.; Trckova, M.; Vondruskova, H.; Zraly, Z.; Pavlik, I. Clay minerals in animal nutrition. *Appl. Clay Sci.* **2011**, *51*, 395–398. [CrossRef]
8.  Pelayo, M. Estudio del Yacimiento de Bentonita de Morrón de Mateo (Cabo de Gata, Almería) como Análogo Natural del Comportamiento de la Barrera de Arcilla de un Almacenamiento de Residuos Radiactivos. Ph.D. Thesis, Universidad Complutense de Madrid, Madrid, Spain, 2013; 311p.
9.  Hassan, A.Z.A.; Abdel, W.M.M. The combined effect of bentonite and natural zeolite on sandy soil properties and productivity of some crops. *Topclass J. Agric. Res.* **2013**, *1*, 22–28.
10. Zhou, C.H.; Zhao, L.Z.; Wang, A.Q.; Chen, T.H.; He, H.P. Current fundamental and applied research into clay minerals in China. *Appl. Clay Sci.* **2016**, *119*, 3–7. [CrossRef]
11. Park, J.H.; Shin, H.J.; Kim, M.H.; Kim, J.S.; Kang, N.; Lee, J.Y.; Kim, K.T.; Lee, J.I.; Kim, D.D. Application of montmorillonite in bentonite as a pharmaceutical excipient in drug delivery systems. *J. Pharm. Investig.* **2016**, *46*, 363–375. [CrossRef] [PubMed]
12. Pandey, S.; Ramontja, J. Natural bentonite clay and its composites for Dye removal: Current state and future potential. *Am. J. Chem. Appl.* **2016**, *3*, 8–19.
13. Luqman, A.; Enobong, R. Bioactivity of quaternary glass prepared from bentonite clay. *J. Adv. Ceram.* **2016**, *5*, 47–53. [CrossRef]
14. Aravindhraj, M.; Sapna, B.T. Influence of Bentonite in Strength and Durability of High Performance Concrete. *Int. Res. J. Eng. Technol.* **2016**, *3*, 5.
15. Masindi, V. Application of cryptocrystalline magnesite-bentonite Clay hybrid for defluoridation of underground water resources: Implication for point of use treatment. *J. Water Reuse Desalination* **2017**, *7*, 3. [CrossRef]
16. Kim, M.J.; Lee, S.R.; Yoon, S.; Jeon, J.S.; Kim, M.S. Optimal initial condition of a bentonite buffer with regard to thermal behavior in a high-level radioactive waste repository. *Comput. Geotech.* **2018**, *104*, 109–117. [CrossRef]
17. Masood, B.; Elahi, A.; Barbhuiya, S.; Ali, B. Mechanical and durability performance of recycled aggregate concrete incorporating low calcium bentonite. *Constr. Build. Mater.* **2020**, *237*, 117760. [CrossRef]
18. Wu, H.L.; Jin, F.; Zhou, A.; Du, Y.J. The engineering properties and reaction mechanism of MgO-activated slag cement-clayey sand-bentonite (MSB) cut-off wall backfills. *Constr. Build. Mater.* **2021**, *271*, 121890. [CrossRef]
19. Wang, Y.; Zhang, H.; Tan, Y.; Zhu, J. Sealing performance of compacted block joints backfilled with bentonite paste or a particle-powder mixture. *Soils Found.* **2021**, *61*, 496–505. [CrossRef]
20. Cunningham, C.; Arribas, A., Jr.; Rytuba, J.; Arribas, A. Mineralized and unmineralized calderas in Spain. Part I: Evolution of the Los Frailes Caldera. *Mineral. Depos.* **1990**, *25*, 21–28. [CrossRef]
21. Arribas, A. Las Mineralizaciones de Metales Preciosos de la Zona Central del Cabo de Gata (Almería) en el Contexto Metalogénico del Sureste de España. Ph.D. Thesis, Universidad de Salamanca, Salamanca, Spain, 1992; pp. 109–148, 186–237.
22. Rytuba, J.; Arribas, A., Jr.; Cunningham, C.; McKee, E.; Podwysocki, M.; Smith, J.; Kelly, W.; Arribas, A. Mineralized and unmineralized calderas in Spain; Part II. Evolution of the Rodalquilar caldera complex and associates gold-alunite deposits. *Mineral. Deposita* **1990**, *25*, 529–535. [CrossRef]
23. Fernández-Soler, J.M. El Volcanismo Calco-Alcalino de Cabo de Gata (Almería). Ph.D. Thesis, Universidad de Granada, Granada, Spain, 1992; 243p.

24. Costafreda, J.L. Geología, Caracterización y Aplicaciones de las Rocas Zeolíticas del Complejo Volcánico de Cabo de Gata (Almería). Ph.D. Thesis, Universidad Politécnica de Madrid, Madrid, Spain, 2008; 515p.
25. Reyes, E. Mineralogía y Geoquímica de las Bentonitas de la Zona Norte de Cabo de Gata (Almería). Ph.D. Thesis, Universidad de Granada, Granada, Spain, 1977; 650p.
26. Leone, G.; Reyes, E.; Cortecci, G.; Pochini, A.; Linares, J. Genesis of bentonitas from Cabo de Gata, Almería, Spain: A stable isotope study. *Clay Miner.* **1983**, *18*, 227–238. [CrossRef]
27. Pelayo, M.; García-Romero, E.; Labajo, M.A.; Pérez del Villar, L. Occurrence of Fe-Mg-rich smectites and corrensite in the Morrón de Mateo bentonite deposit (Cabo de Gata region, Spain): A natural analogue of the bentonite barrier in a radwaste repository. *Appl. Geochem.* **2011**, *26*, 1153–1168. [CrossRef]
28. Google Earth. Available online: https://earth.google.com/web/@36.78396874,-2.07282152,311.26637664a,8201.3672037d,35y,0h,0t,0r (accessed on 17 May 2021).
29. British Standard Institution. *UNE-EN 196-5:2011. Methods of Testing Cement–Part 5: Pozzolanicity Test for Pozzolanic Cement*; British Standard Institution: London, UK, 2011.
30. Martín-Vivaldi, J.L. The Bentonites of Cabo de Gata (Southeast Spain) and of Guelaya Volcanic Province (North Morocco). *Clays Clay Miner.* **1962**, *11*, 327–357. [CrossRef]
31. Linares, J. Chemical evolutions related to the genesis of hydrothermal smectites, Almería, SE Spain. In *Geochemistry and Mineral Formation in the Earth Surface*; Rodríguez-Clemente, R., Tardy, Y., Eds.; CSIC-CNRS: Madrid, Spain, 1987; pp. 567–584.
32. Delgado, A.; Reyes, E. Isotopic study of the diagenetic and hydrothermal origins of the bentonite deposits at Los Escullos (Almería, Spain). In *Current Research in Geology Applied to Ore Deposits*; Fenoll, P., Torres-Ruiz, J., Gervilla, F., Eds.; Universidad de Granada: Granada, Spain, 1993; pp. 657–678.
33. García-Romero, E.; Manchado, E.M.; Suárez, M.; García-Rivas, J. Spanish Bentonites: A review and new data on their geology, mineralogy, and crystal chemistry. *Minerals* **2019**, *9*, 696. [CrossRef]
34. Martínez, J.A.; Caballero, E.; Jiménez, C.; Linares, J. The effect of a volcanic dome over the Cala del Tomate bentonite (Almería). *Cadernos Lab. Xeolóxico Laxe* **2000**, *25*, 67–69.
35. Caballero, E. Quimismo del Proceso de Bentonitización en la Región Volcánica de Cabo de Gata (Almería). Ph.D. Thesis, Universidad de Granada, Granada, Spain, 1985; 328p.
36. Benito, R.; García-Guinea, J.; Valle-Fuentes, F.J.; Recio, P. Mineralogy, geochemistry and uses of mordenite-bentonite ash-tuff beds of Los Escullos, Almería, Spain. *J. Geochem. Explor.* **1998**, *62*, 229–240. [CrossRef]
37. Huertas, F.J.; Carretero, P.; Delgado, J.; Linares, J.; Samper, J. An experimental study on the ion-exchange behaviour of the smectite of Cabo de Gata (Almería, Spain): FEBEX Bentonite. *J. Colloid Interface Sci.* **2001**, *239*, 409–416. [CrossRef] [PubMed]
38. Pérez del Villar, L.; Delgado, A.; Pelayo, M.; Fernández Soler, J.M.; Tsige, A.M.; Cózar, J.S.; Reyes, A. *Natural Thermal Effects Induced on the Bentonite from de Cala del Tomate Deposit (Cabo de Gata, Almería)*; BARRA II Project, Termal Effect; CIEMAT/DIAE/54450/1/04 Report No. 82; Internal Report of CIEMAT: Madrid, Spain, 2004.

*crystals*

*Article*

# Natural Fluorite from Órgiva Deposit (Spain). A Study of Its Pozzolanic and Mechanical Properties

Domingo A. Martín [1,2], Jorge Luis Costafreda [1,*], Esteban Estévez [2], Leticia Presa [1], Alicia Calvo [1], Ricardo Castedo [1], Miguel Ángel Sanjuán [3], José Luis Parra [1] and Rafael Navarro [4,5]

[1] Escuela Técnica Superior de Ingenieros de Minas y Energía, Universidad Politécnica de Madrid, C/Ríos Rosas, 21, 28003 Madrid, Spain; domingoalfonso.martin@upm.es (D.A.M.); leticia.presa.madrigal@upm.es (L.P.); alicia.calvo.paz@alumnos.upm.es (A.C.); ricardo.castedo@upm.es (R.C.); joseluis.parra@upm.es (J.L.P.)

[2] Laboratorio Oficial Para Ensayos de Materiales de Construcción (LOEMCO), C/Eric Kandell, 1, 28906 Madrid, Spain; eestevez@loemco.com

[3] Department of Science and Technology of Building Materials, Civil Engineering School, Technical University of Madrid, 28040 Madrid, Spain; masanjuan@ieca.es

[4] Minera de Órgiva, S.L., Mina Carriles, Polígono 13, Parcela 1, 18400 Órgiva, Spain; rafa.navarro@mineradeorgiva.com

[5] CHARROCK Research Group, University of Salamanca, Plaza de Los Caídos s/n, 37008 Salamanca, Spain

* Correspondence: jorgeluis.costafreda@upm.es; Tel.: +34-609-642-209

**Abstract:** This work presents the results of the partial substitution of Portland cement (PC) by natural fluorite (NF) and calcined fluorite (CF) in mortars, at 10%, 25% and 40%. To meet these objectives, a sample of fluorite was initially studied by XRD, SEM and Raman Spectroscopy (RS). A chemical quality analysis (CQA) and a chemical pozzolanicity test (CPT) at 8 and 15 days were carried out in a second stage to establish the pozzolanic properties of the investigated sample. Finally, a mechanical compressive strength test (MCST) at 7, 28 and 90 days was carried out on specimens made up with PC/NF and PC/CF mixes, at a ratio of 10%, 25% and 40%. XRD, SEM and RS results indicated fluorite as the major mineralogical phase. The CPT and CQA showed an increase in the pozzolanicity of the samples from 8 to 15 days. The MCST showed an increase in compressive strength from 7 to 90 days for both PC/NF and PC/CF specimens. The results obtained establish that fluorite produces positive effects in the mortar and contributes to the gain of mechanical strength over time, being a suitable material for the manufacture of cements with pozzolanic addition with a reduction of $CO_2$ emissions, and by reducing the energy costs of production.

**Keywords:** fluorite; cement; pozzolan; mortar; mechanical strength; reduction of $CO_2$ emissions

**Citation:** Martín, D.A.; Costafreda, J.L.; Estévez, E.; Presa, L.; Calvo, A.; Castedo, R.; Sanjuán, M.Á.; Parra, J.L.; Navarro, R. Natural Fluorite from Órgiva Deposit (Spain). A Study of Its Pozzolanic and Mechanical Properties. *Crystals* **2021**, *11*, 1367. https://doi.org/10.3390/cryst11111367

Academic Editor: Francesco Capitelli

Received: 24 October 2021
Accepted: 9 November 2021
Published: 10 November 2021

**Publisher's Note:** MDPI stays neutral with regard to jurisdictional claims in published maps and institutional affiliations.

## 1. Introduction

The use of fluorite in the making of cement has been known for several decades. In studies such as those presented by Gilvonio and Dominguez [1] the use of fluorite as a flux in the clinkerization process is mentioned; this significantly reduces the clinkerization temperature and obtains cements with greater mechanical strength, which contributes to the reduction of high energy costs during the production of Portland cement [2]. Fluorite has been used in the manufacture of cement with special characteristics, such as fast setting and high initial strength, as demonstrated by Najafi Kanil and Allahverdi [3]. Dominguez et al. [4] and Endzhievskaya et al. [5] have used natural fluorite in the characterization of the clinker sintering process, concluding that phase transformation and, above all, the presence of alite, occur at lower temperatures than usual when this mineral is incorporated. Other researchers, such as Fridrichova et al. [6], have managed to raise the reactivity of the common included of Portland clinker by the controlled addition of natural fluorite. Additionally, Zeta-Garcia et al. [7] have recently shown that the addition of fluorite in the standard clinker with alite, belite and ye'elimite (ABY) causes an increase in the mechanical strength of mortars and favourably affects the production of "ecological cements". Recently, the use of natural fluorite residues as substitutes for traditional aggregates in concrete has been used in proportions of 5%, 10% and 20%, achieving an improvement in mechanical properties [8]. The works of Zhu et al. [9] mention the use of waste from the mining of serpentine and kaolin by means of mixtures with fluorite slurry, obtaining mechanical strength of up to 150 MPa. Achmania et al. [10] detail the uses of fluorite in other fields, such as steel, aluminium metallurgy, the manufacture of plastics, solvents, refrigerants, welding and glass manufacturing, as well as pointing out the use of this mineral in the cement industry. It is pertinent to mention that the fluorite has been used as an index mineral for locating possible deposits of rare earth (REE) elements and niobium, related to carbonatites, as established by Makin et al. [11].

This research aims to demonstrate the efficacy of natural fluorite from a site in the province of Granada, south of the Iberian Peninsula (Spain). For this, natural and calcinated fluorite have been used to partially replace the Portland cement in the mortars, and to establish two fundamental parameters: the proportion of PC/NF and PC/CF mixtures, as well as the ideal calcination temperature threshold of the samples, in order to achieve optimal compression strength.

This work has been structured and developed in two phases: first, natural fluorite has been characterized by XRD, SEM and Raman Spectroscopy (RS) to determine the constitution and morphology of the mineralogical phases present in the sample; secondly, a study was carried out to establish technological quality through calcination tests, chemical analysis, pozzolanicity and compression strength at different ages.

This work is considered novel for several reasons, such as obtaining resistant mortars, the possible manufacture of cements with more environmentally friendly pozzolanic additions, the inclusion of fluorite and its residues within the circular economy cycle and the revitalization of the local industry.

## 2. Materials and Methods

### 2.1. Materials

The materials used in this research consist of natural fluorite, Portland cement and standardized sand. Natural fluorite comes from the Órgiva deposit, located in the Province of Granada, in the south of the Iberian Peninsula. The Órgiva Mining Company exploits the deposit. A volumetric sample of fluorite of 50 kg was taken from the La Candelaria mine, by lithogeochemical fragment sampling.

A Portland cement (PC) type 1, 42.5 N, of normal resistance, with the characteristics and parameters indicated in the standard UNE-EN 197-1:2011 [12], was used in this research.

A standardized sand (NS) of Normsand-CEN EN 196-1 typology, consisting of 98% silica, was used in this work as a fine aggregate.

### 2.2. Methods

#### 2.2.1. Preparation of the Sample

The sample was crushed in two phases. In the first, a reduction in the size of the particles was made up to 3 cm in diameter, with the help of a crusher model Alas. In the second phase, the size of the particles was reduced to 1 cm, by using the Controls brand crusher. The sample was then ground in a Siebtechnick mill model Scheibenschwingmühle TS 1000 in order to obtain two different particle sizes or blaine fineness (BPF): 2000 cm$^2$/g and 5000 cm$^2$/g, respectively. To obtain the BPF of 2000 cm$^2$/g the sample was ground for 50 s, whereas for the fraction 5000 cm$^2$/g the time was 30 s.

#### 2.2.2. Calcination Test

Two subsamples, CF-01 and CF-02, representative of both particle sizes 2000 cm$^2$/g and 5000 cm$^2$/g, were calcinated. The calcination process was carried out at a temperature of 900 °C, for one hour, with the use of a Binder stove, model 9010-0101 ED240; Serial number: 206493; Measuring range and scale division: 1300 °C/1 °C; Electrical power and connection voltage: 100–240 V; Measuring accuracy: +/−3 °C; Made in Germany. After the heating treatment, the samples were crushed again for 40 s, following the procedure described in Section 2.2.1, to remove the compaction produced during the calcination process. All the materials and instruments used in this research belong to the laboratories of the Escuela Técnica Superior de Ingenieros de Minas y Energía and the Laboratorio Oficial para Ensayos de Materiales de Construcción (LOEMCO), both of the Universidad Politécnica de Madrid, Spain.

### 2.2.3. X-ray Diffraction Analysis (XRD)

A study to determine the mineralogical phases present in the natural fluorite sample was performed by X-ray diffraction (XRD) (All the materials and instruments used in this research belong to the laboratories of the Escuela Técnica Superior de Ingenieros de Minas y Energía and the Laboratorio Oficial para Ensayos de Materiales de Construcción (LOEMCO), both of the Universidad Politécnica de Madrid, Madrid, Spain). The equipment used consisted of a Rigaku Miniflex-600, with the capacity to carry out qualitative and quantitative tests. It is equipped with an X-ray tube operating at 600 w, plus a graphite monochromator and a standard scintillation counter. The diffractometer contains a six-position automatic sampler, and has a HyPix-400 MF-2DHPAD, as well as a ShapeFlex Brand sample stand. The software used was the SmartLab Studio II. This equipment has an interface for profile views, data view, 3D views, and view of the crystal structure of the sample. The power used in the analysis process is 1ø, 100–240 V and 50/60 Hz. About 500 mg for each sample were weighed, ground up and screened to a fineness of 74 μm. The standard tablets were then manufactured in their respective moulds and then placed on the sample holder for analysis.

### 2.2.4. Scanning Electron Microscopy (SEM)

A Hitachi S-570 scanning electron microscope from the Laboratorio Centralizado of the Escuela Técnica Superior de Ingenieros de Minas y Energía of the Universidad Politécnica de Madrid was used in the characterization of the morphological properties of the investigated sample.

The equipment has a Kevex-1728 analyser, a Polaron BIORAD, a power supply for evaporation and a Polaron SEM coating system. The equipment attains a resolution of $200 \times 10^3$. Other components are part of the Scanning Electron microscope such as: a lithium-doped silicon, a liquid nitrogen reservoir, an electronic cannon, a filament chamber, control panel for placing the sample in the high vacuum chamber and for varying the angles of the sample position, and an interface module to display the image during electronic scanning of samples.

The equipment operates with two pieces of software, Winshell and Printerface, to handle the information obtained during the study of the analysed sample and to take microphotographs, respectively. To perform the analysis by SEM, the samples in their natural state were previously reduced to a particle diameter of 0.2–0.5 cm; they were then placed on a graphite adhesive tape and later on the sample holder. Next, samples were covered with a layer of vacuum graphite and then placed in the sample holder of the scanning electron microscope.

### 2.2.5. Raman Spectroscopy (RS)

Raman Spectroscopy is a non-destructive analytical technique that is based on the analysis of scattered light and molecular vibrations produced by the incidence of monochromatic light on the surface of the sample, which yields information on the chemical and structural composition.

The sample was analysed using a Spectrometer brand Gemmo Raman-532, Class 1, Laser product, Reference No. 18-Raman/PL Broad Scan with spectral artifacts-Sinhalite, Brownish Yellow. The equipment also has a laser cannon for the concentration of displaced photons; a filter to separate elastic scattered light from inelastic light, a monochromator to separate the wavelengths of inelastically scattered light and a detector that collects and digitizes the signal of the different wavelengths.

### 2.2.6. Chemical Pozzolanicity Test (CPT)

Pozzolanicity is the property of certain materials to react with $Ca(OH)_2$ in solution when these materials have a very fine particle size, usually below 50 microns [13]. Pozzolans can be of natural origin [14] as well as artificial [15]. Two samples of fluorite, one in its natural state (NF) and the other calcinated (CF), were analysed by CPT to determine their pozzolanic properties, following the indications of the standard UNE-EN 196-5:2011 [13] at 8 and 15 days. A portion equivalent to 100 mL of demineralized water was heated to 40 °C for 60 s; then 20 g of a mixture of PC/NF and PC/CF, respectively, were added to this solution. Once the 8 days had elapsed, the solutions were filtered. This procedure was also repeated for a period of 15 days. The concentration of hydroxyl ions [$OH^-$] was calculated using the following equation:

$$[OH^-] = \frac{1000 \times 0.1 \times V_3 \times f_2}{50} = 2 \times V_3 \times f_2 \tag{1}$$

where:

-     [$OH^-$]: is the concentration in hydroxyl ions (mmol/L).
-     $V_3$: is the volume of the hydrochloric acid solution (0.1 mol/L).
-     $f_2$: is the factor of the hydrochloric acid solution (0.1 mol/L).

The concentration of calcium oxide (CaO) was calculated with the following equation:

$$[CaO] = \frac{1000 \times 0.025 \times V_4 \times f_1}{50} = 2 \times V_4 \times f_1 \tag{2}$$

where:

-     [CaO]: is the concentration in calcium oxide (mmol/L).
-     $V_4$: is the volume of EDTA solution used in the titration.
-     $f_1$: is the factor of the EDTA solution.

The pozzolanicity test is positive when the hydroxide concentration of calcium in dissolution is lower than the saturation concentration [13]. The results are seen in the graph, where all samples below the solubility isotherm curve are considered as pozzolan.

### 2.2.7. Chemical Quality Analysis (CQA)

A chemical quality analysis (CQA) was carried out to determine the main major compounds of the natural fluorite (NF) sample analysed in this research; in addition, to determine its quality as a natural aggregate of pozzolanic cement. The different stages of the test were developed following the parameters of the standard UNE-EN 196-2-2014 [16]. The main compounds determined were the following: Total $SiO_2$ (TS), reactive $SiO_2$ (RS), total CaO (TC), reactive calcium (RC), MgO, $Al_2O_3$ and $Fe_2O_3$. In addition, the insoluble residue (I.R.) was calculated in a standardized solution of hydrochloric acid [16].

### 2.2.8. Mechanical Strength Test (MST)

This method was carried out to determine the compression strength of specimens cured at 7, 28 and 90 days, using the procedures indicated in the standard UNE-EN 196-1 [17]. The specimens were developed with standardized mixtures of PC/NF and PC/CF. The percentage of PC replacement by NF and CF, respectively, was 10, 25 and 40%. In addition in the manufacture of the specimens, two blaine particles fineness (BPF) were taken into account: 2000 and 5000 $cm^2$/g, for both NF and CF. A total of 12 samples of specimens were prepared for the mechanical strength test, with three PC/NF-CF replacement ratios (10, 25 and 40%) and two types of BPF (2000 and 5000 $cm^2$/g). All data referring to the preparation and proportion of the components of the specimens are presented in Table 1.

**Table 1.** Proportions of mixtures of PC, NF, CF and NS used in the preparation of the specimens.

| Sample | Proportion (Ratios) | | | | Temperature of Calcination (°C) | Blaine Particle Fineness (BPF) for NF and CF (cm²/g) |
| | PC [1]:NF [2] (%) | PC:CF [3] (%) | NS [4] (g) | DW [5] (g) | | |
|---|---|---|---|---|---|---|
| PC/NF-01 | 90:10 | | | | | |
| PC/NF-02 | 75:25 | | | | | 2000 |
| PC/NF-03 | 60:40 | - | 1350 | 225 | - | |
| PC/NF-04 | 90:10 | | | | | |
| PC/NF-05 | 75:25 | | | | | 5000 |
| PC/NF-06 | 60:40 | | | | | |
| PC/CF-07 | | 90:10 | | | | |
| PC/CF-08 | | 75:25 | | | | 2000 |
| PC/CF-09 | - | 60:40 | 1350 | 225 | 900 | |
| PC/CF-10 | | 90:10 | | | | |
| PC/CF-11 | | 75:25 | | | | 5000 |
| PC/CF-12 | | 60:40 | | | | |

[1] Portlan cement; [2] Natural fluorite; [3] Calcined fluorite; [4] Normalised sand; [5] Distilled water.

## 3. Results and Discussion

### 3.1. X-ray Diffraction (XRD)

The study of the mineralogical phases determined that the analysed sample is composed of a main phase of fluorite, of up to 98.3%; it also has subordinate secondary phases of quartz and calcite, indicating that the ore is practically monomineral (Figure 1). The study detected the three main peaks of fluorite at the following $2\theta$ angular positions: $2\theta = 28.2508$/Intensity Cps = 50.766 (peak 5); $2\theta = 46.9905$/Intensity Cps = 47.552 (peak 17); $2\theta = 55.7479$/Intensity Cps = 14.129 (peak 19) and $2\theta = 26.61$/Intensity Cps = 235 (peak 4) (Figure 1).

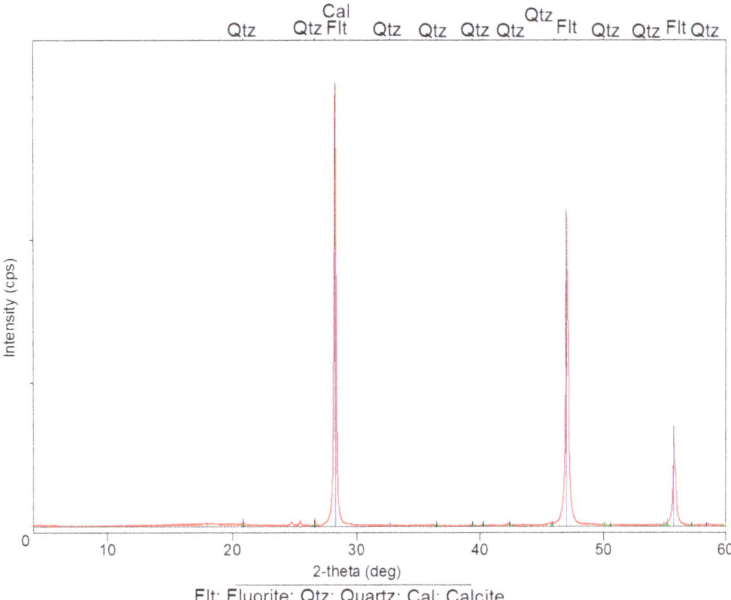


**Figure 1.** X-ray diffraction patterns of the natural fluorite from the research area.

The peaks of quartz and calcite have low intensity and are confined to the background of the X-ray diffraction patterns. The major presence of fluorite in the investigated sample has been confirmed by the analyses of SEM and Raman Spectroscopy (RS) in Sections 3.2 and 3.3.

### 3.2. Scanning Electron Microscopy (SEM)

The analysis of the sample under study shows a majority of fluorite, as shown in Figures 2a–d and 3a–d. In Figure 2a–d the fluorite crystals are large and they usually have a typical orientation that provides an apparent banding. This orientation was possibly caused by tectonics. The direction of accretion and orientation of the fluorite crystals are indicated by arrows in Figure 2a,b.

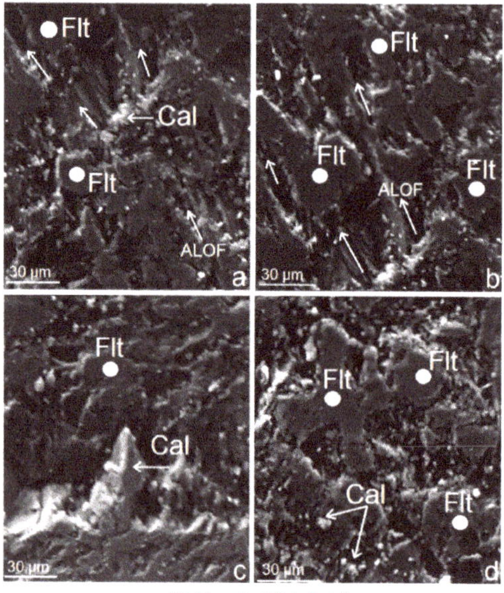


Flt: Fluorite / Cal: Calcite
ALOF: Accretion lines and orientation of fluorite crystals

**Figure 2.** Microphotographs (**a–d**) of the natural fluorite (NF) sample investigated, showing typical banding.

The crystals are enlarged and show an anomalous increase in size, possibly by recrystallization. They form compact crystalline aggregates of zebra appearance that are reflected in the sample as list or bands. Sometimes these aggregates are in the form of impregnations and granular compact masses.

According to the Figures 2a–d and 3a–d, it seems that fluorite comes from two origins: a primary one, prior to tectonic deformation processes (Figure 2a–d), and a postectonic one, with the formation of new generation fluorite, in which the crystals are well preserved (Figure 3a–d).

Calcite forms small aggregates of rhombohedral shapes in empty spaces, as well as druses that develop greater in fractures (Figure 2c). Calcite is very scarce in the samples analysed.

The microphotographs, in Figure 3a–d, show compact crystalline aggregates of fluorite without deformation. Its origin may be related to hydrothermal solutions caused by autometamorphism [18]. All the crystals grow and are arranged in a very tight way, without spaces or pores, it is common to see interpenetrations and minerals twins. Figure 3a–d shows poorly developed calcite crystals.

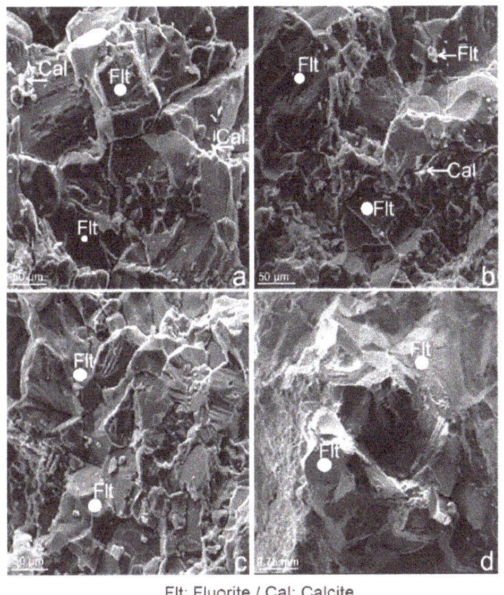



Flt: Fluorite / Cal: Calcite

**Figure 3.** Microphotographs (**a–d**) of the investigated natural fluorite sample without banding.

### 3.3. Raman Spectroscopy (RS)

Figure 4 indicates the presence of virtually pure natural fluorite (NF) in the Raman fingerprint zone to the left of the diagram. The intensity of the peak of the NF exceeds 5500 counts, while the wavelength is 541 nm. The application of the Raman spectroscopy method in this research confirms the results obtained by XRD and SEM, Section 3.1 and Section 3.2, respectively, where the fluorite phase was determined to be the major presence; thus, it was established that the methods used are adequate and complement each other.

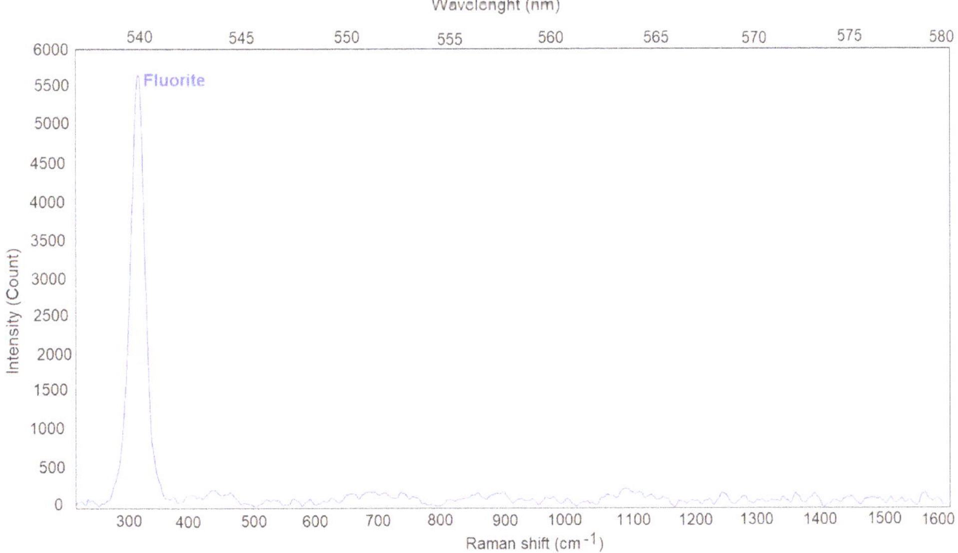



**Figure 4.** Results obtained from the analysis of the NF sample by Raman Spectroscopy.

### 3.4. Pozzolanicity Test (CPT) at 8 and 15 Days

Figure 5a,b demonstrates the pozzolanic character of the natural fluorite (NF) and calcinated (CF) samples investigated. The 8-day test (Figure 5a) fixes both samples immediately below the isothermal solubility curve at 40 °C, indicating an adequate pozzolanic reaction.

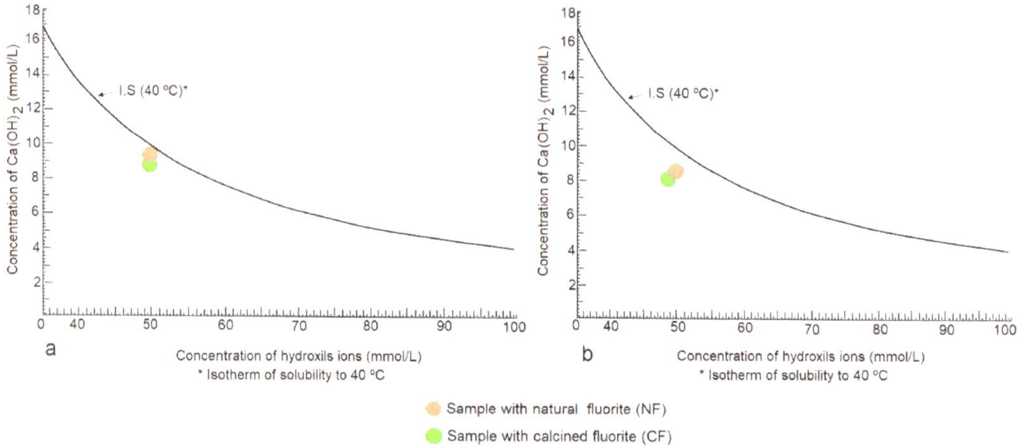


**Figure 5.** Evolution of the pozzolanic behaviour of the NF and CF samples analysed: (**a**) at 8 days; (**b**) at 15 days.

After 15 days of testing, both samples varied their position under the isotherm, which is interpreted as an increase in pozzolanic reactivity over time. It follows that the factors that caused this trend are the percentages of reactive $SiO_2$ and CaO present in the composition of both samples, as shown in Table 2.

**Table 2.** Chemical composition of the NF and CF samples investigated.

| Compounds | Samples | |
|---|---|---|
| | Natural Fluorite (NF) (%) | Calcined Fluorite (CF) (%) |
| Total $SiO_2$ | 53.28 | 56.70 |
| Reactive $SiO_2$ | 51.36 | 55.75 |
| MgO | 0.02 | 0.06 |
| Total CaO | 30.34 | 28.66 |
| Reactive CaO | 28.58 | 28.40 |
| $Fe_2O_3$ | 0.10 | 0.15 |
| $Al_2O_3$ | 0.23 | 0.15 |
| I.R. * | 3.85 | 2.84 |
| $SiO_2/(CaO + MgO)$ | 1.86 | 1.99 |

* Insoluble residue.

Another factor that can determine the pozzolanic capacity of the researched fluorite is its electrical conductivity (EC); in this sense, Liu et al. [19] have determined that the EC is dominated by the fluorine ion, reaching ~0.01 S/m at 650 °C and is potentially important as a carrier of electric charge. Costafreda et al. [20] have demonstrated a similar situation in their study on the electrical conductivity of altered volcanic tuffs. In the works of Rosell et al. [21] it is established that by means of electrical conductivity the index of pozzolanic activity of the materials can be rigorously evaluated.

### 3.5. Chemical Quality Analysis (CQA)

Table 2 shows the result of the chemical analysis of two fluorite samples, one in the natural state (NF), which has already been described in Sections 3.1–3.3, and the other calcinated (CF). Virtually all silica in the NF sample is reactive (51.36%). A similar case occurs with reactive CaO (28.38%). Compounds such as MgO, $Fe_2O_3$ and $Al_2O_3$ are exceptionally low. The contents of I.R. are moderate with a tendency to reach the limit indicated by the standard UNE EN 196-2:2014 [16] and UNE-EN 197-1:2011 [12].

However, the analysis of the CF sample indicates an increase in reactive $SiO_2$ and reactive CaO of 55.70% and 28.40%, respectively; a similar case has been demonstrated in the work of Pei et al. [22]. In addition, the I.R. is comparatively lower in the CF, which is caused by the influence of the increase in temperature and the elimination of the most unstable residues [16].

The percentages of reactive $SiO_2$ and reactive CaO indicated in Table 2 confirm the pozzolanic character of both samples, both calcinated and in their natural state, which is reinforced by what was discussed in Sections 3.4 and 3.6.

### 3.6. Mechanical Strength Test (MST)

The results shown in Figure 6a,b and Figure 7a,b indicate an evident increase in mechanical compressive strength, both for specimens made with PC/NF and those with PC/CF, over the investigation period between 7 and 90 days of curing. Note how the values of compressive strength vary depending on two fundamental factors: the BPF of the NF and CF particles (2000 and 5000 $cm^2/g$) and the degree of substitution of the PC by natural and calcinated fluorite. First, it considers the specific case of the PC/NF-01 to 06 sample series; within this series, the PC/NF-01 to 03 samples replace the PC in the specimens in 10, 25 and 40% with the BPF of the NF particles equal to 2000 $cm^2/g$. The PC/NF-04 to 06 series has the same PC/NF ratios as the above-mentioned series, but the NF BPF is 5000 $cm^2/g$. A substantial increase in mechanical strength is observed in specimens containing 10% NF and BPF 2000 $cm^2/g$, and this increase is even more noticeable in specimens made with BPF of NF equal to 5000 $cm^2/g$, as is the case with the PC/NF-04 to 06 specimen series. According to these results, it can be established that the mechanical strength increases as the portion of NF in the specimen decreases, and when the BPF is reduced from 2000 $cm^2/g$ to 5000 $cm^2/g$ [23].

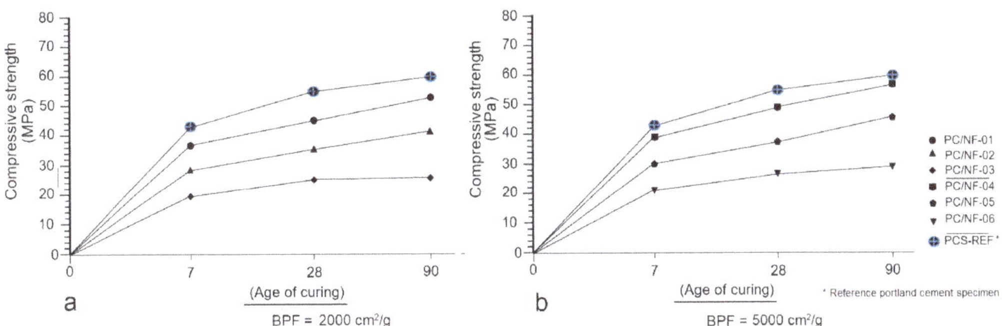


**Figure 6.** Behaviour of the mechanical strength of specimens elaborated with substitution of PC by NF in 10, 25 and 40%. In (**a**) the specimens have BPF of fluorite of 2000 $cm^2/g$, while in (**b**) the BPF is 5000 $cm^2/g$.

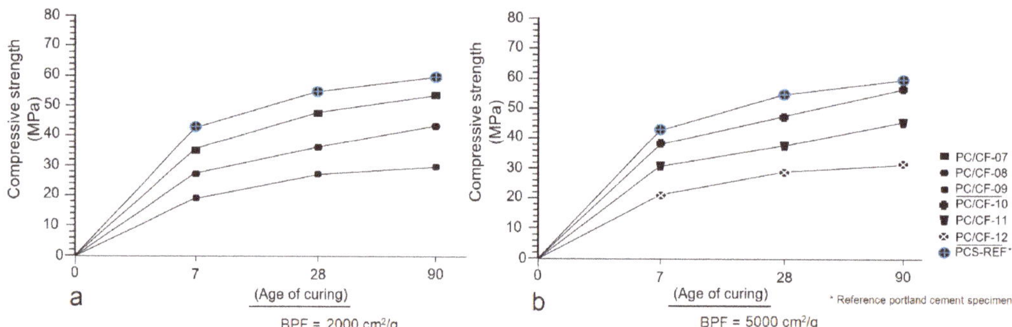



**Figure 7.** Mechanical strengths of specimens elaborated with replacement of PC by CF in 10, 25 and 40%. The BPF of CF in (**a**) is 2000 cm$^2$/g, while in (**b**) it is 5000 cm$^2$/g.

In the specimens made with CF the trend seems to be similar to that seen in Figure 6a,b. An evolution of the compressive strength is observed in the series of specimens PC/CF-07 to 09 and PC/CF-10 to 12, where the highest values correspond to the specimens with the lowest amount of CF and with the lowest BPF (Figure 7a). The analysis carried out in the groups of samples PC/NF-01 to 06 and PC/CF-07 to 12 confirms an increase in mechanical strength from 7 to 90 days in the specimens made with CF and with B.P.F of 2000 cm$^2$/g, with respect to the NF. Similarly, this fact is seen in specimens where the BPF is 5000 cm$^2$/g (Figure 7b).

The comparison of these results with those of the previous subsection shows the relationship that exists between pozzolanicity and mechanical strength [24]. According to Figure 6a,b and Figure 7a,b, in no case were the compressive strength values of the reference Portland cement (PCS-REF) exceeded.

## 4. Conclusions

The researched fluorite sample has a high degree of purity, according to the results of the XRD, SEM and RS analyses.

Both NF and CF have contents of reactive compounds, such as reactive SiO$_2$ and reactive CaO, which cause the pozzolanic reactivity of the studied sample; therefore, this is considered a solid argument to establish the pozzolanic character of the fluorite of the Órgiva deposit.

The mechanical strengths obtained are adequate, and this fact has guaranteed that the tested specimens have reached good compressive strengths at 28 and 90 days. However, an increase in resistance can be forecast beyond 90 days. With a view to large-scale industrial processes, the PC/NF ratio of 10% would be the most effective for the manufacture of pozzolanic cements; even the PC/NF ratio of 25% could also be used, which would lead to a significant decrease in production costs. Depending on the economic solvency of the local company, the PC/NF ratio of 40% could be used as a last resort. In the case of calcinated fluorites, the best PC/CF ratio, 10%, would be the most suitable, followed by a PC/CF of 25%. However, since the difference between the compressive strengths of the PC/NF and PC/CF specimens is not very noticeable, it would be recommended to not initially opt for the large-scale fluorite calcination method, due to the high costs that this represents. In this sense, it should also be considered that excessive grinding in the industrial process, with, for example, a BPF equal to 5000 cm$^2$/g, would increase costs significantly.

The results obtained in this research could serve as a guide for the local Mining Company Órgiva, which would have an impact of added value for its fluorite deposits.

**Author Contributions:** Conceptualization, D.A.M., J.L.C., L.P. and A.C.; methodology, D.A.M., J.L.C. and L.P.; software, D.A.M., J.L.C., L.P., A.C. and R.C.; validation, D.A.M., J.L.C., M.Á.S., J.L.P. and R.N.; formal analysis, D.A.M., J.L.C., E.E., L.P. and A.C.; investigation, D.A.M., J.L.C., E.E., L.P., A.C., R.C., M.Á.S., J.L.P. and R.N.; resources, D.A.M., J.L.C., E.E., L.P., R.C. and R.N.; data curation, D.A.M., J.L.C., L.P., A.C. and R.N.; writing original draft preparation, D.A.M. and J.L.C.; writing review and editing, D.A.M., J.L.C., L.P., R.C., M.Á.S. and J.L.P.; visualization, D.A.M., J.L.C., L.P. and A.C.; supervision, J.L.C. and D.A.M.; project administration, D.A.M. and J.L.C.; funding acquisition, D.A.M., J.L.C. All authors have read and agreed to the published version of the manuscript.

**Funding:** This work has been funded for the Fundación Gómez Pardo and the Laboratorio Oficial para Ensayos de Materiales de Construcción (LOEMCO), Madrid, Spain.

**Institutional Review Board Statement:** Not applicable.

**Informed Consent Statement:** Not applicable.

**Data Availability Statement:** Data is contained within the article.

**Acknowledgments:** The authors wish to thank the Laboratorio Oficial para Ensayos de Materiales de Construcción (LOEMCO) for the preparation of the samples, the performance of the tests and the interpretation of the results, as well as the financial support offered to cover the costs of editing and publication of this work. The authors would also like to acknowledge the financial support provided by the Gómez Pardo Foundation. D.A.M., J.L.C., L.P.M. and A.C.P. thank the Centralised Laboratory of the Escuela Técnica Superior de Ingenieros de Minas y Energía (Universidad Politécnica de Madrid) for the SEM analysis. The authors thank the Laboratorio de Estratigrafía Biomolecular of the Escuela Técnica Superior de Ingenieros de Minas y Energía (Universidad Politécnica de Madrid) for support with X-ray diffraction analysis. The authors are very grateful for the help given by the Laboratorio de Gemología of the Instituto Gemológico de España in the study of fluorite by Raman spectroscopy. The authors are especially grateful to the mining company of Órgiva (Granada, Spain) (Minera de Órgiva, S.L.) for their support with geological work and sampling.

**Conflicts of Interest:** The authors declare no conflict of interest.

# References


1. Gilvonio, R.; Domínguez, F. Ahorro de energía en el proceso de fabricación de clinker de cemento empleando mineral fluorita, (CaF$_2$). *Rev. Soc. Quím. Perú* **2009**, *75*, 303–309.
2. Dahhou, M.; El Hamidi, A.; El Moussaouiti, M.; Azeem Arshadb, M. Synthesis and characterization of belite clinker by sustainable utilization of alumina sludge and natural fluorite (CaF$_2$). *Materialia* **2021**, *20*, 101204. [CrossRef]
3. Najafi Kani, E.; Allahverdi, A. Fast set and high early strength cement from limestone, natural pozzolan, and fluorite. *Int. J. Civ. Eng. Technol.* **2010**, *8*, 362–369. Available online: http://tjce.iust.ac.ir/article-1-306-en.html (accessed on 20 October 2021).
4. Dominguez, O.; Torres-Castillo, A.; Flores-Veleza, L.M.; Torres, R. Characterization using thermomechanical and differential thermal analysis of the sinterization of portland clinker doped with CaF$_2$. *Mater. Charact.* **2010**, *61*, 459–466. [CrossRef]
5. Endzhievskaya, I.G.; Demina, A.V.; Lavorenko, A.A. Synthesis of a mineralizing agent for portland cement from aluminum production waste. *IOP Conf. Ser. Mater. Sci. Eng.* **2020**, *945*. [CrossRef]
6. Fridrichová, M.; Gazdič, D.; Dvoøák, K.; Magrla, R. Optimizing the reactivity of a raw-material mixture for portland clinker firing. *Mater. Technol.* **2017**, *51*, 219–223. [CrossRef]
7. Zea-Garcia, J.D.; Santacruz, I.; Aranda, M.A.G.; De la Torre, A.G. Alite-belite-ye'elimite cements: Effect of dopants on the clinker phase composition and properties. *Cem. Concr. Res.* **2019**, *115*, 192–202. [CrossRef]
8. Güleç, A.; Oğuzhanoğlu, M. Fluorite mineral waste as natural aggregate replacement in concrete. *J. Build. Rehabil.* **2021**, *6*, 26. [CrossRef]
9. Zhu, P.; Wang, L.Y.; Hong, D.; Qian, G.R.; Zhou, M. A study of making synthetic oxy-fluoride construction material using waste serpentine and kaolin mining tailings. *Int. J. Miner. Process.* **2012**, *104–105*, 31–36. [CrossRef]
10. Achmania, J.; Chraibi, I.; Blaise, T.; Barbarand, J.; Brigaud, B.; Bounajma, H. Virtues of fluorite-case of Bou-Izourane fluorite. *Mater. Today Proc.* **2020**, *31*, S114–S121. [CrossRef]
11. Makin, S.A.; Simand, G.J.; Marshall, D. Fluorite and its potential as an indicator mineral for carbonatite-related rare earth element deposits. *Geol. Fieldwork* **2013**, *2014-1*, 207–212.
12. UNE-EN. *Cemento. Parte 1: Composición, Especificaciones y Criterios de Conformidad de Los Cementos Comunes*; Standard UNE-EN 197-1:2011; AENOR: Madrid, Spain, 2011.
13. UNE-EN. *Métodos de Ensayo de Cementos. Parte 5: Ensayo de Puzolanicidad Para Cementos Puzolánicos*; Standard UNE-EN 196-5:2006; AENOR: Madrid, Spain, 2006.
14. Mielenz, R.C.; Greene, K.T.; Schieltz, N.C. Natural pozzolans for concrete. *Econ. Geol.* **1951**, *46*, 311–328. [CrossRef]

15. Moropoulou, A.; Bakolas, A.; Aggelakopoulou, E. Evaluation of pozzolanic activity of natural and artificial pozzolans by thermal analysis. *Thermochim. Acta* **2004**, *420*, 135–140. [CrossRef]
16. UNE-EN. *Métodos de Ensayo de Cementos. Parte 2: Análisis Químico de Cementos*; Standard UNE-EN 196-2:2014; AENOR: Madrid, Spain, 2014.
17. UNE-EN. *Métodos de Ensayo de Cementos. Parte 1: Determinación de Resistencias Mecánicas*; Standard UNE-EN 196-1:2005; AENOR: Madrid, Spain, 2005.
18. Magotra, R.; Namga, S.; Singh, P.; Arora, N.; Srivastava, P.K. A new classification scheme of fluorite deposits. *Int. J. Geosci.* **2017**, *8*, 4. [CrossRef]
19. Liu, H.; Zhu, Q.; Yang, X. Electrical Conductivity of fluorite and fluorine conduction. *Minerals* **2019**, *9*, 72. [CrossRef]
20. Costafreda, J.L.; Martín, D.A.; Presa, L.; Parra, J.L. Altered volcanic tuffs from Los Frailes Caldera. A study of their pozzolanic properties. *Molecules* **2021**, *26*, 5348. [CrossRef] [PubMed]
21. Rosell-Lam, M.; Villar-Cociña, E.; Frías, M. Study on the pozzolanic properties of a natural Cuban zeolitic rock by conductometric method: Kinetic parameters. *Constr. Build. Mater.* **2011**, *25*, 644–650. [CrossRef]
22. Pei, F.; Zhu, G.; Li, P.; Guo, H.; Yang, P. Effects of $CaF_2$ on the sintering and crystallisation of $CaO–MgO–Al_2O_3–SiO_2$ glass-ceramic. *Ceram. Int.* **2020**, *46*, 17825–17835. [CrossRef]
23. Hou, P.; Qian, J.; Cheng, X.; Shahd, S.P. Effects of the pozzolanic reactivity of nanoSiO$_2$ on cement-based materials. *Cem. Concr. Compos.* **2015**, *55*, 250–258. [CrossRef]
24. Grist, E.R.; Paine, K.A.; Heath, A.; Norman, J.; Pinde, H. Compressive strength development of binary and ternary lime–pozzolan mortars. *Mater. Des.* **2013**, *52*, 514–523. [CrossRef]

*Article*

# Mineralization Reaction of Calcium Nitrate and Sodium Silicate in Cement-Based Materials

Isabel Miñano Belmonte [1,*], Mariano Calabuig Soler [1], Francisco J. Benito Saorin [1], Carlos J. Parra Costa [1], Carlos L. Rodríguez López [2], Jorge del Pozo Martin [3], Víctor Martinez Pacheco [4] and Pilar Hidalgo Torrano [4]

[1] Department of Architecture and Building Technologies, Technical/Polytechnic University of Cartagena, Paseo Alfonso XIII, 30203 Cartagena, Spain; mariano.calabuig@upct.es (M.C.S.); francisco.benito@upct.es (F.J.B.S.); carlos.parra@upct.es (C.J.P.C.)

[2] Department of Construction Materials, Centro Tecnológico de la Construcción, C. Sol, 18, Molina de Segura, 30500 Murcia, Spain; crodriguez@ctcon-rm.com

[3] Department of Civil and Environmental Engineering, Universitat Politècnica de Catalunya, 08034 Barcelona, Spain; jorgedelpozo@gmail.com

[4] Department of Innovation and Environment, Cementos La Cruz, S.L., Paraje Tres Santos s/n, 30640 Abanilla, Spain; vmartinez@cementoslacruz.com (V.M.P.); phidalgo@cementoslacruz.com (P.H.T.)

* Correspondence: isabel.minano@upct.es

**Abstract:** The research consists in the design of the new cementitious materials capable of mitigating microfisurative damage through autonomous healing. This lies in the characterization of the materials to employees, study of the expanding agents (sodium silicate and calcium nitrate) and analysis of its mechanical properties and durability. The results revealed that under laboratory conditions, the applied repair agents proved to be powerful in producing an increase in the content of ettringite, favoring the sealing of the fissure. When they heal themselves, they lead to an improvement in durability and mechanical performance.

**Keywords:** self-healing concrete; calcium nitrate; mineralization reaction; cracks

**Citation:** Belmonte, I.M.; Soler, M.C.; Saorin, F.J.B.; Costa, C.J.P.; López, C.L.R.; del Pozo Martin, J.; Pacheco, V.M.; Torrano, P.H. Mineralization Reaction of Calcium Nitrate and Sodium Silicate in Cement-Based Materials. *Crystals* **2022**, *12*, 445. https://doi.org/10.3390/cryst12040445

Academic Editor: Vladislav V. Gurzhiy

Received: 13 February 2022
Accepted: 19 March 2022
Published: 23 March 2022

**Publisher's Note:** MDPI stays neutral with regard to jurisdictional claims in published maps and institutional affiliations.

## 1. Introduction

The deterioration of concrete can happen due to various environmental causes, which can chemically or physically attack the material. Regardless of the nature of the attack, cracks are inevitable in concrete structures, because it is a heterogeneous material. When cracks occur, they allow harmful substances to penetrate the material, without proper maintenance, the propagation of cracks can damage the concrete, putting their safety at risk. In recent years, self-healing techniques have been applied in concrete, these techniques can be identified through various methods, chemical encapsulation; bacterial; mineral mixtures; glass tubes with chemical substances, or by the same cement that was left without hydration.

Design, as well as the production of self-healing concrete, is a topic of great interest today. The great ownership of this type of concrete lies in the ability to mitigate problems related to cracking without external intervention, which leads to prolonging the life of structures and reducing maintenance costs. Healings can be of two types: autogenous and autonomous or autocurative. When the fissures are filled post-secondary hydration of anhydrous particles it is called autogenous healing, while autonomous healing is obtained by adding bacteria, polymer expansive agents, etc. [1–16]. Directly, the use of this type of material contributes to improving the environment, reducing cement consumption by increasing the lifespan of concrete structures.

The expansion agent is another self-healing system, calcium sulfoaluminate-based agents and crystalline additive can provide $Ca^{2+}$ to efficiently improve the self-healing capacity of concrete. Based on this conception, $CaHPO_4 \cdot 2H_2O$ can not only act as a supplier of $Ca^{2+}$ but also as a stabilizer for the healing product. Therefore, the approach of including mineral additives has been developed due to its excellent compatibility with concrete. Mineral additives can react with carbonates that are water-soluble carbon dioxide to form calcium carbonate crystals for crack healing; however, due to the unequal distribution of the concentration of water and carbonate from the surface of the concrete to the interior of the matrix, it is possible that there are fewer ions available within the concrete than on the surface, so the efficiency of self-healing is limited even with the addition of minerals.

This research will analyze the use of expansive agent calcium nitrate and sodium silicate as a repairing agent in cement mixtures. Study involves the implementation of a comprehensive testing program, starting with the individual identification of the main components. Then, in a second stage, self-separating mortar characterization tests are performed to know their mechanical properties, microstructure and durability.

## 2. Materials and Methods

### 2.1. Characterization of Materials

Portland cement type I 52.5R, sodium silicate (MSS) and calcium nitrate (NC), whose chemical composition of the main components, expressed as oxides, was obtained by X-ray fluorescence, the results of which are listed in Table 1.

**Table 1.** Chemical composition of majority components.

| | Cement | MSS | NC |
|---|---|---|---|
| **ÓXIDE** | **COMPOSITION (%)** | | |
| $CO_2$ | 1.21 | 4.36 | |
| $Na_2O$ | 0.372 | 32.13 | |
| $MgO$ | 2.52 | | 0.06 |
| $Al_2O_3$ | 4.09 | 0.12 | 0.017 |
| $SiO_2$ | 16.89 | 27.84 | 0.035 |
| $SO_3$ | 4.061 | 0.022 | 0.0053 |
| $Cl$ | 0.142 | 0.011 | |
| $K_2O$ | 1.33 | 0.147 | |
| $CaO$ | 64.74 | 0.014 | |
| $TiO_2$ | 0.259 | | |
| $Fe_2O_3$ | 3.507 | 0.0209 | 0.0064 |
| $SrO$ | 0.104 | | 0.1577 |
| $Ca(NO_3)_2$ | | | 75.85 |

The analysis of CEM I 52.5R cement to identify the main mineralogical phases is set out in Figure 1, where it is seen that at 11o-2o it has a gypsum peak ($CaSO_4 \cdot 2H_2O$), at peak 12 and 34o-2o it has a peak of calcium manganese aluminum iron oxide ($Ca_2 MnO \cdot 2FeO \cdot 8AlO_5$), and at 23 and 26o-2o you have the peaks of Anhidrita ($CaSO_4$), at 29, 30 and 33o-2o you have the peaks of calcium carbonate ($CaCO_3$). Sodium silicates were produced by melting sodium carbonate ($Na_2CO_3$) at high temperatures with specially selected silica sand. In the diffractogram it can be observed how the majority component appears with various peaks marked red "Sodium Silicate Hydrate" accompanied by hydrogen sodium silicate hydrate ($H_2Na_2[SiO_4][H_2O]_5$) (marked in blue). The peaks were very slender and high intensity, indicating that the sample is made up of particles with an ordered crystalline structure. Finally, in the analysis of the NC by X-ray diffraction we can see that it is a fundamentally amorphous material, by the deviation observed in the baseline. The majority component arises at the peak of nitrocalcite (marked in red) accompanied by calcium nitrate (marked in green) and calcium nitrate hydrate (marked in blue).

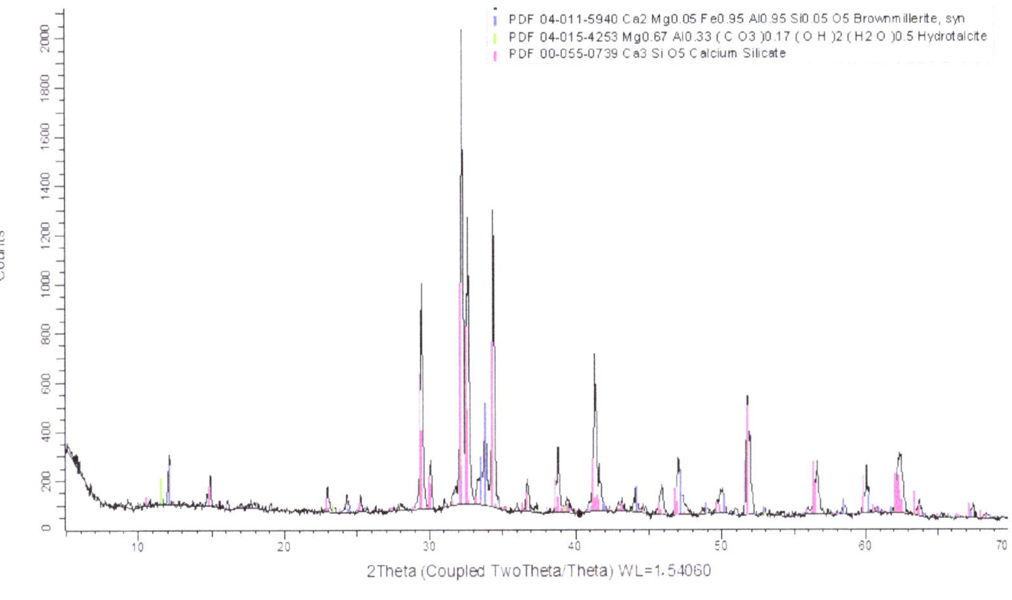

(**a**)

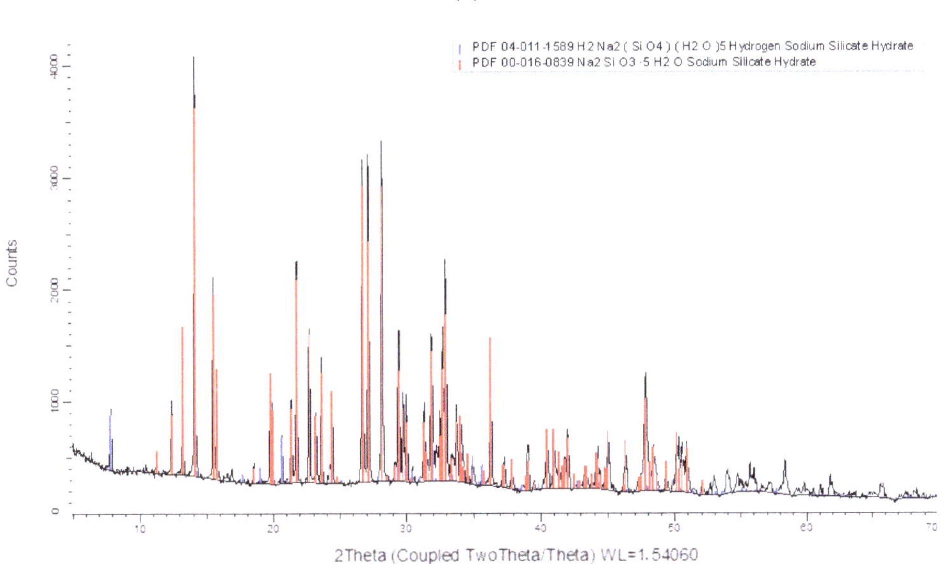

(**b**)

**Figure 1.** *Cont.*

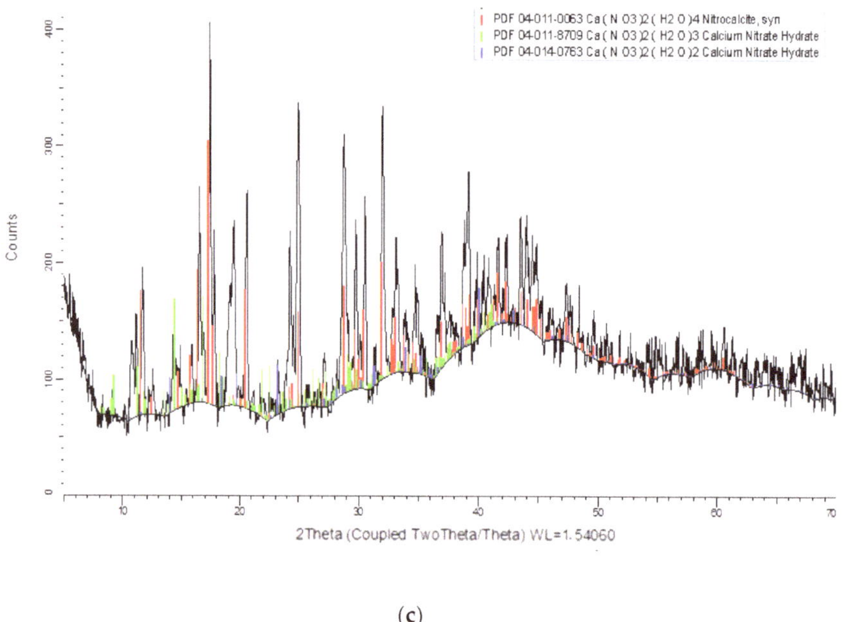

(c)

**Figure 1.** Cement, MSS and NC X-ray diffractogram. (**a**) Cement, (**b**) sodium silicate (MSS) and (**c**) calcium nitrate (NC).

To analyze the behavior of self-healing mortar, the following mixtures were prepared: 450 g of cement, 1350 g of standard sand and 225 g of water. The NC and MSS content relative to cement weight was 3% and 6%. The test to determine compression resistance was conducted as established by UNE-EN196-1, at ages 7, 28 and 90 days. The carbonation depth has been executed in a carbonation chamber consisting of an airtight vessel to which a $CO_2$ bullet with continuous flow has been connected. To speed up the carbonation process, the atmosphere generated in the chamber has been 100% $CO_2$. The temperature has remained between 20 and 25 °C, HR between 60 and 70% and $40 \times 40 \times 80$ mm prismatic specimens were used. After 90 days of curing, the specimens were conditioned at room temperature and humidity for 28 days. Of the six sides of the prism, 4 were protected with aluminum adhesive tape to direct the diffusion only on two sides. Carbonation is measured by contrast of phenolphthalein solution (UNE-112011) by measuring 3 and 7 days. Prior to the spraying of phenolphthalein, the specimens were in two halves. The result will be the average value of at least two distinct sections.

The colorimetric method is used for chloride analysis. The specimens for this test were 40 mm cubic. Five of the faces were painted to be waterproof and watertight to force the spread of chlorides on a single side. These remain submerged in dissolution with 3% sodium chloride until the age of 14 and 28 days.

Microstructure analysis was performed by scanning electron microscope (SEM, FEI Quanta 650 FEG), optical microscope and porosity, using mercury intrusion porosimetry (MIP, AutoPore IV9500) following ASTM D4404-84 (2004).

### 3. Results and Discussion

This section may be divided by subheadings. It should provide a concise and precise description of the experimental results, their interpretation as well as the experimental conclusions that can be drawn.

### 3.1. Effect of NC and MSS on Compression Resistance

The influence of different amounts of NC and MSS on the mechanical properties of cement paste, and the test results are presented in Figure 2. With the addition of 3% by weight and 6% by weight of NC, the compressive resistance of cement paste was improved in relation to the control sample. Compression resistances with 3% NC increased by 14.5% and 2.4% at 7, 28 and 90 days, respectively. Similarly, compression resistances of 6% increased by 6.2% and 5% at 7, 28 and 90 days, respectively.

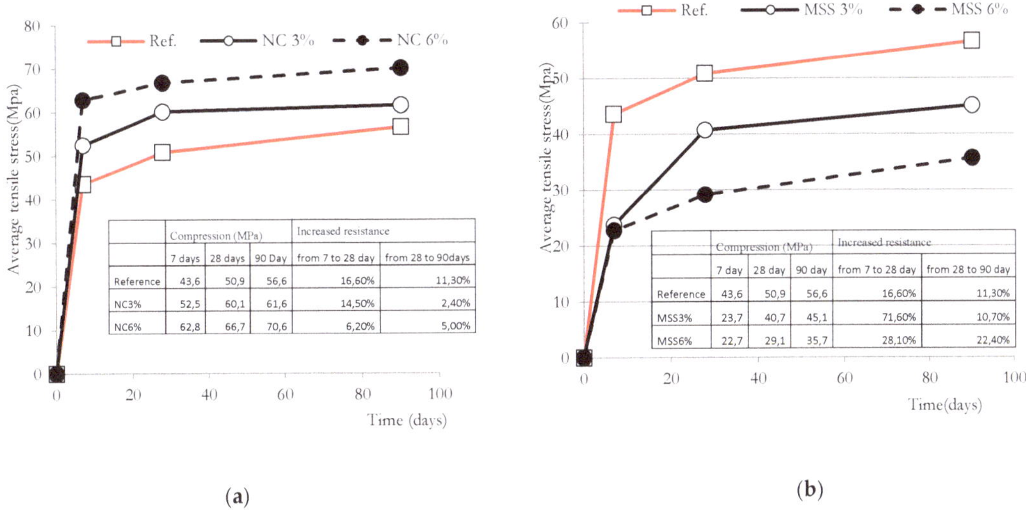

Table for (a):

| | Compression (MPa) | | | Increased resistance | |
|---|---|---|---|---|---|
| | 7 days | 28 days | 90 Day | from 7 to 28 day | from 28 to 90days |
| Reference | 43,6 | 50,9 | 56,6 | 16,60% | 11,30% |
| NC3% | 52,5 | 60,1 | 61,6 | 14,50% | 2,40% |
| NC6% | 62,8 | 66,7 | 70,6 | 6,20% | 5,00% |

Table for (b):

| | Compression (MPa) | | | Increased resistance | |
|---|---|---|---|---|---|
| | 7 day | 28 day | 90 day | from 7 to 28 day | from 28 to 90 day |
| Reference | 43,6 | 50,9 | 56,6 | 16,60% | 11,30% |
| MSS3% | 23,7 | 40,7 | 45,1 | 71,60% | 10,70% |
| MSS6% | 22,7 | 29,1 | 35,7 | 28,10% | 22,40% |

**(a)**　　**(b)**

**Figure 2.** Compression resistance evolution (MPa). (**a**) Calcium nitrate and (**b**) silicate sodium. The comma between the numbers in the figure represents the decimal point.

The resistance of the mortar with sodium silicate is observed to be much lower than that of the reference mortar, the greater the greater the age of the specimens. The results found are consistent with what is reported by various researchers [14], who agree that the use of SS contributes to demean the mechanical behavior of the material but improves the autocurative properties.

### 3.2. Effect of NC and MSS on Carbonation and Chlorides

At 7 days the carbonation depth is greater than 3 days in all the mixtures studied (Figure 3a). According to Papadakis (2000) [13] and Mira et al. (2002) [17] the carbonation depth decreases as calcium components increase. The use of NC (6%) decreases the carbonated thickness, being smaller the higher its content, however it is observed that the differences between the other samples are small compared to that of reference to the age of 7 days. Porosity results presented in successive appliances show that the use of NC can have an impact on carbonation by reducing porosity and improving the capillary network, refining and creating greater tortuousness.

Calcium nitrate reduces the intake of $Cl^-$ (Figure 3b), as we have seen above from $CO_2$, by presenting this a more refined porous structure. Chloride penetration decreases with the lowest porosity presented in samples with higher NC content and/or pore connectivity and decreased chloride binding capacity in the matrix.

### 3.3. Morphology and Microstructure

Figure 4 shows for each mortar the increase in mercury intrusion volume based on the equivalent pore diameter. Highlighting, the shift to the left of the curve black dots, belonging to the NC (6%), which implies a refinement of pores that transform into small capillary pores and that would explain the least detected amount of small pores. The largest critical diameter corresponds to the SS (6%), followed by mortars containing NC (3%) and, finally, by the SS (3%) NC (6%). As mentioned, this reduction in critical diameter entails a general refinement in the porous structure that can be seen by the shift to the left of the different curves.

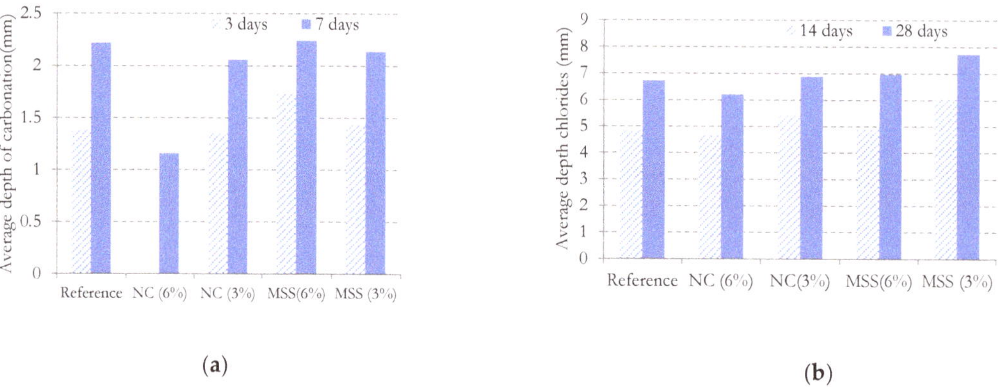

(a)            (b)

**Figure 3.** Carbonation depth and chlorides (mm) (**a**) Carbonation (**b**) Chlorides.

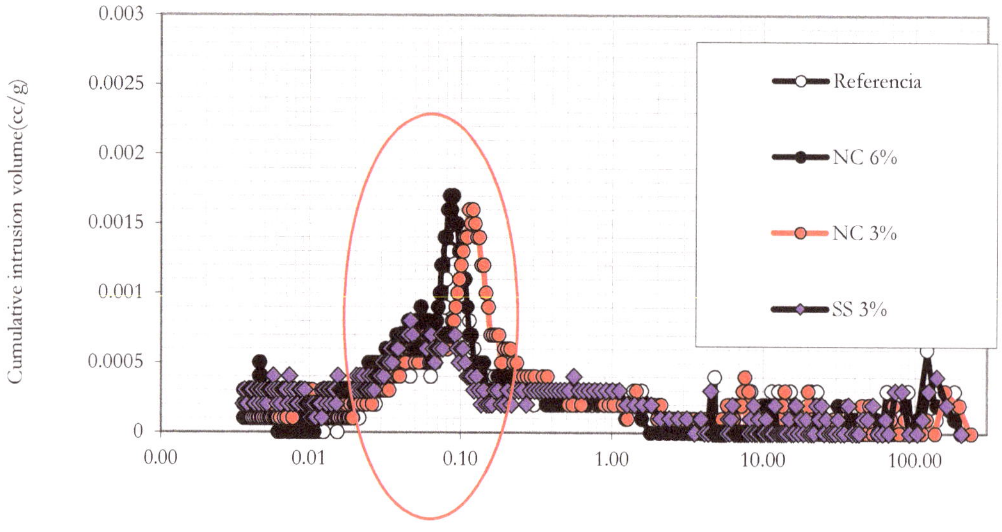

**Figure 4.** Mercury intrusion volume (cc/g).

Summary shows Table 2 reflecting critical diameter values, where the use of healing agents with 6% NC and 3% MSS results in a decrease in diameter size. The NC sample (6%) has lower values of mean pore diameter (am) than the reference sample, the average pore size is 0.086 om and 0.08 om, respectively. This fact is indicative that the NC has a more refined porous structure than the sample without it, which is expected to exhibit better behavior to prevent the penetration of aggressive agents, resulting in greater durability. Another important parameter that would be related to the finesse of the porous structure is the pore size corresponding to the maximum concentration of the pore. The sample of NC

6% and MSS 3% has a smaller threshold diameter (U) and critical diameter (C), which is consistent with a more refined porous microstructure.

**Table 2.** Porosimetric analysis.

|  | Ref | NC (6%) | NC (3%) | MSS (6%) | MSS (3%) |
|---|---|---|---|---|---|
| Total intrusion volume (cc/g) | 0.0626 | 0.0591 | 0.0687 | 0.0750 | 0.0620 |
| Total pore area (m$^2$/g) | 9.1708 | 7.9084 | 8.7333 | 16.0074 | 11.5510 |
| Average pore diameter (^m) | 0.086 | 0.080 | 0.106 | 0.830 | 0.046 |
| Total porosity (%) | 13.6710 | 13.0107 | 14.8482 | 16.0074 | 13.3789 |
| Apparent density (g/mL) | 2.1826 | 2.2019 | 2.1622 | 2.1338 | 2.1587 |

The mercury intrusion porosimetry test also provides total porosity data. Total porosity, with 13.67% for the reference mortar and 13.01% for NC (6%), shows that it decreases with the use of NC. This is due to the densification of the cement matrix by the formation of non-oriented ettringite together with tobermorite, which make the internal structure of the cement denser, because in the spaces between ettringite needles is formed tobermorite, which makes the pores of the internal structure of the cement smaller and increases its durability, this will be best reflected in SEM trials.

The self-healing efficiency of the cracks was characterized by the repair ratio of the area, the permeability coefficient and the healing depth. Meanwhile, the morphologies and polymorphs of the precipitates were analyzed by SEM equipped with an EDS and XRD.

The Figure 5 shows the fissured area with sodium silicate (6%), and the comparative analysis of the elemental composition by dispersed energy. An area with a higher concentration of Ca, Si and Na atoms can be observed. The presence of Na in the cement zone is very significant, since this element is practically absent in the chemical composition of cement [18], as we have seen in Section 2.1.

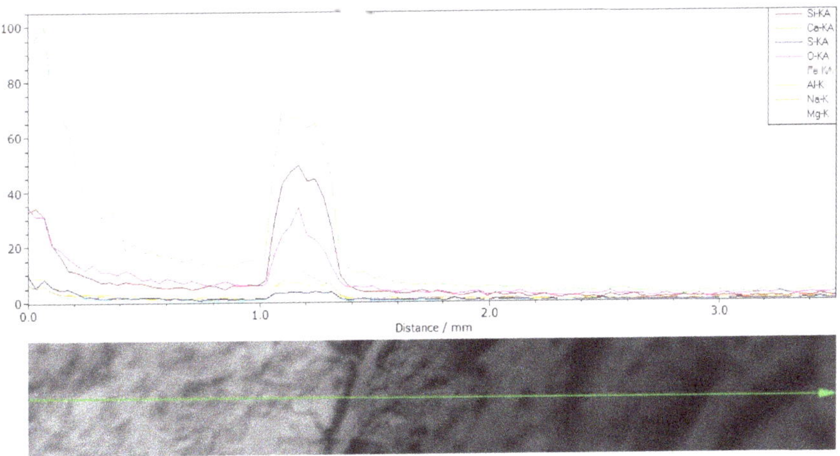


**Figure 5.** EDS sodium silicate (6%).

From the EDS analysis, the main chemical elements of the healing products are Si, O, Ca and Na. It is observed that there is no Ca in the original healing agent "sodium silicate", but there is a transfer of this compound to the cracked area. Since soluble silicates react almost instantaneously with multivalent metal cations to form the corresponding insoluble metal silicate [18], the chemical element Ca in healing products reveals that calcium cations from the cementing matrixes with the sodium silicate solution, and thus, CSH is formed in the cracks. However, there are not enough calcium cations to replace all of the sodium cations in solution. According to the previous discussion, the main resolution mechanism promoted by sodium silicate solution is the reaction of calcium cations with dissolved sodium silicate and the crystallization of available sodium silicate.

That is, in pastes containing sodium silicate repair material ($Na_2SiO_3$) in solid state, it has been observed through SEM images, which react with $Ca(OH)_2$ (portlandite), naturally present in cementing material to form a calcium silicate hydrate (CSH), a compound that works by sealing cracks.

These results coincide with research conducted by Huang and Ye (2011) [19], which also state that the healing products formed in the fissures are the compounds formed by CSH and sodium silicate. Therefore, the main mechanism of self-management that is generated when using a sodium silicate solution is the reaction of calcium cations with dissolved sodium silicate, resulting in the crystallization of sodium silicate.

Figure 6 shows a micrograph of a cement sample with hydrated sodium silicate and on the right its respective EDS analysis at the age of 28 days.

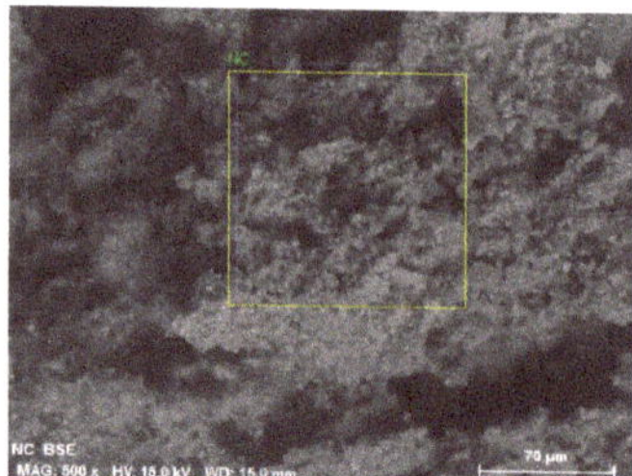
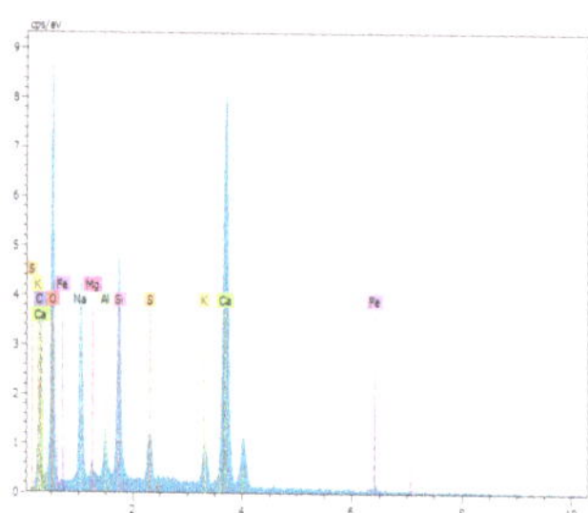

**Figure 6.** The chemical elements of EDS-tested healing products.

Visual inspection of the fissure is also performed through an optical microscope. This consists of sweeping the fissure taking photos so that there is overlap between the previous and subsequent ones for subsequent reconstruction of the fissure. Through Figure 7, the follow-up over time is observed for 79 days of a cracked cement paste with 6% MSS, where the sewn of it is observed due to the presence of new crystals, this coincides with the results obtained by SEM. The lack of cohesion or sewn that is shown in the fissure, reduces as hydration progresses, since much of these cracks are filled with new hydration products improving in this way, the compactness of the internal microstructure by densification of it. At the age of 28 days it was almost cured, the new rehydration and self-healing products between the fissures were clearly observed after 12 days, and after 79 days of rehydration, as shown in the figure the fissure is almost completely sealed (pictured right below). MSS reacts with calcium hydroxide, a product of cement hydration, and produces a hydrated calcium-silica (C-S-H) of gel-a natural bonding material with concrete. The C-S-H gel ($x.(CaO \cdot SiO_2) \cdot H_2O$), partially fills the fissure, and allows some recovery of force.

**Figure 7.** Images of a mortar crack with MSS (6%) at the age of 79 days.

Figure 8 presents by SEM image a mortar with 6% NC, at the age of 28 days of curing micrographs is taken as convention E: ethringite, T: tobermorite and P: portlandita.

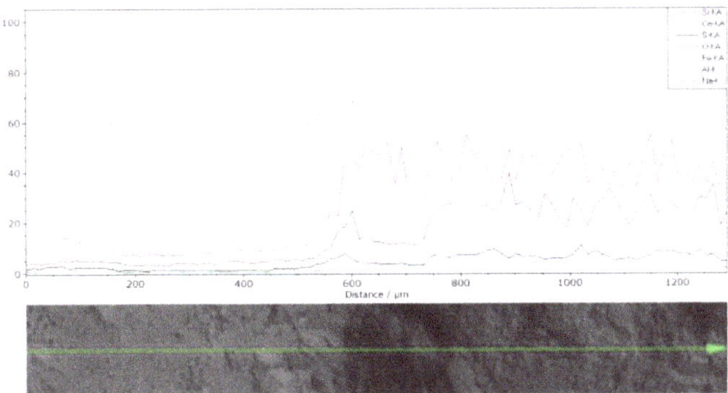

**Figure 8.** SEM images at 1200 magnifications and composition analysis.

The use of NC increases the amount of ettringite present in the structure of the cement, and, also, presents itself in the form of non-oriented needles, making a sewing effect among the other mineral phases. These results explain the improvement in compressive resistance (Section 3.1) where it is reflected that mortars replaced with NC had higher resistances than the Reference samples, especially at early ages, this is due to the shape of non-oriented needles as shown in Figure 9. In the image you can see: ettringite crystals filling a pore (marked in red). Ettringite crystals in air vacuums and fissures are typically two to four micrometers in cross section and twenty to thirty micrometers long. Under conditions of extreme deterioration or decades in a humid environment, white ettringite crystals can completely fill in voids and fissures [20].

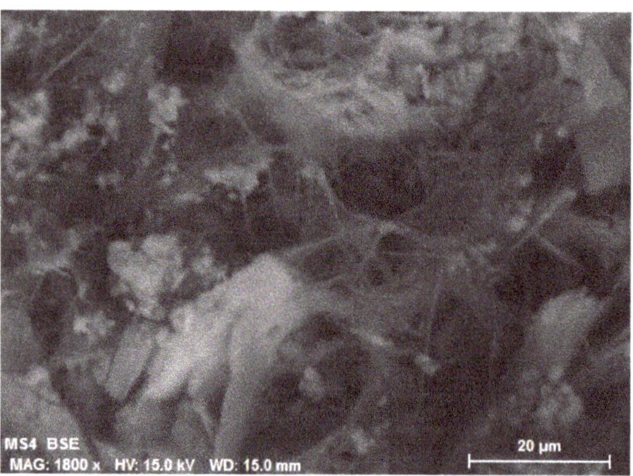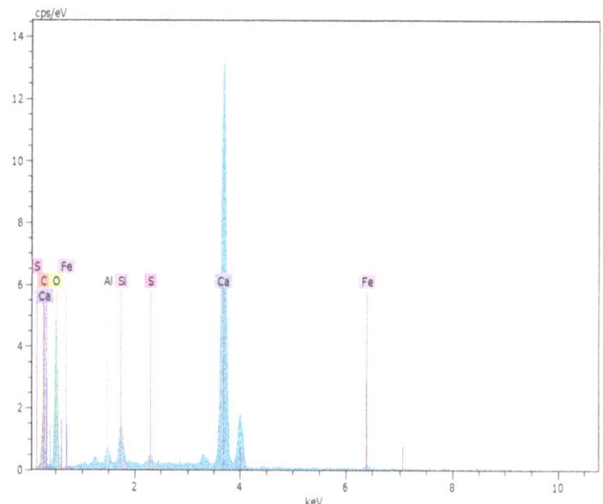

**Figure 9.** Microfissures filling with new ettringite-type hydration products.

In Figure 10 right, you can see bermorite products. Most bermorite has a distinctive morphology, which is described as crystals in the form of leaves or plates found in open spaces where they have enough space to grow. The most bermorite confers excellent mechanical properties. Photomicrography, also, shows a high concentration of ettringite (they look well crystallized and look perfectly in the form of fine elongated needles or "hedgehogs" characteristic of this).

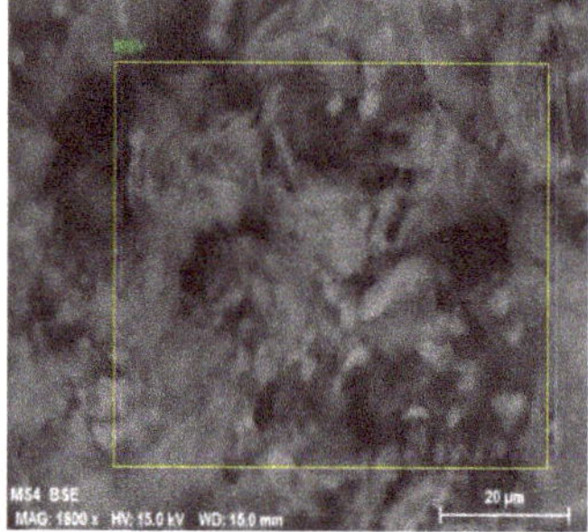

**Figure 10.** SEM image of cement paste with NC 6%.

The use of NC (6%) facilitates the formation of ettringite by penetrating water into the fissure as shown in the optical microscope images (Figure 11). Regarding crystallization, other researchers have tried to take advantage of the formation of crystals with great capacity of expansiveness to achieve the healing of the fissure thanks to its increase in volume or simply generate a large amount of crystals that are deposited and sealed the fissure, such as those obtained in this research with the use of 6% NC. XRD analysis showed that the precipitates in the crack mouth were calcite. When water comes into contact with the non-hydrated cement, greater hydration occurs. In addition, dissolved $CO_2$ reacts with $Ca^{2+}$ to form $CaCO_3$ crystals.

**Figure 11.** Sealed images of mortar cracks with 6% NC.

## 4. Conclusions

- The results showed that although self-adapting properties increase in the case of "sodium silicate", the greatest improvement corresponds to the use of "calcium nitrate" as a self-hoarding agent, as well as its durability properties.
- Micrographs taken at 28 days of SEM curing show that the use of NC increases the amount of ettringite present in the structure of the cement, it is presented in the form of non-oriented needles, making a sewing effect with the other mineral phases The formation of these crystals favors the healing of the fissure by sealing it.
- Use of NC curing agent (6%) in the manufacture of mortars allows obtaining improvements above 20% on the reference mortar, in compression resistance.
- The NC in the cement matrix favors the generation of ettringite and tobermorite, in addition leaves little C3A available for the generation of secondary ettringite, improving durability. The formation of ettringite and tobermorite densifies the cement matrix, resulting in a decrease in its total porosity which results in a reduction in the depth of carbonation and chlorides.
- The results indicated that the rate of calcification using calcium nitrate as a source of calcium had a good efficiency of self-healing in the cracks of the mortar.

**Author Contributions:** All the authors have contributed in the same way in carrying out the research. Conceptualization, I.M.B. and F.J.B.S.; methodology, M.C.S.; validation, P.H.T. and C.J.P.C.; formal analysis, J.d.P.M., V.M.P. and C.L.R.L.; investigation, F.J.B.S., I.M.B. and P.H.T.; resources, V.M.P. and P.H.T. writing—original draft preparation, I.M.B.; writing—review and editing F.J.B.S. All authors have read and agreed to the published version of the manuscript.

**Funding:** The authors of this study would like to thank the Institute for the Promotion of the Region of Murcia (INFO) (Region de Murcia (Spain). for the financing of the 2015.08.ID+I.0028 (http://www.institutofomentomurcia.es/-/caso-cementos-cruz).

**Institutional Review Board Statement:** Not applicable.

**Informed Consent Statement:** Not applicable.

**Data Availability Statement:** Not applicable.

**Acknowledgments:** In this study Cementos La Cruz S.L. and the Polytechnic University of Cartagena have participated.

**Conflicts of Interest:** The authors declare no conflict of interest.

## References

1. Li, V.; Lim, Y.M.; Chan, Y.-W. Feasibility study of a passive smart self-healing cementitious composite. *Compos. Part B* **1998**, *29*, 819–827. [CrossRef]
2. Dry, C.M. Alteration of matrix permeability and associated pore and crack structure by timed release of internal chemicals. *Ceram. Trans.* **1991**, *16*, 729–768.
3. Dry, C.M. Passive tunable fibers matrices. *Int. J. Mod. Phys.* **1992**, *6*, 2763–2771. [CrossRef]

4.      Dry, C.M. Smart building materials which prevent damage or repair themselves. *MRS Online Proc. Libr. (OPL)* **1992**, *276*. [CrossRef]

5.      Dry, C.M. Smart materials which sense, activate and repair damage; hollow porous fibres in composites release chemicals from fibers for self-healing, damage prevention and/or dynamic control. In Proceedings of the 1st European Conference on Smart Structures and Materials, Glasgow, UK, 12–14 May 1992.

6.      White, S.R.; Sottos, N.R.; Geubelle, P.H.; Moore, J.S.; Kessler, M.R.; Sriram, S.R.; Brown, E.N.; Viswanathan, S. Autonomic healing of polymer composites. *Nature* **2001**, *409*, 794797. [CrossRef] [PubMed]

7.      Ahn, T.H.; Kishi, T. The effect of geo-materials on the autogenous healing behavior of cracked concrete. In *Concrete Repair, Rehabilitation and Retrofitting II*; Taylor & Francis Group: London, UK, 2009; ISBN 978-0-415-46850-3.

8.      Li, V. *Autogenous Healing of Engineered Cementitous Composites under Wet-Dry Cycles*; Universidad de Michigan: Ann Arbor, MI, USA, 2009.

9.      Dry, C. Matrix craking repair and filling using active and passive modes for smart timed release of chemicals from fibers into cement matrices. *Smart Mater. Struct.* **1993**, *3*, 118–123. [CrossRef]

10.     Li, V.C.; Herbert, E. Robust Self-Healing Concrete for Sustainable Infrastructure. *J. Adv. Concr. Technol.* **2012**, *10*, 207–218. [CrossRef]

11.     Li, M.; Ranade, R.; Kan, L.; Li, V.C. On improving the infrastructure service life using ECC to mitigate rebar corrosion. In Proceedings of the 2nd International Symposium on Service Life Design for Infrastructure, RILEM PRO 70, Delft, The Netherlands, 4–6 October 2010.

12.     Baroghel-Bouny, V.; Belin, P.; Maultzsch, M.; Henry, D. $AgNO_3$ spray tests: Advantages, weaknesses, and various applications to quantify chloride ingress into concrete. Part 1: Non-steady-state diffusion tests and exposure to natural conditions. *Mater. Struct.* **2007**, *40*, 759. [CrossRef]

13.     Papadakis, V.G. Effect of supplementary cementing materials on concrete resistance against carbonation and chloride ingress. *Cem. Concr. Res.* **2000**, *30*, 291–299. [CrossRef]

14.     Jiang, S.; Lin, Z.; Tang, C.; Hao, W. Preparation and Mechanical Properties of Microcapsule-Based Self-Healing Cementitious Composites. *Materials* **2021**, *14*, 4866. [CrossRef] [PubMed]

15.     Huseien, G.F.; Nehdi, M.L.; Faridmehr, I.; Ghoshal, S.K.; Hamzah, H.K.; Benjeddou, O.; Alrshoudi, F. Smart Bio-Agents-Activated Sustainable Self-Healing Cementitious Materials: An All-Inclusive Overview on Progress, Benefits and Challenges. *Sustainability* **2022**, *14*, 1980. [CrossRef]

16.     Sohail, M.G.; Disi, Z.; Al Zouari, N.; Nuaimi, N.; Al Kahraman, R.; Gencturk, B.; Rodrigues, D.F.; Yildirim, Y. Bio self-healing concrete using MICP by an indigenous Bacillus cereus strain isolated from Qatari soil. *Constr. Build. Mater.* **2022**, *328*, 126943. [CrossRef]

17.     Mira, P.; Papadakis, V.G.; Tsimas, S. Effect of lime putty addition on structural and durability properties of concrete. *Cem. Concr. Res.* **2002**, *32*, 683–689. [CrossRef]

18.     Miñano, I.; Benito, F.J.; Valcuende, M.; Rodríguez, C.; Parra, C.J. Improvements in Aggregate-Paste Interface by the Hydration of Steelmaking Waste in Concretes and Mortars. *Materials* **2019**, *12*, 1147. [CrossRef] [PubMed]

19.     Huang, H.; Ye, G.; Leung, C.; Wan, K. Application of sodium silicate solution as self-healing agent in cementitious materials. In Proceedings of the International RILEM Conference on Advances in Construction Materials Through Science and Engineering, Hong Kong, China, 5–7 September 2011.

20.     Detwiler, R.J.; Powers-Couche, L.J. Effect of Ettringite on Frost Resistance (Efecto de la Etringita Sobre la Resistencia a la Congelación). *Concr. Technol. Today* **1997**, *18*. Available online: https://www.portcement.org/pdf_files/PL973.pdf (accessed on 15 December 2021).

*crystals*

MDPI


*Article*

# Synthesis of Spinel-Hydroxyapatite Composite Utilizing Bovine Bone and Beverage Can


Agus Pramono [1,*], Gerald Ensang Timuda [2,*], Ganang Pramudya Ahmad Rifai [1] and Deni Shidqi Khaerudini [2,*]

[1] Department of Metallurgical Engineering, Sultan Ageng Tirtayasa University, Jl. Jendral Soedirman KM. 3, Cilegon 42435, Banten, Indonesia

[2] Research Center for Physics, National Research and Innovation Agency (BRIN), Building 440-442 Kawasan PUSPIPTEK, Tangerang Selatan 15314, Banten, Indonesia

* Correspondence: agus.pramono@untirta.ac.id (A.P.); gerald.ensang.timuda@brin.go.id (G.E.T.); deni.shidqi.khaerudini@brin.go.id (D.S.K.)



**Abstract:** Spinel-based hydroxyapatite composite (SHC) has been synthesized utilizing bovine bones as the source of the hydroxyapatite (HAp) and beverage cans as the aluminum (Al) source. The bovine bones were defatted and calcined in the air atmosphere to transform them into hydroxyapatite. The beverage cans were cut and milled to obtain fine Al powder and then sieved to obtain three different particle mesh size fractions: +100#, −140# + 170#, and −170#, or Al particle size of >150, 90–150, and <90 μm, respectively. The SHC was synthesized using the self-propagating intermediate-temperature synthesis (SIS) method at 900 °C for 2 h with (HAp:Al:Mg) ratio of (87:10:3 wt.%) and various compaction pressure of 100, 171, and 200 MPa. It was found that the mechanical properties of the SHC are influenced by the Al particle size and the compaction pressure. Smaller particle size produces the tendency of increasing the hardness and reducing the porosity of the composite. Meanwhile, increasing compaction pressure produces a reduction of the SHC porosity. The increase in the hardness is also observed by increasing the compaction pressure except for the smallest Al particle size (<90 μm), where the hardness instead becomes smaller.

**Keywords:** hydroxyapatite composite; spinel; beverage cans; bovine bones



**Citation:** Pramono, A.; Timuda, G.E.; Rifai, G.P.A.; Khaerudini, D.S. Synthesis of Spinel-Hydroxyapatite Composite Utilizing Bovine Bone and Beverage Can. *Crystals* **2022**, *12*, 96. https://doi.org/10.3390/cryst12010096

Academic Editor: Vladislav V. Gurzhiy

Received: 10 December 2021
Accepted: 7 January 2022
Published: 13 January 2022

**Publisher's Note:** MDPI stays neutral with regard to jurisdictional claims in published maps and institutional affiliations.




## 1. Introduction

Hydroxyapatite is an important biomedical implant material. It is the natural bone's inorganic component [1], and has good osteoconductivity, biocompatibility, and bioactivity, thus making it easily incorporated into the bones [1–3]. However, the utilization of chemically synthesized hydroxyapatite for hard tissue replacement is limited because it is expensive, while a large quantity of material is required [3]. Furthermore, synthesized hydroxyapatite has different physicochemical properties (such as strength and chemical composition) than the natural one, leading to lower biological activity [4,5]. Therefore, extracting hydroxyapatites from biological materials has been considered to provide a cheaper and up-scalable alternative synthesis route, and to obtain physicochemical properties as close as those of the natural hydroxyapatite. Several biological materials have been reported as the source of the hydroxyapatite, such as bovine bone [3,6], fish scale [4,7], snail shell [2,8], and eggshell [9,10]. Bovine bone can be considered a biological waste that needs to be recycled. Therefore, processing it into hydroxyapatite is economically and environmentally beneficial.

To further improve the mechanical properties of hydroxyapatite, it can be synthesized into a composite with metal/alloys [11–14], metal oxides [5,15–17], or polymer [1,18]. The composite of hydroxyapatite (SHC) with a single metal element such as in the Mg/HAp system was considered to combine both advantages of biocompatibility and biodegradability in Mg and bioactivity in HAp [11,14]. Improvement of the mechanical properties, such as hardness and compression strength, has been reported in the Mg/HAp composite compared to pure HAp [14]. However, the Mg/HAp system is more prone to corrosion than the alloy matrix [11]. Therefore, the addition of other metals as an alloying component was considered to improve the corrosion properties, such as in Mg-Sn/HAp [11] or Mg-Ca/HAp [13]. Similarly, composite of HAp with metal oxides such as $ZrO_2$, $TiO_2$, $Al_2O_3$ [5], or MgO [17] were considered as the reinforcement to improve the HAp's mechanical stability [5]. Furthermore, bio-inert material such as the $Al_2O_3$ poses no harmful effect on the human body [5].

Several methods to incorporate metals and/or metal oxides into HAp to form a composite have been reported, such as sol-gel [19], reactive spark plasma synthesis [20], stir-assisted squeeze casting [11], and powder metallurgy [13,14]. In this study, the composite of HAp with two metal components, Al and Mg, is reported using a novel approach by the self-propagating intermediate-temperature synthesis (SIS) method. The HAp source is provided by the recycled bovine bone and the Al source is supplied from recycled beverage cans which were reported to contain 93–97% of Al [21–23]. The metal components are prone to oxidation due to the presence of oxygen in the environment during the heating process thus leading to the formation of a HAp-spinel ($MgAl_2O_4$) based composite. The introduction of $MgAl_2O_4$ has been reported to improve the phase stability of HAp during sintering [16]. The effect of Al particle size and compaction pressure on the mechanical properties of the composite is discussed.

## 2. Materials and Methods

### 2.1. Bovine Bones Treatment

Bovine bones (leg part) were used as the hydroxyapatite (HAp) source material and were obtained from local markets in Jakarta, Indonesia. The bones were defatted by twice boiling for 2 h, followed by washing under running water, then basking in the Sun for 8 h. The dried bones were then cut into smaller pieces of bones by a grinder and calcined at 850 °C for 2 h both in an inert and air atmosphere for comparison purposes. Afterward, the air-calcined bones were disk milled for 20 s to obtain the HAp powder.

### 2.2. Aluminum Cans Treatment

Aluminum (Al) and magnesium (Mg) were used as the metal component in the composite. In this study, the Al was provided from recycled beverage cans. The cans were square cut to $1 \times 1$ cm$^2$ then disk milled for 15 min with the on:off interval time of 1:2 (min/min), producing Al powder. Its purity content was then characterized by the X-ray fluorescence (XRF, S2 PUMA Bruker). The powder was then sieved with the sieve sizes of 100#, 140#, and 170# to obtain three different particle size fractions of +100#, −140# + 170#, and −170#, respectively, or Al particle size of >150, 90–105, and <90 μm, respectively.

### 2.3. Self-Propagating Intermediate-Temperature Synthesis (SIS)

A mixture of HAp, Al, and Mg (Merck, 99% purity, particle size 60–300 μm, with the majority at >150 μm, see Figure S1 in Supplementary Information) powders with the ratio of (87:10:3 wt.%) was prepared using a shaker mill for 10 min. The mixture was then pelleted with different compaction pressures of 100, 171, and 200 MPa. The nine different compositions of pellets were prepared from the variation of Al particle sizes and compaction pressures. Afterwards, each pellet was put in the SIS vessel which was then heated at 900 °C for 2 h in an air atmosphere. The SIS vessel is specifically designed so that when undergoing heat treatment in the furnace, the heat propagation can only come from one specific surface instead of from all surfaces in the conventional heat treatment case. In this way, the heat propagation and distribution in the sample became more controllable and homogenous. More detailed information about the SIS method is presented in the Supplementary Information (Figures S2–S4, Table S1).

## 3. Results and Discussions

The hydroxyapatite composite (SHC) synthesized in this study consists of hydroxyapatite (HAp), aluminum (Al), and magnesium (Mg) as the bio-ceramic matrix, reinforcement, and wetting agent/binder [24–28], respectively. The HAp was extracted from the bovine bones. Figure S5 (Supplementary Information) shows the visual appearance of the defatted bovine bones after calcination in the furnace at 850 °C for 2 h. It can be seen that calcination at 850 °C in an inert condition resulted in burnt and charred bones (Figure S5a, Supplementary Information), indicating the formation of carbon instead of the desired HAp. This is because heating in a low or no oxygen atmosphere can lead to pyrolysis, resulting in carbonization of the bones [29]. On the other hand, the bones that are calcined at 850 °C in an air atmosphere have a white color (Figure S5b, Supplementary Information), indicating the formation of the desired HAp [3,30]. The air-calcined bones also lose their previous stickiness feels, indicating the removal of the remaining fat unable to be removed from the previous defatting process. The air-calcined bones were then crushed in the disk milling to obtain the HAp powder used for the next step of the composite synthesis. Figure S6 (Supplementary Information) shows the photograph of the HAp powders from different batches, showing similar white colors. The particle size distribution of the HAp powder is presented in Figure S1 (Supplementary Information) showing that the majority of the HAp particles have sizes at the range of 90–150 µm, in a similar dimension with the Al powders used in this study.

The XRD of the HAp powders from several batches with similar calcination processes is shown in Figure 1, showing a similar pattern indicating good reproducibility. The unit cell results were also compared with the reference HAp peak (COD 96-901-3628). HAp has a hexagonal crystal structure with the space group of P 63/m and the space group number of 176. It reveals that fully crystalline calcined HAp is observed for all the samples with similar reflection peaks. This is also supported by its small difference of lattice constant (unit cell) after Rietveld refinement of each sample (Table 1), which indicates a good reproducibility. Figure 1b shows the diffraction peaks at a small 2θ angle between 30 and 35°, showing a clearer comparison of the three main peaks of (211), (112), and (300). Three out of four samples show almost similar peaks position. Only the HAp-A sample shows a right-shift of the (211) and (300) peaks, as well as the reduction of the (112) peak. One possible reason for the phenomena is the difference in the actual temperature of the calcination process. Ref. [31] reported similar behavior of the HAp derived from the calcined swine bones. A right-shift of some of the peaks and the disappearance tendency of the (112) peak were observed when the calcination was performed at a lower temperature.

**Table 1.** The lattice constant of the HAp powders.

| Sample | HAp-A | HAp-B | HAp-C | HAp-D | Reference * |
|---|---|---|---|---|---|
| a (Å) | 9.387 | 9.425 | 9.423 | 9.426 | 9.421 |
| c (Å) | 6.885 | 6.883 | 6.883 | 6.882 | 6.893 |

* COD 96-901-3628.

The SEM-EDS of the HAp powder is shown in Figure 2. It can be seen that the distribution of Ca and P are almost similar, indicating the formation of hydroxyapatite. The elemental composition derived from the EDS measurements is presented in Table 2. The Ca/P ratio is at 1.47, slightly smaller than the typical hydroxyapatite at 1.67 [32].

**Table 2.** Elemental composition of the HAp powder derived from the EDS measurements.

| Elements | O | P | Ca | Total |
|---|---|---|---|---|
| at. % | 69.05 | 12.54 | 18.41 | 100.00 |

Al is the metal used in the composite here as a reinforcement to improve the mechanical properties of the HAp bio-ceramic. In this study, the Al source was taken from recycled beverage cans. The recycling process includes cutting the cans and then crushing them in the disk milling. After milling, the purity of the crushed powder was characterized using XRF (Table 3). The Al content is at 91.90 ± 1.50%, which is slightly smaller than those reported in the literature (about 93–97% [21–23]). This may come from the paint coating of the can's exterior which was not removed before being recycled, to simplify the recycling process. The impurities are detected in minuscule amounts, where Mn, Si, Fe, and Cu are the four largest impurities percentage at 3.20, 2.80, 1.39, and 0.62%, respectively. Other impurities are less than 0.1% which are relatively negligible. These impurities have been reported to be, instead, intentionally added into HAp for implant materials in many reports to improve the mechanical and biochemical properties of HAp. Mn blended into HAp has been reported to improve the bioactivity of HAp [33]. Mn was also used as a HAp dopant to facilitate its synthesis process and improve its structural and thermal stability [34,35]. Similarly, Fe was also used to dope HAp to induce a superparamagnetic bioactive phase [36], alter its dielectric properties [37], and enhance osteoblast adhesion [35]. Additionally, both Mn and Fe doping show no toxic effect on the osteoblast cell [35]. Si-substituted HAp has also been reported to increase the HAp bioactivity and improve its thermal stability and sinterability [38]. Cu–HAp nanocomposite has also been synthesized to add antibacterial activity with good cytocompatibility [39].

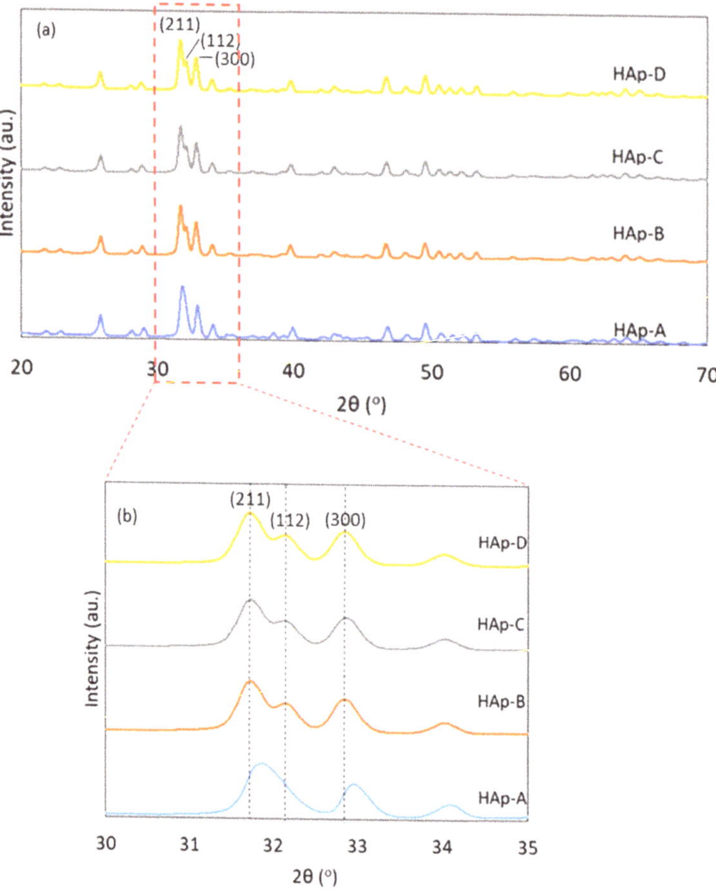


**Figure 1.** (**a**) X-ray diffractogram and (**b**) expanded X-ray diffractogram of the HAp powders taken from several batches of similar calcination process.

**Figure 2.** SEM-EDS of the HAp-based bovine bones after calcination in air atmosphere.

**Table 3.** XRF elemental composition of the Al powder recycled from the beverage can.

| Elements | Percentage (%) | Elements | Percentage (%) |
|---|---|---|---|
| Al | 91.90 ± 1.50 | Ni | 0.06 ± 0.02 |
| Mn | 3.20 ± 0.40 | Zr | 0.03 ± 0.03 |
| Si | 2.80 ± 0.80 | Ti | 0.02 ± 0.09 |
| Fe | 1.39 ± 0.19 | Zn | 0.01 ± 0.01 |
| Cu | 0.62 ± 0.12 | | |

For application as the reinforcement in the composite, the Al powder was further sieved to obtain a finer and homogenous particle size. The smaller particle size is supposed to give a better homogeneity of the particle's distribution, better bonding among the composite's components, and an increase of the composite's density [40]. Therefore, it is a crucial factor that determines the quality of the composite and is further studied in this report.

For the hydroxyapatite-composite (SHC) synthesis, the HAp, Al, and Mg powders (87:10:3 wt.%) were homogenously mixed which were then pelleted before being heated inside the SIS vessel. The compaction pressure is an important parameter in the SIS method [41]. Here, the effect of the variation of the Al particle size and the compaction pressure on the properties of the SHC is studied.

Figure 3 shows the effect of these variations on the hardness and porosity of the SHC produced. The largest hardness (44.92 HV) is obtained by the smallest particle size (<90 μm). The largest porosity (34.36%) is produced by the biggest particle size (>150 μm). Both with the same compaction pressure of 100 MPa. On the other hand, the smallest hardness (8.36 HV) is produced by the biggest particle size (>150 μm, with the compaction pressure of 100 MPa). Meanwhile, the smallest porosity (19.82%) is obtained by the smallest particle size (<90 μm, with the compaction pressure of 200 MPa). In general, for the same value of compaction pressure, there is a tendency that the smaller the Al particle size produces larger hardness and smaller porosity of the composite. This is in accord with the expectation that the smaller particle size can be homogenously distributed in the mixture and increase the composite's density [40], which is shown by the smaller porosity and larger hardness. However, if evaluating the value of each Al particle size group, the tendency is not consistent for all cases. For most of the cases, the larger the compaction pressure tends to give larger hardness and smaller porosity. The exception is for the smallest size-fraction (<90 μm) where the larger the compaction pressure instead gives the smaller hardness (dashed lines in Figure 3a). The trend tendency of the porosity, on the other hand, is more consistent (Figure 3b). The reason for the deviation of the hardness trend is unconfirmed at the present but might be related to the formation of the oxides, which will be discussed later. The largest hardness obtained in this study (44.92 HV) is close to those of the human teeth dentin's (46–64 HV [42,43]) and human cortical bone's (40.38 HV [44]). However, its porosity is at 31.12% while the human cortical bone's one needs to be less than 30% [45]. Some samples fulfilled the porosity criteria for the cortical bone in this study such as (90–105 μm) 171 MPa and (<90 μm) 171 MPa with the porosity of 29.45 and 28.58%, respectively, but their hardness values are much lower than the criteria (at 22.15 and 34.52, respectively). However, judging the trend of the hardness and porosity in the (<90 μm) group, it can be projected that the criteria of hardness at around 40 HV with the porosity of less than 30% is possible to be obtained using a compaction pressure between 100 and 171 MPa.

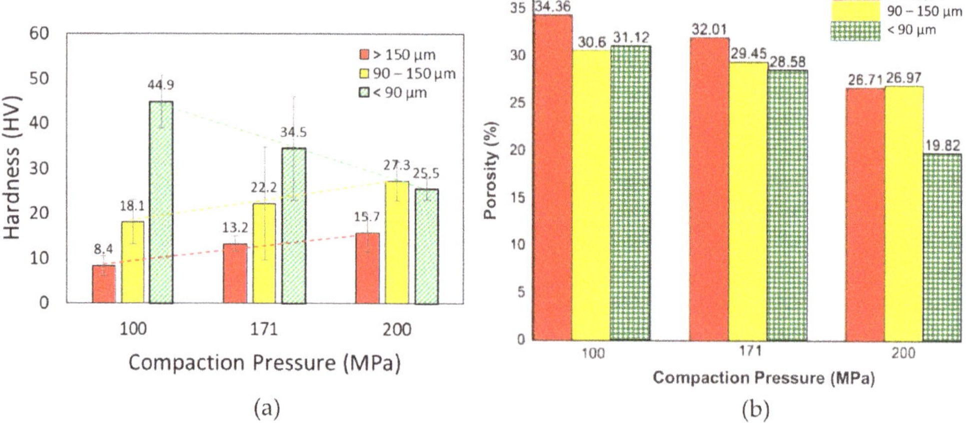

**Figure 3.** Bar charts of the (**a**) hardness and (**b**) porosity of the hydroxyapatite composite as a function of Al particle size-fraction and compaction pressure. Dashed lines are for a vision guide.

Figure 4 shows the XRD of the SHC made by the compaction pressure at 100 MPa with the Al particle size of <90 and >150 μm. Both samples show the formation of oxides in the form of magnesia (MgO), and spinel (MgAl$_2$O$_4$) beside the hydroxyapatite. The formation of the oxides is because the SIS heating was performed in a non-vacuum furnace and without flowing an inert gas as well. Therefore, there was air trapped in the SIS vessel which allows oxidation of the metal components. However, for the Al particle size of <90 μm, there are extra alumina (Al$_2$O$_3$) peaks detected (Figure 4a).

This indicates that in the case of smaller Al particle size (<90 μm), the oxidation of the Al metal into alumina is easier compared to those of spinel. On the other hand, for bigger Al particle size (>150 μm), the oxidation of Al metal is more favorable toward the spinel formation, as shown by the presence of Al and spinel peaks but in the absence of alumina (Figure 4b). Alumina is a known bio-ceramic for medical applications, such as in dental applications (i.e., orthodontic brackets, dental implants, fixed prostheses, and bone cement filler) [46], or joint replacements [47]. The presence of alumina has also been reported to increase the hardness of a composite [48]. The presence of alumina might be responsible for the increase of the hardness observed from 8.36 to 44.92 HV for >150 and <90 μm, respectively. Another possibility of the increase of hardness is coming from the increase of crystallinity of the HAp [49]. The HAp peak at (112) in the >150 μm sample is relatively taller compared to the <90 μm one, indicating its higher crystallinity. The higher the HAp crystallinity instead produces a smaller hardness value in this study, opposite to those reported by Ref. [49]. This indicates that the effect of $Al_2O_3$ formation on the composite's hardness is more dominant.

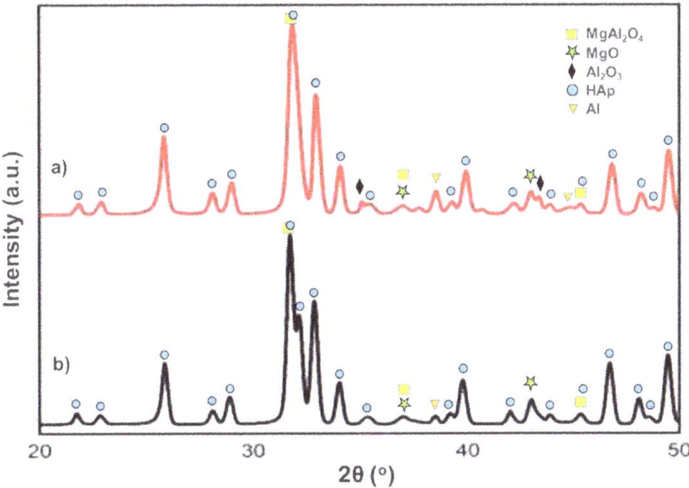


**Figure 4.** XRD of the hydroxyapatite composite made by compaction pressure of 100 MPa and Al particle size of (**a**) <90 and (**b**) >150 μm.

Figure 5 shows the XRD of the hydroxyapatite composite made using the smaller Al particle size (<90 μm) with different compaction pressures of 100 and 200 MPa. As explained above, for the 100 MPa case, the peaks of alumina and Al metal can be observed (Figure 5b). However, increasing the pressure to 200 MPa completely transforms Al metal into the $MgAl_2O_4$ spinel, as shown in the absence of both alumina and Al metal peaks (Figure 5a). This explains the decrease of the hardness of the 200 MPa sample compared to the 100 MPa one (25.49 and 44.92 HV for 200 and 100 MPa samples, respectively), even though the higher pressure is supposed to increase the density (as shown by the smaller porosity in Figure 3b), which eventually increase the hardness. This indicates that for the hydroxyapatite composite made of smaller Al particle size (<90 μm) in this study, the presence of alumina is crucial in increasing its mechanical properties (hence the hardness). The HAp peak at (112) of the 200 MPa sample is relatively taller than those of the 100 MPa sample, indicating its higher crystallinity. Again, like in the above case, the higher HAp crystallinity instead produces smaller hardness, indicating that the effect of $Al_2O_3$ formation on the hardness is more dominant.

Figures 6 and 7 show the SEM-EDS pictures of the SHC made by compaction pressure of 100 MPa and Al particle size of <90 and >150 μm, respectively. The Ca and P distribution of the >150 μm sample seems to be better than the <90 μm one. Nevertheless, for each case, the Ca and P have an almost similar distribution, indicating the formation of the hydroxyapatite. The Ca/P ratio for the two samples, calculated from the elemental composition derived from the EDS measurements (Table 4), is almost similar at 1.47 and 1.46 for <90 and >150 μm, respectively, slightly smaller than the typical hydroxyapatite at 1.67 [32]. For both cases, the distribution of Al and Mg were not necessarily overlapping, indicating the formation of other phases besides the spinel $MgAl_2O_4$. This is supporting the XRD result (Figure 4) where the Al and MgO peaks are also observed besides the spinel $MgAl_2O_4$ one.

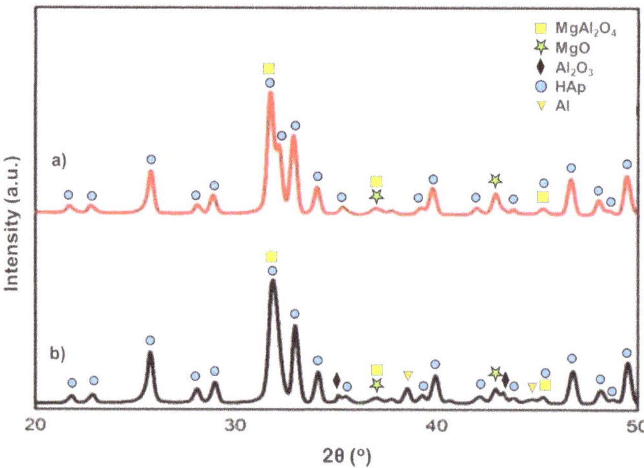

**Figure 5.** XRD of the hydroxyapatite composite made using Al particle size-fraction of <90 μm and compaction pressure of (**a**) 200 and (**b**) 100 MPa.

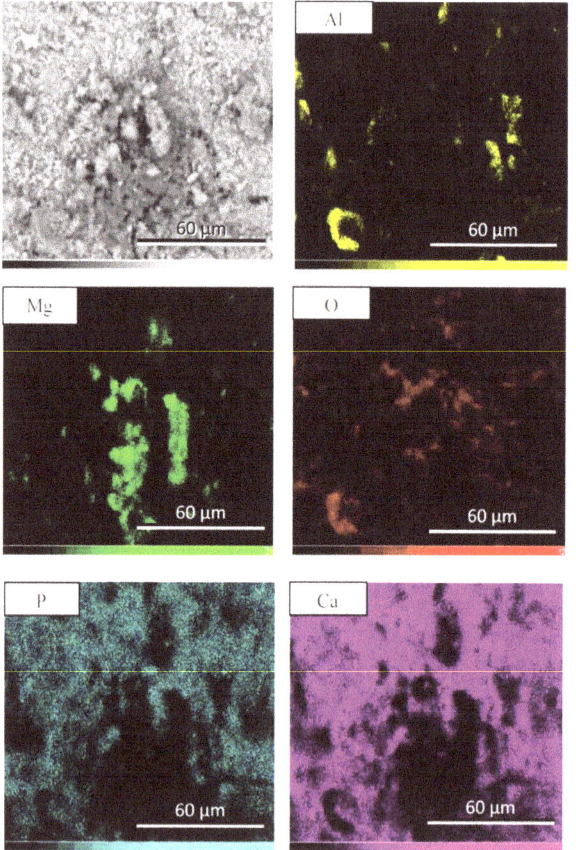

**Figure 6.** SEM-EDX pictures of the hydroxyapatite composite made by Al particle size of <90 μm with compaction pressure of 100 MPa.

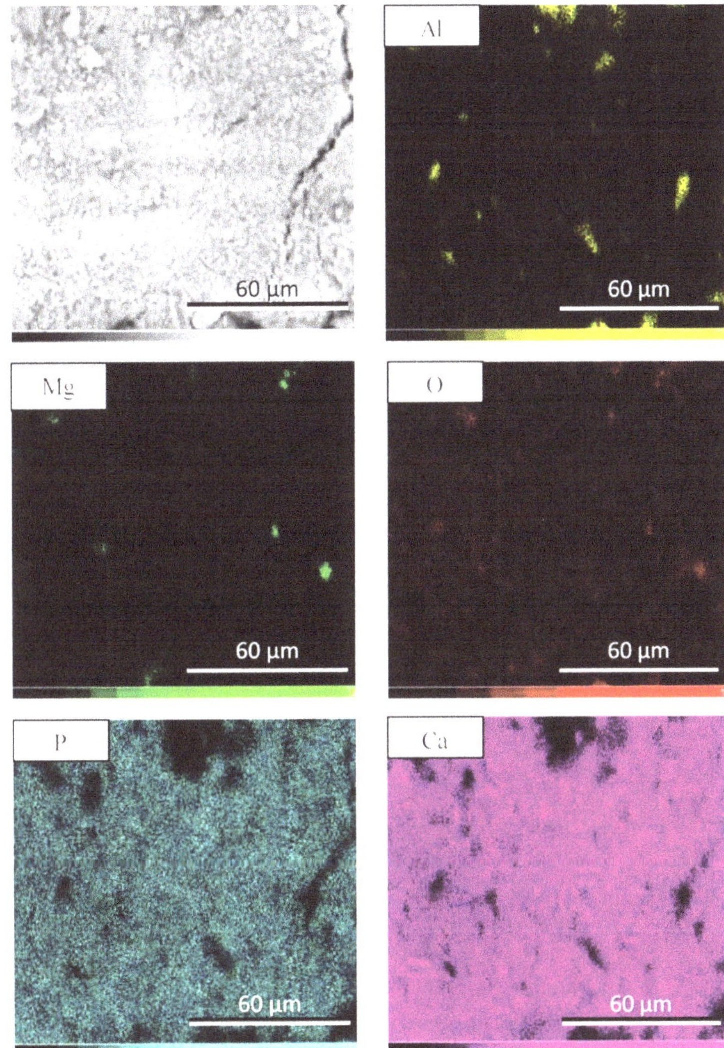

**Figure 7.** SEM-EDS pictures of the hydroxyapatite composite made by Al size-fraction of >150 μm with compaction pressure of 100 MPa.

**Table 4.** Elemental composition derived from the EDS measurements of the hydroxyapatite composite made by Al particle size of <90 μm and >150 μm with the compaction pressure of 100 MPa.

| Elements | at. % | |
|---|---|---|
| | <90 μm | >150 μm |
| O | 66.75 | 66.60 |
| Mg | 2.47 | 2.12 |
| Al | 2.46 | 4.22 |
| P | 11.45 | 10.20 |
| Ca | 16.87 | 14.86 |
| Total | 100.00 | 100.00 |

## 4. Conclusions

We have successfully synthesized spinel-based hydroxyapatite composite (SHC) using bovine bones and beverage cans as hydroxyapatite and aluminum (Al) metal sources, respectively, using the self-propagating intermediate-temperature synthesis (SIS) method. Aluminum particle size and compaction pressure play important roles in determining the mechanical properties of the SHC. Decreasing the Al particle size tends to increase the hardness and reduce the porosity of the SHC. Increasing compaction pressure also tends to decrease the porosity of the SHC. On the other hand, for the hardness value, the tendency of it to increase as the compaction pressure is increased only applies to the bigger Al particle sizes. For the smallest Al particle size, the higher the compaction pressure instead decreases the hardness value. This is probably related to the presence of the alumina in the smaller pressure which contributes to the improvement of the composite's mechanical properties.

**Supplementary Materials:** The following supporting information can be downloaded at https://www.mdpi.com/article/10.3390/cryst12010096/s1: Figure S1. Particle size distribution of (a) the synthesized HAp, and (b) Mg powder. Figure S2. Transient thermal distribution of heat flux by SIS heating. Figure S3. The heat flux plots of SIS heating. Figure S4. Results of Taguchi method optimization evaluated by Minitab. Figure S5. Photograph of bovine bones after calcination in (a) inert nitrogen gas, and (b) air atmosphere. Figure S6. Photograph of the HAp powder from different batches. Table S1. Difference between the minimum and maximum temperature of the sample during SIS heating. After cooling down, the hydroxyapatite-composite (SHC) samples in the SIS vessel were taken out then polished by sandpaper and characterized using Archimedes porosity test [24,25], Vickers hardness test (LECO, LM 100 AT, 3 indentations for each sample), X-ray diffraction (XRD, Smart Lab Rigaku, with the source of Cu-K$\alpha$), and scanning electron microscopy equipped with Energy X-ray Dispersion Spectroscopy (SEM-EDS, SU-3500 Hitachi, Oxford Instruments Aztech One).

**Author Contributions:** Conceptualization, D.S.K.; data curation, G.E.T.; formal analysis, G.P.A.R. and D.S.K.; funding acquisition, A.P.; investigation, G.P.A.R. and D.S.K.; methodology, A.P. and D.S.K.; supervision, A.P., G.E.T. and D.S.K.; validation, A.P., G.E.T. and G.P.A.R.; visualization, A.P.; writing—original draft, G.E.T.; writing—review & editing, A.P. and D.S.K. All authors have read and agreed to the published version of the manuscript.

**Funding:** This research was initiated by Hibah Penelitian Kementrian Pendidikan dan Kebudayaan, part of being funded by Penelitian Terapan Unggulan Perguruan Tinggi (PTUPT), with contract number: B/03/UN43.9/PT.00.03/2020.

**Institutional Review Board Statement:** Not applicable.

**Informed Consent Statement:** Not applicable.

**Data Availability Statement:** Data are contained within the article.

**Acknowledgments:** Authors acknowledge the Mechanical Materials Metallurgical Laboratory I and II, Engineering Faculty of Sultan Ageng Tirtayasa University and Research Center for Physics, National Research and Innovation Agency (BRIN) for research facilities.

**Conflicts of Interest:** The authors declare no conflict of interest.

## References

1. Wu, S.; Wang, J.; Zou, L.; Jin, L.; Wang, Z.; Li, Y. A Three-Dimensional Hydroxyapatite/Polyacrylonitrile Composite Scaffold Designed for Bone Tissue Engineering. *RSC Adv.* **2018**, *8*, 1730–1736. [CrossRef]
2. Sari, Y.W.; Rudianto, R.P.; Nuzulia, N.A.; Sukaryo, S.G. Injectable Bone Substitute Synthesized from Mangrove Snail Shell. *J. Med. Phys. Biophys.* **2017**, *4*, 115–121.
3. Odusote, J.K.; Danyuo, Y.; Baruwa, A.D.; Azeez, A.A. Synthesis and Characterization of Hydroxyapatite from Bovine Bone for Production of Dental Implants. *J. Appl. Biomater. Funct. Mater.* **2019**, *17*, 2280800019836829. [CrossRef]
4. Pon-On, W.; Suntornsaratoon, P.; Charoenphandhu, N.; Thongbunchoo, J.; Krishnamra, N.; Tang, I.M. Hydroxyapatite from Fish Scale for Potential Use as Bone Scaffold or Regenerative Material. *Mater. Sci. Eng. C* **2016**, *62*, 183–189. [CrossRef]
5. Vignesh Raj, S.; Rajkumar, M.; Meenakshi Sundaram, N.; Kandaswamy, A. Synthesis and Characterization of Hydroxyapatite/Alumina Ceramic Nanocomposites for Biomedical Applications. *Bull. Mater. Sci.* **2018**, *41*, 93. [CrossRef]

6.  Budiatin, A.S.; Gani, M.A.; Samirah; Ardianto, C.; Raharjanti, A.M.; Septiani, I.; Putri, N.P.K.P.; Khotib, J. Bovine Hydroxyapatite-Based Bone Scaffold with Gentamicin Accelerates Vascularization and Remodeling of Bone Defect. *Int. J. Biomater.* **2021**, *2021*, 5560891. [CrossRef] [PubMed]

7.  Huang, Y.; Hsiao, P.; Chai, H. Hydroxyapatite Extracted from Fish Scale: Effects on MG63 Osteoblast-like Cells. *Ceram. Int.* **2011**, *37*, 1825–1831. [CrossRef]

8.  Kumar, G.S.; Sathish, L.; Govindan, R.; Girija, E.K. Utilization of Snail Shells to Synthesise Hydroxyapatite Nanorods for Orthopedic Applications. *RSC Adv.* **2015**, *5*, 39544–39548. [CrossRef]

9.  Abdulrahman, I.; Tijani, H.I.; Mohammed, B.A.; Saidu, H.; Yusuf, H.; Ndejiko Jibrin, M.; Mohammed, S. From Garbage to Biomaterials: An Overview on Egg Shell Based Hydroxyapatite. *J. Mater.* **2014**, *2014*, 802467. [CrossRef]

10. Kattimani, V.; Lingamaneni, K.P.; Chakravarthi, P.S.; Sampath Kumar, T.S.; Siddharthan, A. Eggshell-Derived Hydroxyapatite: A New Era in Bone Regeneration. *J. Craniofac. Surg.* **2016**, *27*, 112–117. [CrossRef] [PubMed]

11. Radha, R.; Sreekanth, D. Mechanical and Corrosion Behaviour of Hydroxyapatite Reinforced Mg-Sn Alloy Composite by Squeeze Casting for Biomedical Applications. *J. Magnes. Alloy.* **2020**, *8*, 452–460. [CrossRef]

12. Witte, F.; Feyerabend, F.; Maier, P.; Fischer, J.; Störmer, M.; Blawert, C.; Dietzel, W.; Hort, N. Biodegradable Magnesium-Hydroxyapatite Metal Matrix Composites. *Biomaterials* **2007**, *28*, 2163–2174. [CrossRef]

13. Lim, P.N.; Lam, R.N.; Zheng, Y.F.; Thian, E.S. Magnesium-Calcium/Hydroxyapatite (Mg-Ca/HA) Composites with Enhanced Bone Differentiation Properties for Orthopedic Applications. *Mater. Lett.* **2016**, *172*, 193–197. [CrossRef]

14. Del Campo, R.; Savoini, B.; Muñoz, A.; Monge, M.A.; Garcés, G. Mechanical Properties and Corrosion Behavior of Mg-HAP Composites. *J. Mech. Behav. Biomed. Mater.* **2014**, *39*, 238–246. [CrossRef] [PubMed]

15. Li, J.; Fartash, B.; Hermansson, L. Hydroxyapatite-Alumina Composites and Bone-Bonding. *Biomaterials* **1995**, *16*, 417–422. [CrossRef]

16. Yoon, H.J.; Yoon, J.H.; Park, S.H.; Lee, M.H.; Han, J.S.; Kim, D.J. The Role of MgAl2O4 Powder Packing on Phase Stability of Hydroxyapatite during Sintering. *J. Am. Ceram. Soc.* **2015**, *98*, 1787–1793. [CrossRef]

17. Sreekanth, D.; Rameshbabu, N. Development and Characterization of MgO/Hydroxyapatite Composite Coating on AZ31 Magnesium Alloy by Plasma Electrolytic Oxidation Coupled with Electrophoretic Deposition. *Mater. Lett.* **2012**, *68*, 439–442. [CrossRef]

18. Wei, G.; Ma, P.X. Structure and Properties of Nano-Hydroxyapatite/Polymer Composite Scaffolds for Bone Tissue Engineering. *Biomaterials* **2004**, *25*, 4749–4757. [CrossRef]

19. Papynov, E.K.; Shichalin, O.O.; Apanasevich, V.I.; Plekhova, N.G.; Buravlev, I.Y.; Zinoviev, S.V.; Mayorov, V.Y.; Fedorets, A.N.; Merkulov, E.B.; Shlyk, D.K.; et al. Synthetic Nanostructured Wollastonite: Composition, Structure and "in Vitro" Biocompatibility Investigation. *Ceram. Int.* **2021**, *47*, 22487–22496. [CrossRef]

20. Papynov, E.K.; Shichalin, O.O.; Buravlev, I.Y.; Portnyagin, A.S.; Belov, A.A.; Maiorov, V.Y.; Skurikhina, Y.E.; Merkulov, E.B.; Glavinskaya, V.O.; Nomerovskii, A.D.; et al. Reactive Spark Plasma Synthesis of Porous Bioceramic Wollastonite. *Russ. J. Inorg. Chem.* **2020**, *65*, 263–270. [CrossRef]

21. Risonarta, V.Y.; Anggono, J.; Suhendra, Y.M.; Nugrowibowo, S.; Jani, Y. Strategy to Improve Recycling Yield of Aluminium Cans. *E3S Web Conf.* **2019**, *130*, 01033. [CrossRef]

22. Begum, S. Recycling of Aluminum from Aluminum Cans. *J. Chem. Soc. Pakistan* **2013**, *35*, 1490–1493.

23. Akrom, M.; Marwoto, P. Sugianto Pembuatan Mmc Berbasis Teknologi Metalurgi Serbuk Dengan Bahan Baku Aluminium Dari Limbah Kaleng Minuman Dan Aditif Abu Sekam Padi. *J. Pendidik. Fis. Indones.* **2010**, *6*, 14–19.

24. Slotwinski, J.A.; Garboczi, E.J.; Hebenstreit, K.M. Porosity Measurements and Analysis for Metal Additive Manufacturing Process Control. *J. Res. Natl. Inst. Stand. Technol.* **2014**, *119*, 494–528. [CrossRef] [PubMed]

25. Converse, G.L.; Conrad, T.L.; Merrill, C.H.; Roeder, R.K. Hydroxyapatite Whisker-Reinforced Polyetherketoneketone Bone Ingrowth Scaffolds. *Acta Biomater.* **2010**, *6*, 856–863. [CrossRef] [PubMed]

26. Poovazhagan, L.; Rajkumar, K.; Saravanamuthukumar, P.; Javed Syed Ibrahim, S.; Santhosh, S. Effect of Magnesium Addition on Processing the Al-0.8 Mg-0.7 Si/SiCp Metal Matrix Composites. *Appl. Mech. Mater.* **2015**, *787*, 553–557. [CrossRef]

27. Sangghaleh, A.; Halali, M. Effect of Magnesium Addition on the Wetting of Alumina by Aluminium. *Appl. Surf. Sci.* **2009**, *255*, 8202–8206. [CrossRef]

28. Hasanah, I.U.; Fadhillah, M.F.; Putra, Y.V. Fabrication and Characterization of Aluminum Matrix Composite (AMCs) Reinforced Graphite by Stir Casting Method for Automotive Application. *Mater. Sci. Forum* **2020**, *988*, 17–22.

29. Dahi, E. Optimisation of Bone Char Production Using the Standard Defluoridation Capacity Procedure. *Fluoride* **2015**, *48*, 29–36.

30. Fadhilah, N.; Irhamni, I.; Jalil, Z. Synthesis of Natural Hydroxyapatite from Aceh's Bovine Bone. *J. Aceh Phys. Soc.* **2016**, *5*, 19–21.

31. Londoño-Restrepo, S.M.; Herrera-Lara, M.; Bernal-Alvarez, L.R.; Rivera-Muñoz, E.M.; Rodriguez-García, M.E. In-Situ XRD Study of the Crystal Size Transition of Hydroxyapatite from Swine Bone. *Ceram. Int.* **2020**, *46*, 24454–24461. [CrossRef]

32. Singh, T.P.; Singh, H.; Singh, H. Characterization of Thermal Sprayed Hydroxyapatite Coatings on Some Biomedical Implant Materials. *J. Appl. Biomater. Funct. Mater.* **2014**, *12*, 48–56. [CrossRef]

33. Mahabole, M.; Bahir, M.; Khairnar, R. Mn Blended Hydroxyapatite Nanoceramic: Bioactivity, Dielectric and Luminescence Studies. *J. Biomim. Biomater. Tissue Eng.* **2013**, *18*, 43–59. [CrossRef]

34. Paluszkiewicz, C.; Ślósarczyk, A.; Pijocha, D.; Sitarz, M.; Bućko, M.; Zima, A.; Chróścicka, A.; Lewandowska-Szumieł, M. Synthesis, Structural Properties and Thermal Stability of Mn-Doped Hydroxyapatite. *J. Mol. Struct.* **2010**, *976*, 301–309. [CrossRef]

35. Li, Y.; Widodo, J.; Lim, S.; Ooi, C.P. Synthesis and Cytocompatibility of Manganese (II) and Iron (III) Substituted Hydroxyapatite Nanoparticles. *J. Mater. Sci.* **2012**, *47*, 754–763. [CrossRef]
36. Panseri, S.; Cunha, C.; D'Alessandro, T.; Sandri, M.; Giavaresi, G.; Marcacci, M.; Hung, C.T.; Tampieri, A. Intrinsically Superparamagnetic Fe-Hydroxyapatite Nanoparticles Positively Influence Osteoblast-like Cell Behaviour. *J. Nanobiotechnol.* **2012**, *10*, 32. [CrossRef] [PubMed]
37. Kaygili, O.; Dorozhkin, S.V.; Ates, T.; Al-Ghamdi, A.A.; Yakuphanoglu, F. Dielectric Properties of Fe Doped Hydroxyapatite Prepared by Sol-Gel Method. *Ceram. Int.* **2014**, *40*, 9395–9402. [CrossRef]
38. Bianco, A.; Cacciotti, I.; Lombardi, M.; Montanaro, L. Si-Substituted Hydroxyapatite Nanopowders: Synthesis, Thermal Stability and Sinterability. *Mater. Res. Bull.* **2009**, *44*, 345–354. [CrossRef]
39. Hadidi, M.; Bigham, A.; Saebnoori, E.; Hassanzadeh-Tabrizi, S.A.; Rahmati, S.; Alizadeh, Z.M.; Nasirian, V.; Rafienia, M. Electrophoretic-Deposited Hydroxyapatite-Copper Nanocomposite as an Antibacterial Coating for Biomedical Applications. *Surf. Coat. Technol.* **2017**, *321*, 171–179. [CrossRef]
40. Oh, S.-T.; Sekino, T.; Niihara, K. Effect of Particle Size Distribution and Mixing Homogeneity on Microstructure and Strength of Alumina/Copper Composites. *Nanostruct. Mater.* **1998**, *10*, 327–332. [CrossRef]
41. Halverson, D.C.; Pyzik, A.J.; Aksay, I.A.; Snowden, W.E. Processing of Boron Carbide-Aluminum Composites. *J. Am. Ceram. Soc.* **1989**, *72*, 775–780. [CrossRef]
42. Del Gutiérrez-Salazar, M.P.; Reyes-Gasga, J. Microhardness and Chemical Composition of Human Tooth. *Mater. Res.* **2003**, *6*, 367–373. [CrossRef]
43. Craig, R.G.; Peyton, F.A. The Micro-Hardness of Enamel and Dentin. *J. Dent. Res.* **1958**, *37*, 661–668. [CrossRef]
44. Pramanik, S.; Agarwal, A.K.; Rai, K.N. Development of High Strength Hydroxyapatite for Hard Tissue Replacement. *Trends Biomater. Artif. Organs* **2005**, *19*, 45–49.
45. Keaveny, T.M.; Morgan, E.F.; Yeh, O.C. Bone Mechanics. In *Standard Handbook of Biomedical Engineering and Design*; Kutz, M., Ed.; The McGraw-Hill Companies, Inc.: New York, NY, USA, 2004; pp. 8.1–8.23, ISBN 9780071356374.
46. Al-Sanabani, F.A.; Madfa, A.A.; Al-Qudaini, N. Alumina Ceramic for Dental Applications: A Review Article. *Am. J. Mater. Res.* **2014**, *1*, 26–34.
47. Piconi, C.; Maccauro, G.; Muratori, F.; Brach Del Prever, E. Alumina and Zirconia Ceramics in Joint Replacements. *J. Appl. Biomater. Biomech.* **2003**, *1*, 19–32.
48. Latief, F.H.; Chafidz, A.; Junaedi, H.; Alfozan, A.; Khan, R. Effect of Alumina Contents on the Physicomechanical Properties of Alumina (Al2O3) Reinforced Polyester Composites. *Adv. Polym. Technol.* **2019**, *2019*, 5173537. [CrossRef]
49. Yanagisawa, K.; Zhu, K.; Fujino, T.; Onda, A. Preparation of Hydroxyapatite Ceramics by Hydrothermal Hot-Pressing Technique. *Key Eng. Mater.* **2006**, *309–311*, 57–60. [CrossRef]

*crystals*

*Article*

# Morphology of Barite Synthesized by In-Situ Mixing of Na$_2$SO$_4$ and BaCl$_2$ Solutions at 200 °C

Chunyao Wang [1,2], Li Zhou [3,4,*], Shuai Zhang [1,2], Li Wang [1], Chunwan Wei [5], Wenlei Song [6,7], Liping Xu [8] and Wenge Zhou [1,*]

[1] Key Laboratory of High-Temperature and High-Pressure Study of the Earth's Interior, Institute of Geochemistry, Chinese Academy of Sciences, Guiyang 550081, China; wangchunyao@mail.gyig.ac.cn (C.W.); shuaizhang19950902@163.com (S.Z.); wangli19940204@foxmail.com (L.W.)

[2] College of Earth and Planetary Sciences, University of Chinese Academy of Sciences, Beijing 100049, China

[3] School of Geography and Environmental Science, Guizhou Normal University, Guiyang 550025, China

[4] State Key Laboratory Incubation Base for Karst Mountain Ecology Environment of Guizhou Province, Guiyang 550001, China

[5] Key Laboratory of Orogenic Belts and Crustal Evolution, School of Earth and Space Sciences, Peking University, Beijing 100871, China; cwwei@pku.edu.cn

[6] State Key Laboratory of Continental Dynamics, Department of Geology, Northwest University, Xi'an 710069, China; wlsong@nwu.edu.cn

[7] BIC Brno, Technology Innovation Transfer Chamber, 61200 Brno, Czech Republic

[8] School of Fundamental Science, Zhejiang Pharmaceutical College, No. 666 Siming Road, Ningbo 315500, China; xulp@zjpu.edu.cn

* Correspondence: zhouli@gznu.edu.cn (L.Z.); zhouwenge@mail.gyig.ac.cn (W.Z.)

**Citation:** Wang, C.; Zhou, L.; Zhang, S.; Wang, L.; Wei, C.; Song, W.; Xu, L.; Zhou, W. Morphology of Barite Synthesized by In-Situ Mixing of Na$_2$SO$_4$ and BaCl$_2$ Solutions at 200 °C. *Crystals* **2021**, *11*, 962. https://doi.org/10.3390/cryst11080962

Academic Editor: Vladislav V. Gurzhiy

Received: 8 July 2021
Accepted: 14 August 2021
Published: 16 August 2021

**Abstract:** Barite is an abundant sulfate mineral in nature. Especially, the variety of morphologies of barite is often driven by the mixing of Ba-bearing hydrothermal fluid and sulfate-bearing seawater around hydrothermal chimneys. In order to better understand the factors affecting the morphology and precipitation mechanism(s) of barite in seafloor hydrothermal systems, we synthesized barite by a new method of in-situ mixing of BaCl$_2$ and Na$_2$SO$_4$ solutions at 200 °C while varying Ba concentrations and ratios of Ba$^{2+}$/SO$_4{}^{2-}$, and at room temperature for comparison. The results show that barite synthesized by in-situ mixing of BaCl$_2$ and Na$_2$SO$_4$ solutions at 200 °C forms a variety of morphologies, including rod-shaped, granular, plate-shaped, dendritic, X-shaped, and T-shaped morphologies, while room temperature barites display relatively simple, granular, or leaf like morphologies. Thus, temperature affects barite morphology. Moreover, dendritic barite crystals only occurred at conditions where Ba$^{2+}$ is in excess of SO$_4{}^{2-}$ at the experimental concentrations. The dendritic morphology of barite may be an important typomorphic feature of barite formed in high-temperature fluids directly mixing with excess Ba$^{2+}$ relative to SO$_4{}^{2-}$.

**Keywords:** Barite; hydrothermal synthesis; typomorphic characteristics; in-situ mixing solutions at high temperature

## 1. Introduction

Barite (BaSO$_4$) is a ubiquitous mineral in the earth's crust [1,2]. Early Earth's marine environments were anoxic and sulfate deposits prevailed, resulting in barite as the dominant or the only sulfate mineral within bedded sulfate deposits older than 2.4 Ga [3–7]. Due to its high density and strong resistance to chemical weathering, barite is present throughout Earth's history and has the potential to preserve a record of conditions attending formation, which could be useful for interpretations of Earth's ancient rocks and paleoenvironments [8].

There are many ways of forming natural barite, including magmatic, biogenic, and hydrothermal genesis. Magmatic barite has been found in the Mianning rare earth deposit, Sichuan, Southwestern China [9]. In this deposit, barite is believed to have formed prior to or at the same time as massive quartz and fluorite, in the late stages of magma crystallization [10]. Biogenic barite occurs in various sedimentary environments, generally by combining barium with the skeletons of plankton [11,12]. Hydrothermal barite forms by the reaction of ore-forming hydrothermal fluids with surrounding rock [13,14]. Moreover, hydrothermal barite is the most abundant mineral in seafloor sulfate-sulfide deposits formed in back-arc basins, such as Kuroko, Mariana, and Okinawa deposits [15]. In-depth studies of the deposition mechanisms of sulfate and sulfide minerals in these deposits show that barite is formed by the rapid mixing of $Ba^{2+}$-containing hydrothermal fluid with the surrounding $SO_4^{2-}$-rich seawater [16–19]. For example, barite in the hydrothermal vent of the Endeavour Segment, Juan de Fuca Ridge has a variety of morphologies, including plate, leaf, needle, or radiating cone shapes, and dendritic crystals [19].

On the other hand, many experimental studies on the precipitation of barite from aqueous solution have been carried out to investigate the precipitation kinetics of barite in hydrothermal environments [15,20–25]. The results show that the morphology of barite varies greatly, depending on many factors. However, in previous experiments, barium-bearing solutions and sulfate-bearing solutions were mixed at room temperature, before heating to the desired temperature for crystal growth. This process is quite different from the process of barite precipitation from hydrothermal fluids since barite is believed to be formed by mixing a $Ba^{2+}$-bearing solution with an $SO_4^{2-}$-bearing solution at high temperatures in natural hydrothermal systems [19]. Therefore, the previous experimental results may not be directly applicable to the precipitation of barite in hydrothermal vents.

In order to better simulate the formation process of barite, and better understand the factors affecting the morphology and precipitation mechanisms of barite in seafloor hydrothermal solutions, we synthesized barite by in-situ mixing of barium-bearing and sulfate-bearing solutions of variable concentrations and mixing ratios of $Ba^{2+}/SO_4^{2-}$ at 200 °C and discuss the effects of different mixing ratios of barium and sulfate solutions on the morphology of barite. We also compare the morphologies of barites synthesized by in-situ mixing of $Na_2SO_4$ and $BaCl_2$ solutions at 200 °C with those grown at room temperature.

## 2. Materials and Methods

### 2.1. Materials

The reagents, anhydrous sodium sulfate ($Na_2SO_4$, 99.99%, metals basis) and barium chloride dihydrate ($BaCl_2 \cdot 2H_2O$, 99.99%, metals basis), used in the experiments are all produced by Shanghai Aladdin Biochemical Technology Co., Ltd. Solutions were prepared with ultra-pure water. An $SO_4^{2-}$ concentration similar to that of seawater (0.03 mol/kg) was chosen. In addition, to investigate the effect of concentration on barite precipitation, a higher concentration of $SO_4^{2-}$ (0.1 mol/kg) was also prepared. $Ba^{2+}$ solutions of similar concentration (0.03 mol/kg and 0.1 mol/kg) were also prepared.

### 2.2. Experimental Methods

Hydrothermal synthesis is a common method of synthesis [26–28]. Rather than mixing solutions at room temperature and then heating them to high temperatures for crystal growth, this study synthesized barite by a new method of in-situ mixing of $BaCl_2$ and $Na_2SO_4$ solutions at 200 °C. The experiments utilized conventional stainless steel autoclaves with two Teflon bottle linings (Figure 1a). The height of the stainless steel autoclave is 70 mm and the diameter is 45 mm, while the heights of the outer Teflon bottle and the inner bottle are 55 mm and 30 mm, with diameters of 32 mm and 15 mm, respectively. The filling degree of the whole equipment is 47%. The sodium sulfate solution was first added to the inner Teflon bottle, and the barium chloride solution was added to the outer Teflon bottle, then the Teflon lining with these separated solutions was sealed within the stainless steel autoclave. The stainless steel autoclave with these starting solutions was maintained vertically while in the electric furnace to avoid solutions mixing at temperatures lower than 200 °C. After the temperature was raised to 200 °C and maintained for one hour, the electric furnace was rotated from vertical to horizontal, allowing the solutions in the inner and outer bottles to mix in-situ at 200 °C (Figure 1b). All experiments were kept at 200 °C for 7 days. At the end of the experiment, the power was cut off and the run product was quenched. Note that our pre-experiment results showed that no barite was observed in the solutions in the inner and outer bottles when keeping the temperature at 200 °C for 7 days without rotation. Therefore, the barite produced in the high-temperature rotation experiment was formed by in-situ mixing of solutions at high temperatures.

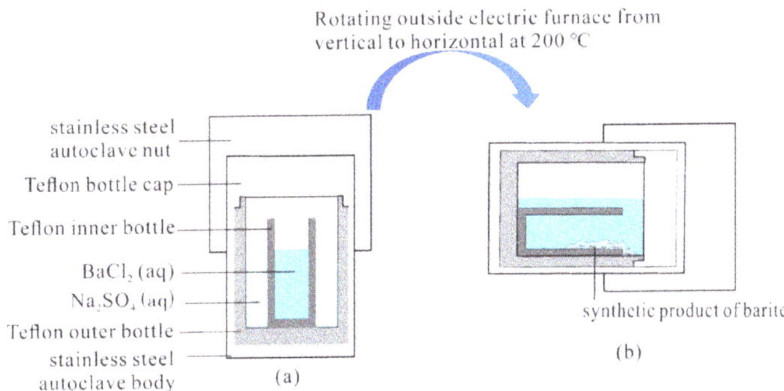


**Figure 1.** Simple diagram of synthetic equipment. (**a**) before mixing at 200 °C (**b**) after mixing at 200 °C.

After quenching and cooling, the solution and products in the Teflon bottle were poured into the centrifugal tube and centrifuged to extract the precipitate. Then the precipitate was washed three times alternately with deionized water and anhydrous ethanol. Finally, the white precipitate was dried at 60 °C to obtain a white powder.

For the mixing experiments at room temperature, different mixing ratios of sodium sulfate solution and barium chloride solution were added into the Teflon bottle and stirred evenly to produce a white precipitate, and then the mixed solution was kept at room temperature for 7 days. The purification of the product used the same procedure as that of the in-situ mixing experiments at 200 °C.

We carried out 10 experiments of in-situ mixing of Na₂SO₄ and BaCl₂ solutions at 200 °C and 10 experiments of mixing solutions at room temperature with 5 solution mixing ratios at two initial concentrations. The experimental conditions are shown in Table 1.

**Table 1.** The experimental conditions.

| Sample | Solution Mixing Ratio (Molar Ratio of Na₂SO₄ to BaCl₂) | $SO_4^{2-}$ (mol/kg) | $Ba^{2+}$ (mol/kg) | $T(°C)$ |
|---|---|---|---|---|
| 1 | 8:2 | 0.03 | 0.03 | 200 |
| 2 | 7:3 | 0.03 | 0.03 | 200 |
| 3 | 1:1 | 0.03 | 0.03 | 200 |
| 4 | 3:7 | 0.03 | 0.03 | 200 |
| 5 | 2:8 | 0.03 | 0.03 | 200 |
| 6 | 8:2 | 0.1 | 0.1 | 200 |
| 7 | 7:3 | 0.1 | 0.1 | 200 |
| 8 | 1:1 | 0.1 | 0.1 | 200 |
| 9 | 3:7 | 0.1 | 0.1 | 200 |
| 10 | 2:8 | 0.1 | 0.1 | 200 |
| 11 | 8:2 | 0.03 | 0.03 | Room temperature |
| 12 | 7:3 | 0.03 | 0.03 | Room temperature |
| 13 | 1:1 | 0.03 | 0.03 | Room temperature |
| 14 | 3:7 | 0.03 | 0.03 | Room temperature |
| 15 | 2:8 | 0.03 | 0.03 | Room temperature |
| 16 | 8:2 | 0.1 | 0.1 | Room temperature |
| 17 | 7:3 | 0.1 | 0.1 | Room temperature |
| 18 | 1:1 | 0.1 | 0.1 | Room temperature |
| 19 | 3:7 | 0.1 | 0.1 | Room temperature |
| 20 | 2:8 | 0.1 | 0.1 | Room temperature |

## 2.3. Analytical Methods

A Renishaw inVia Reflex spectrometer system equipped with a standard confocal microscope was used for Raman spectral analysis at the Institute of Geochemistry, Chinese Academy of Sciences. A Renishaw diode-pumped solid-state laser provided 532 nm laser excitation with 5 mW power at the sample. An 1800 grooves/mm grating was used giving a spectral resolution of 1.2 cm$^{-1}$. Depolarized Raman spectra were obtained using 10 s integration times with 5 accumulations and a 50× Leica long working distance microscope objective, which focused the beam to a spot size of 1.6 μm. Wavenumber calibration was carried out using a silicon standard.

The scanning electron microscope (SEM) images were obtained using an FEI- Scanning Electron Microscope with Scios Dual Beam System at the Institute of Geochemistry, Chinese Academy of Sciences. The instrument was operated at an acceleration voltage of 5–20 kV, a beam current of 0.2–0.4 nA, and a working distance of 7–10 mm. The sample was dispersed in alcohol, dropped onto conductive glue, coated with gold, and then analyzed by a scanning electron microscope.

Powder X-ray diffraction analyses of barite were carried out using an Empyrean diffractometer (Malvern Panalytical) at the Institute of Geochemistry, Chinese Academy of Sciences. The maximum power of the cermet X-ray tube (Cu target) was 2.2 kW; the minimum step of the goniometer was 0.0001°, and the working current was 40 mA, scanning speed was 10°/min, and the scanning range was 10–70°. The data obtained were analyzed with Jade software, and unit cell parameters were obtained.

## 3. Results

### 3.1. Morphology of Barite Synthesized by In-Situ Mixing of Solutions at 200 °C

As shown in Figure 2, barite synthesized using 5 mixing ratios of $Na_2SO_4$ and $BaCl_2$ solutions by in-situ mixing at 200 °C produced six different morphologies, including rod-shaped, granular, plate-shaped, dendritic, X-shaped, and T-shaped crystals. The specific morphologies appearing in the mixed solution of different proportions are shown in Table 2.

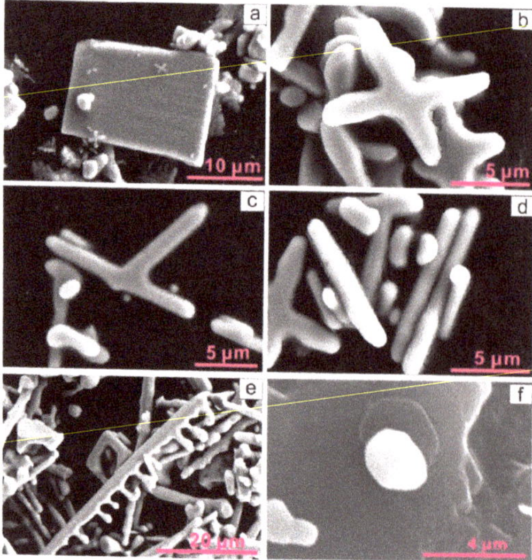

**Figure 2.** Scanning electron microscope (SEM) images of barite synthesized by in situ mixing of $Na_2SO_4$ and $BaCl_2$ solutions at 200 °C. (**a**) plate-shaped (**b**) X-shaped (**c**) T-shaped (**d**) rod-shaped (**e**) dendritic (**f**) granular.

**Table 2.** The morphological statistics of barite appearing in a mixed solution of barium salt and sulfate with different ratios at 200 °C.

| Synthetic Condition | Solution Mixing Ratio (Molar Ratio of $Na_2SO_4$ to $BaCl_2$) | Morphology of Barite | | | | | |
|---|---|---|---|---|---|---|---|
| | | Rod-Shaped | Granular | Plate-Shaped | Dendritic | X-Shaped | T-Shaped |
| 0.03 mol/kg $Na_2SO_4$ and $BaCl_2$ | 8:2 | + | + | - | - | + | + |
| | 7:3 | + | + | - | - | + | + |
| | 1:1 | + | + | + | - | + | + |
| | 3:7 | + | + | + | - | + | + |
| | 2:8 | + | + | + | + | + | + |
| 0.1 mol/kg $Na_2SO_4$ and $BaCl_2$ | 8:2 | + | + | - | - | + | + |
| | 7:3 | + | + | - | - | + | + |
| | 1:1 | + | + | + | - | + | + |
| | 3:7 | + | + | + | + | + | + |
| | 2:8 | + | + | + | + | + | + |

Note: "+" indicates existence and "-" indicates absence.

At concentrations of $Na_2SO_4$ and $BaCl_2$ solutions of 0.03 mol/kg, rod-shaped, granular, X-shaped, and T-shaped crystals appeared in all mixing ratios, while plate-shaped crystals appeared in the mixes of 1:1, 3:7, and 2:8 and dendritic shapes only appeared in mixes of 2:8. In addition, at concentrations of $Na_2SO_4$ and $BaCl_2$ of 0.1 mol/kg, rod-shaped, granular, X-shaped, and T-shaped crystals appeared in all 5 mixing proportions, while plate-shaped crystals appeared in the mixes of 1:1, 3:7, and 2:8, and dendritic appeared in the mixes of 3:7 and 2:8.

The histograms of the barite particle size synthesized in mixes of 0.03 mol/kg $Na_2SO_4$ and $BaCl_2$ with a ratio of 2:8 at 200 °C are shown in Figure 3. The plate-shaped barite with a particle size of 4.0–6.0 μm accounts for about 42% of the total, the rod-shaped barite with a particle size of 1.0–1.5 μm accounts for about 18% of the total, the granular barite with a particle size of 2.0–2.5 μm accounts for about 39% of the total, the dendritic barites with a particle size in the range of 1.5–2.0 μm and 2.0–2.5 μm account for about 26% of the total, respectively, the T-shaped barite with a particle size in the range of 3.0–4.0 μm and 4.0–5.0 μm account for about 33% of the total, respectively, and the X-shaped barite with a particle size of 5.0–6.0 μm accounts for about 33% of the total.

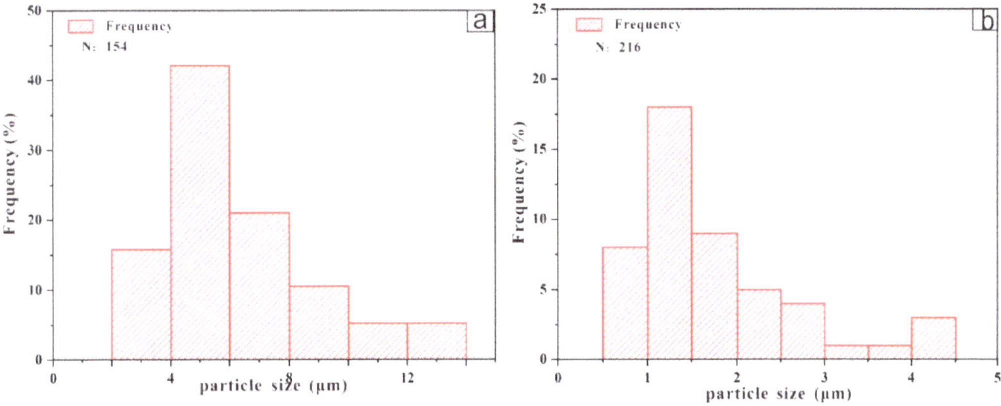


**Figure 3.** *Cont.*

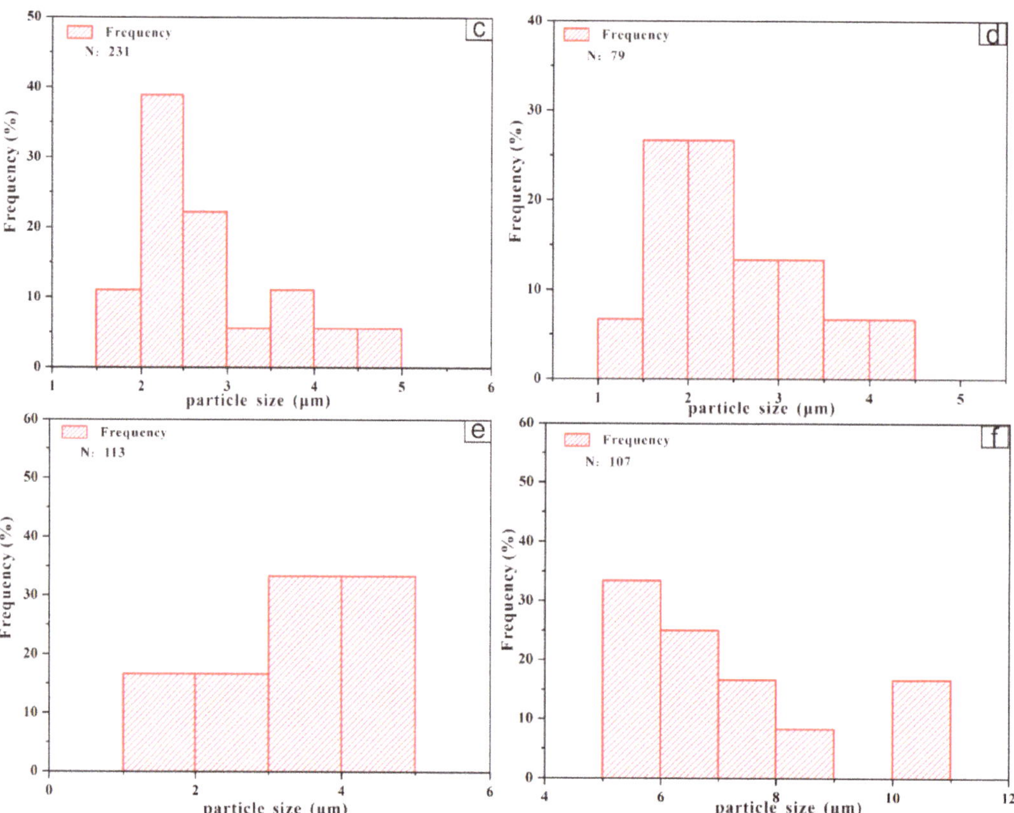

**Figure 3.** Histograms of the barite particle size with different morphology. The barite was synthesized using a mixed solution of 0.03 mol/kg $Na_2SO_4$ and $BaCl_2$ with a ratio of 2:8 at 200 °C. (**a**) plate-shaped (**b**) rod-shaped (**c**) granular (**d**) dendritic (**e**) T-shaped (**f**) X-shaped. N: statistic.

### 3.2. Morphology of Barite Synthesized at Room Temperature

When the initial ion concentration is constant, the morphologies of the barite synthesized at room temperature with different mixing ratios are the same, and the change of the mixing ratio does not change barite morphology (Figure 4). The morphology of barite is leaf-like when the initial concentration of $Na_2SO_4$ and $BaCl_2$ solution is 0.03 mol/kg; the morphology of barite consists of nanospheres when the initial concentration of $Na_2SO_4$ and $BaCl_2$ solution is 0.1 mol/kg. And the particle size distribution of barite synthesized by mixed solutions at room temperature shows a normal distribution (Figure 5).

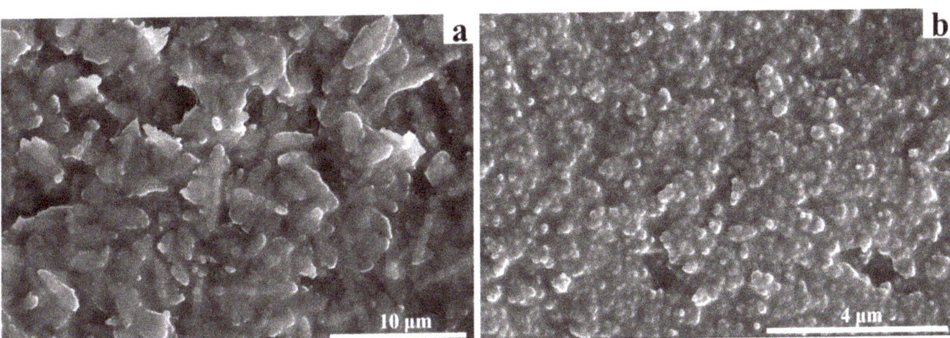

**Figure 4.** Scanning electron microscope (SEM) images of barite synthesized with a mixed solution of 0.03 mol/kg $Na_2SO_4$ and $BaCl_2$ (**a**) and a mixed solution of 0.1 mol/kg $Na_2SO_4$ and $BaCl_2$ (**b**) with a ratio of 1:1 at room temperature.

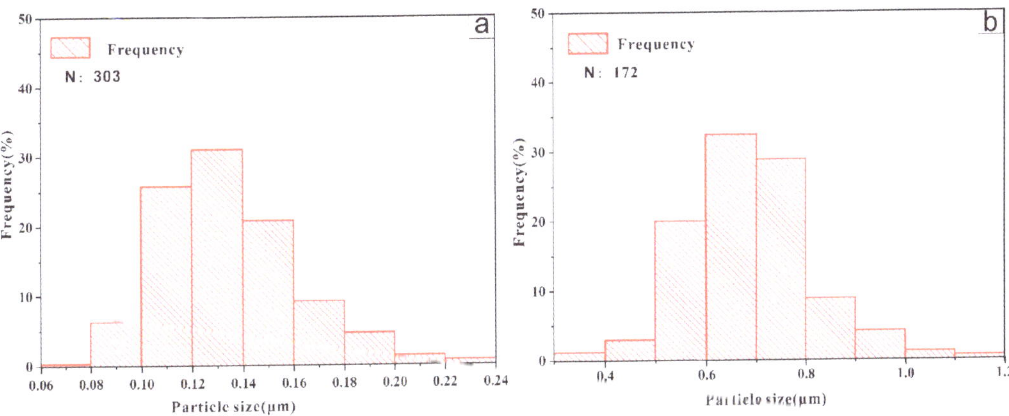

**Figure 5.** Histograms of barite particle size synthesized using a mixed solution of 0.1 mol/kg $Na_2SO_4$ and $BaCl_2$ (**a**) and a mixed solution of 0.03 mol/kg $Na_2SO_4$ and $BaCl_2$ (**b**) with a mixing ratio of 1:1 at room temperature. N: statistic.

### 3.3. Raman Spectra

The Raman spectra of barite synthesized at room temperature and 200 °C are shown in Figure 6. The structure of barite is orthorhombic, of the Pbnm space group [29]. Each S atom coordinates with four oxygen atoms to form $SO_4$ tetrahedra while each $Ba^{2+}$ coordinates with 12 oxygen atoms. The $SO_4$ tetrahedra have Cs site group symmetry which theoretically has 9 degrees of vibrational freedom [30] (i.e., one nondegenerate ($\nu_1$), one doubly degenerate ($\nu_2$), and two triply degenerate modes ($\nu_3$ and $\nu_4$)). Moreover, one additional mode at 1104 $cm^{-1}$ caused by $SO_4$ tetrahedral distortion remained unassigned in this study. The Raman peaks below 400 $cm^{-1}$ are classified as Ba-$O_{12}$ vibrations. The observed patterns are consistent with those reported previously [31,32] and are listed in Table 3.

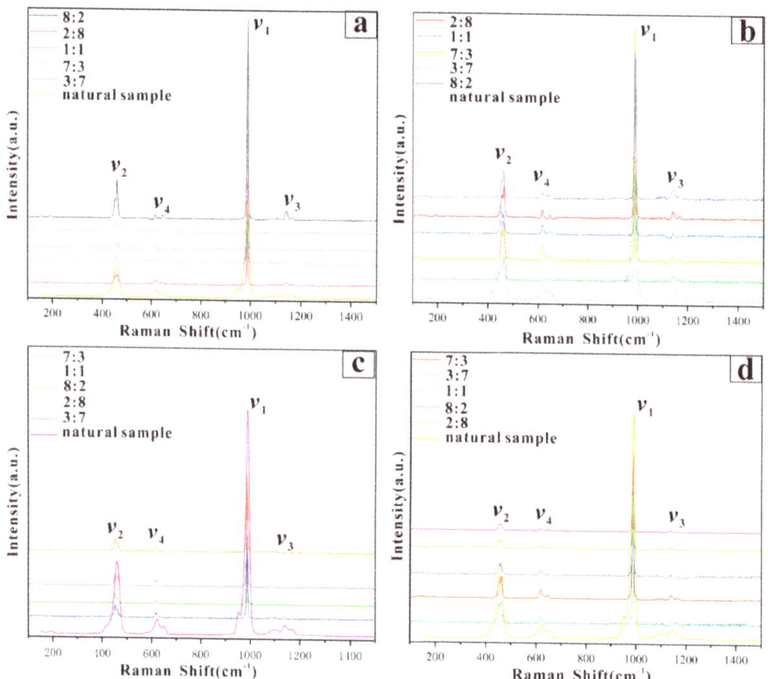

**Figure 6.** Raman spectra of barite synthesized by mixed solutions at room temperature and in-situ mixes of $Na_2SO_4$ and $BaCl_2$ solutions at 200 °C and a natural sample (RRUFFID = R050375). $v_1$, $v_2$, $v_3$, and $v_4$ are assigned to the vibrational modes of the $SO_4$ tetrahedron. (**a**) Synthesized at 200 °C with 0.03 mol/kg $Na_2SO_4$ and $BaCl_2$ solutions, (**b**) Synthesized at 200 °C with 0.1 mol/kg $Na_2SO_4$ and $BaCl_2$ solutions, (**c**) Synthesized at room temperature with 0.03 mol/kg $Na_2SO_4$ and $BaCl_2$ solutions, (**d**) Synthesized at room temperature with 0.1 mol/kg $Na_2SO_4$ and $BaCl_2$ solution. All proportions are molar ratios of $Na_2SO_4$ and $BaCl_2$ solutions.

**Table 3.** Observed Raman vibrational modes (in $cm^{-1}$) of synthesized barite.

| This Study | | | | [31,32] | Mode Assignment |
|---|---|---|---|---|---|
| a | b | c | d | | |
| 453 | 453 | 452 | 453 | 453 | $v_2$ $SO_4$ |
| 462 | 461 | 461 | 462 | 463 | $v_2$ $SO_4$ |
| 618 | 616 | 617 | 616 | | $v_4$ $SO_4$ |
| 622 | 624 | 622 | 622 | 623 | $v_4$ $SO_4$ |
| 649 | 647 | 646 | 647 | 647 | $v_4$ $SO_4$ |
| 988 | 988 | 989 | 987 | 988 | $v_1$ $SO_4$ |
| 1083 | 1084 | 1084 | 1083 | 1083 | $v_3$ $SO_4$ |
| 1103 | 1104 | 1104 | 1104 | 1105 | Unassigned |
| 1139 | 1138 | 1139 | 1138 | | $v_3$ $SO_4$ |
| 1166 | 1167 | 1168 | 1167 | 1167 | $v_3$ $SO_4$ |

Notes: a. Barite synthesized from 0.03 mol/kg $Na_2SO_4$ and $BaCl_2$ solutions with a mixing ratio of 8:2 at 200 °C; b. Barite synthesized by 0.1 mol/kg $Na_2SO_4$ and $BaCl_2$ solution with mixing ratio of 7:3 at 200 °C; c. Barite synthesized by 0.03 mol/kg $Na_2SO_4$ and $BaCl_2$ solution with mixing ratio of 1:1 at room temperature; d. Barite synthesized by 0.1 mol/kg $Na_2SO_4$ and $BaCl_2$ solution with a mixing ratio of 7:3 at room temperature.

The Raman spectra of barite synthesized in these experiments are all similar. Moreover, changes in initial ion concentrations, mixing ratios, and temperature do not change from the positions of characteristic peaks of barite at room temperature (Figure 6). The characteristic Raman peak positions of all synthesized barites are basically consistent with the characteristic peak positions of the pure barite Raman spectrum, and there are no other miscellaneous/unknown peaks. Therefore, Raman hasn't detected any impurities.

### 3.4. Powder XRD

Although the temperatures, concentrations, and mixing ratios are different, the $2\theta$ of the main peak of barite synthesized in our experiments are basically the same as those of the standard barite. Figure 7 shows representative XRD patterns of barites synthesized at room temperature and 200 °C with an initial concentration of 0.03 mol/kg $Na_2SO_4$ and $BaCl_2$ and a mixing ratio of 1:1. The cell parameters of barite synthesized at 200 °C are basically the same as those of barite synthesized at room temperature. The cell parameters of barite synthesized at 200 °C and room temperature are a = 0.8889 nm, b = 0.5454 nm, and c = 0.7160 nm and a = 0.8884 nm, b = 0.5445 nm, and c = 0.7156 nm, respectively, all in good agreement with standard barite (a = 0.8884 nm, b = 0.5456 nm, and c = 0.7157 nm, PDF#83-2053).

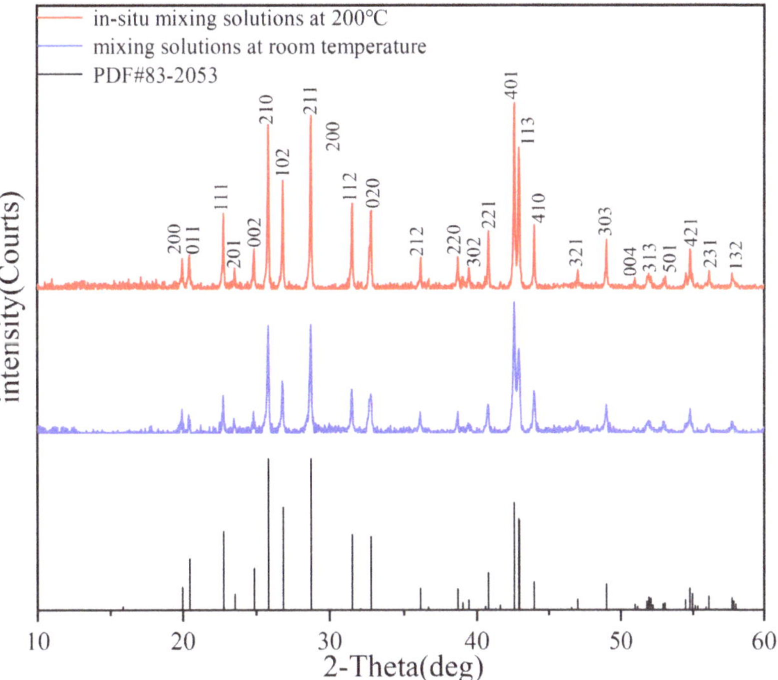


**Figure 7.** XRD diffractograms of barite synthesized by mixed solution at room temperature and in-situ mixing at 200 °C with an initial concentration of 0.03 mol/kg $Na_2SO_4$ and $BaCl_2$ and a mixing ratio of 1:1, along with that of standard barite (PDF#83-2053).

## 4. Discussion

### 4.1. The Effect of Temperature on the Morphology of Barite

Temperature is the key factor affecting the morphology of barite. At elevated temperatures, with all other reaction conditions the same, the morphology of the synthesized barites is different. In this study, the morphology of barite synthesized by mixing solutions at room temperature is relatively simple, which is granular or leaf-like (Figure 4). When the concentration of $Na_2SO_4$ and $BaCl_2$ solution is 0.03 mol/kg, barite is leaf-like, and when the concentrations of $Na_2SO_4$ and $BaCl_2$ solution are 0.1 mol/kg, barite crystals are small and granular. The morphology of the barite synthesized at 25 °C in a previous study [33] is similar to that of leafy barite synthesized at room temperature in this study. Barites synthesized at 150 °C by using barium chloride solutions and native sulfur have the morphologies of rod-shaped and X-shaped [10]. Barites synthesized in this study by in-situ mixing of $Na_2SO_4$ and $BaCl_2$ solutions at 200 °C show six different morphologies, including rod, granular, plate, dendritic, X-shaped, T-shaped crystals, while barites synthesized by mixing barium solutions and sulfate solutions at room temperature and subsequent heating to 200 °C, show near-equiaxed granular crystals of well-developed crystal forms [34]. Therefore, the morphologies of barites change from small granular and leaf-like at 25 °C to rod-shaped and X-shaped at 150 °C, and rod, granular, plate, dendritic, X-shaped, T-shaped crystals at 200 °C, which shows the key control of the morphology of barite is temperature.

### 4.2. Effect of Solution Mixing Ratio on the Morphology of Barite

Precipitation of barite occurs when the product of concentrations of $Ba^{2+}$ and $SO_4^{2-}$ solutions exceed the solubility constant [35]. Varying ionic ratios might be an effective and rather simple way of achieving a degree of crystal morphology control [33,36].

In this study, different solution mixing ratios change the dominant morphology of barite (Table 4). When the concentration of $Na_2SO_4$ and $BaCl_2$ solution is 0.03 mol/kg, the dominant morphology of barite with mixing ratios of 8:2 and 2:8 is rod-shaped, the dominant morphology of barite with a mixing ratio of 7:3 is T-shaped, and the dominant morphology of barite with mixing ratios of 1:1 and 3:7 is granular. However, when the initial concentration of $Na_2SO_4$ and $BaCl_2$ in solution increases from 0.03 mol/kg to 0.1 mol/kg, the dominant morphology of barite is rod-shaped regardless of the mixing ratio. In addition, when the mixing ratios of $Ba^{2+}$ to $SO_4^{2-}$ solutions with the same concentration is greater than 7:3, the average particle size is increased (Figure 8).

**Table 4.** Dominant morphology of barite synthesized by in-situ mixing of $Na_2SO_4$ and $BaCl_2$ solutions with different concentrations and mixing ratios at 200 °C.

| Synthetic Condition | | Dominant Morphology |
|---|---|---|
| **Initial Reactant Concentration** | **Solution Mixing Ratio** $(SO_4^{2-}/Ba^{2+},$ **Molar Ratio)** | |
| 0.03 mol/kg $Na_2SO_4$ and $BaCl_2$ | 8:2 | 28.2% rod-shaped |
| | 7:3 | 29.6% T-shaped |
| | 1:1 | 50.7% granular |
| | 3:7 | 32.8% granular |
| | 2:8 | 41.2% rod-shaped |
| 0.1 mol/kg $Na_2SO_4$ and $BaCl_2$ | 8:2 | 43.5% rod-shaped |
| | 7:3 | 37.7% rod-shaped |
| | 1:1 | 61.7% rod-shaped |
| | 3:7 | 45.8% rod-shaped |
| | 2:8 | 55.4% rod-shaped |

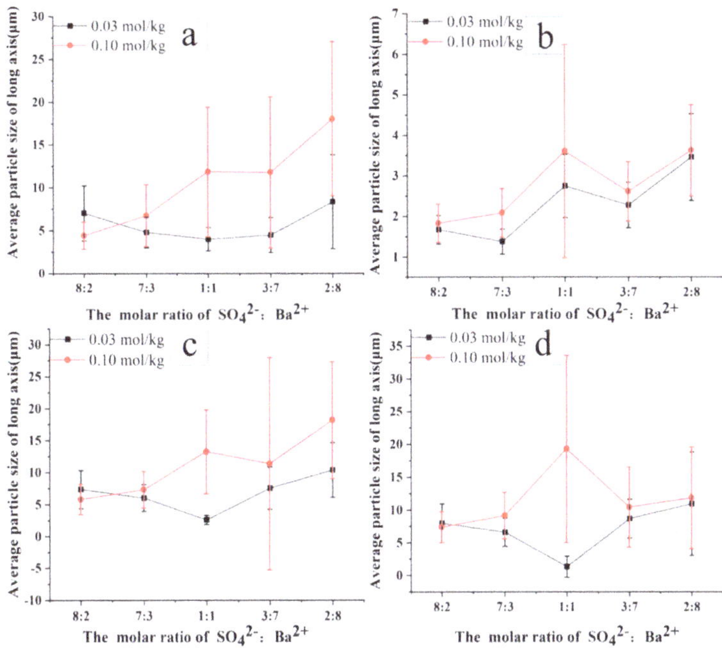

**Figure 8.** Average particle size of barite synthesized with different initial concentration solution and mixing ratio at 200 °C. (**a**) rod-shaped (**b**) granular (**c**) T-shaped (**d**) X-shaped.

For barite synthesized by mixing $Na_2SO_4$ and $BaCl_2$ solutions with the same concentration at room temperature, the morphology does not change with the initial solution mixing ratio (Figure 4).

*4.3. Effect of Supersaturation on the Morphology of Barite*

In many kinetic studies of precipitation, it was found that one of the key parameters for characterizing reacting systems is the supersaturation, S. In the case of barite, the supersaturation can be written as [37]:

$$s = \frac{a\left(Ba^{2+}\right) \cdot a(SO_4^{2-})}{K_{sp}},$$

where $a(Ba^{2+})$ and $a(SO_4^{2-})$ are the activities of $Ba^{2+}$ and $SO_4^{2-}$, respectively, $K_{sp}$ is the solubility product of barite. For barite, $K_{sp}$ is $9.82 \times 10^{-11}$ at 25 °C [8]. Previous work has demonstrated that at high reagent supersaturation ratios (the natural logarithm of supersaturation ratio, lnS = 16.97), a large number of very small spherical nanoparticles will be formed. At low reagent concentrations (lnS = 13.75), the leaf-like products will be formed [37]. Our experimental results are consistent with the previous research results. In our experiments, when the concentrations of $Na_2SO_4$ and $BaCl_2$ are 0.03 mol/kg, barite is leaf-like, and at concentrations of $Na_2SO_4$ and $BaCl_2$ are 0.1 mol/kg, barite is small and granular.

At 200 °C, the solubility product of barite is $4 \times 10^{-10}$ [38], which is larger than that at room temperature. Further, the degree of supersaturation with respect to barite can be changed by varying the barium chloride or sodium sulfate solution concentrations [33]. Dendritic crystals with rough surfaces formed from solutions of high barium chloride concentrations [39]. In our research, dendritic barite was also synthesized when the concentration of barium chloride was high.

Figure 8 shows that although the synthesis temperature and the initial ratios are the same, the particle size of the synthesized barite varies with the initial concentration of $Na_2SO_4$ and $BaCl_2$ in the solution. With increasing ion concentration, barite particle sizes increase, indicating a positive correlation between the initial concentration of $Na_2SO_4$ and $BaCl_2$ solution and the particle size of the synthesized barite. At 0.1 mol/kg $Na_2SO_4$ and $BaCl_2$ in solution, the average particle size of rod-shaped and T-shaped barite crystals increase, while the largest granular and X-shaped barite crystals, form in mixing ratios of 1:1. As plate-shaped and dendritic morphologies only form in mixtures of $Na_2SO_4$:$BaCl_2$ = 1:1 and 2:8, there is no comparison here. Table 4 shows the dominant morphology of barite changing when other reaction conditions are held constant (except for concentration). For example, at higher initial concentrations of $Na_2SO_4$ and $BaCl_2$, the dominant morphology of barite at the mixing ratio of 1:1 and 3:7 changes from granular to rod-like crystal (Table 4). Furthermore, at initial concentrations of $Na_2SO_4$ and $BaCl_2$ of 0.03 mol/kg, dendritic barite only appears at mixing ratios of 2:8. However, dendritic barite appears when the concentration of $Na_2SO_4$ and $BaCl_2$ increases to 0.1 mol/kg and the mixing ratio is 3:7 (Table 2), indicating that increasing $Ba^{2+}$ concentration favors the formation of dendritic barite.

### 4.4. Morphological Development of Barite

The morphological development of barite growing in solution is a complicated process. Previous studies indicate that the final morphology of barite depends on the competition of nucleation and crystal growth at different supersaturation ratios [37]. Moreover, the growth of crystals mainly depends on the growth conditions [40,41]. At lower reagent concentrations, homogeneous nucleation generates numerous nanoparticles, while the growth and heterogeneous nucleation of the particles forms bridges to connect the particles, forming leaf-like crystals [37]. However, at high reagent supersaturation ratios, a short nucleation burst leads to the production of abundant spherical nanoparticles, which exhausts most of the reagent and limits the further growth of particles [42]. The morphological development of barite synthesized at room temperature in this study is consistent with this scenario (Figure 4).

However, the development of the morphology of barite formed by in-situ mixing $BaCl_2$ and $Na_2SO_4$ solutions at 200 °C is more complicated. When mixing in situ at high temperatures, the nucleation and growth of barite occur simultaneously. There is no time to produce perfect crystal nuclei, and supersaturation in the environment around the crystal nucleus is high, so crystal nuclei tend to grow randomly, forming abundant defects. At these defect points, the crystal lattice is disturbed, and new layers begin growing [43]. Moreover, at higher temperatures the fluid state is unstable, and the atomic thermal vibration is strong, so $Ba^{2+}$ and $SO_4^{2-}$ ions can obtain enough energy to overcome the fluid resistance and migrate rapidly. As a result, a wide variety of shapes form, including rod-shaped, granular, plate-shaped, dendritic, X-shaped, and T-shaped.

Moreover, the dendritic barite synthesized by in-situ mixing of 0.03 mol/kg $Na_2SO_4$ and $BaCl_2$ at mixing ratios of 2:8 and 0.1 mol/kg $Na_2SO_4$ and $BaCl_2$ at mixing ratios of 3:7 and 2:8 at 200 °C is very special. The diffusion-limited aggregation model is usually used to explain the formation of such dendritic crystals [40,41]. The core idea of the diffusion-limited aggregation model is that crystal growth is mainly limited by diffusion. During crystal growth, the surrounding growth points diffuse toward the center and "adhere" to the growth center, and all the particles that can diffuse to the crystal nucleus can grow continuously, thus forming a branching epitaxy and dendritic crystal [44]. According to the diffusion-limited aggregation model, the formation process of dendritic barite by in-situ mixing of $BaCl_2$ and $Na_2SO_4$ solutions at mixing ratios of 3:7 and 2:8 at 200 °C can be briefly described as follows (Figure 9).

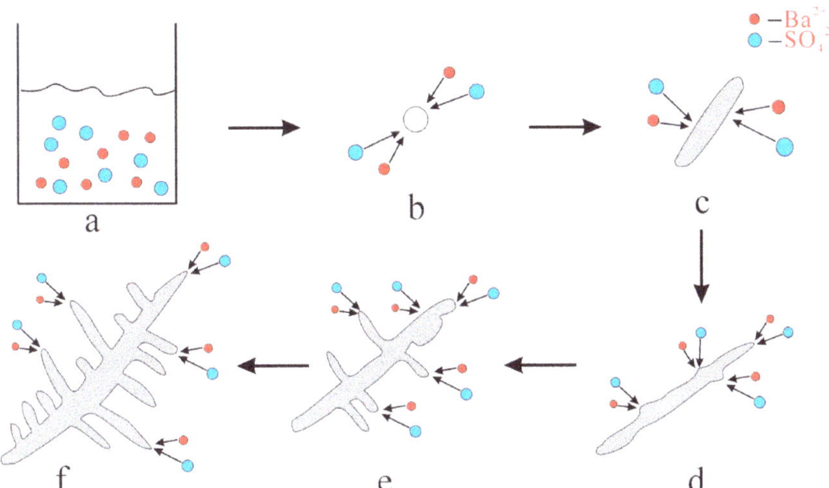



**Figure 9.** A schematic diagram of the formation of dendritic barite at high temperature. (**a**) Mixed solutions of sodium sulfate and barium chloride; (**b**) Granular barite; (**c**) Rod barite; (**d**) develops fine branches; (**e**) Branches continue to develop; (**f**) Dendritic barite.

It is well known that dendritic growth occurs under non-equilibrium conditions leading to morphologies that are not necessarily the most stable in terms of surface energy [45,46]. Due to the high ambient temperature of the crystal nucleus, the unstable fluid state, the strong thermal vibration of atoms, and the high $Ba^{2+}$ concentrations, some crystal nuclei may preferentially grow. Therefore, granular crystals first develop into rod-like crystals (Figure 9b,c). With the progress of growth, on the one hand, the rod-shaped crystals grow along the long axis of the column, however, agglomeration points due to the high local concentration will be formed at some positions on the column surface [47]. Then agglomeration points on the column surface of the rod-shaped barite gradually uplift (Figure 9d). As the nanocrystals continue to gather and grow, new growth fronts are created. Consequently, overgrowth occurs in the new growth front and branches gradually develop on the sides (Figure 9e). Finally, in a manner similar to growth patterns of trees, overgrowth occurs at some point on the branch, resulting in the formation of dendritic barite (Figure 9f).

### 4.5. Geological Implications

The formation and evolution of seafloor hydrothermal vents are often accompanied by barite precipitation [19]. On the seafloor, heat generated by deep magmatic activity drives seawater circulation along mid-ocean ridges, island arcs, and back-arc basins [48]. When the fluids circulate, heated fluids react with the surrounding rock, and vent onto the seafloor, forming hydrothermal chimneys and mounds. Where hydrothermal fluids are discharged, the mixing of hydrothermal fluids with the surrounding seawater drives mineral precipitation [49]. The outer wall of this porous chimney contains a typical series of low-temperature minerals (sphalerite, barite, etc.), which is formed by hydrothermal fluid cooling and/or mixing with seawater [50]. The precipitation of barite is formed by the mixture of $Ba^{2+}$ in hydrothermal fluid and $SO_4^{2-}$ in seawater [35].

Various forms of barite form in hydrothermal vents at Endeavour, from well-formed tabular and bladed crystals to dendritic crystals [19]. Dendritic barite forms where the outer wall of the nozzle is in contact with the seawater, which is very similar to the barite synthesized by in-situ mixing of $BaCl_2$ and $Na_2SO_4$ solutions at 200 °C in this study. Temperature measurements of fluid inclusions in barite from hydrothermal vents show formation temperatures of pressure-corrected inclusions are 180–240 °C, which is also close to the experimental temperature (200 °C) in this study. Combined with field observations and our experimental results, it is concluded that dendritic structures may be an important morphological feature of barite formed by in-situ mixing of seafloor hydrothermal solutions at high temperatures.

However, the mixing ratio of $Na_2SO_4$ and $BaCl_2$ solutions for dendritic barite formation in our experiment are not exactly the same as that for dendritic barite formation in the hydrothermal vent at Endeavour. The dendritic barite is believed to be formed by mixing of $Ba^{2-}$-bearing hydrothermal fluids with the $SO_4^{2-}$-bearing surrounding seawater with a mixing ratio of 4:6 at 180–240 °C in hydrothermal vents at Endeavour [19]. In our experiments, the dendritic barite only occurred under the conditions of excess $Ba^{2+}$ relative to $SO_4^{2-}$. If we regard $Na_2SO_4$ and $BaCl_2$ solutions at 200 °C as $SO_4^{2-}$ rich hot seawater and $Ba^{2+}$-containing hydrothermal fluid, respectively, more than 80% of the hydrothermal components are needed to form dendritic crystals. This hydrothermal component is much larger than that (40%) estimated in the hydrothermal vent of Endeavor [19], which means that the formation of dendritic barite in natural hydrothermal vents may be affected by other species in the fluids and further researches are needed.

## 5. Conclusions

Barite of a range of morphologies was synthesized at 200 °C and room temperature by in-situ mixing of $BaCl_2$ and $Na_2SO_4$ solutions. The synthesized barite was analyzed by SEM, XRD, and Raman spectroscopy. The main conclusions of this study are as follows:

The temperature has an obvious effect on the morphology of synthesized barite. Barite synthesized by in-situ mixing of $BaCl_2$ and $Na_2SO_4$ solutions at 200 °C has rich morphologies, such as rod-shaped, granular, plate-shaped, dendritic, X-shaped, and T-shaped crystals, while the morphology of barite synthesized at room temperature is relatively simple; granular or leaf-like.

When barite was synthesized by in-situ mixing of $BaCl_2$ and $Na_2SO_4$ solutions at 200 °C, the morphology of barite is affected by the initial mixing ratio. Dendritic barites will appear when $Ba^{2+}$ is in excess of $SO_4^{2-}$. In this study, dendritic barite only appeared at $Ba^{2+}/SO_4^{2-}$ molar ratios of 8:2 with initial concentrations of $Na_2SO_4$ and $BaCl_2$ of 0.03 mol/kg or at $Ba^{2+}/SO_4^{2-}$ molar ratios of 8:2 and 7:3 and initial concentrations of $Na_2SO_4$ and $BaCl_2$ of 0.1 mol/kg.

Dendritic textures may be an important typomorphic feature of barite formed under the conditions of high-temperature fluid mixing, such as in seafloor hydrothermal systems. Dendritic barite is formed at the contact between the outer wall of the seafloor hydrothermal vent and seawater, which is very similar to the barite synthesized by in-situ mixing of $BaCl_2$ and $Na_2SO_4$ solutions at 200 °C in this study.

**Author Contributions:** Conceptualization, C.W. (Chunyao Wang) and L.Z.; methodology, L.Z.; software, C.W. (Chunyao Wang), S.Z., L.W., C.W. (Chunwan Wei), W.S. and L.X.; validation, C.W. (Chunyao Wang), S.Z., L.W., C.W. (Chunwan Wei), W.S. and L.X.; investigation, C.W. (Chunyao Wang), S.Z., L.W., C.W. (Chunwan Wei), W.S. and L.X.; writing, C.W. (Chunyao Wang), L.Z., and W.Z. All authors have read and agreed to the published version of the manuscript.

**Funding:** This research was funded by the Natural Science Foundation of China (91962108 and 41773058), the Key Research Program of Frontier Sciences, CAS (QYZDB-SSW-DQC008), EXPRO project of the Czech Science Foundation (No. 19-29124X), and the Research Grant of the Guizhou Normal University (GZNUD2019]4).

**Data Availability Statement:** Not applicable.

**Acknowledgments:** We thank the Editor and three anonymous reviewers whose comments helped improve and clarify this manuscript. We also thank John Mavrogenes and Terry Mernagh for constructive suggestions that helped improve the manuscript.

**Conflicts of Interest:** The authors declare no conflict of interest.

# References


1. Jewell, P.W. Bedded barite in the geologic record. *Mar. Authigenesis Glob. Microb.* **2000**, *64*, 147–161.
2. Bonny, S.M.; Jones, B. Diatom-mediated barite precipitation in microbial mats calcifying at Stinking Springs, a warm sulphur spring system in Northwestern Utah, USA. *Sediment. Geol.* **2007**, *194*, 223–244. [CrossRef]
3. Huston, D.L.; Logan, G.A. Barite, BIFs and bugs: Evidence for the evolution of the Earth's early hydrosphere. *Earth Planet. Sci. Lett.* **2004**, *220*, 41–55. [CrossRef]
4. Senko, J.M.; Campbell, B.S.; Henriksen, J.R.; Elshahed, M.S.; Dewers, T.A.; Krumholz, L. Barite deposition resulting from phototrophic sulfide-oxidizing bacterial activity. *Geochim. Cosmochim. Acta* **2004**, *68*, 773–780. [CrossRef]
5. Bottrell, S.H.; Newton, R.J. Reconstruction of changes in the global sulphur cycling from marine sulphate isotopes. *Earth-Sci. Rev.* **2006**, *75*, 59–83. [CrossRef]
6. Sanz-Montero, M.E.; Rodríguez-Aranda, J.; Cura, M. Bioinduced precipitation of barite and celestite in dolomite microbialites: Examples from Miocene lacustrine sequences in the Madrid and Duero Basins. *Sediment. Geol.* **2009**, *222*, 138–148. [CrossRef]
7. Griffith, E.M.; Paytan, A. Barite in the ocean—Occurrence, geochemistry and palaeoceanographic applications. *Sedimentology* **2012**, *59*, 1817–1835.
8. Widanagamage, I.; Waldron, A.; Glamoclija, M. Controls on Barite Crystal Morphology during Abiotic Precipitation. *Minerals* **2018**, *8*, 480. [CrossRef]
9. Niu, H.; Chen, F. Ree geochemistry of magmatic barite and fluorite. *Acta Mineral. Sin.* **1996**, *16*, 382–388.
10. Niu, H.; Lin, C. The genesis of the mianning rare earth deposit, sichuan province. *Miner. Depos.* **1994**, *013*, 345–353.
11. Bishop, J.K.B. The barite-opal-organic carbon association in oceanic particulate matter. *Nature* **1988**, *332*, 341–343. [CrossRef]
12. Bertram, M.A.; Cowen, J.P. Morphological and compositional evidence for biotic precipitation of marine barite. *J. Mar. Res.* **1997**, *55*, 577–593. [CrossRef]
13. Jin, Z.; Zhu, D.; Hu, W.; Zhang, X.; Wang, Y.; Yan, X.B. Geological and geochemical signatures of hydrothermal activity and their influence on carbonate reservoir beds in Tarim Basin. *Acta Geol. Sin.* **2006**, *80*, 245–253.
14. Vandeginste, V.; Stehle, M.C.; Jourdan, A.L.; Bradbury, H.J.; Manning, C.; Cosgrove, J.W. Diagenesis in salt dome roof strata: Barite-calcite assemblage in Jebel Madar, Oman. *Mar. Pet. Geol.* **2017**, *86*, 408–425.
15. Shikazono, N. Genesis of sulfate mineral in the Kuroko deposits. *Min. Geol.* **1983**, *11*, 229–249.
16. Dill, H.; Carl, C. Sr Isotope Variation in Vein Barites from the NE Bavarian Basement: Relevance for the Source of Elements and Genesis of Unconformity-Related Barite Deposits. *Mineral. Petrol.* **1987**, *36*, 27–39. [CrossRef]
17. Ehya, F. Rare earth element and stable isotope (O, S) geochemistry of barite from the Bijgan deposit, Markazi Province, Iran. *Mineral. Petrol.* **2012**, *104*, 81–93. [CrossRef]
18. Safina, N.P.; Melekestseva, I.Y.; Nimis, P.; Ankusheva, N.N.; Yuminov, A.M.; Kotlyarov, V.A.; Sadykov, S.A. Barite from the Saf'yanovka VMS deposit (Central Urals) and Semenov-1 and Semenov-3 hydrothermal sulfide fields (Mid-Atlantic Ridge): A comparative analysis of formation conditions. *Miner. Depos.* **2015**, *51*, 491–507. [CrossRef]
19. John, J.W.; Mark, H.D.; Margaret, K.T.; Thor, H.; Williamson, N.M.; Margaret, S.; Jan, F.; David, B.; Matthias, F.; Leigh, A.; et al. Precipitation and growth of barite within hydrothermal vent deposits from the Endeavour Segment, Juan de Fura Ridge. *Geochim. Cosmochim. Acta* **2016**, *173*, 64–85.
20. Blount, C.W. Synthesis of barite, celestite, anglesite, witherite, and strontianite from aqueous solutions. *Am. Mineral.* **1974**, *59*, 1209–1219.
21. Liu, S.; Nancollas, G.H. Scanning electron microscopic and kinetic studies of the crystallization and dissolution of barium sulfate crystals. *J. Cryst. Growth* **1976**, *33*, 11–20. [CrossRef]
22. Rizkalla, E.N. Kinetics of the crystallization of barium sulphate. *J. Chem. Soc. Faraday Trans I* **1983**, *79*, 1857–1867. [CrossRef]
23. Uchida, M.; Sue, A.; Yoshioka, T.; Okuwaki, A. Hydrothermal synthesis of needle-like barium sulfate using a barium (II)-EDTA chelate precursor and sulfate ions. *J. Mater. Sci. Lett.* **2000**, *19*, 1373–1374. [CrossRef]
24. Kowacz, M.; Putnis, C.V.; Putnis, A. The effect of cation:anion ratio in solution on the mechanism of barite growth at constant supersaturation: Role of the desolvation process on the growth kinetics. *Geochim. Cosmochim. Acta* **2007**, *71*, 5168–5179. [CrossRef]
25. Ray, D.; Kota, D.; Das, P.; Prakash, S. Microtexture and distribution of minerals in hydrothermal barite-silica chimney from the franklin seamount, sw pacific: Constraints on mode of formation. *Acta Geol. Sin. Engl. Ed.* **2014**, *88*, 213–225.
26. Aymonier, C.; Cansell, F.; Mecking, S.; Moisan, S.; Martinez, V. Synthesis of Particles in Dendritic Structures. U.S. Patent 7,932,311, 26 April 2011.
27. Martin, S.; Espen, D.B.; Mogens, C.; Iversen, B. Size and Morphology Dependence of ZnO Nanoparticles Synthesized by a Fast Continuous Flow Hydrothermal Method. *Cryst. Growth Des.* **2011**, *11*, 4027–4033.
28. Diez-Garcia, M.; Gaitero, J.J.; Dolado, J.S.; Aymonier, C. Ultra-Fast Supercritical Hydrothermal Synthesis of Tobermorite under Thermodynamically Metastable Conditions. *Angew. Chem.* **2017**, *129*, 3210–3215. [CrossRef]
29. Miyake, M.; Minato, I.; Morikawa, H.; Iwai, S. Crystal structures and sulphate force constants of barite, celestite, and anglesite. *Am. Mineral.* **1978**, *63*, 506–510.
30. Ross, S.D. Inorganic infrared and raman spectra. *J. Mol. Struct.* **1973**, *15*, 468–469.
31. Griffith, W.P. Raman studies on rock-forming minerals. Part II. Mineral containing MO3, MO4, and MO6 groups. *J. Chem. Soc.* **1970**, 286–291. [CrossRef]
32. Griffith, W.P. Advances in the Raman and Infrared spectroscopy of minerals. *Adv. Spectrosc.* **1987**, *14*, 119–186.

33. Wong, D.C.Y.; Jaworski, Z.; Nienow, A.W. Effect of Ion Excess on Particle Size and Morphology during Barium Sulphate Precipitation: An Experimental Study. *Chem. Eng. Sci.* **2001**, *56*, 727–734. [CrossRef]

34. Wang, L.; Zhou, L.; Zhang, S.; Wang, C.; Zhou, W. Hydrothermal synthesis of barite using the in-situ high temperature mixing method. *Acta Mineral. Sin.* **2021**, *41*, 139–149.

35. Blount, C.W. Barite solubilities thermodynamic and quantities up to 300 °C and 1400 bars. *Am. Mineral.* **1977**, *62*, 942–957.

36. Zhang, H.; Zhu, L.; Chen, J.; Chen, L.; Liu, C.; Yuan, S. Morphologically Controlled Synthesis of $Cs_2SnCl_6$ Perovskite Crystals and Their Photoluminescence Activity. *Crystals* **2019**, *9*, 258. [CrossRef]

37. Li, S.; Xu, J.; Luo, G. Control of crystal morphology through supersaturation ratio and mixing conditions. *J. Cryst. Growth* **2007**, *304*, 219–224. [CrossRef]

38. Strtibel, G. Zur Kenntnis und genetischen Bedeutung des Systems $BaSO_4$-NaCl-$H_2O$. *Neues Jahrb. Mineral. Mon.* **1967**, *4*, 223–234.

39. Shikazono, N. Precipitation mechanisms of barite in sulfate-sulfide deposits in back-arc basins. *Geochim. Cosmochim. Acta* **1994**, *58*, 2203–2213. [CrossRef]

40. Paul, M. Formation of fractal clusters and networks by irreversible diffusion-limited aggregation. *Phys. Rev. Lett.* **1983**, *51*, 1119–1122.

41. Sander, L.M. Diffusion-Limited Aggregation, a Kinetic Critical Phenomenon. *Contemp. Phys.* **2000**, *41*, 203–218. [CrossRef]

42. Murray, C.B.; Kagan, C.R.; Bawendi, M.G. Synthesis and characterization of monodisperse nanocrystals and close-packed nanocrystal assemblies. *Annu. Rev. Mater. Sci.* **2000**, *30*, 545–610. [CrossRef]

43. Judat, B.; Kind, M. Morphology and internal structure of barium sulfate–derivation of a new growth mechanism. *J. Colloid Interface Sci.* **2004**, *269*, 341–353. [CrossRef]

44. Huang, Q.; Zhao, S. Development of dentrite study. *J. Synth. Cryst.* **2002**, *31*, 486.

45. Chernov, A.A.; Lewis, J. computer model of crystallization of binary systems; kinetic phase transitions. *J. Phys. Chem. Solids* **1967**, *28*, 2185–2198. [CrossRef]

46. Kobayashi, T.; Kuroda, T. *Morphology of Crystals*; Sunagawa, I., Ed.; Terra Scientific Publishing: Tokyo, Japan, 1987.

47. Angerhöfer, M. Untersuchungen zur Kinetik der Fällungskristallisation von Bariumsulfat. Ph.D. Thesis, Technical University Munich, Munich, Germany, 1995.

48. Baker, E.T.; German, C.R.; Elderfield, H. Hydrothermal Plumes over Spreading-Center Axes: Global Distributions and Geological Inferences. *Am. Geophys. Union* **1995**, *91*, 47–71.

49. Hannington, M.D.; Jonasson, I.R.; Herzig, P.M.; Petersen, S. Physical and chemical processes of seafloor mineralization at mid-ocean ridges. *Seafloor Hydrotherm. Syst. Phys. Chem. Biol. Geol. Interact.* **1995**, *91*, 115–157.

50. Tivey, M.K.; Stakes, D.S.; Cook, T.L.; Hannington, M.D.; Petersen, S. A model for growth of steep-sided vent structures on the Endeavour Segment of the Juan de Fuca Ridge: Results of a petrologic and geochemical study. *J. Geophys. Res. Solid Earth* **1999**, *104*, 22859–22883. [CrossRef]

*Article*

# Effect of Gold Nanoparticles on the Crystallization and Optical Properties of Glass in ZnO-MgO-Al$_2$O$_3$-SiO$_2$ System

Georgiy Shakhgildyan [1,*], Veniamin Durymanov [2], Mariam Ziyatdinova [1,3], Grigoriy Atroshchenko [1], Nikita Golubev [1], Alexey Trifonov [4,5], Olga Chereuta [1], Leon Avakyan [2], Lusegen Bugaev [2] and Vladimir Sigaev [1]

[1] Department of Glass and Glass-ceramics, Mendeleev University of Chemical Technology, Miusskaya Sq., 9, 125047 Moscow, Russia; ziiatdinova.m.z@muctr.ru (M.Z.); grigsmith@yandex.ru (G.A.); golubev.n.v@muctr.ru (N.G.); 203896@muctr.ru (O.C.); sigaev.v.n@muctr.ru (V.S.)
[2] Physical Department, Southern Federal University, Zorge Street 5, 344090 Rostov-on-Don, Russia; shakior_5454@mail.ru (V.D.); laavakyan@sfedu.ru (L.A.); bugaev@sfedu.ru (L.B.)
[3] P. N. Lebedev Physical Institute, Russian Academy of Sciences, Leninskiy Av., 53, 119333 Moscow, Russia
[4] Institute of Physics and Applied Mathematics, National Research University of Electronic Technology (MIET), Shokin Square, Bld. 1, Zelenograd, 124498 Moscow, Russia; trif123456@yandex.ru
[5] Scientific Research Institute of Physical Problems Named after F.V. Lukin, Pass. 4806, Bld. 6, Zelenograd, 124498 Moscow, Russia
* Correspondence: shakhgildian.g.i@muctr.ru

**Abstract:** Gold nanoparticles precipitated in transparent glass-ceramics could pave the way for the development of multifunctional materials that are in demand in modern photonics and optics. In this work, we explored the effect of gold nanoparticles on the crystallization, microstructure, and optical properties of ZnO-MgO-Al$_2$O$_3$-SiO$_2$ glass containing TiO$_2$ and ZrO$_2$ as nucleating agents. X-ray diffraction, transmission electron microscopy, Raman, and optical spectroscopy were used for the study. We showed that gold nanoparticles have no effect on the formation of gahnite nanocrystals during the glass heat treatments, while optical properties of the glass-ceramics are strongly dependent on the gold addition. A computational model was developed to predict optical properties of glass during the crystallization, and the possibility for adjusting the localized surface plasmon resonance band position with the heat treatment temperature was shown.

**Keywords:** glass-ceramics; phase separation; nucleation; crystallization; microstructure; gahnite; gold nanoparticles; surface plasmon resonance; plasmonics

**Citation:** Shakhgildyan, G.; Durymanov, V.; Ziyatdinova, M.; Atroshchenko, G.; Golubev, N.; Trifonov, A.; Chereuta, O.; Avakyan, L.; Bugaev, L.; Sigaev, V. Effect of Gold Nanoparticles on the Crystallization and Optical Properties of Glass in ZnO-MgO-Al$_2$O$_3$-SiO$_2$ System. *Crystals* **2022**, *12*, 287. https://doi.org/10.3390/cryst12020287

Academic Editors: Vladislav V. Kharton and Xiang-Hua Zhang

Received: 30 January 2022
Accepted: 17 February 2022
Published: 18 February 2022

**Publisher's Note:** MDPI stays neutral with regard to jurisdictional claims in published maps and institutional affiliations.



## 1. Introduction

Research and development in the field of new optical materials are becoming especially important today, in an era of challenges in the field of photonics and integrated optics [1]. In addition to optical materials based on glasses, crystals, and polymers, transparent glass-ceramics, which combine the properties of various materials, are beginning to play an increasingly important role in optical materials science [2,3]. Transparent glass-ceramics are multiphase materials that consist of crystallites distributed in a glass matrix; the unique properties of transparent glass-ceramics are achieved due to the small size of crystallites, their chemical content and structure, and chemical composition of the residual glassy phase also plays an important role [4].

The phases formed in glass-ceramics can be oxide and non-oxide crystallites, metal nanoparticles (NPs), clusters, or semiconductor quantum dots [4]. The precipitation of different crystallites in transparent glass-ceramics is being actively studied to develop new light-emitting and laser media [5–7], materials with controlled values of the thermal expansion coefficient [8], and high-strength transparent cover materials [9–11]. Similarly, glass-ceramics based on metal NPs (mainly silver or gold) are in great demand as new nonlinear optical and plasmonic materials [12–16]. Moreover, glass-ceramics with metal

NPs are promising systems for the implicating of the effect of rare-earth ion fluorescence enhancement [17], and different combinations of the interaction between NPs and rare-earth ions have been demonstrated to modify the spectral properties of glass-ceramics [18–21].

However, the study of glass-ceramics, in which crystalline phases of various types are precipitated simultaneously, has received quite little attention, despite the fact that the combination of various crystallites in a glass matrix can lead to the creation of glass-ceramics with a wide range of properties [22,23]. Moreover, it is known that metal NPs can also act as crystallization catalysts in the production of glass-ceramics during heat treatment [24]. Few works have been devoted to the study of gold NPs on crystallization, microstructure, and thermal and mechanical properties of glass-ceramics. M. Garai et al. showed that microstructural variation caused by gold NPs in silicate glass-ceramics significantly affects the thermal and mechanical properties [25]. The strong effect of gold NPs on the crystallization kinetics as well as microstructure and thermal and mechanical properties was also observed during the preparation of mica glass-ceramics [26,27]. Kochetkov et al. reported that precipitation of gold NPs plays the main role in the process of the heterogeneous lithium disilicate crystallization in glass; it was found that regions with an elevated concentration of lithium ions formed around colloidal gold particles [28]. Accelerated crystal growth in the gold-doped glass-ceramics was observed by Thieme et al. during heat treatments of zinc-silicate glass [29]. Gold NPs were also introduced in the silicate glass-ceramics to establish their solid oxide fuel cell sealing ability, while no effect of NPs on the crystallization kinetics was reported [30]. Kracker et al. reported on the detailed study of the optical properties of gold NPs precipitated in the silicate glass-ceramics with low thermal expansion, whereas no data were provided on the influence of gold NPs on the microstructure of the glass-ceramics [31]. This implies that gold NPs definitely have an effect on the crystallization and properties of glass-ceramics, but more work is needed to further explore this area for the development of new multifunctional transparent glass-ceramics.

Recently, we showed that gold NPs precipitated in the phase-separated glasses and glass-ceramics of the $ZnO-MgO-Al_2O_3-SiO_2$ system containing $TiO_2$ and $ZrO_2$ demonstrate tunable position of the localized surface plasmon resonance (LSPR) [32]. The present study aims to explore the effect of gold NPs on the crystallization, microstructure, and optical properties of the mentioned glass system. Appropriate investigation techniques were used to evaluate the role of gold-doping in the glass-ceramics production process.

## 2. Materials and Methods

### 2.1. Glass Synthesis

In this work we synthesized glasses in the $ZnO-MgO-Al_2O_3-SiO_2$ system containing $SnO_2$, $Na_2O$, $TiO_2$ and $ZrO_2$ oxides with the chemical composition described in the previous work [32]. We added 0.2 g of $HAuCl_4$ to the glass batch for the Au-doping of glass (designated as Au-doped) and did not add gold for the Au-free glass (designated as Matrix). Glass batch was calculated to prepare 1000 g of bulk glass.

The process of the glass synthesis was described previously [32] and represents traditional melt-quenching technique with the subsequent annealing to reduce residual stress. Obtained glass samples were transparent and free of defects.

### 2.2. Glass Characterization

The visual appearance of the samples was captured by digital camera. For the density determination of studied samples Archimedes method with distilled water was used.

For the determination of the glass transition temperature ($T_g$) and the crystallization temperature ($T_C$) differential scanning calorimetry (DSC) was used. Bulk glass samples of about 20 mg in weight were loaded in the platinum crucible and heated in the simultaneous thermal analyzer NETZSCH STA 449 F3 Jupiter (NETZSCH-Gerätebau, Selb, Germany) with a dynamic flow atmosphere of Ar. The temperature range was from room temperature to 950 °C with a heating rate of 10 °C/min.

X-ray diffraction patterns (XRD) of powdered samples were recorded by means of a diffractometer Bruker D2 Phaser (Bruker AXS GmbH, Karlsruhe, Germany) employing nickel-filtered CuK$\alpha$ radiation. Crystal phases were identified by comparing the peak position and relative intensities in the XRD pattern with the ICDD PDF-2 database (release 2011). The mean crystallite size was estimated from broadening of the XRD peak at about 37° according to Scherrer's equation:

$$D = \frac{K\lambda}{\Delta \cos \theta},$$ (1)

where $\lambda$ is the wavelength of the X-ray radiation (1.5406 Å), $\theta$ is the diffraction angle, $\Delta$ is the width of the peak at half of its maximum and $K$ is the constant assumed to be 1 [33]. The crystallized fraction was evaluated as $100 \cdot (A_p / A_x)$, where $A_x$ and $A_p$ are the area of the whole XRD pattern (without background) and the area of the peaks considered as the area outside of the broad amorphous XRD pattern, respectively. Indicated areas were calculated (in cps × degrees) using DIFFRAC.EVA software [1].

The microstructure of the samples was studied by high-resolution transmission electron microscopy (HRTEM) with the transmission electron microscope FEI Tecnai G2 20 S-Twin (FEI, Hillsboro, OR, USA), in 200 kV mode. Bulk glass samples were grounded in an agate mortar to fine powders and dispersed in ethanol. The obtained solution was dropped on a microscope grid which was dried for 20 min. The HRTEM images were analyzed with the ImageJ software (version 1.53n. https://imagej.nih.gov/ij/ (accessed on 20 January 2022)).

Spectroscopy studies were performed using Raman and optical spectroscopy. For the first NTEGRA Spectra spectrometer (NT-MDT, Zelenograd, Moscow, Russia) with the Ar laser beam (488 nm excitation wavelength) was used and for the last Shimadzu UV-3600 spectrophotometer (Shimadzu, Kyoto, Japan) was used. Double-sided polished samples were utilized for the spectroscopy studies. The glass refractive index was determined by an ATAGO DR-M4 Abbe refractometer (ATAGO Co., Tokyo, Japan).

## 3. Results and Discussion

### 3.1. Physicochemical Properties

The DSC curves for the two glass samples under study are similar, showing the glass transition temperature, $T_g$, is about 740 °C for both glasses, as well as the temperature of the crystallization $T_C$ about 872 °C (Figure 1a). To study the influence of the gold NP precipitation on the phase-separation, crystallization, and optical properties of glass, we performed a series of heat treatments at temperatures below and above the $T_g$ and the $T_C$ both for the Matrix and Au-doped samples.

The density variation for the Matrix and Au-doped glass samples with the heat treatment temperature is shown in Figure 1b. It demonstrates a complex behavior typical for density variation of spinel-based glass-ceramics and was reported previously for similar glass systems [34,35]. One can see that density variations for both samples are very close. After the heat treatment in the 650–750 °C range, the density rapidly increases and reaches the maximum values up to 2.98 g/cm$^3$ at 875 °C heat treatment temperature. Further increase of the heat treatment temperature leads to the decrease of the density down to 2.92 g/cm$^3$, which can be explained by the precipitation of mixed crystal phases affecting the overall density of the composite material. The observed increase of the density after the heat treatment at 750 °C could suggest amorphous phase separation in the glass. This assumption is confirmed below using TEM, Raman, and optical spectroscopy data.

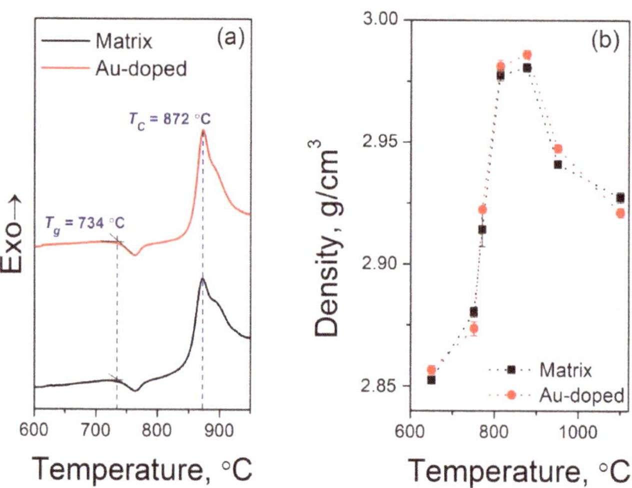

**Figure 1.** (**a**) DCS curves of the glass samples ($T_g$—glass transition temperature, $T_C$—the temperature of the crystallization peak); (**b**) variation of density with the heat treatment temperature for the glass samples. The heat treatment duration was 20 h. The dashed lines in (**b**) serve as a guide for the eye. The error bars match the size of the symbols.

## *3.2. XRD Studies*

According to the XRD data of the Matrix series (Figure 2), the glass sample after the heat treatment at 750 °C is likely amorphous, showing a broad bump in the 20–30° range, as does glass sample before the heat treatment. When analyzing the XRD patterns, what draws attention is a slight shift of the amorphous halo position to smaller angles for treated samples, which suggests a change in the composition of the residual glass. Since the position of the amorphous halo becomes close to that of the fused silica, the residual glass seems to be more enriched in silica compared to the composition of the parent glass [34].

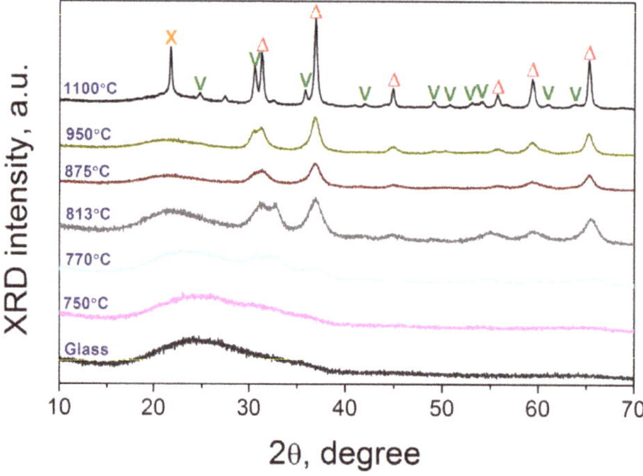

**Figure 2.** XRD patterns of the glass samples within the Matrix series after treatment at the indicated temperatures for 20 h compared to the XRD pattern of the parent glass. Symbols are indicated the following crystal phases: $\Delta$—#01-074-1136 ($ZnAlO_4$), V—#01-074-9434 ($ZrTiO_4$), X—#00-060-0061 ($Al_2Mg_3O_{18}Si_6$). The patterns are shifted for the convenience of observation.

Crystal phases start to brightly appear on the XRD patterns after the heat treatments at 770 °C and higher temperatures. The main crystal phase for the studied samples is gahnite ($ZnAlO_4$), which is believed to be the single-phase up to the heat treatment at 950 °C. At this temperature, $ZrTiO_4$ crystallites start to precipitate, and further increase of the heat treatment temperature up to 1100 °C leads to the opacity of the sample and sharpening of the XRD peaks that is caused by a significant increase of the crystallized fraction and crystallite size. We suggest that $ZrTiO_4$ crystallites could start to precipitate already at 770 °C since the extremely broadened peak at 2θ ~32 deg. can be observed on the XRD pattern. Moreover, we need to underline that $Mg^{2+}$ and $Zn^{2+}$ ions have nearly the same ionic radii, thus the incorporation of Mg in the crystal phase will not lead to a drastic change in the lattice constant. Since the glass composition contains nearly the same amount of ZnO and MgO, we can suggest formation of the $(Mg, Zn)Al_2O_4$ crystal phase. In this work we use gahnite to describe the precipitated crystal phase; future studies using the TEM elemental analysis will help to clarify the exact composition of the precipitated crystals.

Figure 3 shows the data for the crystallized fraction and crystallite size determined from the abovementioned XRD patterns. As can be seen, the crystallized fraction rapidly grows after treatment at temperatures higher than 770 °C and exceeds 65% for the sample heat treated at 1100 °C. Calculated crystallite size seems to be increased linearly from about 3–4 nm after the treatment at 770 °C to 23–25 nm for the opaque sample treated at 1100 °C.

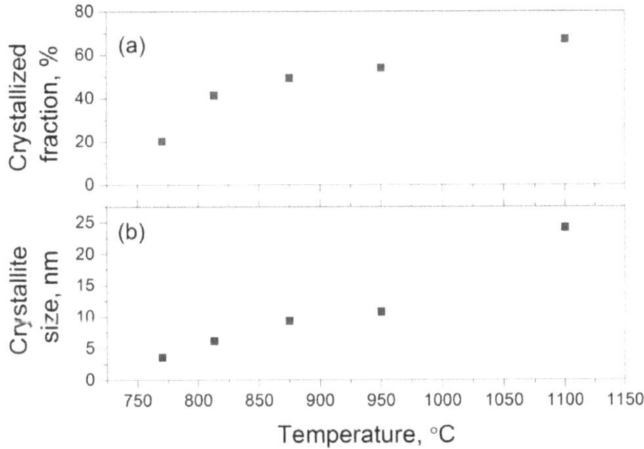

**Figure 3.** (**a**) Crystallized fraction and (**b**) crystallite size of the gahnite determined from the XRD studies for the glass samples heat treated up to 1100 °C for 20 h for the Matrix series.

These findings along with the data on sample density suggest that during the low-temperature heat treatment of the glass samples, amorphous phase separation could occur while the further increase of the treatment temperature leads to the precipitation of gahnite crystals, the sizes and fraction of which rapidly grow with the temperature. In order to analyze the influence of gold addition on the crystallization process of the glass samples, we compare XRD patterns of samples from the Matrix and Au-doped series (Figure 4). One can see that the detailed comparison of the XRD patterns does not allow to commit any changes: both at 770 and 950 °C the XRD patterns are practically identical. Based on these data, we can propose that gold NPs precipitated in the glass samples do not affect the crystallization of gahnite.

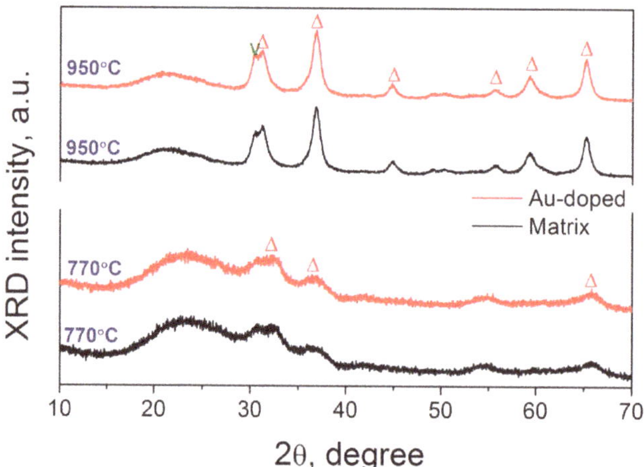



**Figure 4.** XRD patterns of the glass samples within the Matrix and Au-doped series after treatment at 770 or 950 °C for 20 h. Symbols are indicated the following crystal phases: $\Delta$—#01-074-1136 ($ZnAlO_4$), V—#01-074-9434 ($ZrTiO_4$).

### 3.3. TEM/HRTEM Studies

To study the structure of the Au-doped samples in more detail, we performed a series of TEM measurements. The TEM characterization of the Au-doped glass sample heat treated at 750 °C (Figure 5) indicated the presence of inhomogeneous contrast regions about 10 nm in size uniformly dispersed in the glass matrix. These areas are not crystalline since no crystalline fringes can be observed in the HRTEM images. We propose that these regions are formed by amorphous droplets due to liquid-liquid phase separation and enriched with $TiO_2$ and $ZrO_2$. The study of TEM images does not lead to the detection of gold NPs. This is possibly due to a very low concentration of gold in the glass.

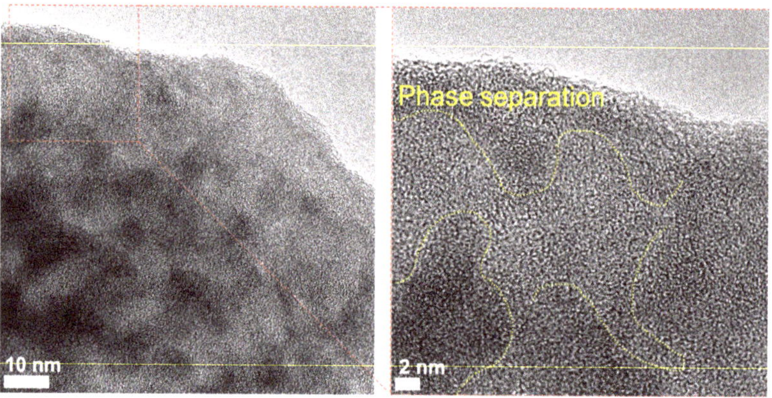



**Figure 5.** TEM image of the Au-doped glass sample heat treated at 750 °C for 20 h. The right image depicts magnified region in red.

Figure 6a shows the microstructure of the glass sample after the heat treatment at 770 °C. One can see the appearance of the abovementioned inhomogeneous regions indicating that phase separation still proceeds at this treatment temperature. At once, the HRTEM image depicts single nanocrystals about 3–5 nm in size (Figure 6b); the observed lattice fringes prove that the particles indeed are crystallites. The interplanar spacing of the crystallites is 0.29 nm which is consistent with the spacing of gahnite (220) crystallographic planes [36].

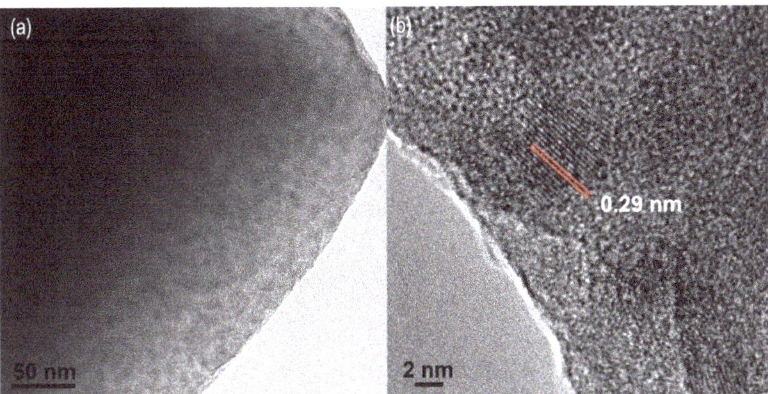

**Figure 6.** (**a**) TEM and (**b**) HRTEM images of the Au-doped glass sample heat treated at 770 °C for 20 h.

In order to further analyze the observed crystals, we need once again to underline the fact of the total similarity between XRD patterns for glasses from the Matrix and Au-doped series (Figure 4). In this regard, we can compare the data (crystallinity and crystallite size) calculated from the XRD patterns for the Matrix series (Figure 3) and data obtained from the TEM studies for the Au-doped series. Hence, based on the XRD results, we can conclude that the nanocrystals observed in the HRTEM image refer to gahnite. Moreover, the calculated crystallite size of 3–4 nm (Figure 3b) perfectly matches with HRTEM data.

As shown by XRD analysis, the increase of the heat treatment temperature up to 950 °C results in the growth of both the nanocrystal size and the crystallized fraction, which one can see also on the TEM image (Figure 7a). The estimated crystallized fraction of more than 50% matches well with the visual analysis of the observed image, while calculated nanocrystal size (10–11 nm) perfectly correlates with the mean value from the size distribution histogram (inset on Figure 7). Analysis of the HRTEM image and corresponding selected area diffraction (SAED) pattern confirms the formation of gahnite ($ZnAl_2O_4$) nanocrystals, which correlates with the XRD data indicating that gahnite is the dominant crystal phase in the investigated sample. Thus, we can preliminarily confirm our assumption about the lack of gold NP influence on the glass crystallization.

### 3.4. Raman Spectra

Analysis of the Raman spectra provides more insights into the glass structure change during the heat treatment. Figure 8 shows the Raman spectra of the glass samples for the Matrix series. One can see that the spectrum for the initial glass as well as for the glass sample after the treatment at 650 °C formed by three bands: one broadband with maximum at ~450 cm$^{-1}$, the bands at ~800, and at ~900 cm$^{-1}$ [37,38]. The band at ~460 cm$^{-1}$ is associated with the vibrations of [$SiO_4$] tetrahedrons of a glass matrix, the band at ~900 cm$^{-1}$ can be assigned to [$TiO_4$] tetrahedrons built into the glass matrix, while the band at ~800 cm$^{-1}$ has more complex nature: such band is superimposed by the band associated with the vibrations of [$SiO_4$] tetrahedra and with the vibrations of the Ti–O bonds in [$TiO$]$_5$ and in [$TiO$]$_6$ polyhedrons, which could be formed during the phase separation process [39].

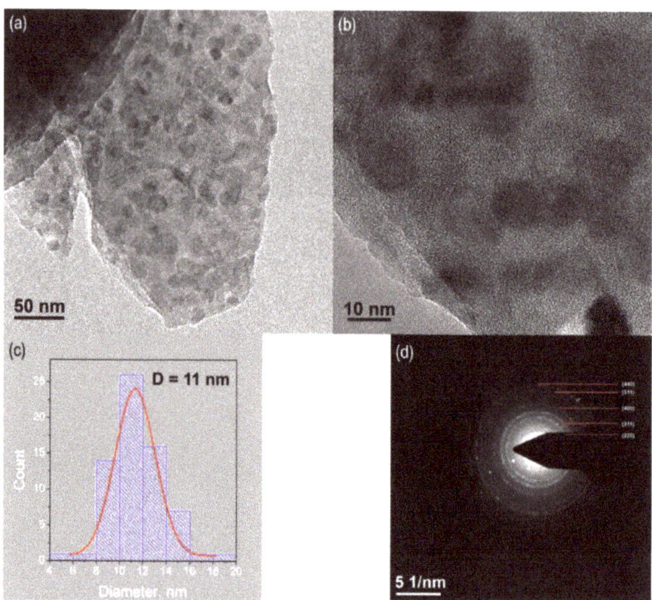

**Figure 7.** (**a**) TEM and (**b**) HRTEM images of the Au-doped glass sample heat treated at 950 °C for 20 h. (**c**) the size distribution of nanocrystals and (**d**) corresponding SAED patterns of the area in (**b**).

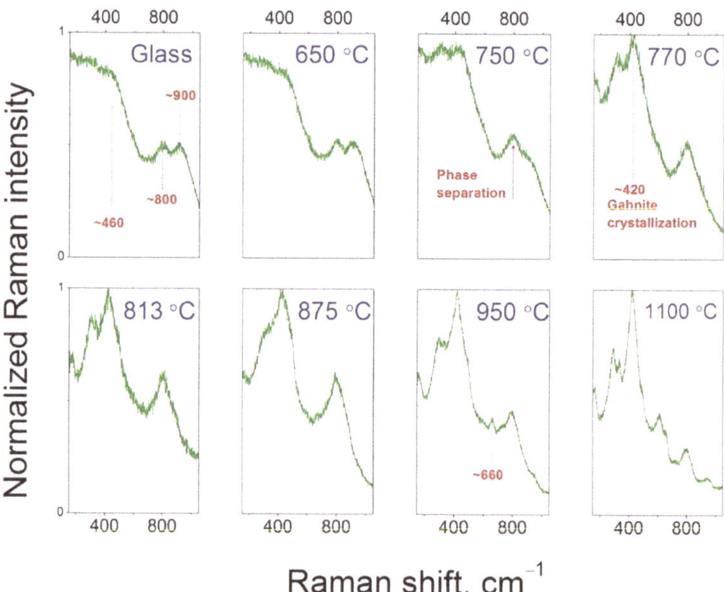


**Figure 8.** Raman spectra of the glass samples of the Matrix series before and after the heat treatment in the 650–1100 °C range for 20 h.

It is known that the increase of the ~800 cm$^{-1}$ band intensity with the decrease of the ~900 cm$^{-1}$ band could be the indicator of the phase separation development in the glass during the heat treatment [34]. Hence, we analyze the evolution of the Raman bands in the high-frequency part of the spectra with the heat treatment temperature increase.

Figure 8 clearly shows that the ~800 cm$^{-1}$/~900 cm$^{-1}$ intensity ratio rapidly rises from 650 to 750 °C and higher temperatures. This observation means that essentially all titania enters into the zinc aluminotitanate inhomogeneous regions, which confirms that the phase separation occurs in the glass samples treated at 750 °C and higher temperatures.

The first band related to the gahnite formation (at ~420 cm$^{-1}$) appears in the spectrum of the sample treated at 770 °C and continuously sharpens with the temperature increase. The second characteristic gahnite band at ~660 cm$^{-1}$ appears in the spectrum after the heat treatment at 950 °C. The comparison of the characteristic Raman spectra for the Matrix and Au-doped series is shown in Figure 9. One can see that spectra for all treatment temperatures are similar for the Matrix and Au-doped series. This finding once again indicates the negligible influence of the Au NP formation on the structure change of the studied glasses. Nevertheless, optical properties of the Au-doped glass samples undergo crucial changes during the heat treatment, as is shown in the next section.

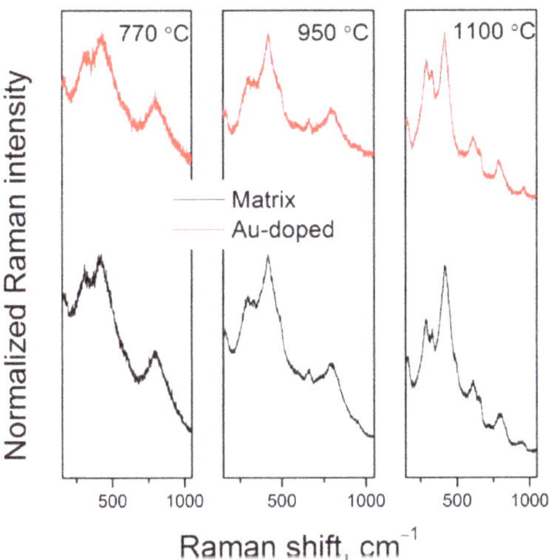


**Figure 9.** Raman spectra of the glass samples heat treated in the 770–1100 °C range for 20 h for the Matrix and Au-doped series.

### 3.5. Optical Properties

Figure 10 displays optical absorption spectra of the Matrix glass samples before and after the heat treatment in the 650–950 °C range as well as photos of the samples. One can see that both the initial glass and samples heat treated up to 950 °C are transparent, while the sample treated at 1100 °C is translucent (absorption spectrum presented in Figure S1 of the Supplementary Materials). There is no coloration in the initial glass and samples treated up to 813 °C. Higher temperature treatment leads to light yellowish coloration for the sample treated at 875 °C and pronounced yellow color for the sample treated at 950 °C.

The observed color changes are related to the distinct shift of the UV absorption edge upon the heat treatment: for the initial glass, the UV absorption edge is observed at 375 nm and for the glass treated at 950 °C the edge redshifted to 422 nm. These changes are likely to have complex nature, and different factors can be responsible for the observed shift. The shift (from 3.3 to 2.9 eV) can be due to the onset of band-to-band excitations of precipitated gahnite crystals in the glass matrix, the increase of crystallized fraction, and the consequent increase of intensity of the UV absorption tail. Moreover, some absorption contribution could also be caused by $O^{2}$–[$Ti^{4+}$] charge transfer transitions [34].

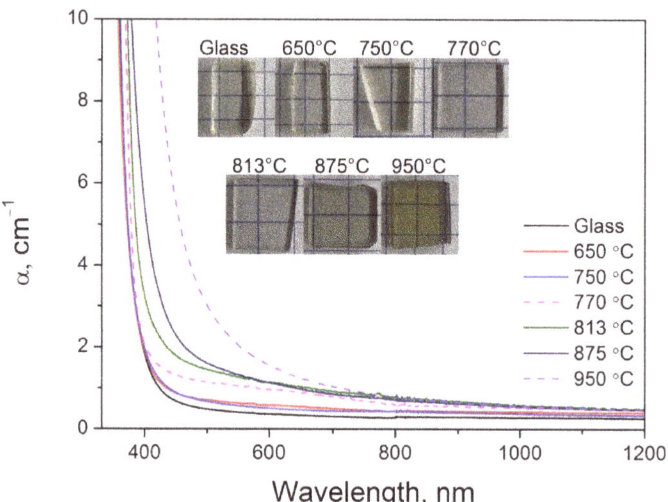



**Figure 10.** Absorption spectra of the Matrix glass samples before and after treatment at the indicated temperatures for 20 h. Inset—photos of the samples.

Analysis of the absorption spectra of Au-doped glasses reveals the strong differences in the optical properties and crucial effect of the gold NPs. Figure 11a presents absorption spectra of the Au-doped glass samples before and after the heat treatment. It is obvious that the spectrum and appearance of the initial Au-doped glass and the spectrum of the initial Matrix glass are the same. However, all spectra of the heat treated Au-doped glass samples contain the absorption band with the repositionable maxima (Figure 11b). This band is characteristic of the LSPR of gold NPs and resulted from the resonant oscillation of conduction electrons at the surface of the NPs stimulated by incident light [40]. All heat-treated Au-doped samples are transparent and colored (Figure 11c). The color of the glass samples undergoes distinct change with the heat treatment temperature rise: from deep-blue at 750–770 °C, to violet at 813 °C, crimson at 875 °C, and finally bloody red at 950 °C.

Observed color changes are related to the blue shift of the LSPR band with the heat treatment (Figure 11b). The overall plasmonic shift obtained in this work is 117 nm which is the largest value for the gold NPs precipitated in glass to the best of our knowledge. Hence, despite there being an absence of any effect of gold NPs precipitated in the glass under study on the structure change and crystallization kinetics, they play a crucial role in the change of optical properties.

The origin of the LSPR shift and the observed color change we previously reported in [32]. We suggested that upon the low-temperature heat treatment, gold NPs precipitate in the phase-separated regions which probably have high values of the refractive index. Such an environment of the NPs with high refraction provides a redshift of the LSPR band. Further increase of the heat treatment temperature leads to the formation of gahnite nanocrystals from the phase separation sites. This results in the decline of the overall refractive index around the gold NPs and the blue shift of the LSPR band.

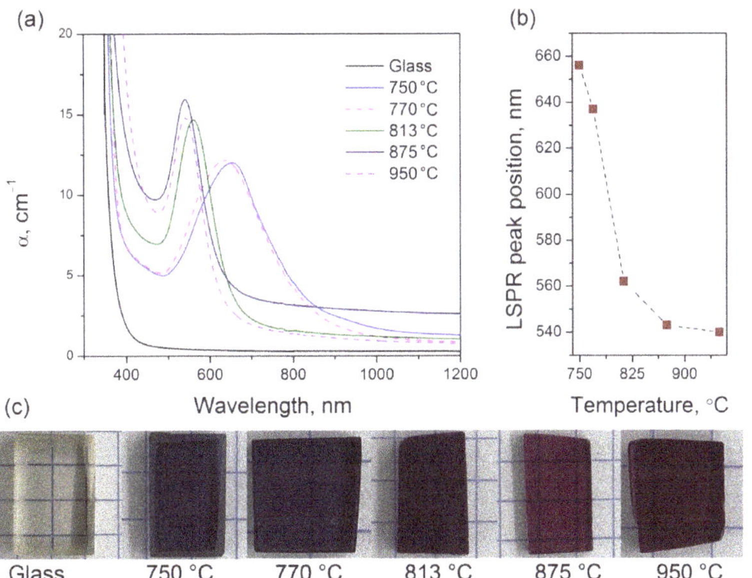



**Figure 11.** (**a**) Absorption spectra of the Au-doped glass samples before and after the heat treatment in the 650–950 °C range for 20 h; (**b**) change of the LSPR peak wavelength with a variation of the heat treatment temperature; (**c**) photos of the samples.

In order to verify the applicability of the previous suggestions to the glasses under study, we performed computer simulations of the absorption spectra of the Au-doped samples with the evolution of the heat treatment temperature. To take into account changes in the sizes of NPs and the refractive index of the glass, we consider an ensemble of nanoparticles described by a two-dimensional truncated normal distribution [41]:

$$f(D,n) = \frac{1}{2\pi\sigma_D\sigma_n\sqrt{1-r^2}}exp\left[-\frac{1}{1-r^2}\left(\frac{(D-\mu_D)^2}{2\sigma_D^2} + \frac{(n-\mu_n)^2}{2\sigma_n^2} - r\frac{(D-\mu_D)(n-\mu_n)}{\sigma_D\sigma_n}\right)\right] \tag{2}$$

where the variables $D$ (nanoparticle diameter) and $n$ (local refractive index of the glass) were limited: $2 < D < 15$ nm, $1.5 < n < 3.0$. The distribution parameters $\mu_D, \sigma_D, \mu_n, \sigma_n,$ and $r$ were selected so as to ensure the best agreement between the calculated spectrum and the experimental one. For an uncut normal distribution, the parameters $\mu$ and $\sigma$ determine the mean and variance, respectively.

The spectrum of each nanoparticle was calculated within the framework of the Mie theory [40], taking into account size-dependent corrections to the dielectric function of bulk gold, as we did earlier in [32] using the MSTM-Studio program [42]. The theoretical absorption spectrum of the glass with NPs was calculated in the diluted limit as a linear combination of contributions from each nanoparticle with weights determined by Equation (2).

The achieved theoretical description of the experimental curves and the resulting two-dimensional distributions are shown in Figure 12. One can see a very good agreement between the experimental and fitted spectra. This confirms the applicability of the proposed approach for the description of the LSPR shift in the glass during the heat treatment.

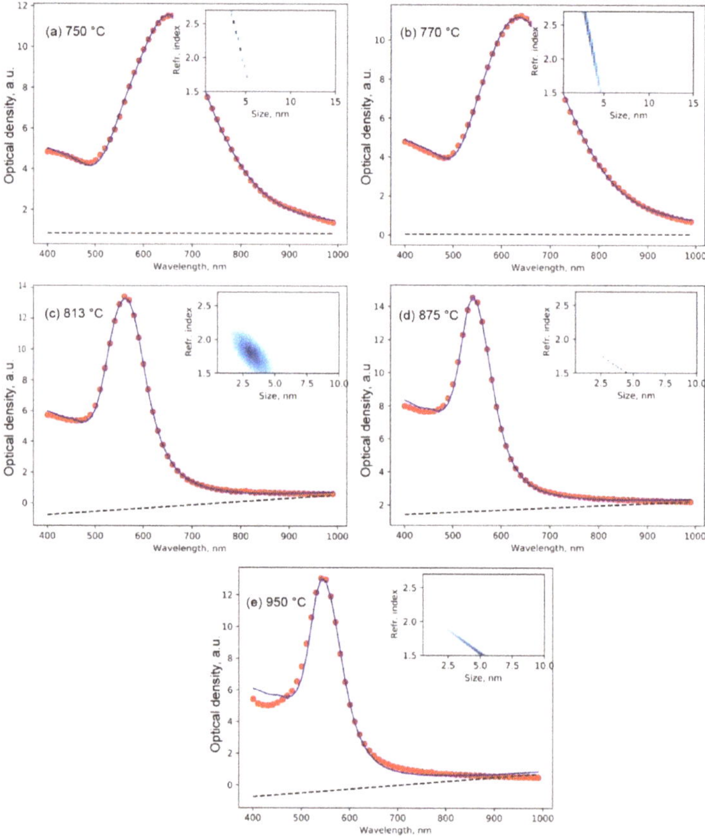

**Figure 12.** (a–e) Fitting quality of the absorption spectra of the Au-doped glass samples upon the heat treatment in the 750–950 °C range. Red dots—experimental spectrum, solid blue line—theoretical spectrum, dashed line—background contribution. The inset illustrates the obtained 2D normal distribution of NPs over the sizes and glass refraction index defined by Equation (2).

Figure 13a,b illustrates the dependence of the distribution of NPs by size and refractive index, respectively, of regions with NPs from the evolution of heat treatment temperature. With increasing of the temperature, the average NP size $\mu_D$ increases (lines in Figure 13a), while the size dispersion $\sigma_D$ (width of the filled regions in Figure 13a) is retained. This behavior is expected and indicates the continued formation of gold NPs as the temperature increases. Despite the proposed increase of the NP size the overall quantity of the NPs is too low, which is why we cannot detect it using the TEM.

The highest refractive index with the average value $\mu_n \approx 2.5$ of the glass regions containing gold NPs is achieved at a treatment temperature of 750 °C. Heat treatment at higher temperatures leads to a decrease in the average refractive index of the material, down to the values of the initial glass, $n = 1.5$–1.6.

The evidence from the study of the optical properties of the Matrix and Au-doped glasses subjected to heat treatments suggests that gold NPs have a great effect on the color and absorption spectra of glasses. Moreover, phase separation plays a crucial role in the location of the LSPR band in the visible part of the spectrum. We propose that a possible area of future research would be to investigate the plasmonic enhancement of the rare-earth ions' photoluminescence by the tunable LSPR in the co-doped Au/rare-earth glass systems.

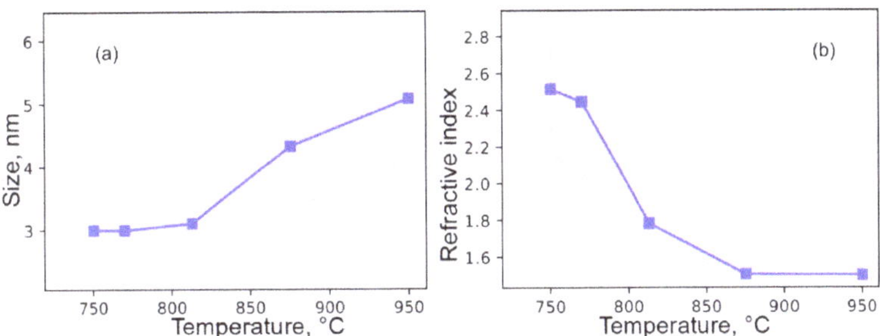

**Figure 13.** Illustration of the obtained size distributions (**a**) and glass refractive index (**b**) for gold NPs in samples obtained at heat treatment temperatures from 750 to 950 °C. The solid line corresponds to the parameter μ, and the shaded areas correspond to the deviation from μ by ±0.5σ.

## 4. Conclusions

The effect of Au-doping on structure, crystallization, and optical properties of ZnO-MgO-$Al_2O_3$-$SiO_2$ glass containing $TiO_2$ and $ZrO_2$ as nucleating agents were studied. We showed that thermally-induced precipitation of gold NPs has no effect on the structure and crystallization of the studied glass. Taken together, obtained results suggest that regardless of the Au-doping, heat treatment of the glass above the crystallization peak (872 °C) leads to the volume formation of spinel-type gahnite crystallites from 10 to 25 nm in size. Heat treatment below the crystallization peak results in the development of the amorphous phase-separated regions up to 10 nm in size and the formation of 3–4 nm crystallites.

Conversely, the findings from the performed studies on the optical properties suggest that Au-doping has a crucial effect on the color and electronic structure of the glass. Gold NPs precipitated in the glass during the heat treatments cause the appearance of the absorption band due to the LSPR of NPs. Change of the heat treatment temperature allows adjusting the LSPR band position with the possible shift of 117 nm in the visible part of the spectrum. Using the computational methods, we showed that one possible implication of this is a change of the local refractive index around the NPs in the 1.5–2.6 range occurring during the phase separation and crystallization process of glass. We propose that a possible area of future research would be to investigate the plasmonic enhancement of the rare-earth ions' photoluminescence by the tunable LSPR in the co-doped Au/rare-earth glass systems.

**Supplementary Materials:** The following are available online at https://www.mdpi.com/article/10.3390/cryst12020287/s1, Figure S1: Absorption spectrum of the glass sample heat-treated at 1100 °C for 20 h. Inset shows the photo of the glass sample.

**Author Contributions:** Conceptualization, G.S. and V.D.; Funding acquisition, G.S.; Investigation, M.Z., G.A., N.G., A.T. and O.C.; Methodology, G.A. and N.G.; Supervision, V.S.; Visualization, M.Z., A.T. and L.A.; Writing—original draft, G.S. and L.A.; Writing—review and editing, N.G., L.A., L.B. and V.S. All authors have read and agreed to the published version of the manuscript.

**Funding:** This research was funded by Mendeleev University of Chemical Technology, project number 2020-012.

**Institutional Review Board Statement:** Not applicable.

**Informed Consent Statement:** Not applicable.

**Data Availability Statement:** The data presented in this study are available on request from the corresponding author. The data are not publicly available due to privacy restriction.

**Conflicts of Interest:** The authors declare no conflict of interest.

# References


1. Soukoulis, C.M.; Wegener, M. Past Achievements and Future Challenges in the Development of Three-Dimensional Photonic Metamaterials. *Nat. Photonics* **2011**, *5*, 523–530. [CrossRef]
2. Komatsu, T.; Honma, T. Optical Active Nano-Glass-Ceramics. *Int. J. Appl. Glass Sci.* **2013**, *4*, 125–135. [CrossRef]
3. Gorni, G.; Velázquez, J.; Mosa, J.; Balda, R.; Fernández, J.; Durán, A.; Castro, Y. Transparent Glass-Ceramics Produced by Sol-Gel: A Suitable Alternative for Photonic Materials. *Materials* **2018**, *11*, 212. [CrossRef]
4. Liu, X.; Zhou, J.; Zhou, S.; Yue, Y.; Qiu, J. Transparent Glass-Ceramics Functionalized by Dispersed Crystals. *Prog. Mater. Sci.* **2018**, *97*, 38–96. [CrossRef]
5. Veber, A.; Lu, Z.; Vermillac, M.; Pigeonneau, F.; Blanc, W.; Petit, L. Nano-Structured Optical Fibers Made of Glass-Ceramics, and Phase Separated and Metallic Particle-Containing Glasses. *Fibers* **2019**, *7*, 105. [CrossRef]
6. Golubev, N.V.; Ignat'eva, E.S.; Mashinsky, V.M.; Kozlova, E.O.; Sigaev, V.N.; Monguzzi, A.; Paleari, A.; Lorenzi, R. Pre-Crystallization Heat Treatment and Infrared Luminescence Enhancement in $Ni^{2+}$-Doped Transparent Glass-Ceramics. *J. Non-Cryst. Solids* **2019**, *515*, 42–49. [CrossRef]
7. Lorenzi, R.; Golubev, N.V.; Ignat'eva, E.S.; Sigaev, V.N.; Ferrara, C.; Acciarri, M.; Vanacore, G.M.; Paleari, A. Defect-Assisted Photocatalytic Activity of Glass-Embedded Gallium Oxide Nanocrystals. *J. Colloid Interface Sci.* **2022**, *608*, 2830–2838. [CrossRef]
8. Venkateswaran, C.; Sreemoolanadhan, H.; Vaish, R. Lithium Aluminosilicate (LAS) Glass-Ceramics: A Review of Recent Progress. *Int. Mater. Rev.* **2021**, 1–38. [CrossRef]
9. Shakhgil'dyan, G.Y.; Savinkov, V.I.; Shakhgil'dyan, A.Y.; Alekseev, R.O.; Naumov, A.S.; Lopatkina, E.V.; Sigaev, V.N. Effect of Sitallization Conditions on the Hardness of Transparent Sitalls in the System $ZnO–MgO–Al_2O_3–SiO_2$. *Glass Ceram.* **2021**, *77*, 426–428. [CrossRef]
10. Sant'Ana Gallo, L.; Célarié, F.; Bettini, J.; Rodrigues, A.C.M.; Rouxel, T.; Zanotto, E.D. Fracture Toughness and Hardness of Transparent $MgO–Al_2O_3–SiO_2$ Glass-Ceramics. *Ceram. Int.* **2021**. [CrossRef]
11. Wang, Z.; Wang, Z.; Gan, L.; Zhang, J.; Wang, P. Structure/Property Nonlinear Variation Induced by Gamma Ray Irradiation of Boroaluminosilicate Transparent Glass Ceramic Containing Gahnite Nanocrystallite. *J. Non-Cryst. Solids* **2022**, *578*, 121346. [CrossRef]
12. Chakraborty, P. Metal Nanoclusters in Glasses as Non-Linear Photonic Materials. *J. Mater. Sci.* **1998**, *33*, 2235–2249. [CrossRef]
13. Savinkov, V.I.; Shakhgil'Dyan, G.Y.; Paleari, A.; Sigaev, V.N. Synthesis of Optically Uniform Glasses Containing Gold Nanoparticles: Spectral and Nonlinear Optical Properties. *Glass Ceram. (Engl. Transl. Steklo Keram.)* **2013**, *70*, 143–148. [CrossRef]
14. Shakhgildyan, G.Y.; Lipatiev, A.S.; Fedotov, S.S.; Vetchinnikov, M.P.; Lotarev, S.V.; Sigaev, V.N. Microstructure and Optical Properties of Tracks with Precipitated Silver Nanoparticles and Clusters Inscribed by the Laser Irradiation in Phosphate Glass. *Ceram. Int.* **2021**, *47*, 14320–14329. [CrossRef]
15. Shakhgil'dyan, G.Y.; Lipat'ev, A.S.; Vetchinnikov, M.P.; Popova, V.V.; Lotarev, S.V.; Sigaev, V.N. Femtosecond Laser Modification of Zinc-Phosphate Glasses with High Silver Oxide Content. *Glass Ceram. (Engl. Transl. Steklo Keram.)* **2017**, *73*, 420–422. [CrossRef]
16. Lipat'ev, A.; Shakhgil'dyan, G.; Lipat'eva, T.; Lotarev, S.; Fedotov, S.; Vetchinnikov, M.; Ignat'eva, E.; Golubev, N.; Sigaev, V.; Kazanskii, P. Formation of Luminescent and Birefringent Microregions in Phosphate Glass Containing Silver. *Glass Ceram.* **2016**, *73*, 277–282. [CrossRef]
17. Drexhage, K.H. Influence of a Dielectric Interface on Fluorescence Decay Time. *J. Lumin.* **1970**, *1–2*, 693–701. [CrossRef]
18. Shakhgil'dyan, G.Y.; Ziyatdinova, M.Z.; Kovgar, V.V.; Lotarev, S.V.; Sigaev, V.N.; Prusova, I.V. Effect of Gold Nanoparticles on the Spectral Luminescence Properties of $Eu^{3+}$-Doped Phosphate Glass. *Glass Ceram. (Engl. Transl. Steklo I Keram.)* **2019**, *76*, 121–125. [CrossRef]
19. Shakhgildyan, G.Y.; Ziyatdinova, M.Z.; Vetchinnikov, M.P.; Lotarev, S.V.; Savinkov, V.I.; Presnyakova, N.N.; Lopatina, E.V.; Vilkovisky, G.A.; Sigaev, V.N. Thermally-Induced Precipitation of Gold Nanoparticles in Phosphate Glass: Effect on the Optical Properties of $Er^{3+}$ Ions. *J. Non-Cryst. Solids* **2020**, *550*, 120408. [CrossRef]
20. Jiménez, J.A.; Smith, S. Gold-Assisted Enhancement of the Luminescence of $Mn^{2+}$ Ions Induced by Silicon in Phosphate Glass. *Phys. Lett. A* **2020**, *384*, 126776. [CrossRef]
21. Danmallam, I.M.; Ghoshal, S.K.; Ariffin, R.; Bulus, I. Europium Luminescence in Silver and Gold Nanoparticles Co-Embedded Phosphate Glasses: Judd-Ofelt Calculation. *Opt. Mater.* **2020**, *105*, 109889. [CrossRef]
22. Deubener, J.; Allix, M.; Davis, M.J.; Duran, A.; Höche, T.; Honma, T.; Komatsu, T.; Krüger, S.; Mitra, I.; Müller, R.; et al. Updated Definition of Glass-Ceramics. *J. Non-Cryst. Solids* **2018**, *501*, 3–10. [CrossRef]
23. Sakamoto, A.; Yamamoto, S. Glass-Ceramics: Engineering Principles and Applications. *Int. J. Appl. Glass Sci.* **2010**, *1*, 237–247. [CrossRef]
24. Stookey, S.D. Catalyzed Crystallization of Glass in Theory and Practice. *Ind. Eng. Chem.* **1959**, *51*, 805–808. [CrossRef]
25. Garai, M.; Sasmal, N.; Molla, A.R.; Tarafder, A.; Karmakar, B. Effects of In-Situ Generated Coinage Nanometals on Crystallization and Microstructure of Fluorophlogopite Mica Containing Glass-Ceramics. *J. Mater. Sci. Technol.* **2015**, *31*, 110–119. [CrossRef]
26. Garai, M.; Murthy, T.S.R.C.; Karmakar, B. Microstructural Characterization and Wear Properties of Silver and Gold Nanoparticle Doped K-Mg-Al-Si-O-F Glass-Ceramics. *Ceram. Int.* **2018**, *44*, 22308–22317. [CrossRef]
27. Garai, M.; Reka, A.A.; Karmakar, B.; Molla, A.R. Microstructure–Mechanical Properties of $Ag^0/Au^0$ Doped K–Mg–Al–Si–O–F Glass-Ceramics. *RSC Adv.* **2021**, *11*, 11415–11424. [CrossRef]

28. Kochetkov, D.A.; Nikonorov, N.V.; Sycheva, G.A.; Tsekhomskii, V.A. The Effect of Gold Nanoparticles on Crystallization Processes in Photostructured Lithium-Silicate Glass. *Glass Phys. Chem.* **2013**, *39*, 351–357. [CrossRef]
29. Thieme, C.; Kracker, M.; Patzig, C.; Thieme, K.; Rüssel, C.; Höche, T. The Acceleration of Crystal Growth of Gold-Doped Glasses within the System BaO/SrO/ZnO/SiO$_2$. *J. Eur. Ceram. Soc.* **2019**, *39*, 554–562. [CrossRef]
30. Garai, M.; Molla, A.R.; Reka, A.A.; Karmakar, B. Wide Thermal Expansion in Ag$^0$/Au$^0$ Nanoparticle Doped SiO$_2$-MgO-Al$_2$O$_3$-B$_2$O$_3$-K$_2$O-MgF$_2$ Glass-Ceramics. *Mater. Today Proc.* **2021**, *50*, 134–138. [CrossRef]
31. Kracker, M.; Thieme, C.; Thieme, K.; Patzig, C.; Berthold, L.; Höche, T.; Rüssel, C. Redox Effects and Formation of Gold Nanoparticles for the Nucleation of Low Thermal Expansion Phases from BaO/SrO/ZnO/SiO$_2$ Glasses. *RSC Adv.* **2018**, *8*, 6267–6277. [CrossRef]
32. Shakhgildyan, G.; Avakyan, L.; Ziyatdinova, M.; Atroshchenko, G.; Presnyakova, N.; Vetchinnikov, M.; Lipatiev, A.; Bugaev, L.; Sigaev, V. Tuning the Plasmon Resonance of Gold Nanoparticles in Phase-Separated Glass via the Local Refractive Index Change. *J. Non-Cryst. Solids* **2021**, *566*, 120893. [CrossRef]
33. Langford, J.I.; Wilson, A.J.C. Scherrer after Sixty Years: A Survey and Some New Results in the Determination of Crystallite Size. *J. Appl. Crystallogr.* **1978**, *11*, 102–113. [CrossRef]
34. Basyrova, L.; Bukina, V.; Balabanov, S.; Belyaev, A.; Drobotenko, V.; Dymshits, O.; Alekseeva, I.; Tsenter, M.; Zapalova, S.; Khubetsov, A.; et al. Synthesis, Structure and Spectroscopy of Fe$^{2+}$:MgAl$_2$O$_4$ Transparent Ceramics and Glass-Ceramics. *J. Lumin.* **2021**, *236*, 118090. [CrossRef]
35. Loiko, P.A.; Dymshits, O.S.; Skoptsov, N.A.; Malyarevich, A.M.; Zhilin, A.A.; Alekseeva, I.P.; Tsenter, M.Y.; Bogdanov, K.V.; Mateos, X.; Yumashev, K.V. Crystallization and Nonlinear Optical Properties of Transparent Glass-Ceramics with Co:Mg(Al,Ga)$_2$O$_4$ Nanocrystals for Saturable Absorbers of Lasers at 1.6–1.7. *Mm. J. Phys. Chem. Solids* **2017**, *103*, 132–141. [CrossRef]
36. Kurajica, S.; Šipušić, J.; Zupancic, M.; Brautović, I.; Albrecht, M. ZnO-Al$_2$O$_3$-SiO$_2$ Glass Ceramics: Influence of Composition on Crystal Phases, Crystallite Size and Appearance. *J. Non-Cryst. Solids* **2020**, *553*, 120481. [CrossRef]
37. Golubkov, V.V.; Dymshits, O.S.; Petrov, V.I.; Shashkin, A.V.; Tsenter, M.Y.; Zhilin, A.A.; Kang, U. Small-Angle X-Ray Scattering and Low-Frequency Raman Scattering Study of Liquid Phase Separation and Crystallization in Titania-Containing Glasses of the ZnO–Al$_2$O$_3$–SiO$_2$ System. *J. Non-Cryst. Solids* **2005**, *351*, 711–721. [CrossRef]
38. Loshmanov, A.A.; Sigaev, V.N.; Khodakovskaya, R.Y.; Pavlushkin, N.M.; Yamzin, I.I. Small-Angle Neutron Scattering on Silica Glasses Containing Titania. *J. Appl. Crystallogr.* **1974**, *7*, 207–210. [CrossRef]
39. Dymshits, O.S.; Zhilin, A.A.; Petrov, V.I.; Tsenter, M.Y.; Chuvaeva, T.I.; Shashkin, A.V.; Golubkov, V.V.; Kang, U.; Lee, K.-H. A Raman Spectroscopic Study of Phase Transformations in Titanium-Containing Magnesium Aluminosilicate Glasses. *Glass Phys. Chem.* **2002**, *28*, 66–78. [CrossRef]
40. Kreibig, U.; Genzel, L. Optical Absorption of Small Metallic Particles. *Surf. Sci.* **1985**, *156*, 678–700. [CrossRef]
41. Tong, Y.L. *The Multivariate Normal Distribution*; Springer: New York, NY, USA, 1990; ISBN 978-1-4613-9657-4.
42. Avakyan, L.A.; Heinz, M.; Skidanenko, A.V.; Yablunovski, K.A.; Ihlemann, J.; Meinertz, J.; Patzig, C.; Dubiel, M.; Bugaev, L.A. Insight on Agglomerates of Gold Nanoparticles in Glass Based on Surface Plasmon Resonance Spectrum: Study by Multi-Spheres T-Matrix Method. *J. Phys. Condens. Matter* **2018**, *30*, 045901. [CrossRef] [PubMed]

MDPI

St. Alban-Anlage 66

4052 Basel

Switzerland

Tel. +41 61 683 77 34

Fax +41 61 302 89 18

www.mdpi.com

*Crystals* Editorial Office

E-mail: crystals@mdpi.com

www.mdpi.com/journal/crystals